AF739035

Ascorbate-Glutathione Pathway and Stress Tolerance in Plants

Naser A. Anjum • Shahid Umar • Ming-Tsair Chan
Editors

Ascorbate-Glutathione Pathway and Stress Tolerance in Plants

Springer

Editors
Naser A. Anjum
Centre for Environmental and Marine Studies (CESAM) and
Department of Chemistry
University of Aveiro
3810–193 Aveiro
Portugal
anjum@ua.pt; naser.a.anjum@gmail.com

Shahid Umar
Department of Botany
Faculty of Science
Hamdard University
New Delhi
India
s_umar9@hotmail.com

Ming-Tsair Chan
Academia Sinica Biotechnology Center in Southern Taiwan
Sinshih Township
Tainan County 74146
Taiwan
mbmtchan@gate.sinica.edu.tw

ISBN 978-90-481-9403-2 e-ISBN 978-90-481-9404-9
DOI 10.1007/978-90-481-9404-9
Springer Dordrecht Heidelberg London New York

Library of Congress Control Number: 2010934293

© Springer Science+Business Media B.V. 2010
No part of this work may be reproduced, stored in a retrieval system, or transmitted in any form or by any means, electronic, mechanical, photocopying, microfilming, recording or otherwise, without written permission from the Publisher, with the exception of any material supplied specifically for the purpose of being entered and executed on a computer system, for exclusive use by the purchaser of the work.

Printed on acid-free paper

Springer is part of Springer Science+Business Media (www.springer.com)

To

A Great Philanthropist, Thinker, Visionary

&

The Founder of Jamia Hamdard

(Hamdard University), New Delhi, INDIA

Janab Hakeem Abdul Hameed
(1908–1999)

Foreword

Anti-scorbutic factor (Vitamin C), ascorbic acid, was isolated from paprika by A. Szent-Gyorgyi, and its anti-scorbutic effect was confirmed in guinea-pigs (Biochem. J. 27: 278-285 (1933). Occurrence of anti-scorbutic factor in fruits, vegetables and adrenal glands of vertebrates had been deduced from their protective effects against scurvy. However, its isolation from either lemon or adrenal glands was a very heavy work in 1920-30, which required several tons (!) of the materials and several years, because of its low contents and disturbing components. In 1933, Szent-Gyorgyi in Szeged, Hungary, established a simple way for preparation of 450 g crystalline ascorbic acid within a month from local paprika (Hungarian red pepper, *Capscicum annuum*). This isolation was initiated by his finding of high reducing potentials in paprika juice, as deduced by disappearance of blue dibromophenolindophenol. Even in the current data book on food nutrients, the content of vitamin C in paprika is one of the highest ones among vegetables and fruits, indicating a sharp sense of Szent-Gyorgyi focusing to paprika as the starting material for the isolation of anti-scorbutic factor. In those days, a large amount of ascorbate was required for the determination of its structure, which was accomplished by W. Haworth in England. The revealed structure opened a gate to synthesize ascorbate, which was soon done by W. Haworth and T. Reichstein, independently. Thus, ascorbate is the first vitamin which is chemically synthesized.

Based on these works and the following many works and surveys on the antioxidants in plants, World Cancer Research Fund (2007) recommended; "Eat at least five servings (total at least 400 g) of a variety of non-starchy vegetables and of fruits every day, to protect from cancer", as one of the personal recommendations for foods to escape from cancer. This is a very reasonable one considering that plants are always exposed to most stressful environments among organisms; strong sun light, highest oxygen concentration in leaf tissues, and low homeostasis in respects of temperature and other environmental factors. Thus, plants are expected to contain very effective antioxidants at high contents for survival under natural environments, therefore, both vegetables and fruits are very rich in the antioxidants which are able to protect DNA from ROS.

Even though ascorbate isolated from plants has contributed so much to understand its nutritional effect for human, the physiological and biochemical functions of ascorbate in plants themselves have remained obscure for many years. In plants, it had been supposed that ascorbate (AsA), with glutathione (GSH), plays a role as general antioxidants as in animals. However, ascorbate-specific reaction or enzyme in plants was not identified, except for ascorbate oxidase, up to 1980.

When chloroplasts are exposed to higher photon intensities over that required for CO_2-fixation, excess photons induce the reduction of oxygen in place of CO_2. Mehler found the photoproduction of H_2O_2 in chloroplasts and then, in 1970s, this H_2O_2 is shown to be produced through the SOD-catalyzed disproportionation of superoxide, which is the primary photoreduced product of oxygen in PS I (Asada). Because even very low amounts of H_2O_2 inactivate the CO_2-fixation enzymes, plant chloroplasts should equip the effective scavenging system of H_2O_2. Chloroplasts do not contain any catalase, then, peroxidase is a possible scavenger of H_2O_2. In 1980-81, ascorbate-specific peroxidase (APX) was found in *Euglena* cells (Shigeoka et al.) which do not contain catalase, and also in plant chloroplasts (Nakano and Asada). Further, the regeneration of AsA from the oxidized ascorbates (monodehydroascorbate, MDHA radical & dehydroascorbate, DHA) by respective reductases has been characterized including glutathione peroxidase (GPX) and glutathione reductase (GR), all of them are required to scavenge ROS in chloroplasts. Thus, ascorbate-glutathione (AsA-GSH) and related enzymes are essential to scavenge effectively ROS generated in chloroplasts under environmental stress.

In addition to chloroplasts, APX and other related enzymes for either scavenging or adjustment of ROS levels have been found in other cell organelles, such as mitochondria, peroxisomes, plasma membranes, apoplasts and other cell compartments. Further, the superoxide-generating, plasma membrane-bound NADPH oxidase has been found and plays a role in the generation for physiologically functional ROS. These systems have an intimate relation with AsA-GSH, as in chloroplasts.

The present monograph, edited by eminent scientists Drs. Naser A. Anjum, Shahid Umar and Ming-Tsair Chan, covers the current progress on the physiologically functional ROS, in relation to AsA-GSH. I believe that the revealed mechanisms of signaling and other functions of ROS with intimate correlations of AsA-GSH pathway in plants, and its further progress should contribute to understand the ROS-signal systems in responses to under biotic and biotic stresses. Further, they should provide the focusing points to be analyzed further to support higher crop yields. Furthermore, the ROS-related signaling system mediate signaling system mediated or adjusted by AsA-GSH system in plants should give a clue to understand the similar systems for human health, as have been done since AsA was isolated from paprika.

July 29th 2010

Kozi Asada
Kyoto University (Emeritus)
Kyoto, JAPAN
e-mail: k.asada@ks5.ecs.kyoto-u.ac.jp

Preface

Since last several decades, increasing agricultural productivity has been a challenging task to fulfill the requirements of enough food to feed rapidly increasing world population in the changing environment. Both biotic and abiotic stress factors are continued to negatively affect various aspects of plant growth and development leading to relative decrease in the potential maximum yields by more than fifty percent. These stress factors have been shown to affect various aspects of plant system including the acceleration in the formation of reactive oxygen species (ROS). Although, reactive oxygen species are important signal molecules that regulate plant responses to environmental stress factors but these must be rapidly processed and/or detoxified if oxidative damage is to be averted in cells. The ascorbate (AsA)-glutathione (GSH) pathway is a key part of the network of reactions involving enzymes and metabolites with redox properties for the detoxification of ROS, and thus to avert the ROS-accrued oxidative damage in plants.

Both AsA and GSH are intimately linked in terms of their major physiological functions in AsA-GSH pathway and many of these processes are correlated with endogenous AsA-GSH levels especially under stress conditions. In addition to having major role during vital phases of the plant life cycle, AsA and GSH determine the lifetime of reactive oxygen species within the cellular environment and provide crucial protection against oxidative damage. While research into the responses of individual components of plant antioxidant defense system has benefited greatly from advances in molecular technology, the cross-talks and inter-relationships studies on the physiological, biochemical and molecular aspects of the cumulative response of various components of AsA-GSH pathway to stress factors and their significance in plant stress tolerance have received comparatively very little or no attention.

The present book has concentrated more on cumulative responses of the components of AsA-GSH pathway in plant stress tolerance with emphasis on the unique insights and advances gained by molecular exploration than whole plant antioxidant defense system. In fact, these studies/reports based on inter-relationships and/or cross-talks are expected to lead to understand and improve the mechanisms of stress tolerance in plants. Therefore, the present volume would definitely be an ideal source of scientific information to the advanced students, junior researchers, faculty and scientists involved in agriculture, plant sciences, molecular biology, biochemistry, biotechnology and related areas.

We are thankful to contributors for their interests, significant contributions and cooperation that eventually made the present volume possible. Thanks are also due to all the well-wishers, teachers, seniors, research students and affectionate family members. Without their unending support, motivation and encouragements the present grueling task would have never been accomplished.

We would like to offer our sincere thanks to Mr. Jacco Flipsen, Ineke Ravesloot, Purushothaman Saravanan and their team at Springer and SPi for their continuous support which made our efforts successful.

Last but not least, the financial supports to our research from Foundation for Science & Technology (FCT), PORTUGAL (SFRH/BPD/64690/2009), Council of Scientific & Industrial Research (CSIR), New Delhi, INDIA [(9/112(0401)2K8-EMR-I (312208/2K7/1)], Agricultural Biotechnology Research Center (ABRC), Academia Sinica, TAIWAN, Potash Research Institute of India (PRII), Gurgaon, INDIA and International Potash Institute (IPI), SWITZERLAND are gratefully acknowledged.

August 6th 2010

Aveiro, Portugal — Naser A. Anjum
New Delhi, India — Shahid Umar
Tainan, Taiwan — Ming-Tsair Chan

Contents

1 **Regulatory Role of Components of Ascorbate–Glutathione Pathway in Plant Stress Tolerance** 1
Dariusz Latowski, Ewa Surówka, and Kazimierz Strzałka

2 **Ascorbate and Glutathione in Organogenesis, Regeneration and Differentiation in Plant In vitro Cultures** 55
Jarosław Tyburski and Andrzej Tretyn

3 **Role of Ascorbate Peroxidase and Glutathione Reductase in Ascorbate–Glutathione Cycle and Stress Tolerance in Plants** 91
Cai-Hong Pang and Bao-Shan Wang

4 **The Ascorbate–Gluathione Cycle and Related Redox Signals in Plant–Pathogen Interactions** 115
Elżbieta Kuźniak

5 **Regulation of the Ascorbate–Glutathione Cycle in Plants Under Drought Stress** 137
Adriano Sofo, Nunzia Cicco, Margherita Paraggio, and Antonio Scopa

6 **Glutathione and Herbicide Resistance in Plants** 191
Zornitsa Ivanova Katerova and Lyuba Petar-Emil Miteva

7 **Ascorbate and Glutathione: Protectors of Plants in Oxidative Stress** 209
Qaisar Mahmood, Raza Ahmad, Sang-Soo Kwak, Audil Rashid, and Naser A. Anjum

8 Changes in the Glutathione and Ascorbate Redox State Trigger Growth During Embryo Development and Meristem Reactivation at Germination 231
Claudio Stasolla

9 A Winning Two Pair: Role of the Redox Pairs AsA/DHA and GSH/GSSG in Signal Transduction 251
Günce Şahin and Mario C. De Tullio

10 Involvement of AsA/DHA and GSH/GSSG Ratios in Gene and Protein Expression and in the Activation of Defence Mechanisms Under Abiotic Stress Conditions 265
Vasileios Fotopoulos, Vasileios Ziogas, Georgia Tanou, and Athanassios Molassiotis

11 Ascorbate–Glutathione Cycle: Enzymatic and Non-enzymatic Integrated Mechanisms and Its Biomolecular Regulation 303
Juan Pablo Martínez and Héctor Araya

12 Coordinate Role of Ascorbate–Glutathione in Response to Abiotic Stresses 323
Imran Haider Shamsi, Sisi Jiang, Nazim Hussain, Xianyong Lin, and Lixi Jiang

13 Regulation of Genes Encoding Chloroplast Antioxidant Enzymes in Comparison to Regulation of the Extra-plastidic Antioxidant Defense System 337
Margarete Baier, Nicola T. Pitsch, Marina Mellenthin, and Wei Guo

14 The Peroxisomal Ascorbate–Glutathione Pathway: Molecular Identification and Insights into Its Essential Role Under Environmental Stress Conditions 387
Sigrun Reumann and Francisco J. Corpas

15 Identification of Potential Gene Targets for the Improvement of Ascorbate Contents of Genetically Modified Plants 405
Adebanjo A. Badejo and Muneharu Esaka

Index 429

Contributors

Raza Ahmad
Department of Environmental Sciences, COMSATS Institute of Information Technology, Abbottabad 22060, Pakistan

Naser A. Anjum
Centre for Environmental and Marine Studies (CESAM) & Department of Chemistry, University of Aveiro, 3810-193 Aveiro, Portugal
Academia Sinica, Agricultural Biotechnology Research Center (ABRC), Taiwan and
Department of Botany, Division of Plant Physiology and Biochemistry, Faculty of Life Sciences, Aligarh Muslim University, Aligarh-202 002, UP, India
anjum@ua.pt; naser.a.anjum@gmail.com

Héctor Araya
Faculty of Medicine, Department of Nutrition, University of Chile, Independencia 1027, Santiago, Chile and
Faculty of Pharmacy, Department of Food and Nutrition, University of Valparaíso, Avda Gran Bretaña 1111, Valparaíso, Región de Valparaíso, Chile and
Regional Center for the Study of Healthy Foods (CREAS), Blanco 1623, Oficina 1402, Edificio Torres Mar del Sur 2-, Valparaíso

Adebanjo A. Badejo
Faculty of Life and Environmental Science, Shimane University, Matsue, Shimane 690-8504, Japan
badejoaa1@yahoo.com

Margarete Baier
Plant Science, Heinrich-Heine-University, Universitaetsstrasse 1, 40225 Düsseldorf, Germany and
Plant Physiology, Free University Berlin, Königin-Luise-Strasse 12-16, 14195 Berlin, Germany
margarete.baier@uni-duesseldorf.de

Nunzia Cicco
Istituto di Metodologie per l'Analisi Ambientale, Consiglio Nazionale delle Ricerche, C. da S. Loia, Zona Industriale, 85050 Tito Scalo (PZ), Italy

Francisco J. Corpas
Departamento de Bioquímica, Biología Celular y Molecular de Plantas, Estación Experimental del Zaidín (EEZ), CSIC, Granada, Spain

Mario C. De Tullio
Plant Biology and Pathology, Aldo Moro University, Bari, Italy
detullio@botanica.uniba.it

Muneharu Esaka
Graduate School of Biosphere Sciences, Hiroshima University, Kagamiyama, Higashi-Hiroshima, Hiroshima 739-8528, Japan
mesaka@hiroshima-u.ac.jp

Vasileios Fotopoulos
Department of Agricultural Sciences, Biotechnology and Food Science, Cyprus University of Technology, PC 3036 Limassol, Cyprus

Wei Guo
Plant Science, Heinrich-Heine-University, Universitaetsstrasse 1, 40225 Düsseldorf, Germany

Nazim Hussain
Institute of Crop Science, College of Agriculture and Biotechnology, Zhejiang University, 268 Kaixuan Road, Hangzhou 310029, PR of China

Lixi Jiang
Institute of Crop Science, College of Agriculture and Biotechnology, Zhejiang University, 268 Kaixuan Road, Hangzhou 310029, PR of China
jianglx@zju.edu.cn

Sisi Jiang
Institute of Agricultural Chemistry, College of Environmental and Resource Sciences, 288 Kaixuan Roas, Hangzhou 310029, PR of China

Zornitsa Ivanova Katerova
Acad. M. Popov Institute of Plant Physiology, Bulgarian Academy of Sciences, Acad. G. Bonchev Str., Bl. 21, 1113 Sofia, Bulgaria
katerovazor@yahoo.com

Elżbieta Kuźniak
Department of Plant Physiology and Biochemistry, University of Łódź, 90-237 Łódź, Banacha 12/16, Poland
elkuz@biol.uni.lodz.pl

Sang-Soo Kwak
Environmental Biotechnology Research Center, Korea Research Institute of Bioscience and Biotechnology (KRIBB), Eoeun-dong 52, Yuseong, Daejon, 305-806, South Korea

Dariusz Latowski
Faculty of Biochemistry, Biophysics and Biotechnology, Department of Plant Physiology and Biochemistry, Jagiellonian University, Gronostajowa 7, 30-387 Krakow, Poland

Xianyong Lin
Institute of Agricultural Chemistry, College of Environmental and Resource Sciences, 288 Kaixuan Roas, Hangzhou 310029, PR of China

Qaisar Mahmood
Department of Environmental Sciences, COMSATS Institute of Information Technology, Abbottabad 22060, Pakistan
mahmoodzju@gmail.com

Juan Pablo Martínez
Institute of Agriculture Research (INIA), Chorrillos 86, La Cruz, Regiӧn de Valparaíso, Chile and
Regional Center for the Study of Healthy Foods (CREAS), Blanco 1623, Oficina 1402, Edificio Torres Mar del Sur 2, Valparaíso
jpmartinez@inia.cl; jpg_martinez@hotmail.com

Marina Mellenthin
Plant Science, Heinrich-Heine-University, Universitaetsstrasse 1, 40225 Düsseldorf, Germany

Lyuba Petar-Emil Miteva
Resbiomed EOOD, 81 B, Bulgaria Blvd, Sofia, Bulgaria

Athanassios Molassiotis
Faculty of Agriculture, Aristotle University of Thessaloniki, 54 124 Thessaloniki, Greece
amolasio@agro.auth.gr

Cai-Hong Pang
Shandong Academy of Forestry, Jinan 250014, Shandong, P.R. China

Margherita Paraggio
Istituto di Metodologie per l'Analisi Ambientale, Consiglio Nazionale delle Ricerche, C. da S. Loia, Zona Industriale, 85050 Tito Scalo (PZ), Italy

Nicola T. Pitsch
Plant Science, Heinrich-Heine-University, Universitaetsstrasse 1, 40225 Düsseldorf, Germany

Audil Rashid
Department of Environmental Sciences, PMAS Arid Agriculture University, Rawalpindi, Pakistan

Sigrun Reumann
Faculty of Science and Technology, Centre for Organelle Research (CORE), University of Stavanger, Stavanger, Norway
sigrun.reumann@uis.no

Günce Şahin
Abant Izzet Baysal University, Bolu, Turkey

Antonio Scopa
Dipartimento di Scienze dei Sistemi Colturali, Forestali e dell'Ambiente, Università degli Studi della Basilicata, Via dell'Ateneo Lucano 10, 85100 Potenza, Italy

Imran Haider Shamsi
Institute of Crop Science, College of Agriculture and Biotechnology, Zhejiang University, 268 Kaixuan Road, Hangzhou 310029, PR of China

Adriano Sofo
Dipartimento di Scienze dei Sistemi Colturali, Forestali e dell'Ambiente, Università degli Studi della Basilicata, Via dell'Ateneo Lucano 10, 85100 Potenza, Italy
adriano.sofo@unibas.it

Kazimierz Strzałka
Faculty of Biochemistry, Biophysics and Biotechnology, Department of Plant Physiology and Biochemistry, Jagiellonian University, Gronostajowa 7, 30-387 Krakow, Poland
kazimierzstrzalka@gmail.com

Claudio Stasolla
Department of Plant Science, University of Manitoba, Winnipeg R3T 2N2, Manitoba, Canada
stasolla@ms.umanitoba.ca

Ewa Surówka
Institute of Plant Physiology, Polish Academy of Sciences, Niezapominajek 21, 30-239 Krakow, Poland

Georgia Tanou
Faculty of Agriculture, Aristotle University of Thessaloniki, 54 124 Thessaloniki, Greece

Andrzej Tretyn
Department of Biotechnology, Institute of General and Molecular Biology, Nicolaus Copernicus University, Gagarina 9, 87-100 Toruń, Poland

Jarosław Tyburski
Department of Biotechnology, Institute of General and Molecular Biology, Nicolaus Copernicus University, Gagarina 9, 87-100 Toruń, Poland
tybr@umk.pl

Bao-Shan Wang
Key Lab of Plant Stress Research, College of Life Sciences,
Shandong Normal University, Jinan 250014, Shandong, P.R. China
bswang@sdnu.edu.cn

Vasileios Ziogas
Faculty of Agriculture, Aristotle University of Thessaloniki,
54 124 Thessaloniki, Greece

Chapter 1
Regulatory Role of Components of Ascorbate–Glutathione Pathway in Plant Stress Tolerance

Dariusz Latowski, Ewa Surówka, and Kazimierz Strzałka

Abstract Ascorbate (Asc) and glutathione (GSH) are important molecules functioning in several vital processes in plant cells, including the ascorbate–glutathione cycle. They are involved in basic metabolic reactions normally occurring in plants as well as in those evoked by abiotic or biotic stresses. Asc and GSH are localized in most of cell compartments such as cytoplasm, mitochondria, peroxisomes and chloroplasts, the Asc being additionally also found in the apoplast. These small molecular weight compounds protect cells against oxidative stress and damage by detoxifying reactive oxygen species (ROS) and ROS-generated toxic metabolic products. They do this either directly by scavenging or indirectly through the activation of defense mechanisms. Asc and GSH are engaged in maintaining cellular redox homeostasis, being at the same time involved in redox signaling. They may interact with different molecules and signaling pathways during the resistance responses. Besides the total level of Asc and GSH in the cell the ratio between reduced and oxidized forms of these molecules play an important role in the activation of various defense mechanisms. Furthermore, glutathione is involved in specific functions such as detoxification of heavy metals, transfer and storage of sulfur, regulation of expression of defense-related genes and protein activity, while ascorbate acts as a signal-transducing molecule, cofactor of some enzymes and biosynthetic precursor of oxalic and L-tartaric acids. Moreover all kinds of xanthophylls cycle, considered as the most important photoprotective mechanism in plants, are strongly dependent on the acid form of ascorbate. The biochemical, physiological and genetic aspects of the involvement of ascorbate and glutathione, localized in different cell compartments, in controlling cellular redox state, plant stress tolerance, and defense mechanisms will be discussed.

D. Latowski and K. Strzałka (✉)
Faculty of Biochemistry, Biophysics and Biotechnology, Department of Plant Physiology and Biochemistry, Jagiellonian University, Gronostajowa 7, 30-387 Krakow, Poland
e-mail: kazimierzstrzalka@gmail.com

E. Surówka
Institute of Plant Physiology, Polish Academy of Sciences, Niezapominajek 21, 30-239 Krakow, Poland

N.A. Anjum et al. (eds.), *Ascorbate-Glutathione Pathway and Stress Tolerance in Plants*,
DOI 10.1007/978-90-481-9404-9_1, © Springer Science+Business Media B.V. 2010

Keywords Ascorbate • Glutathione • Antioxidant compartmentalisation • Cellular redox state • Defence mechanisms • Signal transduction pathways • Stress tolerance • Xanthophyll cycle

1 Introduction

Exposure of plants to stresses such as high light stress (HL), ultraviolet radiation, drought stress and desiccation, salt stress, chilling, heat shock, heavy metals, air pollutants, mechanical stress, nutrient deprivation and pathogen attack and can give rise to excess accumulation of reactive oxygen species (ROS) at the cellular level (Stohs and Bagchi 1995; Noctor and Foyer 1998; Alscher et al. 2002; Schützendübel and Polle 2002). In plant cells ROS such as superoxide radical ($O_2{\cdot}^-$), hydrogen peroxide (H_2O_2), hydroxyl radical (·OH), and singlet oxygen ($^1O_2$) are mainly produced in chloroplasts, mitochondria, and peroxisomes as well as by the plasma membrane–bound NADPH oxidases and the cell wall–bound NAD(P)H oxidase–peroxidase (Foyer and Noctor 2000; Jiang and Zhang 2001; del Río et al. 2002; Sweetlove et al. 2002; Foyer and Noctor 2003; Apel and Hirt 2004; Kristensen et al. 2004; Mittler et al. 2004; Møller and Kristensen 2004; Papadakis and Roubelakis-Angelakis 2005; Papadakis and Roubelakis-Angelakis 2005; Halliwell 2006; Ślesak et al. 2007). The enhanced production of ROS during stress can pose a threat to cells but it is also thought that ROS act as signals for the activation of stress-response and defense pathways (Karpinski et al. 1999; Desikan et al. 2001; Knight and Knight 2001; Mittler 2002). ROS can be viewed as cellular indicators of stress and as secondary messengers involved in the stress-response signal transduction pathway thus, their level has to be kept under tight control by ROS-scavenging mechanisms, including enzymatic and non-enzymatic antioxidants. Under normal growth conditions, the antioxidative defence system provides the adequate protection against reactive oxygen species (Foyer and Halliwell 1976; Fridovich 1986; Asada and Takahashi 1987). However, under abiotic and biotic stresses the balance between ROS production and scavenging them by antioxidants may be changed and then ROS are generated in excess leading to cell death (Fig. 1) (Apel and Hirt 2004; Mittler et al. 2004; Dietz 2003; Gechev and Hille 2005).

The extent to which ROS accumulate is determined by the antioxidative system, which enables organisms to maintain proteins and other cellular components in an active state for metabolism and provide a dynamic metabolic interface between stress perception by plant cell and physiological responses (Mahalingam and Fedoroff 2003; Foyer and Noctor 2005, 2009; Mittler et al. 2004; Van Breusegem and Dat 2006; Ślesak et al. 2007; Niewiadomska and Borland 2008). Variable contents and concentrations of antioxidant between individual cell compartments enable plants to generate ROS-related signaling involving fundamental processes in plant cells. Moreover rapid, compartment-specific changes in redox state and ROS, related to redox signaling, can be achieved either by repression/activation of the antioxidant defense system or by modifying ROS (especially $O_2{\cdot}^-$, H_2O_2) production or by both

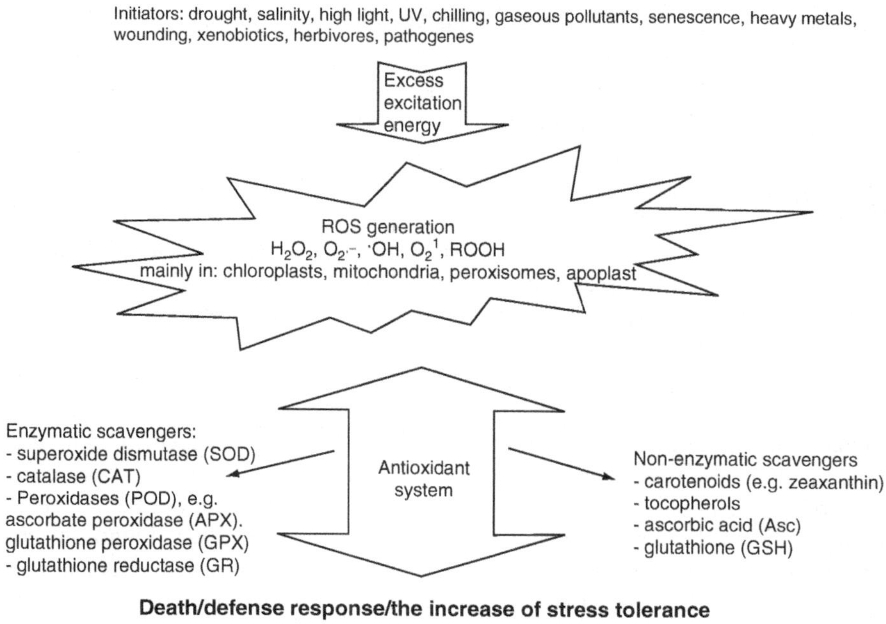

Fig. 1 The response of plants to different abiotic and biotic stresses. The antioxidant system enables plants to regulate reactive oxygen species (ROS) level and influence on ROS-dependent signal induction

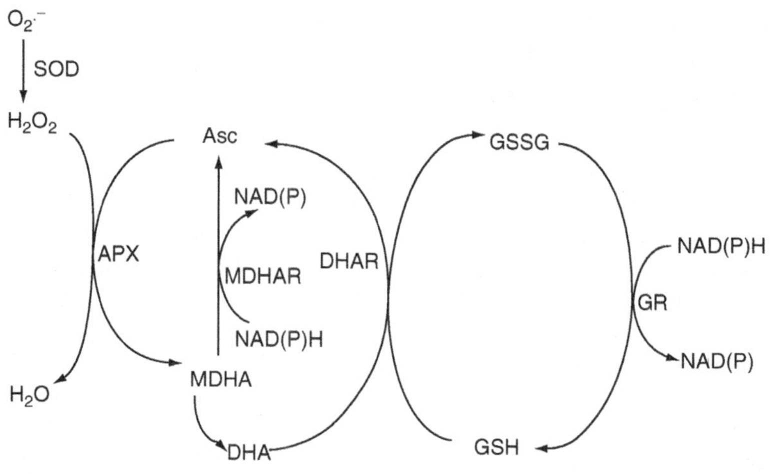

Fig. 2 The scheme of ascorbate–glutathione cycle. Asc – ascorbate, APX – ascorbate peroxidase, DHA – dehydroascorbate, DHAR – dehydroascorbate reductase, MDHA – monodehydroascorbate, GR – glutathione reductase, GSH – reduced glutathione, GSSG – oxidized glutathione

in individual organellas (Alscher 1997; Foyer and Noctor 2000; Pastori and Foyer 2002; Xiong et al. 2002; Foyer and Noctor 2005; Kuźniak and Skłodowska 2005; Fernandez-García et al. 2009). Superoxide dismutase (SOD) – the enzyme known as the first line of defence, catalyzes the dismutation of $O_2\cdot^-$ radicals to molecular oxygen and H_2O_2 (Fridovich 1986; Alscher et al. 2002). H_2O_2 scavenging is accomplished by catalase, various peroxidases, and the ascorbate–glutathione pathway, also known as the Halliwell–Asada cycle (Fig. 2), which is catalyzed by a set of four enzymes (Willekens et al. 1997; Wojtaszek 1997; Noctor 1998; Kingston-Smith and Foyer 2000; Jung 2003; Asada 2006; Queval et al. 2008; Almagro et al. 2009; Cosio and Durand 2009).

First, the hydrogen peroxide is scavenged via the oxidation of ascorbate (2,3-endiol-L-gulonic acid-γ-lactone; L-ascorbic acid; Asc) by ascorbate peroxidase (APX). This enzyme is involved in the oxidation of Asc to monodehydroascorbate (MDHA), which can be converted back to Asc via monodehydroascorbate reductase (MDHAR). MDHA is further rapidly converted to dehydroascorbate (DHA), which is converted back to Asc by the action of dehydroascorbate reductase (DHAR). DHAR utilizes glutathione (γ-L-glutamyl-L-cysteinyl glycine; GSH), which is regenerated by glutathione reductase (GR) from its oxidized form, glutathione disulfide (GSSG). The ascorbate–glutathione cycle operates in chloroplasts, mitochondria, peroxisomes, cytosol, and in the apoplast (Smirnoff 2000; Mittler 2002, 2004; Pignocchi and Foyer 2003; Asada 2006). Apart from the implication of ascorbate and glutathione in Halliwell–Asada cycle, these water soluble compounds could act independently. The specificity of functions of Asc and GSH are related to their compartmentalization (Veljovic-Jovanovic et al. 2001).

2 The Importance of Ascorbate–Glutathione Cycle Compounds in Different Cell Comparments Under Stress Conditions

Majority of plant cells contain in most compartments very large quantities of ascorbate (vitamin C) in milimolar concentrations (10–300 mM) and glutathione (0.1–25 mM) (Noctor and Foyer 1998; Smirnoff 2000, 2002; Noctor et al. 2002; Ruiz and Blumwald 2002; Mou et al. 2003; Ball et al. 2004; Freeman et al. 2004; Gomez et al. 2004; Shao et al. 2008). Hence, they have the capacity to deal with very high fluxes of H_2O_2 production and with the complex network of reactions surrounding ROS and keeping redox homeostasis (Noctor et al. 2002).

The important roles of Asc and GSH for tolerance towards environmental stresses were emphasize in investigations involvement mutants and transgenic plants lines with altered levels of these antioxidants (Pastori et al. 2003; Ball et al. 2004).

Seven ascorbate-deficient (*vtc*) mutants for *Arabidopsis thaliana* have been identified. These mutants represent four different VTC loci (*VTC1, VTC2, VTC3, VTC4*) involved in the maintenance of the Asc pool in the cell (Table 1).

Enzymes encoded by genes localized in these loci involved in major pathway of Asc biosynthesis in all chlorophyll-containing plants, in which GDP-D-mannose, formed from D-mannose-1-phosphate, is successively converted to GDP-L-galactose, L-galactose-1-phosphate, L-galactose, L-galactono-1,4-lactone, and finally to

Table 1 Comaparison of phenotype of ascorbate (vtc) deficient mutant plants

Genotype of *Arabidopsis thaliana* mutants	Published phenotype characteristics	References
vtc1-1	No phenotypic effects from increased oxidative stress increased sensitivity to ozone; the ascorbate levels 25–30% of wild-type levels; slower growth rate; no effect on photosynthesis rate or hydrogen peroxide levels under moderate light intensity; sensitive to UV-B and UV-C radiation; reduced basal thermotolerance microarrays analysis: 171 genes differentially expressed when compare to WT; 12.9% genes involved in cell defence – increase of pathogenesis related (PR) proteins; 60% higher concentration of ABA; increased resistance to infection by virulent pathogens; increased activity of peroxidase other than APX	Conklin et al. 1996, 2000, Veljovic-Jovanovic et al. 2001, Gatzek et al. 2002, Pastori et al. 2003, Barth et al. 2004, Filkowski et al. 2004, Müller-Moulé et al. 2004, Larkindale et al. 2005, Pavet et al. 2005, Colville and Smirnoff 2008; Gao and Zhang 2008
vtc1-2	Increased sensitivity to ozone; the ascorbate levels 25–30% of wild-type levels; reduced basal thermotolerance	Conklin et al. 2000, Larkindale et al. 2005
vtc2-1	Increased sensitivity to ozone; the ascorbate levels 25–30% of wild-type levels; slower growth rate; no effect on photosynthesis rate or hydrogen peroxide levels under moderate light intensity; reduced ability to zeaxanthin synthesis under high light; reduced basal thermotolerance; greater thermoinduced photon emission; increased resistance to infection by virulent pathogens reduced acclimation to high light demonstrated as the increase of lipid peroxidation and bleaching; in spite of 30% higher levels of GSH than the wild type	Conklin et al. 2000, Veljovic-Jovanovic et al. 2001, Havaux 2003, Pastori et al. 2003, Barth et al. 2004, Müller-Moulé et al. 2004, Larkindale et al. 2005, Pavet et al. 2005

(continued)

Table 1 (continued)

Genotype of *Arabidopsis thaliana* mutants	Published phenotype characteristics	References
vtc2-2	The ascorbate levels 25–30% of wild-type levels; slightly sensitive to ozone than wild-type plants; reduced ability to synthesise zeaxanthin under high ligh reduced basal thermotolerance	Conklin et al. 2000, Müller-Moulé et al. 2004, Larkindale et al. 2005
vtc2-3	The ascorbate levels 50% of wild-type levels in vtc2-3; slightly sensitive to ozone than wild-type plants; reduced ability to synthesise zeaxanthin under high light; reduced basal thermotolerance	Conklin et al. 2000, Müller-Moulé et al. 2004 Larkindale et al. 2005
vtc3	The ascorbate levels 50% of wild-type levels; slightly sensitive to ozone than wild-type plants	Conklin et al. 2000
vtc4	The ascorbate levels 50% of wild-type levels; slightly sensitive to ozone than wild-type plants	Conklin et al. 2000

L-ascorbate (Wheeler et al. 1998; Conklin et al. 1999, 2000, 2006; Dowdle et al. 2007; Smirnoff et al. 2004; Ishikawa et al. 2006; Colville and Smirnoff 2008; Linster et al. 2007; Linster and Clarke 2008).

Plants possess also alternative Asc biosynthetic pathway, which is suggested to play an important role under certain conditions, at particular developmental stages, and it seems to be engaged in salvage mechanism for carbon resulting from, i.e., the breakdown of cell walls (Davey et al. 1999; Agius et al. 2003). This Asc biosynthetic pathway includes the conversion of the uronic acids, D-glucuronic acid (D-GlcUA) and a precursor of L-galactono-1,4-lactone (L-GalL) such as D-galacturonic acid, to L-ascorbic acid (Davey et al. 1999; Agius et al. 2003; Lorence et al. 2004). In both biosynthetic pathway, the synthesis of L-galactono-1,4-lactone is rate-limiting step for Asc biosynthesis (Bartoli et al. 2000; Millar et al. 2002; Foyer and Noctor 2005; Bartoli et al. 2006). The regulation of the expression genes related to Asc metabolism in leaves of *Arabidopsis* plants was shown to be closely related to the photosynthetic electron transport (Yabuta et al. 2007, 2008) as well as jasmonic acid (Turner et al. 2002; Wolucka et al. 2005; Bartoli et al. 2006; Dowdle et al. 2007; Maruta et al. 2008). Plants make Asc via de novo synthesis pathways (Wheeler et al. 1998) and carbon skeleton re-cycling networks (Smirnoff et al. 2004).

The four glutathione-deficient mutants for *A. thaliana* have been characterized: *cad2-1* (Cobbett et al. 1998), *rax1* (Ball et al. 2004) and *pad2* (Parisy et al. 2007) and *rml1* (Vernoux et al. 2000) and applied in the studies of GSH importance in plant stress response (Table 2).

Table 2 Comaparison of genotype and phenotype of glutathione deficient mutant plants

Genotype of *Arabidopsis thaliana* mutants	Gene locus and enzyme of mutation	Published phenotype characteristics	References
cad2-1	Mutations in the gene At4g23100 which encodes γ-glutamylcysteine synthetase (γ-ECS, GSH1)	75–80% Reduction in GSH content when compared to wild type; two amino acid deletion (P238, K239); no phenotype in the absence of stress; the reduced ability to produce sufficient amounts of heavy metal binding phytochelatins from the precursor GSH; cadium hypersensitive	Cobbett et al. 1998
rax1	Mutations in the gene At4g23100 which encodes γ-ECS	R229K amino acid change; 50–80% decrease in the GSH level; lost of ability to properly control expression of cytosolic APX2	Ball et al. 2004
pad2	Mutations in the gene At4g23100 which encodes γ-ECS, GSH1	80% Reduction in GSH content; pathogen-sensitive mutant; increased susceptibility to *Phytophthora brassicae* infection; phytoalexin deficient	Glazebrook and Ausubel 1994, Glazebrook et al. 1996, 1997; Parisy et al. 2007
rml1	Mutations in the gene At4g23100, which encodes γ-ECS; *GSH1*	Shows a D259N amino acid change; not capable of maintaining cell division in the root meristem; arrested plant development even under optimal growth conditions	Vernoux et al. 2000, Cairns et al. 2006
GSH1 knockout mutant		Embryonic lethal phenotype; embryos produce no GSH and develop to their full size but just after greening, mutant embryos bleach	Cairns et al. 2006
GSH2 knockout mutants		Exhibit a seedlings; lethal phenotype	Pasternak et al. 2008

These mutants contain the mutation in the gene *At4g23100*, which encodes γ-glutamylcysteine synthetase (γ-ECS, GSH1) (Meyer 2008). The knockout of GSH1 through T-DNA insertions causes an embryonic lethal phenotype (Cairns et al. 2006). Similarly, knockout of glutathione synthetase (GSH2) through T-DNA insertions also causes lethal phenotype. So far, it is the state of the art in that the only isolated mutants of GSH2 are T-DNA insertion mutants (Kopriva and Rennenberg 2004; Mullineaux and Rausch 2005; Meyer 2008; Pasternak et al. 2008; Szalai et al. 2009). These two enzymes: GSH1 and GSH2 catalyse ATP-dependent reactions in the glutathione synthesis pathway. First enzyme catalyses the formation of γ-glutamylcysteine (γ-EC) from L-glutamate and cysteine and the second one bound a glycine to the dipeptide (γ-glutamylcysteine). Some authors point out that γ-glutamylcysteine synthetase is the controlling enzyme for overall flux through the pathway and synthesis of γ-EC is important step in the pathway (Jez et al. 2004; Hothorn et al. 2006; Hicks et al. 2007; Rennenberg et al. 2007; Meyer et al. 2008). GSH1 is exclusively located in the plastids, whereas GSH2, is located in both plastids and cytosol (Wachter 2004; Wachter and Rausch 2005; Mullineaux and Rausch 2005; Wachter et al. 2005; Hothorn et al. 2006; Hicks et al. 2007; Pasternak et al. 2008). In *A. thaliana*, *GSH1* and *GSH2* are present as single genes (May and Leaver 1994; Ullmann et al. 1996; The Arabidopsis Genome Initiative 2000; Mullineaux and Rausch 2005).

2.1 Apoplast and Symplast

The apoplast is the extraprotoplastic matrix of plant cells, consisting of all compartments from the external face of the plasmalemma to the cell wall (Dietz 1996). It represents the first line of contact between the plant and its external environment (Wojtaszek 1997; Bolwell et al. 1999). The apoplast plays crucial role in the perception of abiotic and biotic stresses as well as in plant defense and survival. Apoplast functions also as a highly sensitive and flexible compartment in signal-transducing processes throughout the plasmalemma. Through the cross-talk between apoplast and cytoplasm (symplast) the level as well as redox status of Asc in the apoplast are regulated (Fig. 3) (Diaz-Vivancos et al. 2006; Pignocchi and Foyer 2003; Pignocchi et al. 2003, 2006).

Asc is regenerated though Asc–GSH cycle or several mechanisms that include MDHAR in the symplast (Fig. 3) (Horemans et al. 2000; Pignocchi et al. 2003; Foyer 2004). Several types of Asc and DHA specific carriers localised in plasmalemma such as, i.e., Asc-dependent cytochrome b_{561} transporters are engaged in these processes (Foyer 2004; Preger et al. 2005). Asc-dependent cytochrom b_{561} was shown to be reduced by Asc on one face of plasmalemma and oxidized on the other by MDHA or other substrates for MDHA reductase (MDHAR, Preger et al. 2005). However, the preferably transported from the apoplastic to the cytoplasmic side of the membrane is DHA and this process is dependent upon plasma membrane proton gradient (Kollist et al. 2008).

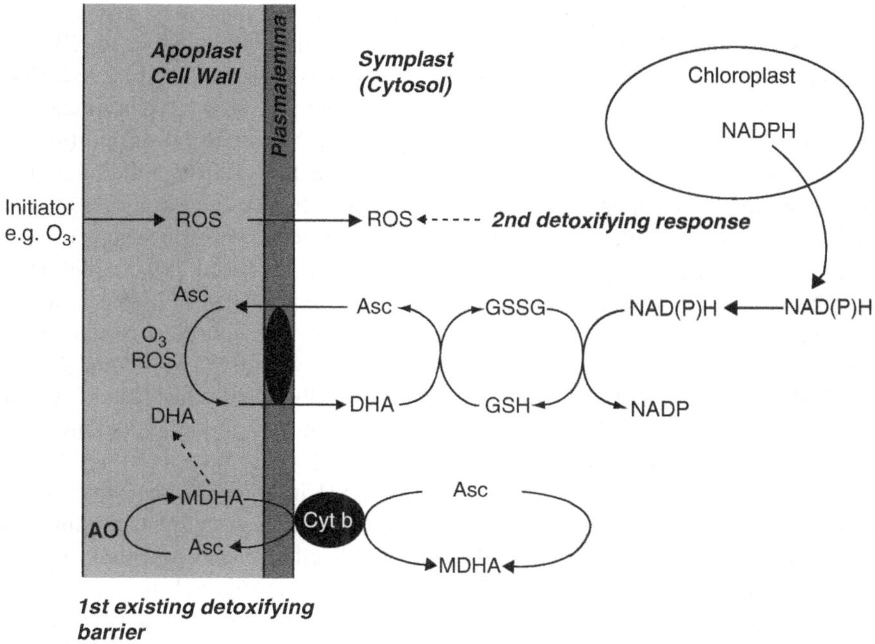

Fig. 3 Scavenging of ROS by ascorbate in apoplastic space of the cell. ROS – reactive oxygen species, Asc – ascorbate, DHA – dehydroascorbate, MDHA – monodehydroascorbate, GSH – reduced glutathione, GSSG – oxidized glutathione, AO – ascorbate oxidase

The capacity of redox buffering in the apoplast is much weaker than inside the cell (Horemans et al. 2000; Pignocchi et al. 2003), and the homeostasis in the apoplast is easily perturbed and this is why environmental factors can be perceived by the cell throughout the components of the apoplastic space (Lyons et al. 1999; Sattelmacher 2001; Bolwell et al. 2002; Pignocchi and Foyer 2003). The apoplast matrix does not contain NAD(P)H as well as GSH and its derivatives (Noctor 2006; Zechmann et al. 2006, 2008) and it is equiped with antioxidative enzymes such as SOD, MDHAR as well as Asc. Asc is considered as its major antioxidant (Hernández et al. 2001; Foyer and Noctor 2005, 2009; Diaz-Vivancos et al. 2006) and it is present in the apoplast at millimolar concentrations (up to about 10% of the of the whole cell Asc content). These all apoplastic antioxidants are inter-linked with oxidants like ($O_2^{\cdot -}$) and H_2O_2, which are also present in the apoplast especially under stress conditions (Hernández et al. 2001; Pignocchi and Foyer 2003). It is well known that abiotic and biotic stress factors like: pathogens, drought, salinity, air pollutants or strong light cause oxidation of Asc directly or indirectly (Mittler 2002, 2004; Foyer and Noctor 2003, 2005, 2009). The oxidation of Asc occurring in the apoplast and the related alteration in its redox state are the primary symptoms of oxidative burst. However, with the other hand these alterations are related to defense responses through the regulation of signal transduction cascades and gene expression patterns. The ascorbate redox state in the apoplast is largely independent of that in the symplasm which is relatively constant throughout the life of a cell. Moreover, the sensors located in the apoplast may have different properties compared with those operating in the symplast (Delauney et al. 2002).

The redox state of the Asc pool in the apoplast is also regulated by ascorbate oxidase (AO), which is a cell wall-localized enzyme that uses oxygen to catalyse the oxidation of Asc to the unstable radical MDHA that rapidly disproportionates to yield DHA and Asc (Fotopoulos et al. 2006). AO activity, through modulation of the redox state of the apoplastic AA pool, strongly influences the responses of plant cells to external and internal stimuli. As it was presented on transgenic tobacco (*Nicotiana tabacum*) plants with modified cell wall-localized AO, oxidation of apoplastic Asc was associated with the loss of the auxin response, higher mitogen-activated protein kinase activities, and susceptibility to a virulent strain of the pathogen *Pseudomonas syringae* (Pignocchi and Foyer 2003; Foyer and Noctor 2006; Pignocchi et al. 2006). The control of AO by light and other factors such as gibberellic acid (GA), salicylic acid (SA) and auxin may modulate defense processes (Pignocchi and Foyer 2003; Pignocchi et al. 2006).

In addition to Asc, plants also accumulate GSH in compartments such as the chloroplast, cytosol and mitochondria. Plants maintain most cytoplasmic thiols in the reduced (2SH) state because of the low thioldisulfide redox potential imposed by millimolar amounts of the thiol buffer, glutathione. Although the amounts of GSH which is present inside the cell are probably small (Gukasyan et al. 2002), the intracellular ratio GSHred/GSHox, which determines the redox potential, could have been high and sufficient to reduce ROS.

2.2 Chloroplasts

Chloroplasts are frequently exposed to ROS in the light, due to the presence of high redox potential components, excited states of pigments, and generation of free electrons during photosynthetic electron transport in the thylakoid membrane. The energy sensing and signalling pathways initiated in chloroplasts in relation to the photochemical and biochemical dysfunctions of photosynthesis extend beyond these organelles such as mitochondria and affect also processes other than photosynthesis (Mullineaux and Karpinski 2002; Rossel et al. 2002; Ball et al. 2004; Mateo et al. 2004; Mittler et al. 2004; Kuźniak et al. 2009). Therefore ROS generated in excess have to be effectively scavenged, and this can be done by several systems/mechanisms like the xanthophyll cycle, photorespiration and other changes in metabolic activity (Demmig-Adams and Adams 1996; Kozaki and Takeba 1996; Eskling et al. 1997; Osmond et al. 1997), and a number of other nonenzymatic and enzymatic antioxidants such as i.e. carotenoids and α-tocopherol (vitamin E) (Smirnoff 1993; Niyogi 2000; Mullineaux and Karpinski 2002; Apel and Hirt 2004; Mittler et al. 2004; Foyer and Noctor 2005, 2009; Hofius et al. 2004; Munné-Bosch 2004; Munné-Bosch and Falk 2004; Plaxton and Podestá 2006).

Within the chloroplastic mechanisms of protection against the potentially deleterious effects of ROS, a pivotal role play Asc and GSH. The reduced forms of Asc and

GSH, together with antioxidant enzymes: APX, MDHAR, DHAR, glutathione peroxidise (GPx) and glutathione reductase (GR) make up the ascorbate–glutathione cycle (Fig. 2), and they are involved in ROS scavenging, similarly to water–water cycle that enables the electron flow through the photosynthetic electron transport carriers at limited availability of electron acceptors, such as $NADP^+$ and/or CO_2 (Asada et al. 1999; Asada 2006).

Asc and GSH are water-soluble antioxidants and they are able to act directly. They scavenge lipid hydroperoxide (LOOH) and remove H_2O_2 generated during photosynthetic processes in the light. Asc and GSH serve both as cofactors for peroxidases and as reductants. Asc is the main substrate for H_2O_2 reduction by thylakoid-bound and stromal ascorbate peroxidases (Melhorn et al. 1993). Ascorbate peroxidase reduces H_2O_2 to water producing two molecules of oxidised ascorbate, monodehydroascorbate (MDHA), which rapidly disproportionates to Asc and DHA (Asada 2006). The MDHAR and DHAR are responsible for regenerating Asc, the later using GSH as reducing substrate. Glutathione reductases regenerate GSH in a NADPH-dependent reaction (Foyer and Halliwell 1976). The low antioxidant buffering could allow oxidative signals to accumulate within the lumen, and this may be important in redox signal transduction leading to programmed cell death (PCD) (Van Breusegem and Dat 2006).

Asc and GSH help to maintain the integrity of the photosynthetic membranes under oxidative stress (Havaux 2003; Noctor and Foyer 1998; Asada 2006; Smirnoff and Wheeler 2000; Munné-Bosch and Alegre 2002). Asc may indirectly protect α-tocopherol by scavenging ROS, and it may participate in the recycling of α-tocopheroxyl radicals to α-tocopherol (Veljovic-Jovanovic et al. 2001; Pastori and Foyer 2002), thus resulting in DHA accumulation under stress (Munné-Bosch and Alegre 2002, 2003, 2007; Packer et al. 1979; Smirnoff and Wheeler 2000). In the lumen Asc is the electron donor in the enzymatic conversion of violaxanthin to zeaxanthin by the lumenal violaxanthin deepoxidase. In experiments with Asc-deficient *Arabidopsis thaliana* mutants it was demonstrated that the expose of *vtc2-2* mutant to high light was accompanied with reduced nonphotochemical quenching (NPQ) related to the alteration in xanthophyl cycle pigments pattern (Conklin et al. 2000; Veljovic-Jovanovic et al. 2001; Müller-Moulé et al. 2002, 2003, 2004), while in other ascorbate deficient mutants follow oxidative stress responses (Kiddle et al. 2003; Pastori et al. 2003; Rossel et al. 2002; Davletova et al. 2005; Mahalingam et al. 2005). Some authors (Rautenkranz et al. 1994; Horemans et al. 2000) presented data supporting the increased transport of Asc from the cytosol to chloroplasts in stressed plants, and thylakoid membranes are able to transport Asc by diffusion (Foyer and Lelandais 1996; Horemans et al. 2000).

The unique status of chloroplasts is also related to the relationship of C, N and S assimilation pathways in these organelles that are influenced by a broad spectrum of environmental conditions (Paul and Foyer 2001). The complexity of these chloroplast regulations may be emphasize by GSH and salicylic acid (SA) biosynthesis (Kuźniak et al. 2009). The high concentrations of GSH found in the chloroplast could be to some extent dependent on import from the cytosol, although the contributions of the two compartments to GSH formation may vary between different cells and these processes also influence physiological status (Foyer et al. 2001). The wheat chloroplasts are able to import glutathione and this activity is not affected by either light or ATP (Noctor et al. 2002). Transcripts for one of the transporters with high specificity to GSH, cloned in *Arabidopsis* have been shown to be induced by xenobiotic exposure, and not by H_2O_2 or cadmium (Cagnac et al. 2004).

Another key player in glutathione content is Cys availability, which concentration is a limiting factor for γ-ECS activity (Harms et al. 2000; Droux 2004). Bick et al. (2001) shown that ozone exposure leads to up-regulation of sulphur assimilation in response to decrease in GSH:GSSG, via activation of adenosine 5′-phosphosulphate reductase and a total tissue glutathione level is typically 5–20 times more abundant than free Cys and is highly mobile throughout the plant (Herschbach and Rennenberg 1995). Moreover, glycine supply for biosynthesis of glutathione from γ-glutamyl-cysteine (γ-EC) is to some extent light dependent, and has been suggested to come from the photorespiratory cycle (Kopriva and Rennenberg 2004). Together with metabolites such as *O*-acetylserine (Hirai et al. 2003), GSH may act as an indicator of sulphur status. A simple regulatory mechanism has been proposed with regard to the regulation of demand driven sulphur assimilation by positive signals such as *O*-acetylserine and negative signals such as GSH (Kopriva and Rennenberg 2004).

2.3 *Mitochondria and Peroxisomes*

Besides chloroplasts, mitochondria are also an important cellular site for production of ROS and are themselves a target of oxidative damage. Thus, these organelles are equipped with various antioxidant defences that help to inhibit any enhancement of reactive oxygen species production. The constituents of the Asc–GSH cycle have also been localized in these organelles and GSH and GPx represent the most important antioxidant compounds in mitochondria (Jiménez et al. 1997). The Asc synthesis is related to the level of the mitochondrial electron transport chain through cytochrome c pathway and non-phosphorylating pathway related to alternative oxidase (AOX). Since these both mitochondrial pathways are involved in the regulation of photosynthesis it could be supposed that the Asc synthesis is to some extend the respond of cells to stress factors. Jiménez et al. (1998) demonstrated on pea leaves that the ascorbate–glutathione cycle compounds are also present in peroxisomes. In peroxisomes the increase of glutathione pools is a result of senescence when the strong oxidative damage takes place. The differential response to senescence of the mitochondrial and peroxisomal ascorbate–glutathione cycle compounds suggests that mitochondria could be affected by oxidative damage earlier than peroxisomes. Willekens et al. (1997) have demonstrated the specific oxidation of the glutathione pool in catalase-deficient barley suggested that the increase of GSH content may be a compensatory mechanism. Kopriva and Rennenberg (2004) pointed out that photorespiration pathway in peroxisomes may supply glycine for glutathione synthesis in normal metabolism or under stress conditions. Changes in the redox state of GSH in mitochondria correlate with changes in oxidative damage of mitochondrial DNA. Thus, the GSH redox status plays an important role in controlling critical functions for mitochondrial and cell survival. In lines with altered aconitase and malate dehydrogenase (MDH) activities revealed a specific increase of ascorbate in leaves (Carrari et al. 2003; Nunes-Nesi et al. 2005), what is coupled to a general up-regulation of levels of transcripts encoding genes associated with photosynthesis (Urbanczyk-Wochniak et al. 2006). Sweetlove et al. (2007) suggest that redox signals emanating from the mitochondria are important in setting the cellular machinery to maintain the overall redox balance, and experiments on transgenic/mutant plants hint at important role for ascorbate in the co-ordination of major pathways of energy metabolism in the leaf (Nunes-Nesi et al. 2007). Chew et al. (2003) stated that the key components of the ascorbate–glutathione cycle in *Arabidopsis* cell organelles are encoded in nuclear genome by single organellar targeted isoforms that are dual localized in the chloroplast stroma and the mitochondria. The defined set of proteins has dual activities in the intermembrane space and matrix and many of these proteins are dual-targeted to chloroplasts and Moreover, changes in transcript level of APX, MDHAR, and GR genes were induced by oxidative stresses imposed on chloroplasts and/or mitochondria and elevated transcript levels were maintained during photosynthetic operation in the light.

3 The Action of Ascorbate and Glutathione as Important Components of the Antioxidative System

The effects of oxidative stress depend on the type of cell, the level of oxidative stress experienced, and the protective mechanisms functioning in that cell type. The sensitivity of cells to oxidative stress is determined by their intrinsic antioxidant systems, in particular the levels of Asc and GSH (Banki et al. 1996). The level of antioxidants is related to the potential extent of antioxidative protection and the balance between their synthesis, oxidation, and regeneration (Boo et al. 2000; Robinson and Bunce 2000; Mittler et al. 2001; Tausz et al. 2001; Herbinger et al. 2002). Positive correlations between high antioxidant capacity and environmental stress tolerance have frequently been found. The relationship between the dehydroascorbate/ascorbate redox couples and the glutathione disulfide/glutathione (GSSG/2GSH) ratio are frequently used as markers of plant stress (Noctor and Foyer 1998).

The most prominent and best established functions of Asc and GSH are those of crucial antioxidants in the Asc–GSH cycle (Arrigoni 1994; Cordoba and Gonzales-Reyes 1994; Noctor and Foyer 1998; Asada 1999). The Asc–GSH cycle serves the removal of H_2O_2, which are inevitably formed as by-products of the normal metabolism or as a consequence of environmental stresses. This cycle comprises three interdependent redox couples: Asc/DHA, GSH/GSSG, and NADPH/$NADP^+$ (May et al. 1998).

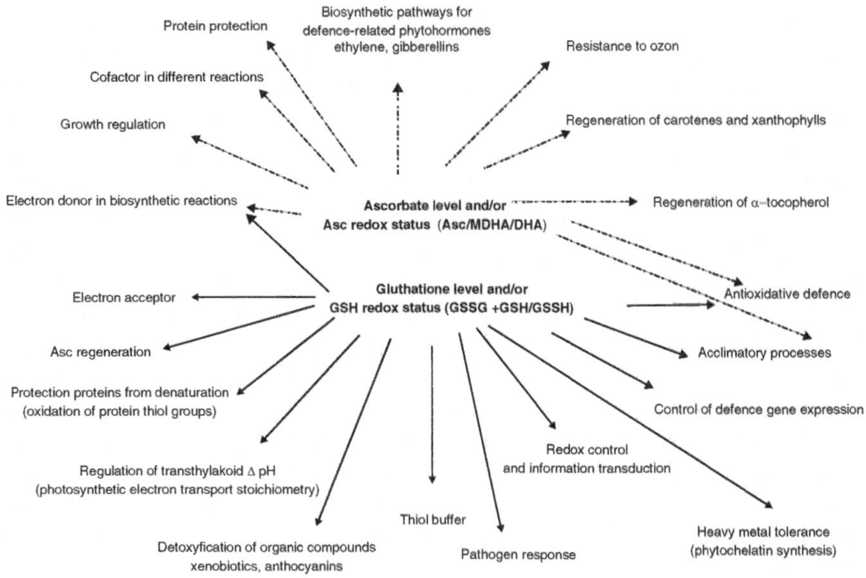

Fig. 4 The metabolic processes influenced by ascorbate and glutathione

Plant cells synthesize Asc and use it as a hydrophilic redox buffer to provide protection against oxidative challenge (Fig. 4).

Asc is known as a major low molecular antioxidant in plants protecting them by directly reaction with ROS such as $O_2{\cdot}^-$, $\cdot OH$, $^1O_2$ and organic radicals. Indirectly Asc protect cells against ROS by participation in:

- Ascorbate photonic energy dissipation mechanisms, such as the water–water cycle (Asada 1999), ascorbate–glutathione and xanthophyll cycles
- Recycling of lipid soluble α-tocopherol by reduction of its oxidised form (Munné-Bosch and Alegre 2002)
- Protection of enzymes (Padh 1990)

Arabidopsis thaliana plants grown in high light (HL) accumulate more Asc and have higher ascorbate peroxidase activity than plants grown in low light (LL) (Tabata et al. 2002; Müller-Moulé et al. 2004; Bartoli et al. 2006; Dowdle et al. 2007). The Asc levels in leaves show a diurnal rhythm (Dutilleul et al. 2003; Pignocchi et al. 2003; Tamaoki et al. 2003), vary with the daylight conditions associated with long-term seasonal variations (Grace and Logan 1996) and the maintain of a constant Asc pool level is dependent on the rate of its biosynthesis (Pallanca and Smirnoff 2000; Chen et al. 2003). It was shown that leaves of tropical plants with lacking DHAR activity under HL turn yellow and accumulated high level of flavonoid pigments (Yamasaki et al. 1999). The overexpression of DHAR increases total tissue ascorbate pools (Chen et al. 2003; Chen and Gallie 2004, 2005).

An increased content of Asc protects proteins and lipids against oxidative damage in plants subjected to water and salt stress (Tambussi et al. 2000; Shalata and

Neumann 2001). The protective effects of Asc was documented also in plants exposed to ozone (Sanmartin et al. 2003). Retsky et al. (1993) demonstrated that plants accumulate also DHA during stress. DHA is able to act as an antioxidant. The effectivity of DHA was even higher than Asc when low-density lipoprotein from copper ion induced oxidation was investigated.

Parallel to Asc, glutathione, plays an important functions in cellular defence and protection. Its reduced form, GSH, exists interchangeably with the oxidized form, GSSG. Schafer and Buettner (2001) pointed out that the GSSG/2GSH couple is the major cellular redox buffer and GR may be a central determinant of antioxidant capacity by controlling the GSH/GSSG ratio. Due to the stability of the disulfide bridge in GSSG, it is more stable than Asc and tocopherol, particularly in dry biological systems (Kranner et al. 2006). In most tissues, glutathione is maintained in GSH form. However, under stressful conditions its redox status: [GSH/(GSSG + GSH)] changes and the relative amount of GSSG may be enhanced considerably (Noctor and Foyer 1998; Wingsle et al. 1999). The efficiency with which GSSG can be re-converted to GSH during the reductive inactivation of peroxides contributes to the centrality of GSH in antioxidant defences. The significant decrease in GSH:GSSG ratio, followed by glutathione pool accumulation after exposure to ozone (Gupta et al. 1991; Pasqualini et al. 2001; Bick et al. 2001). A similar response precedes leaf bleaching in maize subjected to chilling (Gomez et al. 2004). The involvement of GR in the maintenance of a high GSH/GSSG ratio in plants during stress conditions was observed in tomato under salinity, desiccated *Myrothamnus flabellifolia* and in wheat under drought stress (Shalata et al. 2001; Kranner et al. 2002; Kocsy et al. 2004). May et al. (1998) have also suggested that the pool size of GSH in plants enable to percept changes from the environment and it is able to up-regulate defense mechanisms involved other redox active components such as i.e., Asc, NADPH pools during environmental stress. Thus, the size of the reduced glutathione pool shows marked alterations in response to a number of environmental conditions and the activation of GSH synthesis and accumulation are a general feature of enhanced oxidation processes (Foyer and Noctor 2003, 2005, 2009).

The response of the glutathione pool is also highly dependent on stress location. The concentration and/or redox state of glutathione correlates also with the adaptation of plants to abiotic and biotic stresses and tolerance of xenobiotics. The considerable increases in the pool of GSH have been measured in response to chilling, heat shock, pathogen attack, air pollution, drought, extreme temperatures and heavy metal stress (Dietz 2003; Edwards et al. 1991; Foyer and Noctor 1998; Gupta et al. 1991; Kocsy et al. 1996, 2000, 2004a,b; May et al. 1998; Madamanchi and Alscher 1991; Nieto-Sotelo and Ho 1986; Saito 2004; Parisy et al. 2007). It can react directly with a range of ROS, and indirectly throughout enzyme-catalysed reactions link GSH to the detoxification of H_2O_2 in the ascorbate–glutathione cycle. GSH, is the reducing co-factor for several enzymes involved in ROS detoxification, in particular peroxidases, as well as for enzymes such as formaldehyde dehydrogenase that may be important in processing products of lipid peroxidation. Additionally, in reactions catalyzed by glutathione S-transferases

(GSTs) GSH is involved in removing lipid peroxides, methylglyoxal, and herbicides (Moons 2005; Rausch et al. 2007; Yadav et al. 2008; Szalai et al. 2009). Different types of GSTs catalyse conjugation of GSH to xenobiotics or endogenous metabolites, while others have peroxidase or isomerase activities (Edwards et al. 2000; Noctor 2006).

Exposure of *Arabidopsis* cell cultures to oxidative stress (by the addition of aminotriazole, menadione or fenchlorazole) lead to the increase of cellular glutathione content and γ-ECS at the activity but not mRNA level in the bundle sheath exposed to chilling. This phenomenom, however was not observed in mesophyll cells, what suggested different GSH content in the two types of cells (Gómez et al. 2004). The increase of γ-ECS activity and GSH level was related to cadmium tolerance in tomato and chilling tolerance in maize (Chen and Goldsbrough 1994; Kocsy et al. 1996). Transgenic poplars overexpressing γ-ECS has the higher, than wild-type, possibility to accumulate Cd and parallel presented the increase tolerance to chloracetanilide herbicides (Gullner et al. 2001; Koprivova et al. 2002). The enhanced synthesis of GSH together with phytochelatins in various plant species exposed to cadmium stress was also described (Mendoza-Cózatl et al. 2005). In *Arabidopsis* the abundance of *GSH1* transcripts coded γ-ECS was increased after treatment with heavy metals such as cadmium and copper (Xiang and Oliver 1998), while in maize as a response to chilling stress (Gomez et al. 2004). In *Brassica juncea* the abundance of *GSH1* and *GSH2* transcripts was increased by cadmium (Schäfer et al. 1998). Ozone and catalase deficiency are both known to trigger production of jasmonic acid (JA), which has been shown to increase *GSH1* and *GSH2* transcripts (Xiang and Oliver 1998; Harada et al. 2000). Moreover, the expression of the gene coding γ-ECS in *Arabidopsis thaliana* plants exposed to chilling stress and in maize fumigated with ozone were also described (Gómez et al. 2004; Sasaki-Sekimoto et al. 2005). The expression of the GR, GST, and GPX genes and resulting in increases in the activities of the corresponding enzymes was stated in *Euphorbia esula* plants exposed to xenobiotic (diclofop-methyl) (Anderson and Davis 2004). However, in pea plants affected by heat stress increased GR gene expression was not accompanied by the higher GR activity (Kurganova et al. 1999; Escaler et al. 2000). Kocsy et al. (2000, 2004) have demonstrated in $_{35}$S-labeling experiments that osmotic stress, high temperature, and cold treatments induced a greater increase in GSH synthesis in tolerant wheat genotypes than in sensitive ones.

Furthermore, some authors (Durner et al. 1999; Díaz et al. 2003) presented that glutathione may be required as a relatively stable carrier of nitric oxide (NO), in the form of *S*-nitrosoglutathione (GSNO). Barroso et al. (2006) presented that GSNO participate in response of plants to Cd stress. Barker et al. (1996) demonstrated that the oxidative damage (i.e., by reactive nitrogen species – RNS) only occurs in the presence of a marked GSH deficiency.

GSH, together with Asc are generally considered to constitute an antioxidant buffer that is involved in control of ROS level. Mittova et al. (2003) suggested that the enzymes of GSH synthesis and metabolism are induced together in response to stress which, in turn affects the level of ascorbate and glutathione.

4 The Role of Ascorbate–Glutathione Cycle in Regulation of the Xanthophyll Cycles Activity

In almost all species of plants exists a photoprotection mechanism, strongly dependent on ascorbate, called common name "xanthophyll cycle". Term "xanthophyll cycle" means cyclic enzymatic removal of epoxy groups from xanthophylls under strong light condition (de-epoxidation) and reverse reaction, during which the de-epoxidised xanthophylls are epoxidised. First reaction is catalyzed by de-epoxidases and the second one by epoxidases. The xanthophyll cycles take place in the thylakoid membranes. De-epoxidases are located in the thylakoid lumen and bind to membrane when the lumen becomes acidified, while epoxidases are postulated as stromal enzymes, connected with thylakoid membrane and they operate at rather neutral pH. Asc is necessary for all known types of de-epoxidases. Three xanthophylls which are de-epoxidated in nature are violaxanthin (Vx), diadinoxanthin (Ddx) and lutein-epoxide (Lx). Depending on what kind of xanthophyll is de-epoxidased three different xanthophyll cycles are distinguished (Fig. 5) (Jahns et al. 2009).

When Vx, undergoing de-epoxidation, the cycle is called violaxanthin cycle (Fig. 5a). Since it is widely distributed among all higher plants, ferns, mosses, some green algae like Phaeophyta, Chlorophyta or some Rhodophyta (Siefermann-Harms 1985) and it was the first xanthophyll cycle which has been described in the literature (Sapozhnikov et al. (1957) it is called xanthophyll cycle. Pioneering work by the groups of Yamamoto and Hager in the 1960s and 1970s established the basic biochemical characteristics of the Vx cycle (Hager 1969) in which Vx is reversibly converted to Zx via the intermediate antheraxanthin (Ax) (Yamamoto 1979). These two-step reactions, during which two epoxy groups are stepwise removed (in the de-epoxidation reactions) or inserted (in the epoxidation reactions), are catalyzed by the Vx de-epoxidase (VDE) and the Zx epoxidase (ZE), respectively (Fig. 5a).

In one of marine algae species, called *Mantoniella squamata*, a modification of the Vx cycle was found (Goss et al. 1998). The most intriguing feature of this type of the cycle is that the second step of the de-epoxidation i.e. conversion of Ax to Zx is very slow and, as a consequence, and contrary to typical Vx cycle, Ax is accumulated in place of Zx. These observations can lead to the conclusion, that VDE from Mantoniella has less affinity to Ax than VDEs from other plants. Up till now this is the only known modification of the xanthophylls cycle involved Vx.

In diatoms and some other groups of algae, the xanthophyll cycle is called diadinoxanthin cycle (Fig. 5b) because it comprises the reversible conversion of monoepoxide, diadinoxanthin (Ddx) and epoxide-free, diatoxanthin (Dtx) (Stransky and Hager 1970; Olaizola et al. 1994; Coesel et al. 2008). Analysis of genome of one of diatoms, *Phaeodactylum tricornutum*, indicates one *VDE* gene and two *VDE*-like genes, designated as *PtVDE* and *PtVDL1* and *2*, respectively. The first was proposed to be involved in the Vx cycle, which is also present in these organisms, while the latter were suggested to be more specialized in the chromist-specific Ddx-cycle. Proteins coded by *VDL* genes contain larger lipocalin domain than typical VDE and they were found only in chlorophyll c-containing chromist algae. This

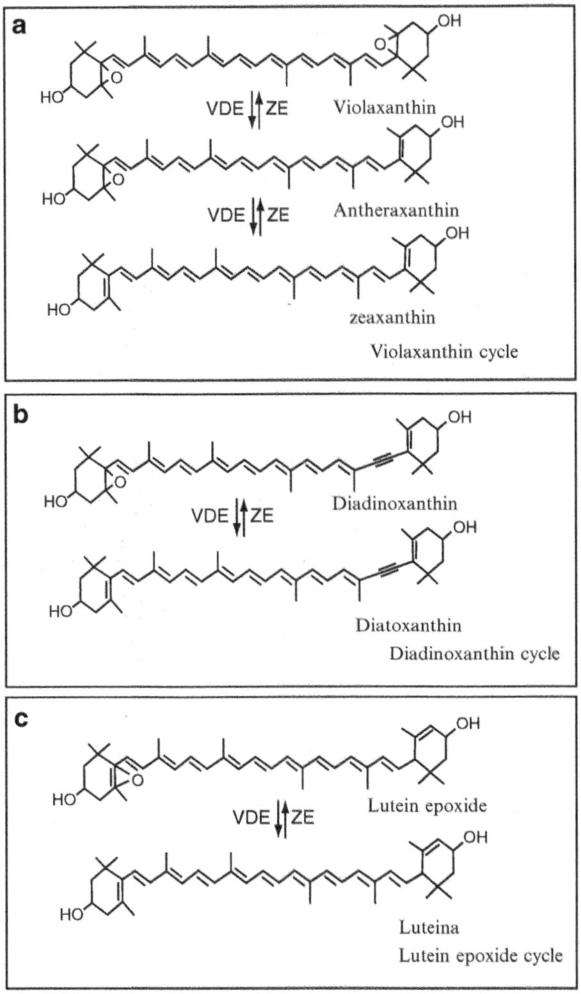

Fig. 5 Three different types of xanthophyll cycles: (**a**) violaxanthin cycle, called also commonly xanthophyll cycle, (**b**) diadinoxanthin cycle, (**c**) epoxy lutein cycle

is why these enzymes were suggested to be able to more efficiently bind, and eventually de-epoxidise, brown algal-specific molecules such as diadinoxanthin. They were called diadinoxanthin de-epoxidases (DDE) (Coesel et al. 2008).

The third type of xanthophyll cycle is lutein epoxide cycle which has been identified in green tomato fruits (Rabinowitch et al. 1975) and recently in holoparasitic *Cuscuta reflexa* (Bungard et al. 1999), hemiparasitic plants (Matsubara et al. 2001, 2003), *Quercus species* (Garcia-Plazaola et al. 2002), a tropical tree *Inga sapindoides*, and others (Garcia-Plazaola et al. 2007). This cycle comprises reversible light-induced changes at the level of lutein epoxide (Lx) and lutein (L) (Fig. 5c). These reactions are probably catalysed by the same enzymes like Vx-cycle (Yamamoto and Higashi 1978; Goss 2003; Garcia-Plazaola et al. 2003, 2007).

The main postulated function of the xanthophyll cycle, connected with production of de-epoxidised xanthophylls, is to allow harmless energy dissipation by the NPQ process. NPQ is recognised as a central regulatory mechanism for protecting plants from photodamage (Arsalane et al. 1994; Olaizola and Yamamoto 1994; Frank et al. 1996; Grouneva et al. 2008, 2009; Matsubara et al. 2001, 2005; Müller et al., 2001; Garcìa-Plazaola et al. 2003). Additional postulated functions of the xanthophyll cycles are:

- Protection against oxidative stress of lipids (Havaux et al. 1991; Sarry et al. 1994)
- Photoreception of blue light (Quinones and Zeiger 1994; Srivastava and Zeiger 1995)
- Modulation of the physical properties of the thylakoid membrane (Gruszecki and Strzałka 1991; Tardy and Havaux 1997)
- Part of abscisic acid synthesis pathway (Marin et al. 1996)

All known enzymes of the xanthophyll cycles have almost the same properties and requirements. Factors that influence the de-epoxidases activity include the availability of substrates, redox potential, lipid composition, temperature and pH. At low light intensities (lumenal pH around 7) the enzymes are water soluble. Upon shift to high light conditions the pH of the lumen drops, with the overall effect that de-epoxidation commences and shows an optimum around pH 5 for VDE or 5,5 for DDE. Vx de-epoxidation starts at pH-values below 6.5 while conversion of Ddx to Dtx starts at pH 7 and the de-epoxidation state increases rapidly with increasing acidification (Jakob et al. 2001). The drop in pH is necessary for binding of enzymes to the membrane but also for the protonation of ascorbate (pKa 4.1) to create acid form of reduced ascorbate (AscH) (Fig. 6) (Bratt et al. 1995).

All known types of de-epoxidases require ascorbate (Asc) as reductant to carry out de-epoxidation (Hager 1969; Yamamoto 1979). The possible role of Asc availability in regulation of de-epoxidases activity was discovered when inhibition of VDE by competitive consumption of Asc by ascorbate peroxidase (APX) upon addition of hydrogen peroxide was noticed (Neubauer and Yamamoto, 1994). Bratt et al. (1995) showed that the optimum Asc concentration for VDE activity is strongly pH-dependent. At pH 4.5–5.5 the enzyme becomes saturated at 10–20 mM Asc (Table 3), whereas at pH 6.0 it is not saturated even at 100 mM.

The Asc concentration in chloroplasts is supposed to be in the range of 10–50 mM (Gillham and Dodge 1986; Schöner and Krause 1990; Foyer 1993) although it may increase in response to stress factors such as high light irradiances and chilling (Gillham and Dodge 1987; Schöner and Krause 1990). The Asc-dependent pH optimum is similar for both VDE located in the thylakoids and partially purified enzyme (Yamamoto 1979), albeit the partially purified VDE shows a broader pH optimum and a shift to higher pH in comparison with VDE in thylakoids (Table 3).

These differences may be caused by the limitation of Asc passage across the thylakoid membrane at higher pH, due to the negative charge of the base form of Asc (Asc^-) (Fig. 6). These results suggest that VDE has a pH-dependent K_M for Asc

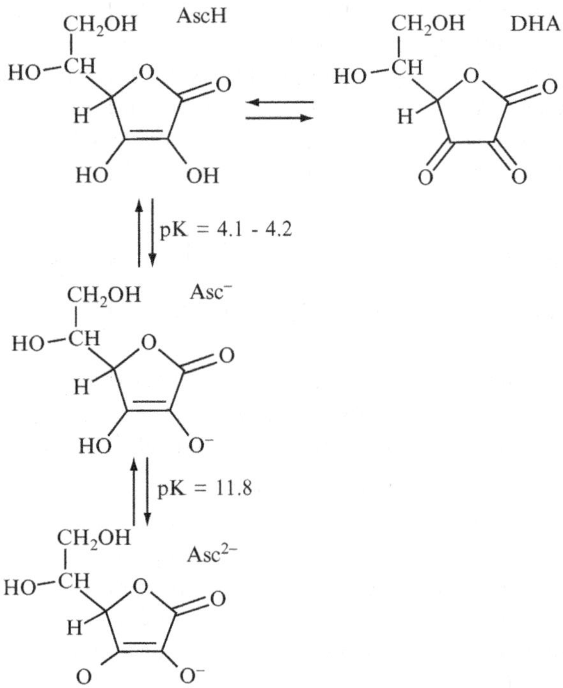

Fig. 6 Different forms of ascorbate. Three reduced forms (AscH, Asc^{-}, Asc^{2-}) and dehydroascorbate (DHA) as oxidised form of ascorbate (Asc)

Table 3 The pH-optimum of VDE or DDE activity in dependence on the Asc concentration

Asc concentration [mM]	[a]Optimum pH of VDE activity in intact thylakoids	[a]Optimum pH of partially purified VDE activity	Optimum pH of partially purified DDE[b]
0.5	4.7	4.88	5.00
3.0	4.8–4.9	5.15	5.30
30.0	5.0	5.25	5.58

[a]Bratt et al. 1995
[b]Grouneva et al. 2006

(Table 4) and that Asc is not simply a cofactor but a co-substrate for VDE (Bratt et al. 1995; Eskling et al. 1997).

Interestingly, all determined values of the K_M for Asc fell within the range of 0.10 ± 0.02 mM when Asc concentrations are expressed for the reduced acid form of Asc (AscH) and assuming a pKa of 4.1. This suggests that not the negatively charged Asc but rather the acidic form AscH is the substrate for VDE or DDE (Fig 6).

AscH as a protonated form of Asc is an endogenous electron and proton donor for de-epoxidation and activates VDE and DDE (Yamamoto 1979; Bratt et al. 1995; Sokolove and Marsho 1976; Neubauer and Yamamoto 1994; Eskling et al. 1997). These de-epoxidases catalyze electron and proton transfer from AscH to one or two

Table 4 Apparent K_M values (mM) of violaxanthin de-epoxidase (VDE) and diadinoxanthin de-epoxidase (DDE) for Asc at different pH values. ND - not determined

	K_M VDE (mM)		
pH	[a]	[b]	K_M DDE**
4.5	0.3	ND	0.6 ± 0.05
5.0	1.0	2.3 ± 0.09	0.7 ± 0.04
5.5	2.5	5.0 ± 0.56	1.2 ± 0.05
6.0	10	13.8 ± 0.91	6.6 ± 0.34
6.5	–	ND	8.7 ± 0.79

[a]Bratt et al. 1995
[b]Grouneva et al. 2006

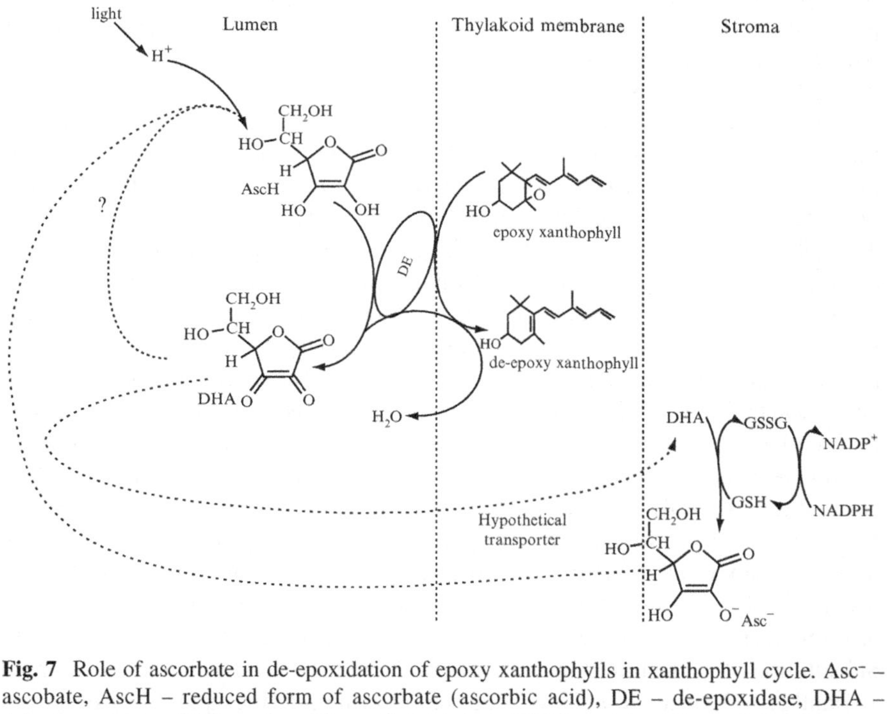

Fig. 7 Role of ascorbate in de-epoxidation of epoxy xanthophylls in xanthophyll cycle. Asc^- – ascobate, AscH – reduced form of ascorbate (ascorbic acid), DE – de-epoxidase, DHA – dehydroascorbate, GSH – reduced form of glutathione, GSSG – oxidised form of glutathione

epoxide groups of epoxy-xanthophylls by that creating completely de-epoxidised xanthophylls like Zx or Dtx, water and dehydroascorbate (DHA) (Fig. 7) (Hager 1969). AscH is oxidised during de-epoxidation to DHA, but no mechanism engaged in rereduction of DHA into Asc in thylakoid lumen is known and therefore a transport system for the DHA in the thylakoid membrane has been postulated (Bratt et al. 1995) although such system has never been shown.

Foyer and Lelandais (1996) suggested presence of a carrier for Asc in the plasma membrane and chloroplast envelope but no transport system for Asc across the thylakoid membrane. Because the negatively charged form of Asc may not penetrate the membrane by simple diffusion, the transport of Asc into thylakoid lumen might be facilitated by a postulated DHA transporter in exchange for DHA (Eskling et al. 1997). This putative system may allow the rereduction of DHA to Asc in the stroma via the GSH cycle consuming NADPH and GSH and protonation of Asc to create the AscH required for de-epoxidation would finally be possible in lumen at low pH as given in the light (Fig. 7). The concentration of the reduced acid form of ascorbate would be determined by the light-driven acidification of the thylakoid lumen.

It was reported that plants grown at high light intensities contained higher levels of Asc and oxygen radical scavenging enzymes than plants grown under low light (Gillham and Dodge 1987; Schöner and Krause 1990). Both the Asc concetration and the amount of V, A, Z increase when plants are grown at high light intensities (Demmig-Adams et al. 1995; Logan et al. 1996). The requirement of an Asc transporter in the thylakoid membrane and the possible limitation of VDE activity by Asc consuming reactions in the stroma have further been supported by studies with isolated chloroplasts (Neubauer and Yamamoto 1994).

Grouneva et al. (2006) found that partially purified DDE has a three to four times higher affinity for the AscH than VDE. Similarly to VDE (Bratt et al. 1995) the ascorbate affinity of DDE strongly depends on the pH value but the K_M value of DDE for Asc at pH 5 was determined to be 0.7 mM while the K_M for VDE was 2.3 mM (Table 4) (Grouneva et al. 2006). At high Asc concentrations a strong shift of the pH optimum towards higher pH values was observed and DDE was found to be still active at almost neutral pH values, showing very fast and strong response in activity to small pH changes in the thylakoid lumen. It was also shown that the ascorbate affinity is an intrinsic feature of the respective enzyme and does not depend on the nature of the substrate being de-epoxidized. The high affinity of DDE for ascorbate indicates that, even at a limited availability of reduced ascorbate, high enzyme activity is possible at low pH values. On the other hand, at high ascorbate concentrations, DDE activity can be shifted towards neutral pH values, thereby facilitating a very fast and strong response to small pH changes in the thylakoid lumen (Grouneva et al. 2006). The different Asc affinities of VDE and DDE may result from differences in their amino acid sequences, where four highly conserved histidine residues have been shown to be important for the Asc binding by VDE. Substitution of the histidines in positions H121 and H124, as well as in positions H 167 and H 173, resulted both in changes of enzyme activities and Asc affinities of the modified spinach de-epoxidases (Emanuelsson et al. 2003). It should be emphasized that the two features do not correlate, meaning that a lower de-epoxidase activity does not necessarily result in a lower affinity for Asc. This indicates that the respective histidine residues might be responsible for the binding of Asc as well as for enzymatic activity. It is suggested that a modification of the histidine-enriched protein region might have taken place in the diatom de-epoxidase, leading to the enhanced affinity of DDE for its co-substrate Asc.

The higher Asc affinity of DDE may also reflect physiological differences between diatoms and higher plants. At a limited availability of reduced form of ascorbic acid (AscH), high enzyme activity at low pH values is possible. This situation can occur in vivo under high light illumination, where high Asc consumption by APX and a strongly stimulated Mehler reaction are taking place (Claquin et al. 2004). The high Asc affinity of DDE would ensure efficient Ddx de-epoxidation even if large amounts of Asc were used for other photoprotective processes. Besides, a first series of measurements of the Asc content of diatom cells suggests that diatoms, in contrast to higher plants and green algae, contain a highly oxidized ascorbate pool under high light illumination (Grouneva et al. 2006).

At high Asc concentrations, DDE activity can be moreover shifted towards neutral pH values, thereby facilitating a very fast and strong response to small pH changes in the thylakoid lumen. This situation is also observed in intact diatom cells that are dark incubated for longer time periods. In these dark-adapted cells, slight pH changes in the thylakoid lumen, because of chlororespiratory electron flow, have been reported to induce significant de-epoxidation of Ddx to Dtx (Jakob et al. 2001).

As shown above, availability of AscH is the main mechanism by which the ascorbate–glutathione pathway controls all types of the xanthophyll cycles, hence also all protective functions regulated by these cycles.

5 The Role of Compounds of Ascorbate–Glutathione Cycle in the Controlling Cellular Redox State and Their Involvement in Signal Transduction Pathways

The redox homeostasis in plant cells is defined by the presence of pools of antioxidant compounds that have possibility to absorb and buffer reductants and oxidants. The coordinated action of different components of the antioxidative systems maintain the cellular redox homeostasis and redox metabolism as well as it is related to signal generation and signal transmission. Venis and Napier (1997) defined signal transduction as a sequence of events that is initiated by a stimulus and culminates in a characteristic physiological or biochemical response.

Noctor (2005) distinguished, at least two types of stress signaling defined as "dynamic" and "static" act at a physiological level linked to redox metabolites. It is suggested that both types of signalling through redox couples may coexist. A 'dynamic' signaling is postulated to involved a change in redox status or concentration of redox active compounds, while 'static' signalling could simply draw on existing pools of redox active compounds or it is related to sensing of a threshold concentration or redox potential. Dietz (2003) grouped redox signals on three types based on their origin and mode of action in signal transmission. Type-I signals derive specifically from single pathways, type-II signals integrate redox information from various pathways and type-III redox signals indicate more extreme redox imbalances and their transmission depends on cross-talk with other signaling cascades. Similarly, Pfannschmidt et al. (2001, 2003) defined three classes of redox signals based on perturbation of the photosynthetic electron transfer depending on the light intensity. Class 1 signals originate from specific redox pairs in the photosynthetic electron transport chain, e.g. the reduced and oxidized PQ. Class 2 signals depend on the redox state of stromal thioredoxins, NAD(P)H as well as GSH and Asc, whereas Class 3 signals are mediated by ROS and other signaling molecules. Each of these signals has the potential to function as developmental and environmental stimulus and to change the metabolic state of a given plant cell by short term metabolic control within seconds to minutes or long term genetic mechanisms on the time scale of hours. Baier and Dietz (2005) pointed out that signals transmitted by redox shifts in cellular redox components or by ROS are distinguished from signals transmitted by second messengers synthesized inside chloroplasts.

The potent signalling effects of ROS require their concentrations to be controlled by a battery of antioxidants that determine the lifetime and intensity of the signal (Mittler 2002; Mittler et al. 2004). Thus, this flexible and complex regulatory system is involved not only in maintaining the prooxidant/antioxidant equilibrium but also in a specific redox-based stress sensing and signalling (Bräutigam et al. 2009).

The individual plant cell compartments like chloroplasts, mitochondria, peroxisomes, cytosol and apoplast posses different antioxidant buffering capacities determined by differences in synthesis, transport, and/or degradation that allow to control redox-sensitive signals independently. On the other hand, the interactions between these cell compartments as well as the complex network of reactions surrounding ROS generation and consumption, are crucial for redox homeostasis in metabolism (Scheibe et al. 2005), and ROS signaling (Noctor 2006; Foyer and Noctor 2005, 2009; Joo et al. 2005). Local perturbation of redox buffering system is likely an important process in the transmission of ROS signals, particularly those related to stress conditions. Specific compartment-based signaling can be achieved, among others, via differential changes in the amounts and relative reduction states of the ascorbate and glutathione pools, which are part of the cellular information-rich redox buffers (Schafer and Buettner 2001; Dietz 2003; Noctor 2006). The interactions between the redox couples such as NAD(P)/NAD(P)H, GSSG/GSH or additionally in plants DHA/AscH are involved in maintaining of cellular redox homeostasis (Fig. 8).

The regeneration of oxidised glutathione and the reduction of monodehydroascorbate to ascorbate by monodehydroascorbate reductase are NAD(P)H-dependent, demonstrating a metabolic link between the NAD(P)-dependent redox system and the low molecular weight antioxidants. The Asc and GSH, in addition to pyridine nucleotides such as NADH or NADPH, are the key mediators of redox transfer in the soluble phase of plant cells. This function requires that their reduced forms react slowly with oxygen and its more reactive derivatives so that oxidation is enzyme dependent (Noctor 2006; Wormuth et al. 2007). In photosynthetically active cells which are not strongly affected by stress factors, balance between the amounts of NADPH, ROS, Asc and GSH is kept. However, under stress conditions leading to ROS production, the redox balance at a cellular level may be affected.

Stress signaling linked to the three redox metabolites: NAD(P), GSH and Asc have been found to be crucial to set the appropriate defence responses and to achieve a successful acclimation by plant. The redox active elements, i.e. ferredoxin, NADPH and GSH, are dependent to some extent, on the photosynthetic electron transport chain or other metabolic processes and feed electrons into the network (Dietz 2003; Kuźniak et al. 2009). The $NAD(P)^+$/NAD(P)H, GSSG/GSH,

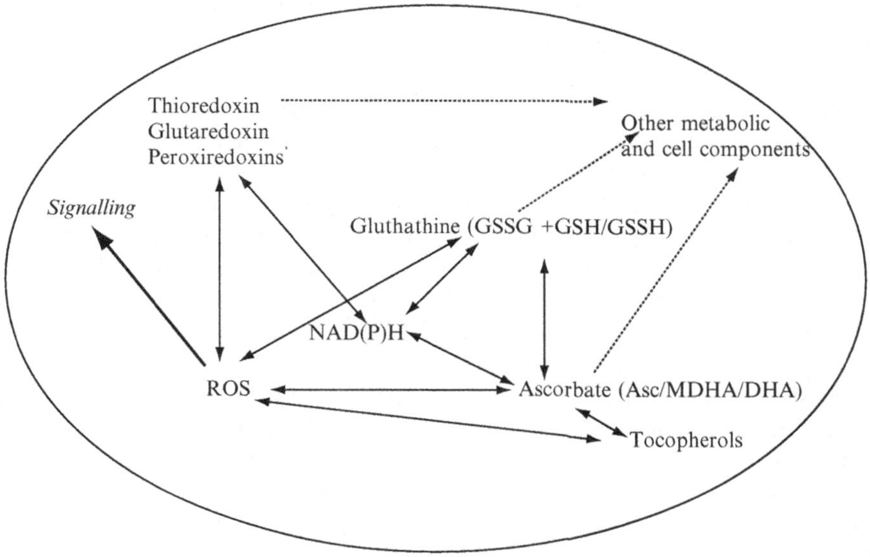

Fig. 8 Potential plant cells for influence on cellular redox level and signal induction. The relationship between some components of the cell redox buffers

and DHA/AscH redox couples, integrated with the redox regulatory network, could undergo a distinct physiological functions.

Redox systems involved in sensing, transferring and multiplying signals are responsible for coordination of respiration, photorespiration and alternative respiration to keep photosynthesis working at high level. These events can be viewed in terms of the interacting pathways contributing to the adjustment of metabolic processes to a new state of homeostasis, usually referred to as acclimation.

Both oxidants and antioxidants fulfill signaling roles to provide information on plant defense, using kinase-dependent and independent pathways that are initiated by redox-sensitive receptors modulated by thiol status (Kuźniak et al. 2009). Thiol-based regulation may be important in plant acclimation to environmental change, particularly where redox interactions play a key role in the orchestration of the abiotic stress response (Desikan et al. 2005). In plants, thiol containing of proteins are oxidized by ROS, either directly or indirectly, to give relatively stable oxidation products with modified physical conformations or biochemical activities. The major low molecular weight antioxidants such as Asc and GSH determine the specificity of the signal. Low molecular antioxidant are a part of network associated with the mechanisms by which plant cells sense the environment and make appropriate adjustments to gene expression, metabolism and physiology (Foyer and Noctor 2005).

Asc can act as a signal transducing molecule in plants (Pastori et al. 2003; Fotopoulos et al. 2008) participating in the interaction with the environment, for instance to ozone (Sanmartin et al. 2003), pathogens and oxidizing agents (Fotopoulos et al. 2006), and water stress (Fotopoulos et al. 2008). Pastori et al. (2003)

demonstrated that the leaves of the *vtc1* mutant (Table 1), which have only 30% of the Asc content of the wild type, show extensive re-adjustments in gene expression that can be reversed by Asc feeding. This finding suggests that Asc is involved in metabolic cross-talk between redox-regulated pathways. However, when Asc concentrations are not supplemented by feeding, ROS do not accumulate in *vtc1* mutants (Veljovic-Jovanovic et al. 2001). Moreover, it could be noted that also DHA may act as a potential factor in the signaling pathway. The chemical structure of DHA, with its vicinal carbonyl group, gives the molecule its peculiar reactivity (Deutsch 2000) with, amongst other compounds, thiols, whereby it induces oxidative protein folding (Banhegyi et al. 2003). Reversible modification of specific proteins by DHA could be important in cell signaling. It seems that Asc oxidation to DHA constitutes a vital signal transduction module governing plant reaction to stressful environmental conditions. A large fraction of total leaf DHA is probably located in the apoplast (Veljovic-Jovanovic et al. 2001; De Pinto and De Gara 2004). The Asc/DHA ratio in this compartment is postulated to be crucial in plant signalling (Pignocchi and Foyer 2003). Redox reactions within the apoplast can influence signal transduction cascades. The apoplast is also a major site of NO synthesis as ascorbate reductant (Bethke et al. 2004). It was also demonstrated that the interaction of GSH with NO the S-nitrosoglutathione (GSNO) is formed and it may interconnect the ROS- and reactive nitrogen-based signaling pathways (Neill et al. 2002) as well as it is thought to function in plants as a mobile reservoir of NO bioactivity (Durner et al. 1999; Díaz et al. 2003).

Glutathione, the next component of ascorbate–glutathione cycle is an abundant metabolite in plants that functions as modulator or signal transducer (Noctor et al. 2002; Gomez et al. 2004; Noctor and Foyer 1998). GSH and GSSH may also function as signal molecules in many cellular processes. It shows strong interplay between concentration and the redox state. Glutathione pool level might be regulated, at least partially, by hydrogen peroxide and on the redox state of the plastochinon pool (Karpinski et al. 1997, 1999). Conditions that trigger accumulation of GSSG often also lead to the subsequent increase in total glutathione and are related, among others, to the stress-induced changes in the H_2O_2 content (Neill et al. 2002; Foyer and Noctor 2003, 2005, 2009; Noctor 2006; Dietz 2008; Quan et al. 2008). This phenomenon is observed in catalase-deficient barley (Smith et al. 1985), during exposure of poplar to ozone (Gupta et al. 1991), following chilling of maize (Gomez et al. 2004), during incompatible interactions between barley and powdery mildew (Vanacker et al. 2000) and during the developing of photoperiod-dependent strategies in the acclimation of *Arabidopsis thaliana* plants (Queval et al. 2007).

Many observations suggest that changes in glutathione status may be as important as enhanced ROS pools in redox signaling (Creissen et al. 1999; Vanacker et al. 2000; Mou et al. 2003; Ball et al. 2004; Gomez et al. 2004; Evans et al. 2005). When GSH levels are low the cell environment will be oxidising and the functioning of enzymes, particularly those with thiol groups, will be altered. Moreover, the GSH/GSSG ratio as well as enzymes such as GR involved in controlling the GSH/GSSG ratio are able to determined of antioxidant capacity and thus may be involved in redox signaling. The GSH/GSSG couple is able to modify the activity of various compounds (enzymes, regulatory proteins) directly through the reduction/oxidation of their disulfide bridges/sulfhydryl groups and through the (de)glutathionylation of sulfhydryl groups. With the other hand GSH/GSSG couple is an indicator of the general cellular thiol-disulphide redox balance (Fig. 9).

The regulation of proteins by the GSH/GSSG couple may occur also due to cross-talk between GSH/GSSG and other redox systems like glutathionylation or thiol-disulfide transition, which may have a role in signaling and plant responses to stress conditions (Rausch et al. 2007). Protein glutathionylation (i.e., the formation of a reversible mixed-disulfide bond between a small-molecular-weight thiol such as glutathione and specific cysteine residues of various proteins) seems to play an important role in perception of glutathione status by the cell. Therefore, accumulation of GSSG similar like ROS-catalyzed generation of protein thiol radicals may be sufficient to trigger protein glutathionylation. Glutathionylation may also occur independently of enhanced ROS production or redox perturbation of the glutathione pool. As a target for glutathionylation two Calvin cycle enzymes such as aldolase

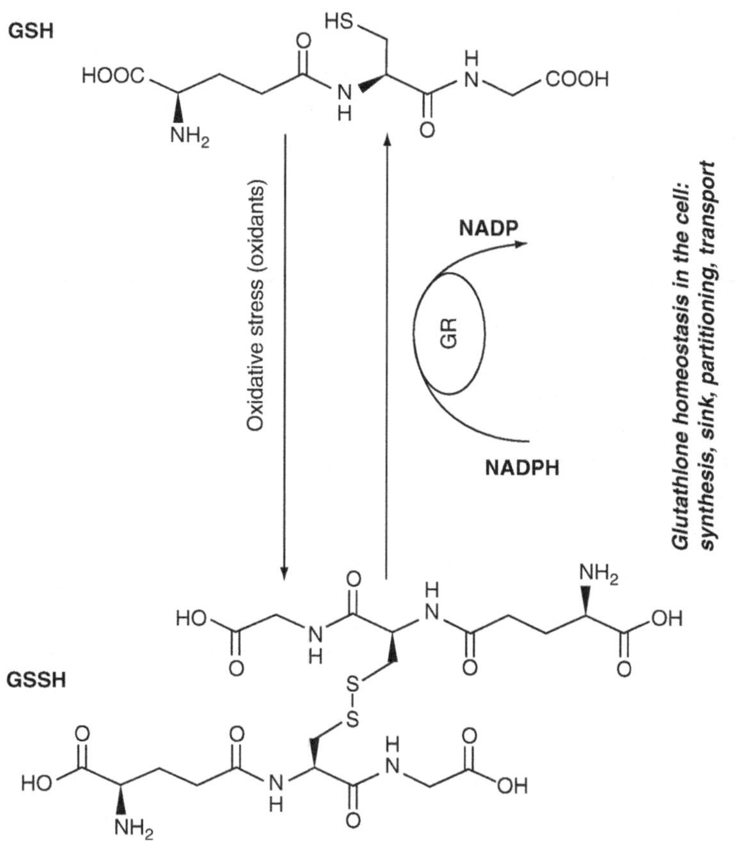

Fig. 9 The glutathione in the reduced and oxidized form. The glutathione pool avability and the redox state of glutathione pool are related to ROS level and metabolic processes

and triose phosphate isomerase have been found (Ito et al. 2003). In Arabidopsis, oxidative stress-induced glutathionylation was described for a number of other proteins, including several GSTs and glutaredoxins (Grx) (Dixon et al. 2005). Glutaredoxins take part in transfer electrons reversibly between GSH and target proteins and they are involved in redox-regulated processes participating in stress responses (Hisabori et al. 2007; Vieira et al. 2006; Meyer 2008; Kuźniak et al. 2009). Glutathionylation could be an important redox signaling mechanism allowing cells to sense and signal harmful stress conditions and trigger appropriate responses and adaptation, and has been shown to be involved in the regulation of several signal transduction pathways (Ghezzi 2005; Hurd et al. 2005; Michelet et al. 2006; Dalle-Donne et al. 2007, 2008; Gallogly and Mieyal 2007; Ghezzi and Di Simplicio 2007; Townsend 2007; Rouhier et al. 2008, Gao et al. 2008). This posttranslational modification can protect cysteine thiols against irreversible oxidation to sulfinic or sulfonic acid, modulate enzyme activity by modification of catalytic site Cys residues or affect biological activity by competing with other thiol modifications (Foyer and Noctor 2005, 2009; Noctor 2006; Gao et al. 2008). This modification apparently occurs mainly under oxidative/nitrosative stress conditions but may also be important under normal conditions, especially for regeneration of several thiol peroxidases (Rouhier et al. 2008).

Glutathione together with H_2O_2 and/or $O_2{\cdot}^-$ can act both as messenger molecules in cellular signal transduction pathways in several organelles and as factors in plant defense responses (del Río et al. 1996; Karpinski et al. 1997).

6 Influence of Ascorbate and Glutathione on Gene Expression Associated with Biotic and Abiotic Stress Response

Apart from the well documented role of antioxidative enzymes and antioxidants in removing H_2O_2 and other ROS, there is also a strong interaction between oxidants and antioxidants, at the level of gene expression and translation (May et al. 1998; Karpinski et al. 1999).

Asc and GSH as well as Asc/DHA or GSH/GSSG redox couples have been demonstrated as a transcriptional regulators and being regulated by metabolic processes (Baier et al. 2000; Noctor et al. 2000; Veljovic-Jovanovic et al. 2001) (Fig. 10).

The changes in the cellular redox balance constitute an early event in H_2O_2 signal transduction as reduction of the cellular redox buffer and thus the cell's ability to maintain a high GSH/GSSG ratio potentiated the plant's antioxidant response. The perturbs cellular redox balance may serve as an inducer for the set of defense-related genes such as: H_2O_2-induced expression of glutathione-S-trasferase1 (*GST1*) gene (Rentel and Knight 2004) or genes of pathogenesis-related (PR) proteins (Foyer et al. 1997; Durner et al. 1998; Foyer and Noctor 2003, 2005, 2009). PR proteins are crucial components of defense mechanism and they may contribute to the innate immunity of plants. They are induced in response to pathogen infection and/or related to stress situations and are used widely as marker genes/proteins

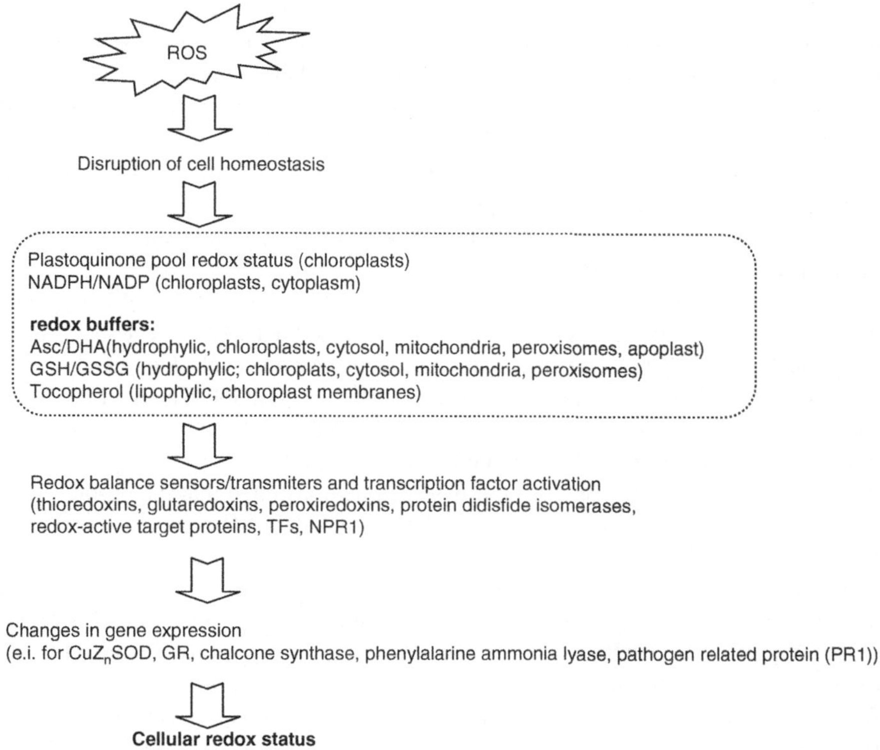

Fig. 10 The role of the redox active components in the regulation of gene expression

to study the plant self-defense mechanism(s). The PR proteins are classified into 14 groups (PR-1 to PR-14) based on genes sequence or predicted sequence of amino acids, serological relationships and biological activity of their structural homologues within the groups (van Loon and van Strien 1999). The low levels of Asc, can act as an elicitor of pathogen resistance responses and leads to the induction of PR genes (PR-1, PR-2 and PR-5), as shown by the analysis of the *vtc1* and *vtc2* mutants (Pastori et al. 2003; Conklin and Barth 2004; Barth et al. 2004; Pavet et al. 2005). Feeding Asc to leaves of Asc deficient *vtc1* mutant represses the accumulation of PR-1 transcripts, whereas the levels of PR-5 and PR-2 transcripts were not significantly changed (Pastori et al. 2003) and these effects are accompanied by enhanced glutathione level (Mou et al. 2003; Gomez et al. 2004).

Horling et al. (2003) demonstrated that the treatment of Arabidopsis plants with ascorbate declined in shoots of these plants the level of transcript of the other proteins engaged in stress response such as the chloroplast-targeted peroxiredoxins namely the two-cysteine (2-Cys) peroxiredoxins (*prx Q* and *prx II E*). These enzymes are involved in protecting photosynthesis. Baier et al. (2000) point out that the decrease of 2-Cys peroxiredoxins in transgenic *A. thaliana* plants caused up-regulation of genes related to ascorbate metabolism such as stromal ascorbate peroxidase and particularly monodehydroascorbate reductase, while gene products like glutathione reductase involved in glutathione metabolism were unaltered similarly to catalase, and superoxide dismutase. On the hand Lamkemeyer et al. (2006) observed alter transcripts of glutathione-related genes in Prx Q knockout plants.

The expression of defense-related gene is also related to the absolute amounts of glutathione as well as the ratio of GSH to GSSG. The oxidation of even a small part of the GSH in a cell can shift significantly the GSH:GSSG ratio, with consequent implications for patterns of gene expression (Hwang et al. 1992; Sen 2000). Such changes allow the organism to respond to oxidative stress caused by altered cellular or environmental conditions.

As it was presented by some authors changes in the redox state of GSH in plants exposed to HL stress are associated with the increased activity of heat shock transcription factors (HSFs) and raised abscisic acid (ABA) level (Jabs et al. 1996; Karpinski et al. 1997, 1999; Fryer et al. 2003; Ball et al. 2004; Chang et al. 2004; Mateo et al. 2006; Kuźniak et al. 2009). Moreover, both GSH and GSSG elicit phenylalanine ammonia-lyase (Phe ammonia lyase) enzyme activity and phytoalexin accumulation (Edwards et al. 1991). It has been demonstrated that the promoters of genes involved in the synthesis of phytoalexins include GSH-responsive elements (Levine et al. 1994). In carrot, inhibition of glutathione synthesis induced phytoalexin accumulation (Guo et al. 1993). Exogenous application of GSH increased level of Phe ammonia lyase and chalcone synthase transcripts in bean cell suspensions (Wingate et al. 1988). Chalcone synthase (CHS) is pivotal for the biosynthesis of flavonoid antimicrobial phytoalexins and anthocyanin pigments, catalyzes the first and key regulatory step in the branch of phenylpropanoid biosynthesis pathway specific for synthesis of flavonoid pigments and UV protectants (Ferrer et al. 1999; Ryder et al. 1987). Parisy et al. (2007) demonstrated that the low GSH concentration in the *pad2-1* mutant (Table 2) did not affect the transcript abundance of γ-ECS and GSHS. Hovewer, after inoculation of this mutant with *Phytophthora brassicae* gene expression for these enzymes was much more strongly induced than in the wild type. Ball et al. (2004) determined the transcriptional response of the allelic glutathione biosynthetic mutants *rax1-1* and *cad2-1* (Table 2). Typical target genes with altered transcript accumulation under stress are those for PR proteins, e.g. PR1, chitinase, and PAL (encoding phenylalanine ammonium lyase), and antioxidant enzymes as peroxidases, dehydroascorbate reductase and CuZn-superoxide dismutase. The sets of target genes widely overlap with the transcripts induced by pathogens and wounding (Cheong et al. 2002; Mahalingam et al. 2003).

It is also observed that GSH as a one of thiols similarly to thioredoxin and Cys enhance nuclear transcription factor kappa (NF-κ) in mesophyll cells of maize leaves and this suggests that redox control of translation is a key determinant of protein abundance in this plant (Pastori et al. 2000). The demonstration of thiol modulation of nonexpressor of pathogen related genes (NPR1) and its interacting transcription factors (Després et al. 2003) potentially resolves some of the issues surrounding pathogen-induced changes in glutathione status (Edwards et al. 1991, Vanacker et al. 2000) as well as induction of PR genes in plants in which glutathione pools are enhanced by γ-ECS overexpression or in response to the decrease of antioxidative enzyme capacity (Creissen et al. 1999; Rizhsky et al. 2002). In *Medicago trunculata* roots GSH synthesis pathway is regulated by NO-donors (sodium nitroprusside and nitrosoglutathione) treatment and *γ-ESC* and *GSHS* genes for gamma-glutamylcysteine synthetase (γ-ECS), glutathione synthetase (GSHS), respectively are up regulated (Innocenti et al. 2007).

Asc and GSH and Asc/DHA or GSH/GSSG redox couples may regulate the expression of genes involved in plants stress response with the one side, but synthesis of these molecules as well as metabolism is affected by the abiotic and biotic stress factors. Enzymes involved in Asc and GSH metabolism (Szalai et al. 2009) are regulated at different levels in various plant species and the control of the corresponding enzymes may depend on the organ and cell type and on the developmental stage.

7 Relationship Between Different Components of Ascorbate–Glutathione Cycle and Metabolic Processes Involved in Plant Defense Response

The induction of defence pathways is a key feature of response of all organisms to stress. The dynamic interactions between cell compartments are important for effective stress signal relay and the induction of defense mechanisms. Under stress conditions plants regulate their resource allocation in a way that increases stress tolerance (Mateo et al. 2004; Feys et al. 2005).

Plant defense may be modulated simultaneously in response to changes in Asc availability via changes in the ABA and ABA/giberelin acid (GA) signaling pathways. The abundance of key transcripts is modulated by Asc availability in leaves. Thus, plants sense not only changes in the redox state of their Asc pool but also the absolute amount of this major redox buffer (Foyer 2004). Asc plays a crucial role in protection and regulation of photosynthesis, acting among others in the Mehler peroxidase reaction with ascorbate peroxidase to regulate the redox state of photosynthetic electron carriers (Foyer and Allen 2003). Futhermore, higher Asc levels may support the growth of plants under conditions of high salinity (Shalata and Neumann 2001). In opposite to these data, it was demonstrated that the deficient in Asc mutant *vtc1* (Table 1) is more resistant to bacterial and fungal pathogens (Barth et al. 2004). Millar et al. (2003) showed that Asc synthesis and hence stress responses in plants is controlled by respiration. Asc and its metabolic precursors give rise to oxalic acid (OxA) found in calcium oxalate crystals in specialized crystal idioblast cells in plants as well as for soluble oxalate accumulated in leaf vacuoles (Wagner 1981; Kostman et al. 2001). The accumulation of OA crystals in plant tissues is suggested to be involved in regulation of cellular calcium levels and sequestration of toxic metals, and to confer resistance to herbivory (Deutsch 2000; Horner et al. 2000; Keates et al. 2000; Franceschi and Nakata 2005). Asc participates also in synthesis of ethylene, gibberellins and anthocyanins (Smirnoff and Wheeler 2000; Smirnoff 2000). Asc, functions also as a biosynthetic precursor of L-tartaric acids (TA). TA plays a critical role in determining the suitability of grapes for use in winemaking – berry. TA is largely responsible for controlling juice pH and through TA addition during vinification, the winemaker can minimize oxidative and microbial spoilage, thereby promoting both organoleptic and ageing potentials of the finished wine (DeBolt et al. 2006). Moreover, Asc, GSH and H_2O_2 function as upstream/downstream components of hormone-mediated signal transduction. Some authors (Chen and Gallie 2006; Fotopoulos et al. 2008) described altered stomatal behaviour in ascorbate oxidase (AO) over-expressing tobacco plants, and they suggested that DHA may act as an early activator of stomatal closure thus improving the control over transpiration. Under stressful conditions Asc oxidation to DHA take place and, in turn, this molecule can modulate plant responses to stress by regulating ABA synthesis and increasing hydrogen peroxide production (López-Carbonell et al. 2006).

Involvement of apoplastic Asc, in stress responses enable plants to cope with unfavorable conditions by modulating plant growth. Asc and AO in the apoplast are key players in both cell division and cell elongation. Many environmental and metabolic factors influence the rate and direction of cell elongation (Gonzalez-Reyes et al. 1994). DHA is believed to signal the redox state of the apoplastic environment, and hence to allow the cell to perceive stress in the environment. DHA accumulation in the apoplast may trigger the arrest of cell division (Potters et al. 2000).

Glutathione is abundant and ubiquitous thiol with proposed roles in the storage and transport of reduced sulphur, the synthesis of proteins and nucleic acids and as a modulator of enzyme activity. Stable protein disulphide bonds are relatively rare except in quiescent tissues such as seeds, where GSSG is allowed to accumulate. Moreover, GSH is a substrate for several reductive enzymes, including enzymes that reduce peroxides (Foyer and Noctor 2005a,b). Glutathione is used also in detoxification of reactive ketoaldehydes such as methylglyoxal what is a potential target for engineering tolerance to stresses such as high salinity (Singla-Pareek et al. 2003; Noctor 2006, Noctor and Foyer 2005). Other indirect roles of glutathione are in heavy metal detoxification as a precursor to phytochelatins (Cobbett and Goldsborough 2002) and as a transport form of Cys (Kopriva and Rennenberg 2004). Cold treatments and the certain metals such cadmium or copper that favour ROS production can induce transcripts for the enzymes of glutathione synthesis (Rodríguez-Serrano et al. 2009).

The redox changes in the glutathione pool as well as changes in hydrogen peroxide levels are genetically and functionally interconnected with the SA signaling pathway regulating light acclimatory processes (Karpinska et al. 2000, 2003; Mateo et al. 2004, 2006). Glutathione is able to induce PR transcript induction (Gómez et al. 2004a,b), whereas localized cell death occurs in Asc deficient plants (Pastori et al. 2003). These effects point to opposing functions for GSH and Asc in redox signal transduction. A striking relationship between glutathione oxidation and mitochondrial DNA damage during aging has also been reported (Yen et al. 1994; Esteve et al. 1999). In addition, accumulation of GSSG is often associated with tissue death or quiescence (Wachter et al. 2005). Kranner et al. (2006) demonstrated that the alteration in the half-cell reduction potential of the GSH/GSSG couple ($E_{GSSG/2GSH}$), a major cellular antioxidant and redox buffer is part of the signaling cascade that initiates programmed cell death (PCD), finally causing internucleosomal DNA fragmentation in the final, or execution phase, of PCD (Kranner et al. 2006; Schafer and Buettner 2001). Changes of the redox state of glutathione was demonstrated, among others, as the response to excess light stress (Karpinski et al. 1997, 1999; Panchuk et al. 2002; Fryer et al. 2003; Ball et al. 2004; Chang et al. 2004).

Acknowledgement This work was supported by the Polish Ministry of Science and Higher Education (project No. 50/N-DFG/2007/0).

References

Alhagdow M, Mounet F, Gilbert L, Nunes-Nesi A, Garcia V, Just D, Petit J, Beauvoit B, Fernie AR, Rothan C, Baldet P (2007) Changes in ascorbate redox state through silencing of L-galactono-1, 4-lactone dehydrogenase (L-GalLDH) affect plant and fruit development in tomato. Plant Physiol 145:1408–1422

Allen JF (2003) Cyclic, pseudocyclic and noncyclic photophosphorylation: new links in the chain. Trends Plant Sci 8:15–19

Allen JF, Pfannschmidt T (2000) Balancing the two photosystems: photosynthetic electron transfer governs transcription of reaction centre genes in chloroplasts. Philos Trans R Soc Lond B Biol Sci 355:1351–1357

Almagro L, Ros LVG, Belchi-Navarro S, Bru R, Barceló AR, Pedreño MA (2009) Class III peroxidases in plant defence reactions. J Exp Bot 60:377–390

Alscher RG, Donahue JL, Cramer CL (1997) Reactive oxygen species and antioxidants: Relationships in green cells. Physiol Plant 100:224–233

Alscher RG, Erturk N, Heath LS (2002) Role of superoxide dismutases (SODs) in controlling oxidative stress in plants. J Exp Bot 53:1331–1341

Anderson JV, Davis DG (2004) Abiotic stress alters transcript profiles and activity of glutathione S-transferase, glutathione peroxidase, and glutathione reductase in Euphorbia esula. Physiol Plant 120:421–433

Apel K, Hirt H (2004) Reactive oxygen species: metabolism, oxidative stress, and signal transduction. Annu Rev Plant Biol 55:373–399

Aro EM, Ohad I (2003) Redox regulation of thylakoid protein phosphorylation. Antioxid Redox Signal 5:55–67

Arsalane W, Rousseau B, Duval J-C (1994) Influence of the pool size of the xanthophyll cycle on the effects of light stress in a diatom: competition between photoprotection and photoinhibition. Photochem Photobiol 60:237–243

Asada K (1999) The water–water cycle in chloroplasts: scavenging of active oxygen and dissipation of excess photons. Annu Rev Plant Physiol Plant Mol Biol 50:601–639

Asada K (2006) Production and scavenging of reactive oxygen species in chloroplasts and their functions. Plant Physiol 141:391–396

Asada K, Takahashi M (1987) Production and scavenging of active oxygen in chloroplasts. In: Kyle DJ, Osmond CB, Arntzen CJ (eds) Photoinhibition. Elsevier, Amsterdam, pp 227–287

Baier M, Dietz KJ (2005) Chloroplasts as source and target of cellular redox regulation: a discussion on chloroplast redox signals in the context of plant physiology. J Exp Bot 56:1449–1462

Baier M, Ströher E, Dietz KJ (2004) The acceptor availability at photosystem I and ABA control nuclear expression of 2-Cys peroxiredoxin-A in *Arabidopsis thaliana*. Plant Cell Physiol 45:997–1006

Baier M, Noctor G, Foyer C, Dietz KJ (2000) Antisense suppression of 2-Cys peroxiredoxin in Arabidopsis thaliana specifically enhances the activities and expression of enzymes associated with ascorbate metabolism, but not glutathione metabolism. Plant Physiol 124:823–832

Ball L, Accotto G-P, Bechtold U, Creissen G, Funck D, Jimenez A, Kular B, Leyland N, Mejia-Carranza J, Reynolds H, Karpinski S, Mullineaux PM (2004) Evidence for a direct link between glutathione synthesis and stress defence gene expression in Arabidopsis. Plant Cell 16:2448–2462

Banhegyi G, Csala M, Szarka A, Varsanyi M, Benedetti A, Mandl J (2003) Role of ascorbate in oxidative protein folding. Biofactors 17:37–46

Banki K, Hutter E, Colombo E, Gonchoroff NJ, Perl A (1996) Glutathione levels and sensitivity to apoptosis are regulated by changes in transaldolase expression. J Biol Chem 271:32994–33001

Barroso JB, Corpas FJ, Carreras A, Rodríguez-Serrano M, Esteban FJ, Fernández-Ocaña A, Chaki M, Romero-Puertas MC, Valderrama R, Sandalio LM del Río LA (2006) Localization of S-nitrosoglutathione and expression of S-nitrosoglutathione reductase in pea plants under cadmium stress. J Exp Bot 57:1785–1793

Barker JE, Heales SJR, Cassidy A, Bolaños JP, Land JM, Clarka JB (1996) Depletion of brain glutathione results in a decrease of glutathione reductase activity; an enzyme susceptible to oxidative damage. Brain Res 716:118–122

Barth C, Moeder W, Klessig DF, Conklin PL (2004) The timing of senescence and response to pathogens is altered in the ascorbate-deficient Arabidopsis mutant vitamin c-11. Plant Physiol 134:1784–1792

Bartoli CG, Guiamet JJ, Kiddle G, Pastori G, Di Cagno R, Theodoulou FL, Foyer CH (2005) The relationship between L-galactono-1,4-lactone dehydrogenase (GalLDH) and ascorbate content in leaves under optimal and stress conditions. Plant, Cell and Environment 28:1073–1081

Bartoli CG, Pastori GM, Foyer C (2000) Ascorbate biosynthesis in mitochondria is linked to the electron transport chain between complexes III and IV. Plant Physiol 123:335–344

Bartoli CG, Yu J, Gómez F, Fernández L, McIntosh L, Foyer CH (2006) Inter-relationships between light and respiration in the control of ascorbic acid synthesis and accumulation in Arabidopsis thaliana leaves. J Exp Bot 57:1621–1631

Bartoli C, Tambussi E, Diego F, Foyer CH (2009) Control of ascorbic acid synthesis and accumulation and glutathione by the incident light red/far red ratio in *Phaseolus vulgaris leaves*. FEBS Letters 583:118–122

Beligni MV, Fath A, Bethke PC, Lamattina L, Jones RL (2002) Nitric oxide acts as an antioxidant and delays programmed cell death in barley aleurone layers. Plant Physiol 129:1642–1650

Bethke PC, Murray RB, Russell LJ (2004) Apoplastic synthesis of nitric oxide by plant tissues. Plant Cell 16:332–341

Bick JA, Setterdahl AT, Knaff DB, Chen Y, Pitcher LH, Zilinskas BA, Leustek T (2001) Regulation of the plant-type 5′-adenylysulfate reductase by oxidative stress. Biochem 40:9040–9048

Bode S, Quentmeier CC, Liao PN, Hafi N, Barros T, Wilk L, Bittner F, Walla PJ (2009) On the regulation of photosynthesis by excitonic interactions between carotenoids and chlorophylls PNAS 106:12311–12316

Bolwell GP, Blee KA, Butt VS, Davies DR, Gardner SL, Gerrish C, Minibayeva F, Rowntree EG, Wojtaszek P (1999) Recent advances in understanding the origin of the apoplastic oxidative burst in plant cells. Free Radical Research 31:S137–145

Bolwell GP, Bindschedler LV, Blee KA, Butt VS, Davies DR, Gardner SL, Gerrish C, Minibayeva F (2002) The apoplastic oxidative burst in response to biotic stress in plants: a tree-component system. J Exp Bot 53:1367–1376

Bonardi V, Pesaresi P, Becker T, Schleiff E, Wagner R, Pfannschmidt T, Jahns P, Leister D (2005) Photosystem II core phosphorylation and photosynthetic acclimation require two different protein kinases. Nature 437:1179–1182

Bratt CE, Arvidsson PE, Carlsson M, Åkerlund HE (1995) Regulation of violaxanthin de-epoxidase activity by pH and ascorbate concentration. Photosyth Res 45:165–175

Bräutigam K, Dietzel L, Kleine T, Ströher E, Wormuth D, Dietz K-J, Radke D, Wirtz M, Hell R, Dörmann P, Nunes-Nesi A, Schauer N, Fernie AR, Oliver SN, Geigenberger P, Leister D, Pfannschmidt T (2009) Dynamic plastid redox signals integrate gene expression and metabolism to induce distinct metabolic states in photosynthetic acclimation in Arabidopsis. Plant Cell 21:2715–2732

Van Breusegem F, Dat JF (2006) Reactive oxygen species in plant cell death. Plant Physiol 141:384–390

Bright J, Desikan R, Hancock JT, Weir IS, Neill SJ (2006) ABA induced NO generation and stomatal closure in Arabidopsis are dependent on H_2O_2 synthesis. Plant J 45:113–122

Bungard RA, Ruban AV, Hibberd JM, Press MC, HortonP SJD (1999) Unusual carotenoid composition and a new type of xanthophyll cycle in plants. Proc Natl Acad Sci USA 96:1135–1139

Bulley S, Rassam M, Hoser D, Otto W, Schünemann N, Wright M, MacRae E, Gleave A, Laing W (2009) Gene expression studies in kiwifruit and gene over-expression in Arabidopsis indicates that GDP-L-galactose guanyltransferase is a major control point of vitamin C biosynthesis. J Exp Bot 60:765–778

Burkey KO, Eason G, Fiscus EL (2003) Factors that affect leaf extracellular ascorbic acid content and redox status. Physiol Plant 117:51–57

Cagnac O, Bourbouloux A, Chakrabarty D, Zhang MY, Delrot S (2004) AtOPT6 transports glutathione derivatives and is induced by primisulfuron. Plant Physiol 135:1378–1387

Cairns NG, Pasternak M, Wachter A, Cobbett CS, Meyer AJ (2006) Maturation of Arabidopsis seeds is dependent on glutathione biosynthesis within the embryo. Plant Physiol 141:446–455

Carrari F, Nunes-Nesi A, Gibon Y, Lytovchenko A, Loureiro ME, Fernie AR (2003) Reduced expression of aconitase results in an enhanced rate of photosynthesis and marked shifts in carbon partitioning in illuminated leaves of wild species tomato. Plant Physiol 133:1–14

Castillo FJ, Greppin H (1988) Extracellular ascorbic acid and enzyme activities related to ascorbic acid metabolism in *Sedum album* L. leaves after ozone exposure. Environ Exp Bot 28:231–238

Chang CC-C, Ball L, Fryer MJ, Baker NR, Karpinski S, Mullineaux PM (2004) Induction of ASCORBATE PEROXIDASE 2 expression in wounded Arabidopsis leaves does not involve known wound-signalling pathways but is associated with changes in photosynthesis. Plant J 38:499–511

Chen Z, Gallie DR (2004) The ascorbic acid redox state controls guard cell signaling and stomatal movement. Plant Cell 16:1143–1162

Chen Z, Gallie DR (2005) Increasing tolerance to ozone by elevating foliar ascorbic acid confers greater protection against ozone than increasing avoidance. Plant Physiol 138:1673–1689

Chen Z, Gallie DR (2006) Dehydroascorbate reductase affects leaf growth, development, and function. Plant Physiol 142:775–787

Chen J, Goldsbrough PB (1994) Increased activity of [gamma]-glutamylcysteine synthetase in tomato cells selected for cadmium tolerance. Plant Physiol 106:233–239

Chen Z, Young TE, Ling J, Chang SCh, Gallie DR (2003) Increasing vitamin C content of plants through enhanced ascorbate recycling. Proc Natl Acad Sci USA 100:3525–3530

Cheong YH, Chang HS, Gupta R, Wang X, Zhu T, Luan S (2002) Transcriptional profiling reveals novel interactions between wounding, pathogen, abiotic stress, and hormonal responses in Arabidopsis. Plant Physiol 129:661–677

Chew O, Whelan J, Millar AH (2003) Molecular definition of the ascorbate-glutathione cycle in *Arabidopsis* mitochondria reveals dual targeting of antioxidant defenses in plants. JBC – J Biol Chem 278:46869–46877

Claquin P, Kromkamp JC, Martin-Jezquel V (2004) Relationship between photosynthetic metabolism and cell cycle in a synchronized culture of the marine alga *Cylindrotheca fusiformis* (Bacillariophyceae). Eur J Phycol 39:33–41

Cobbett C, Goldsbrough P (2002) PHYTOCHELATINS AND METALLOTHIONEINS: Roles in heavy metal detoxification and homeostasis. Annu Rev Plant Biol 53:159–182

Cobbett CS, May MJ, Howden R, Rolls B (1998) The glutathionedeficient, cadmium-sensitive mutant, cad2-1, of *Arabidopsis thaliana* is deficient in c-glutamylcysteine synthetase. Plant J 16:73–78

Coesel S, Obornik M, Varela J, Falciatore A, Bowler C (2008) Evolutionary origins and functions of the carotenoid biosynthetic pathway in marine diatoms. PLoS ONE 3(8):e2896. doi:10.1371/journal.pone.0002896

Colville L, Smirnoff N (2008) Antioxidant status, peroxidase activity, and PR protein transcript levels in ascorbate-deficient Arabidopsis thaliana vtc mutants. J Exp Bot 59:3857–3868

Conklin PL, Barth C (2004) Ascorbic acid, a familiar small molecule intertwined in the response of plants to ozone, pathogens, and the onset of senescence. Plant Cell Environ 27:959–970

Conklin PL, Pallanca JE, Last RL, Smirnoff N (1997) L-ascorbic acid metabolism in the ascorbate-deficient Arabidopsis mutant *vtc1*. Plant Physiol 115:1277–1285

Conklin PL, Williams EH, Last RL (1996) Environmental stress tolerance of an ascorbic acid-deficient Arabidopsis mutant. Proc Natl Acad Sci USA 93:9970–9974

Conklin PL, Saracco SA, Norris SR, Last RL (2000) Identification of ascorbic acid-deficient *Arabidopsis thaliana* mutants. Genetics 154:847–856

Conklin PL, Gatzek S, Wheeler GL, Dowdle J, Raymond MJ, Rolinski S, Isupov M, Littlechild JA, Smirnoff N (2006) Arabidopsis thaliana VTC4 encodes L-galactose-1-P phosphatase, a plant ascorbic acid biosynthetic enzyme. J Biol Chem 281:15662–15670

Conklin PL, Norris SR, Wheeler GL, Williams EH, Smirnoff N, Last RL (1999) Genetic evidence for the role of GDP-mannose in plant ascorbic acid (vitamin C) biosynthesis. Proc Natl Acad Sci USA 96:4198–4203

Conklin PL, Saracco SA, Norris SR, Last RL (2000) Identification of ascorbic acid-deficient *Arabidopsis thaliana* mutants. Genetics 154:847–856

Conklin PL, Williams EH, Last RL (1996) Environmental stress sensitivity of an ascorbic acid-deficient Arabidopsis mutant. Proc Natl Acad Sci USA 93:9970–9974

Cosio C, Durand Ch (2009) Specific functions of individual class III peroxidase genes. J Exp Bot 60:391–408

Creissen G, Firmin J, Fryer M, Kular B, Leykand N, Reynolds H, Pastori G, Wellburn F, Baker H, Wellburn A (1999) Elevated glutathione biosynthesis causes increased oxidative stress. Plant Cell 11:1277–1291

Cruz JA, Avenson TJ, Kanazawa A, Takizawa K, Edwards GE, Kramer DM (2005) Plasticity in light reactions of photosynthesis for energy production and photoprotection. J Exp Bot 56:395–406

Dalle-Donne I, Milzani A, Gagliano N, Colombo R, Giustarini D, Rossi R (2008) Molecular mechanisms and potential clinical significance of S-glutathionylation. Antioxid Redox Signal 10:445–473

Dalle-Donne I, Rossi R, Giustarini D, Colombo R, Milzani A (2007) S-glutathionylation in protein redox regulation. Free Radic Biol Med 43:883–898

Darie CC, De Pascalis L, Mutschler B, Haehnel W (2006) Studies of the Ndh complex and photosystem II from mesophyll and bundle sheath chloroplasts of the C4-type plant *Zea mays*. J Plant Physiol 163:800–808

Davey MW, Gilot C, Persiau G, Østergaard J, Han Y, Bauw GC, Van Montagu MC (1999) Ascorbate biosynthesis in Arabidopsis cell suspension culture. Plant Physiol 121:535–543

Davletova S, Schlauch K, Coutu J, Mittler R (2005) The zinc-finger protein Zat12 plays a central role in reactive oxygen and abiotic stress signaling in Arabidopsis. Plant Physiol 139:847–856

DeBolt S, Cook DR, Ford CM (2006) L-Tartaric acid synthesis from vitamin C in higher plants. PNAS 103:5608–5613

Delaunay A, Pflieger D, Barrault MB, Vinh J, Toledano MB (2002) A thiol peroxidase is an H_2O_2 receptor and redox-transducer in gene activation. Cell 111:471–481

Demmig-Adams B, Adams WWIII, Logan BA, Verhoeven AS (1995) Xanthophyll cycle-dependent energy dissipation and flexible photosystem II efficiency in plants acclimated to light stress. Aust J Plant Physiol 22:249–260

Demmig-Adams B, Adams WW III (1996) The role of xanthophyll cycle carotenoids in the protection of photosynthesis. Trend Plant Sci 1:21–26

Davey MW, Gilot C, Persiau G, Østergaard J, Han Y, Bauw GC, Van Montagu MC (1999) Ascorbate biosynthesis in Arabidopsis cell suspension culture. Plant Physiol 121:535–543

Depège N, Bellafiore S, Rochaix JD (2003) Role of chloroplast protein kinase Stt7 in LHCII phosphorylation and state transition in Chlamydomonas. Science 299:1572–1575

Desikan R, Hancock JT, Bright J, Harrison J, Weir I, Hooley R, Neill SJ (2005) A Role for ETR1 in hydrogen peroxide signaling in stomatal guard cells. Plant Physiol 137:831–834

Desikan R, Mackerness SAH, Hancock JT, Neill SJ (2001) Regulation of the Arabidopsis transcriptome by oxidative stress. Plant Physiol 127:159–172

Després C, Chubak C, Rochon A, Clark R, Bethune T, Desveaux D, Fobert PR (2003) The Arabidopsis NPR1 disease resistance protein is a novel cofactor that confers redox regulation of DNA binding activity to the basic 452 domain/leucine zipper transcription factor TGA1. Plant Cell 15:2181–2191

Deutsch JC (1998) Ascorbic acid oxidation by hydrogen peroxide. Anal Biochem 255:1–7

Deutsch JC (2000) Dehydroascorbic acid. J Chromatogr A 881:299–307

Dìaz M, Achkor H, Titarenko E, Martínez MC (2003) The gene encoding glutathione-dependent formaldehyde dehydrogenase/GSNO reductase is responsive to wounding, jasmonic acid and salicylic acid. FEBS Letters 543:136–139

Díaz-Vivancos P, Barba-Espín G, Clemente-Moreno MJ, Hernández JA (2010) Characterization of the antioxidant system during the vegetative development of pea plants Biol Plant 54:76–82

Diaz-Vivancos P, Rubio M, Mesonero V, Periago PM, Barceló AR, Martínez-Gómez P, Hernández JA (2006) The apoplastic antioxidant system in Prunus: response to long-term plum pox virus infection. J Exp Bot 57:3813–3824

Dietz KJ (1996) Functions and responses of the leaf apoplast under stress. Prog Bot 58:221–254

Dietz KJ (2005) Plant thiol enzymes and thiol homeostasis in relation to thiol-dependent redox regulation and oxidative stress. In: Smirnoff N (ed) Antioxidants and reactive oxygen species in plants. Blackwell, Oxford Publishers, pp 25–52

Dietz KJ (2003) Redox control, redox signaling, and redox homeostasis in plant cells. Int Rev Cytol 228:141–193

Dietz KJ (2008) Redox signal integration: from stimulus to networks and genes. Physiol Plant 133:459–468

Dietz KJ, Jacob S, Oelze ML, Laxa M, Tognetti V, de Miranda SM, Baier M, Finkemeier I (2006) The function of peroxiredoxins in plant organelle redoc metabolism. J Exp Bot 57:1697–1709

Dietz KJ, Stork T, Finkemeier I, Lamkemeyer P, Li WX, El-Tayeb MA, Michel KP, Pistorius E, Baier M (2004) The role of peroxiredoxins in oxygenic photosynthesis of cyanobacteria and higher plants: peroxide detoxification or redox sensing? In: Demmig-Adams B, Adams W, Mattoo A (eds) Photoprotection, photoinhibition, gene regulation, and environment. Kluwer, Dordrecht, The Netherlands

Dixon DP, Skipsey M, Grundy NM, Edwards R (2005) Stress-induced protein S-glutathionylation in Arabidopsis. Plant Physiol 138:2233–2244

Dowdle J, Ishikawa T, Gatzek S, Rolinski S, Smirnoff N (2007) Two genes in Arabidopsis thaliana encoding GDP-l-galactose phosphorylase are required for ascorbate biosynthesis and seedling viability. Plant J 52:673–689

Dron M, Clouse SD, Dixon RA, Lawton MA, Lamb CJ (1988) Glutathione and fungal elicitor regulation of a plant defense promoter in electroporated protoplasts. Proc Natl Acad Sci USA 85:6738–6742

Droux M (2004) Sulfur assimilation and the role of sulfur in plant metabolism: a survey. Photosynth Res 79:331–348

Drozdowicz YM, Jones RL (1995) Hormonal regulation of organic and phosphoric acid release by barley aleurone layers and scutella. Plant Physiol 108:769–776

Durner J, Wendehenne D, Klessig DF (1998) Defence gene induction in tobacco by nitric oxide, cyclic GMP, and cyclic ADP-ribose. Proc Natl Acad Sci USA 95:10328–10333

Dutilleul C, Garnier M, Noctor G, Mathieu C, Chétrit P, Foyer CH, de Paepe R (2003) Leaf mitochondria modulate whole cell redox homeostasis, set antioxidant capacity, and determine stress resistance through altered signaling and diurnal regulation. Plant Cell 15:1–16

Edreva A (2005) Pathogenesis-related proteins: research progress in the last 15 years. Gen Appl Plant Physiol 31:105–124

Edwards R, Blount JW, Dixon RA (1991) Glutathione and elicitation of the phytoalexin response in legume cell cultures. Planta 184:403–409

Edwards R, Dixon DP, Walbot V (2000) Plant glutathione S-transferases: enzymes with multiple functions in sickness and in health. Trends Plant Sci 5:193–198

Emanuelsson A, Eskling M, Åkerlund HE (2003) Chemical and mutational modification of histidines in violaxanthin de-epoxidase from Spinacia oleracea. Physiol Plant 84:218–224

Escaler M, Aranda MA, Roberts IM, Thomas CL, Maule AJ (2000) A comparison between virus replication and abiotic stress (heat) as modifiers of host gene expression in pea. Mol Plant Pathol 1:159–167

Eskling M, Arvidsson PO, Åkerlund HE (1997) The xanthophyll cycle, its regulation and components. Physiol Plant 100:806–816

Esteve JM, Mompo J, De la Asunción JG, Sastre J, Asensi M, Boix J, Vinã JR, Vinã J, Pallardo FV (1999) Oxidative damage to mitochondrial DNA and glutathione oxidation in apoptosis: studies *in vivo* and *in vitro*. FASEB J 13:1055–1064

Fernandez-García N, Martí MC, Jimenez A, Sewilla F, Enrique Olmos E (2009) Sub-cellular distribution of glutathione in an Arabidopsis mutant (*vtc1*) deficient in ascorbate. J Plant Physiol 166:2004–2012

Feys BJ, Wiermer M, Bhat RA, Moisan LJ, Medina-Escobar N, Neu Ch, Cabral A, Parker JE (2005) Arabidopsis SENESCENCE-ASSOCIATED GENE101 stabilizes and signals within an ENHANCED DISEASE SUSCEPTIBILITY1 complex in plant innate immunity. Plant Cell 17:2601–2613

Ferrer JL, Jez JM, Bowman ME, Dixon RA, Noel JP (1999) Structure of chalcone synthase and the molecular basis of plant polyketide biosynthesis. Nat Struci Biol 6:775–784

Filkowski J, Kovalchuk O, Kovalchuk I (2004) Genome stability of vtc1, tt4, and tt5 *Arabidopsis thaliana* mutants impaired in protection against oxidative stress. Plant J 38:60–69

Foyer CH (1993) Ascorbic acid. In: Alscher RG, Hess JL (eds) Antioxidants in higher plants. CRC Press, Boca Raton, FL, pp 312–358

Foyer C, Lelandais M (1996) A comparision of the relative rates of transport of ascorbate and glucose across the thylakoid, chloroplast and plasmalemma membranes of pea leaf mosophyll cells. J Plant Physiol 148:391–398

Fotopoulos V, Sanmartin M, Kanellis AK (2006) Effect of ascorbate oxidase over-expression on ascorbate recycling gene expression in response to agents imposing oxidative stress. J Exp Bot 57:3933–3943

Fotopoulos V, De Tullio MC, Barnes J, Kanellis AK (2008) Altered stomatal dynamics in ascorbate oxidase over-expressing tobacco plants suggest a role for dehydroascorbate signaling. J Exp Bot 59:729–737

Foyer CH, Allen JF (2003) Lessons from redox signaling in plants. Antioxid Redox Signal 5:3–5

Foyer CH, Fletcher JM (2001) Plant antioxidants: colour me healthy. Biologist 48:115–120

Foyer CH, Halliwell B (1976) The presence of glutathione and glutathione reductase in chloroplasts: a proposed role in ascorbic acid metabolism. Planta 133:21–25

Foyer CH, Lelandais M (1996) A comparison of the relative rates of transport of ascorbate and glucose across the thylakoid, chloroplast and plasma membranes of pea leaf mesophyll cells. J Plant Physiol 148:391–398

Foyer CH, Lopez-Delgrado H, Dat JF, Scott IM (1997) Hydrogen peroxide- and glutathione-associated mechanisms of acclimatory stress tolerance and signalling. Physiol Plant 100:241–254

Foyer CH, Noctor G (1998) Ascorbate and gluthathione: keeping active oxygen under control. Annu Rev Plant Physiol Plant Mol Biol 49:249–279

Foyer CH, Noctor G (2000) Oxygen processing in photosynthesis: regulation and signalling. New Phytol 146:359–388
Foyer CH, Noctor G (2003) Redox sensing and signaling associated with reactive oxygen in chloroplasts, peroxisomes and mitochondria. Physiol Plant 119:355–364
Foyer CH, Noctor G (2005a) Oxidant and antioxidant signalling in plants: a re-evaluation of the concept of oxidative stress in a physiological context. Plant Cell Environ 28:1056–1071
Foyer CH, Noctor G (2005b) Redox homeostasis and antioxidant signaling: a metabolic interface between stress perception and physiological responses. Plant Cell 17:1866–1875
Foyer CH, Noctor G (2009) Redox regulation in photosynthetic organisms: signalling, acclimation, and practical implications. Antioxid Redox Signal 11:861–905
Foyer CH, Theodoulou FL, Delrot S (2001) The functions of interand intracellular glutathione transport systems in plants. Trends Plant Sci 6:486–492
Franceschi VR, Nakata PA (2005) CALCIUM OXALATE IN PLANTS: Formation and Function. Annu Rev Plant Biol 56:41–71
Frank HA, Cua A, Chynwat V, Young A, Gosztola D, Wasielewski MR (1996) The lifetimes and energies of the first excited singlet states of diadinoxanthin and diatoxanthin: the role of these molecules in excess energy dissipation in algae. Biochim Biophys Acta 1277:243–252
Freeman JL, Persans MW, Nieman K, Albrecht C, Peer W, Pickering IJ, Salt DE (2004) Increased glutathione biosynthesis plays a role in nickel tolerance in Thlaspi nickel hyperaccumulators. Plant Cell 16:2176–2191
Fridovich I (1986) Superoxide dismutases. Adv Enzymol Relat Areas Mol Biol 58:61–97
Fryer MJ, Ball L, Oxborough K, Karpinski S, Mullineaux PM, Baker NR (2003) Control of ascorbate peroxidase 2 expression by hydrogen peroxide and leaf water status during excess light stress reveals a functional organisation of Arabidopsis leaves. Plant J 33:691–705
De Gara (2004) L: Ascorbate and plant growth: from germination to cell death. In Vitamin C Functions and Biochemistry in Animals and Plants. Asard H, May JM, Smirnoff N (eds) Oxford: BIOS Scientific Publishers, vol 1, pp 83–89
Gallogly MM, Mieyal JJ (2007). Mechanisms of reversible protein glutathionylation in redox signaling and oxidative stress. Curr Opin Pharmacol 7:381–391
Gao XH, Bedhomme M, Veyel D, Zaffagnini M, Lemaire SD (2009) Methods for analysis of protein glutathionylation and their application to photosynthetic organisms. Mol Plant 2:218–235
Gao Q, Zhang LX (2008) Ultraviolet-B-induced oxidative stress and antioxidant defense system responses in ascorbate-deficient vtc1 mutants of *Arabidopsis thaliana*. J Plant Physiol 165:138–148
Garcia-Plazaola JI, Hernandez A, Errasti E, Becerril JM (2002) Occurrence and operation of the lutein epoxide cycle in Quercus species. Funct Plant Biol 29:1075–1080
Garcia-Plazaola JI, Hernandez A, Olano JM, Becerril JM (2003) The operation of the lutein epoxide cycle correlates with energy dissipation. Funct Plant Biol 30:319–324
Garcia-Plazaola JI, Matsubara S, Osmond CB (2007) Review: The lutein epoxide cycle in higher plants: its relationships to other xanthophyll cycles and possible functions. Funct Plant Biol 34:759–773
Gatzek S, Wheeler GL, Smirnoff N (2002) Antisense suppression of L-galactose dehydrogenase in *Arabidopsis thaliana* provides evidence for its role in ascorbate synthesis and reveals light modulated L-galactose synthesis. Plant J 30:541–553
Gechev TS, Hille J (2005) Hydrogen peroxide as a signal controlling plant programmed cell death. J Cell Biol 168:17–20
Ghezzi P (2005) Oxidoreduction of protein thiols in redox regulation. Biochem Soc Trans 33:1378–1381
Ghezzi P (2005a) Regulation of protein function by glutathionylation. Free Radic Res 39:573–580
Ghezzi P (2005b) Oxidoreduction of protein thiols in redox regulation. Biochem Soc Trans 33:1378–1381
Ghezzi P, Di Simplicio P (2007) Glutathionylation pathways in drug response. Curr Opin Pharmacol 7:398–403

Giacomelli L, Rudella A, van Wijk KJ (2006) High light response of the thylakoid proteome in Arabidopsis wild type and the ascorbate-deficient mutant vtc2-2. A comparative proteomics study. Plant Physiol 141:685–701
Gillham DJ, Dodge AD (1986) Hydrogen-peroxide scavenging systems within pea chloroplasts. A quantitative study. Planta 167:246–251
Gillham DJ, Dodge AD (1987) Chloroplast superoxide and hydrogen-peroxide scavenging systems from pea leaves. Seasonal variations. Plant Sci 50:105–109
Giovannoni JJ (2007) Completing a pathway to plant vitamin C synthesis. PNAS 104:9109–9110
Glazebrook J, Ausubel FM (1994) Isolation of phytoalexindeficient mutants of *Arabidopsis thaliana* and characterization of their interactions with bacterial pathogens. Proc Natl Acad Sci USA, 91:8955–8959
Glazebrook J, Rogers EE, Ausubel FM (1996) Isolation of Arabidopsis mutants with enhanced disease susceptibility by direct screening. Genetics 143:973–982
Glazebrook J, Zook M, Merrit F, Kagan I, Rogers EE, Crute IR, Holub EB, Hammerschmidt R, Ausubel FM (1997) Phytoalexin-deficient mutants of Arabidopsis reveal that PAD4 encodes a regulatory factor and that four PAD genes contribute to downy mildew resistance. Genetics 146:381–392
Gomez LD, Vanacker H, Buchner P, Noctor G, Foyer CH (2004a) Intercellular distribution of glutathione synthesis in maize leaves and its response to short-term chilling. Plant Physiol 134:1662–1671
Gomez LD, Noctor G, Knight MR, Foyer CH (2004b) Regulation of calcium signalling and gene expression by glutathione. J Exp Bot 55:1851–1859
Gonzalez-Reyes JA, Hidalgo A, Caler JA, Palos R, Navas P (1994) Nutrient-uptake changes in ascorbate free radical-stimulated onion roots. Plant Physiol 104:271–276
Goss R, Böhme K, Wilhelm C (1998) The xanthophylls cycle of Mantoniella squamata converts violaxanthin into antheraxanthin but not to zeaxanthin: consequences for the mechanism of enhanced non-photochemical energy dissipation. Planta 205:613–621
Goss R (2003) Substrate specificity of the violaxanthin de-epoxidase of the primitive green alga Mantoniella squamata (Prasinophyceae). Planta 217:801–812
Grace SC, Logan BA (2000) Energy dissipation and radical scavenging by the plant phenylopropanoid pathway. Phil Trans R Soc Lond 355:1499
Grant JJ, Loake GJ (2000) Role of reactive oxygen intermediates and cognate redox signalling in plant disease resistance. Plant Physiol 124:21–29
Green MA, Fry SC (2005) Vitamin C degradation in plant cells via enzymatic hydrolysis of 4-O-oxalyl-L-threonate. Nature 433:83–87
Grouneva I, Jakob T, Wilhelm C, Goss R (2006) Influence of ascorbate and pH on the activity of the diatom xanthophyll cycle-enzyme diadinoxanthin de-epoxidase. Physiol Plant 126:205–211
Grouneva I, Jakob T, Wilhelm C, Goss R (2008) A new multicomponent NPQ mechanism in the diatom Cyclotella meneghiniana. Plant Cell Physiol 49(8):1217–25
Grouneva I, Jakob T, Wilhelm C, Goss R (2009) The regulation of xanthophyll cycle activity and of non-photochemical fluorescence quenching by two alternative electron flows in the diatoms Phaeodactylum tricornutum and Cyclotella meneghiniana. Biochim Biophys Acta 1787(7):929–938
Gruszecki WI, Strzałka K (1991) Does the xanthophyll cycle takes part in the regulation of the fluidity of the thylakoid membrane? Biochim Biophys Acta 1060:310–314
Gukasyan HJ, Lee VHL, Kim KJ, Kannan R (2002) Net glutatione secretion across primary cultured rabbit conjunctival epithelial cell layers. Investigative Ophthalmology Visual Sci 43:1154–1161
Gullner G, Kömives T, Rennenberg H (2001) Enhanced tolerance of transgenic poplar plants overexpressing gamma-glutamylcysteine synthetase towards chloroacetanilide herbicides. J Exp Bot 52:971–979
Guo Z-J, Lamb Ch, Dixon RA (1998) Potentiation of the oxidative burst and isoflavonoid phytoalex in accumulation by serine protease inhibitors. Plant Physiol 118:1487–1494
Gupta AS, Alscher RG, McCune D (1991) Response of photosynthesis and cellular antioxidants to ozone in populus leaves. Plant Physiol 96:650–655

Halliwell B (2006) Reactive species and antioxidants: redox biology is a fundamental theme of aerobic life. Plant Physiol 141:312–322

Hancock RD, Viola R (2005) Biosynthesis and catabolism of L-ascorbic acid in plants. Crit Rev Plant Sci 24:167–188

Harada E, Yamaguchia Y, Koizumia N, Hiroshia S (2002) Cadmium stress induces production of thiol compounds and transcripts for enzymes involved in sulfur assimilation pathways in Arabidopsis. J Plant Physiol 159:445–448

Harms K, Von Ballmoos P, Brunold C, Höfgen R, Hesse H (2000) Expression of a bacterial serine acetyltransferase in transgenic potato plants leads to increased levels of cysteine and glutathione. Plant J 22:335–343

Hager A (1969) Lichtbedingte pH-Erniedrigung in einem Chloroplastenkompartiment als Ursache der enzymatischen Violaxanthin- Zeaxanthin-Umwandlung; Beziehungen zur Photophosphorylierung. Planta 89:224–243

Hartung W, Radin JW, Hendrix DL (1988) Abscisic-acid movement into the apoplastic solution of water-stressed cotton leaves – role of apoplastic pH. Plant Physiol 86:908–913

Havaux M (2003) Spontaneous and thermoinduced photon emission: New methods to detect and quantify oxidative stress in plants. Trends Plant Sci 8:409–413

Havaux M, Dall'Osto L, Bassi R (2007) Zeaxanthin has enhanced antioxidant capacity with respect to all other xanthophylls in Arabidopsis leaves and functions independent of binding to PSII antennae. Plant Physiol 145:1506–1520

Havaux M, Gruszecki WI, Dupont I, Leblanc RM (1991) Increased heat emission and its relationship to the xanthophyll cycle in pea leaves exposed to strong light stress. J Photochem Photobiol B Biol 8:361–370

Herbinger K, Tausz M, Wonisch A, Soja G, Sorger A, Grill D (2002) Complex interactive effects of drought and ozone stress on the antioxidant defence systems of two wheat cultivars. Plant Physiol Biochem 40:691–696

Hernández JA, Ferrer MA, Jiménez A, Barceló AR, Sevilla F (2001) Antioxidant systems and $O_2^{\cdot-}/H_2O_2$ production in the apoplast of pea leaves. Its relation with salt-induced necrotic lesions in minor veins. Plant Physiol 127:817–831

Herschbach C, Rennenberg H (1995) Long-distance transport of 35S-sulphur in 3-year-old beech trees (*Fagus sylvatica*). Physiol Plant 95:379–386

Herschbach C, Scheerer U, Ronnenberg H (2009) Redox states of glutathione and ascorbate in root tips of poplar (Populus tremulaxP. alba) depend on Pohlem transport from the shoot to the roots. J Exp Bot 61(4):1–10

Hicks LM, Cahoon RE, Bonner ER, Rivard RS, Sheffield J, Jez JM (2007) Thiol-based regulation of redox-active glutamate-cysteine ligase from Arabidopsis thaliana. Plant Cell 19:2653–2661

Hidalgo A, Gonzalez-Reyes JA, Navas P (1989) Ascorbate free radical enhances vacuolisation in onion root meristems. Plant Cell Environ 12:455–460

Hirai MY, Fujiwara T, Awazuhara M, Kimura T, Noji M, Saito K (2003) Global expression profiling of sulphur-starved Arabidopsis by DNA macroarray reveals the role of O-acetyl-l-serine as a general regulator of gene expression in response to sulphur nutrition. Plant J 33:651–663

Hisabori T, Motohashi K, Hosoya-Matsuda N, Ueoka-Nakanishi H, Romano PGN (2007) Towards a functional dissection of thioredoxin networks in plant cells. Photochem Photobiol 83: 145–151

Hofius D, Hajirezaei M-R, Geiger M, Tschiersch H, Melzer M, Sonnewald U (2004) RNAi-mediated tocopherol deficiency impairs photoassimilate export in transgenic potato plants. Plant Physiol 135:1256–1268

Horemans N, Foyer CH, Asard H (2000) Transport and action of ascorbate at the plant plasma membrane. Trends Plant Sci 5:263–267

Horling F, Lamkemeyer P, König J, Finkemeier I, Kandlbinder A, Baier M, Dietz KJ (2003) Divergent light-, ascorbate-, and oxidative stress-dependent regulation of expression of the peroxiredoxin gene family in Arabidopsis. Plant Physiol 131:317–325

Horner HT, Kausch AP, Wagner BL (2000) Ascorbic acid: a precursor of oxalate in crystal idioblasts of *Yucca torreyi* in liquid root culture. Int J Plant Sci 161:861–868

Hothorn M, Wachter A, Gromes R, Stuwe T, Rausch T, Scheffzek K (2006) Structural basis for the redox control of plant glutamate cysteine ligase. J Biol Chem 281:27557–27565
Howden R, Andersen CR, Goldsbrough PB, Cobbett CS (1995) Cadmium sensitive, glutathione deficient mutant of Arabidopsis thaliana. Plant Physiol 107:1067–1073
Hurd TR, Filipovska A, Costa NJ, Dahm CC, Murphy MP (2005) Disulphide formation on mitochondrial protein thiols. Biochemical Society Transactions 33:1390–1393
Hwang C, Sinskey AJ, Lodish HF (1992) Oxidized redox state of glutathione in the endoplasmic reticulum. Science 257:1496–1502
Innocenti G, Pucciariello C, Le Gleuher M, Hopkins J, Stefano M, Delledonne M, Puppo A, Baudouin E, Frendo P (2007) Glutathione synthesis is regulated by nitric oxide in *Medicago truncatula* roots. Planta 225:1597–1602
Ishikawa T, Dowdle J, Smirnoff N (2006) Progress in manipulating ascorbic acid biosynthesis and accumulation in plants. Physiol Plant 126:343–355
Ishikawa T, Shigeoka S (2008) Recent advances in ascorbate biosynthesis and the physiological significance of ascorbate peroxidase in photosynthesizing organisms. Biosci Biotechnol Biochem 72:1143–1154
Ito H, Iwabuchi M, Ogawa K (2003) The sugar-metabolic enzymes aldolase and triose-phosphate isomerase are targets of glutathionylation in *Arabidopsis thaliana*: detection using biotinylated glutathione. Plant Cell Physiol 44:655–660
Jabs T, Dietrich RA, Dangl JL (1996) Initiation of runaway cell death in an Arabidopsis Mutant by extracellular superoxide. Science 273:1853–1856
Jahns P, Latowski D, Strzałka K (2009) Mechanism and regulation of the violaxanthin cycle: the role of antenna proteins and membrane lipids. Biochim Biophys Acta 1787:3–14
Jiang M, Zhang J (2001) Effect of abscisic acid on active oxygen species, antioxidative defence system and oxidative damage in leaves of maize seedlings. Plant Cell Physiol 42:1265–1273
Jakob T, Goss R, Wilhelm C (2001) Unusual pH-dependence of diadinoxanthin de-epoxidase activation causes chlororespiratory induced accumulation of diatoxanthin in the diatom Phaeodactylum tricornutum. J Plant Physiol 158:383–390
Jez JM, Cahoon RE, Chen S (2004) *Arabidopsis thaliana* glutamatecysteine ligase. Functional properties, kinetic mechanism, and regulation of activity. J Biol Chem 279:33463–33470
Jiménez A, Hernández JA, Pastori G, del Río LA, Sevilla F (1998) Role of the ascorbate-glutathione cycle of mitochondria and peroxisomes in the senescence of pea leaves. Plant Physiol 118:1327–1335
Jiménez A, Hernández JA, del Río LA, Sevilla F (1997) Evidence for the presence of the ascorbate–glutathione cycle in mitochondria and peroxisomes of pea leaves. Plant Physiol 114:275–284
Joo JH, Wang S, Chen JG, Jones AM, Fedoroff NV (2005) Different signalling and cell death roles of heterotrimeric G protein alpha and beta subunits in the Arabidopsis oxidative stress response to ozone. Plant Cell 17:957–970
José A, Hernández FMA, Jiménez A, Barceló AR, Sevilla F (2001) Antioxidant systems and O_2^{-}/H_2O_2 production in the apoplast of pea leaves. Its relation with salt-induced necrotic lesions in minor veins. Plant Physiol 127:817–831
Karpinska B, Wingsle G, Karpinski S (2000) Antagonistic effects of hydrogen peroxide and glutathione on acclimation to excess excitation energy in Arabidopsis. IUBMB Life 50:21–26
Karpinski S, Escobar C, Karprinska B, Creissen G, Mullineaux PM (1997) Photosynthetic electron transport regulates the expression of cytosolic ascorbate peroxidase genes in *Arabidopsis* during excess light stress. Plant Cell 9:627–640
Karpinski S, Gabrys H, Mateo A, Karpinska B, Mullineaux PM (2003) Light perception in plant disease defense signaling. Curr Opin Plant Biol 6:390–396
Karpinska B, Karlsson M, Schinkel H, Streller S, Suss KH, Melzer M, Wingsle G (2001) A novel superoxide dismutase with a high isoelectric point in higher plants. Expression, regulation, and protein localization. Plant Physiol 126:1668–1677
Karpinski S, Reynolds H, Karpinska B, Wingsle G, Creissen G, Mullineaux P (1999a) Systemic signaling and acclimation in response to excess excitation energy in Arabidopsis. Science 284:654–657

Kasahara M, Kagawa T, Oikawa K, Suetsugu N, Miyao M, Wada M (2002) Chloroplast avoidance movement reduces photodamage in plants. Nature 420:829–832

Keates SE, Tarlyn NM, Loewus FA, Franceschi VR (2000) L-Ascorbic acid and L-galactose are sources for oxalic acid and calcium oxalate in *Pistia stratiotes*. Phytochem 53: 433–440

Kiddle G, Pastori GM, Bernard S, Pignocchi C, Antoniw J, Verrier PJ, Foyer CH (2003) Effects of leaf ascorbate content on defense and photosynthesis gene expression in Arabidopsis thaliana. Antioxid Redox Signal 5:23–32

Kingston-Smith AH, Foyer CH (2000) Bundle sheath proteins are more sensitive to oxidative damage than those of the mesophyll in maize leaves exposed to paraquat or low temperatures. J Exp Bot 51:123–130

Knight H, Knight MR (2001) Abiotic stress signaling pathways: specificity and cross-talk. Trends Plant Sci 6:262–267

Kocsy G, Brunner M, Riiegsegger A, Stamp P, Brunold Ch (1996) Glutathione synthesis in maize genotypes with different sensitivities to chilling. Planta 198:365–370

Kocsy G, Szalai G, Galiba G (2004a) Effect of osmotic stress on glutathione and hydroxymethyl-glutathione accumulation in wheat. J Plant Physiol 161:785–794

Kocsy G, Szalai G, Vágújfalvi A, Stéhli L, Orosz G, Galiba G (2000) Genetic study of glutathione accumulation during cold hardening in wheat. Planta 210:295–301

Kocsy G, Kobrehel K, Szalai G, Duviau M-P, Buzás Z, Galiba G (2004b) Abiotic stress-induced changes in glutathione and thioredoxin h levels in maize. Environ Exp Bot 52:101–112

Kollist H, Moldau H, Oksanen E, Vapaavuori E (2008) Ascorbate transport from the apoplast to the symplast in intact leaves. Physiol Plant 113:377–383

Koprivova A, Kopriva S, Jäger D, Will B, Jouanin L, Rennenberg H (2002) Evaluation of transgenic poplar over-expressing enzymes of glutathione synthesis for phytoremediation of cadmium. Plant Biol 4:664–670

Koprivova A, North KA, Kopriva S (2008) Complex signaling network in regulation of sulfate assimilation by salt stress in Arabidopsis roots. Plant Physiol. 146:1408–1420

Kopriva S, Wiedemann G, Reski R (2007) Sulfate assimilation in basal land plants – what does genomic sequencing tell us? Plant Biol 9:556–564

Kopriva S, Rennenberg H (2004) Control of sulphate assimilation and glutathione synthesis: interaction with N and C metabolism. J Exp Bot 55:1831–1842

Kostman TA, Tarlyn NM, Loewus FA, Franceschi VR (2001) Biosynthesis of L-Ascorbic Acid and conversion of carbons 1 and 2 of L-Ascorbic Acid to oxalic acid occurs within individual calcium oxalate crystal idioblasts. Plant Physiol 125:634–640

Kovtun Y, Chiu WL, Tena G, Sheen J (2000) Functional analysis of oxidative stress-activated mitogen-activated protein kinasecascade in plants. Proc Natl Acad Sci USA 97:2940–2945

Kranner I, Beckett RP, Wornik S, Zorn M, Pfeifhofer HW (2002) Revival of a resurrection plant correlates with its antioxidant status. Plant J 31:13–24

Kranner I, Birtić S, Anderson KM, Pritchard HW (2006) Glutathione half-cell reduction potential: universal stress marker and modulator of programmed cell death? Free Radic Biol Med 40:2155–2165

Kristensen BK, Askerlund P, Bykova NV, Egsgaard H, Moller IM (2004) Identification of oxidised proteins in the matrix of rice leaf mitochondria by immunoprecipitation and two-dimensional liquid chromatographytandem mass spectrometry. Phytochemistry 65:1839–1851

Kurganova LN, Veselov AP, Sinitsina YV, Elikova EA, Kulaeva ON (1999) Lipid peroxidation products as possible mediators of heat stress response in plants. Russ J Plant Physiol 46:181–185

Kuźniak E, Niewiadomska E, Miszalski Z, Karpinski S (2009) The role of chloroplasts and redox status in holistic regulation of stress responses in plants. In: Maksymiec W (ed) Compartmentation of responses to stresses in higher plants, true or false. Transworld Research Network (ISBN: 978-81-7895-422-6), pp 163–192

Kuźniak E, Skłodowska M (2005) Compartment-specific role of the ascorbate–glutathione cycle in the response of tomato leaf cells to *Botrytis cinerea* infection. J Exp Bot 56:921–933

Laing WA, Wright MA, Cooney J, Bulley SM (2007) The missing step of the L-galactose pathway of ascorbate biosynthesis in plants, an L-galactose guanyltransferase, increases leaf ascorbate content. PNAS, USA 104:9534–9539

Lamkemeyer P, Laxa M, Collin V, Li W, Finkemeier I, Schottler MA, Holtkamp V, Tognetti VB, Issakidis-Bourguet E, Kandlbinder A, Weis E, Miginiac-Maslow M, Dietz KJ (2006) Peroxiredoxin Q of *Arabidopsis thaliana* is attached to the thylakoids and functions in context of photosynthesis. Plant J 45:968–981

Larkindale J, Hall JD, Knight MR, Vierling E (2005) Heat stress phenotypes of Arabidopsis mutants implicate multiple signaling pathways in the acquisition of thermotolerance. Plant Physiol 138:882–897

Lee SK, Kader AA (2000) Preharvest and postharvest factors influencing vitamin C content of horticultural crops. Post Harv Biol Technol 20:207–220

Levine A, Tenhaken R, Dixon R, Lamb C (1994) H_2O_2 from the oxidative burst orchestrates the plant hypersensitive disease resistance response. Cell 79:583–593

Lin LS, Varner JE (1991) Expression of ascorbate oxidase in zucchini squash (Cucurbita pepo L.). Plant Physiol 96:159–165

Linster CL, Gomez TA, Christensen KC, Adler LN, Young BD, Brenner C, Clarke SG (2007) Arabidopsis VTC2 encodes a GDP-L-galactose phosphorylase, the last unknown enzyme in the Smirnoff–Wheeler pathway to ascorbic acid in plants. J Biol Chem 282: 18879–18885

Linster CL, Clarke SG (2008) L-Ascorbate biosynthesis in higher plants: the rope of VTC2. Trends Plant Sci 13:567–573

Logan BA, Barker DH, Demmig-Adamas B, Adams WW III (1996) Acclimation of leaf carotenoid composition and ascorbate levels to gradients in the light environment within an Australian rainforest. Plant Cell Environ 19:1083–1090

Van Loon LC, Pierpoint WS, Boller T, Conejero V (1994) Recommendations for naming plant pathogenesis-related proteins. Plant Mol Biol Rep 12:254–264

Van Loon LC, Van Strien EA (1999) The families of pathogenesis-related proteins, their activities, and comparative analysis of PR-1 type proteins. Physiol Mol Plant Pathol 55:85–97

López-Carbonell M, Munné-Bosch S, Alegre L (2006) The ascorbate-deficient vtc-1 Arabidopsis mutant shows altered ABA accumulation in leaves and chloroplasts. J Plant Growth Regul 25:137–144

Lyons T, Ollerenshaw J, Barnes JD (1999) Impacts of ozone on Plantago major: apoplastic and symplastic antioxidant status. New Phytol 141:253–263

Madamanchi NR, Alscher RG (1991) Metabolic bases for differences in sensitivity of two Pea cultivars to sulfur dioxide. Plant Physiol 97:88–93

Mahalingam R, Fedoroff NV (2003) Stress response, cell death and signaling: the many faces of ROS. Physiol Plant 119:56–68

Mahalingam R, Gomez-Buitrago A, Eckardt N, Shah N, Guevara-Garcia A, Day P, Raina R, Fedoroff NV (2003) Characterizing the stress/defense transcriptome of Arabidopsis. Genome Biol 4:R20. Epub 2003

Marin E, Nussaume L, Quesada A, Gonneau M, Sotta B, Hugueney P, Frey A, Marion-Poll A (1996) Molecular identification of zeaxanthin epoxidase of Nicotiana plumbaginifolia, a gene involved in abscisic acid biosynthesis and coresponding to the ABA locus of *Arabidopsis thaliana*. EMBO J 15:2331–2342

Marrè MT, Albergoni FG, Moroni A, Marrè E (1989) Light-induced activation of electrogenic HR extrusion and KR uptake in Elodea densa depends on photosynthesis and is mediated by the plasma membrane ATPase. J Exp Bot 40:343–352

Maruta T, Yonemitsu M, Kabuta Y, Tamoi M, Ishikawa T, Shigeoka S (2008) Arabidopsis phosphomannose isomerase 1, but not phosphomannose isomerase 2, is essential for ascorbic acid biosynthesis. J Biol Chem 283:28842–28851

Mateo A, Funck D, Mühlenbock P, Kular B, Mullineaux PM, Karpinski S (2006) Controlled levels of salicylic acid are required for optimal photosynthesis and redox homeostasis. J Exp Bot 57:1795–1807

Matsubara S, Gilmore AM, Osmond CB (2001) Diurnal and acclimatory responses of violaxanthin and lutein epoxide in the Australian mistletoe *Amyema miquelii*. Aust J Plant Physiol 28:792–800

Matsubara S, Morosinotto T, Bassi R, Christian AL, Fischer-Schliebs E, Luttge U, Orthen B, Franco AC, Scarano FR, Forster B, Pogson BJ, Osmond CB (2003) Occurrence of the lutein-epoxide cycle in mistletoes of the Loranthaceae and Viscaceae. Planta 217:868–879

Matsubara S, Naumann M, Martin R, Nichol C, Rascher U, Morosinotto T, Bassi R, Osmond B (2005) Slowly reversible de-epoxidation of luteinepoxide in deep shade leaves of a tropical tree legume may 'lock in' lutein-based photoprotection during acclimation to strong light. J Exp Bot 56:461–468

May MJ, Leaver CJ (1994) *Arabidopsis thaliana* c-glutamylcysteine synthetase is structurally unrelated to mammalian, yeast, and *Escherichia coli* homologs. Proc Natl Acad Sci USA 91:10,059–10,063

May MJ, Vernoux T, Leaver C, Van Montagu M, Inzé D (1998) Glutathione homeostasis in plants: implications for environmental sensing and plant development. J Exp Bot 49:649–667

Mehlhorn H, Lelandais M, Korth HG, Foyer CH (1996) Ascorbate is the natural substrate for plant peroxidases? FEBS Lett 378:203–206

Melhorn H (1990) Ethylene-promoted ascorbate peroxidase activity protects plants against hydrogen peroxide, ozone and paraquat. Plant Cell Environ 13:971–976

Mendoza-Cózatl D, Loza-Tavera H, Hernández-Navarro A, Moreno-Sánchez R (2005) Sulphur assimilation and glutathione metabolism under cadmium stress in yeast, protists and plants. FEMS Microbiol Rev 29:653–671

Meyer AJ (2008) The integration of glutathione homeostasis and redox signaling. J Plant Physiol 165:1390–1403

Meyer AJ, Hell R (2005) Glutathione homeostasis and redoxregulation by sulfhydryl groups. Photosynth Res 86:435–457

Michelet L, Zaffagnini M, Massot V, Keryer E, Vanacke H, Miginiac-Maslow M, Issakidis-Bourguet E, Lemaire SD (2006) Thioredoxins, glutaredoxins, and glutathionylation: new crosstalks to explore. Photosynth Res 89:225–245

Mieda T, Kabuta Y, Rapolu M, Motoki T, Takeda T, Yoshimura K, Ishikawa T, Shigeoka S (2004) Feedback inhibition of spinach L-galactose dehydrogenase by L-ascorbate. Plant Cell Physiol 45:1271–1279

Millar AH, Mittova V, Kiddle G, Heazlewood JL, Bartoli CG, Theodoulou FL, Foyer CH (2003) Control of ascorbate synthesis by respiration and its implications for stress responses. Plant Physiol 133:443–447

Mittler R (2002) Oxidative stress, antioxidants and stress tolerance. Trends Plant Sci 7:405–410

Mittler R, Merquiol E, Hallk HE, Rachmilevitch S, Kaplan A, Cohen M (2001) Living under a "dormant" canopy: a molecular acclimation mechanism of the desert plant *Retama raetam*. Plant J 25:407–416

Mittler R, Vanderauwera S, Gollery M, Van Breusegem F (2004) Reactive oxygen gene network of plants. Trends Plant Sci 9:490–498

Mittova V, Guy G, Tal M, Volokita M (2004) Salinity up-regulates the antioxidative system in root mitochondria and peroxisomes of the wild salt-tolerant tomato species *Lycopersicon pennellii*. J Exp Bot 55:1105–1113

Møller IM, Kristensen BK (2004) Protein oxidation in plant mitochondria as a stress indicator. Phytochem Photobiol Sci 3:730–735

Moons A (2005) Regulatory and functional interactions of plant growth regulators and plant glutathione S-transferases (GSTS). Plant Hormones 72:155–202

Mou Z, Fan W, Dong X (2003) Inducers of plant systemic acquired resistance regulate NPR1 function through redox changes. Cell 113–935

Müller P, Li XP, Niyogi KK (2001) Non-photochemical quenching. A response to excess light energy. Plant Physiol 125:1558–1566

Müller-Moulé P, Conklin PL, Niyogi KK (2002) Ascorbate deficiency can limit violaxanthin de-epoxidase activity *in vivo*. Plant Physiol 128:970–977

Mullineaux P, Karpinski S (2002) Signal transduction in response to excess light: getting out of the chloroplast. Curr Opin Plant Biol 5:43–48

Mullineaux PM, Karpinski S, Baker NR (2006) Spatial dependence for hydrogen peroxide-directed signaling in light-stressed plants. Plant Physiol 141:346–350

Mullineaux PM, Rausch T (2005) Glutathione, photosynthesis and the redox regulation of stress-responsive gene expression. Photosynth Res 86:459–474

Munné-Bosch S (2005) The role of α-tocopherol in plant stress tolerance. J Plant Physiol 162:743–748

Munné-Bosch S, Alegre L (2002) The function of tocopherols and tocotrienols in plants. Crit Rev Plant Sci 21:31–57

Munné-Bosch S, Alegre L (2003) Drought-induced changes in the redox state of α-tocopherol, ascorbate, and the diterpene carnosic acid in chloroplasts of Labiatae Species differing in carnosic acid contents. Plant Physiol 131:1816–1825

Munné-Bosch S, Falk J (2004) New insights into the function of tocopherols in plants. Planta 218:323–326

Munné-Bosch S, Weiler EW, Alegre L, Müller M, Düchting P, Falk J (2007) a-Tocopherol may influence cellular signaling by modulating jasmonic acid levels in plants. Planta 225:681–691

Murata Y, Pei ZM, Mori IC, Schroeder J (2001) Abscisic acid activation of plasma membrane Ca2R channels in guard cells requires cytosolic NAD(P)H and is differentially disrupted upstream and downstream of reactive oxygen species production in abi1-1 and abi2-1 protein phosphatase 2C mutants. Plant Cell 13:2513–2523

Müller P, Li X-P, Niyogi KK (2001) Non-photochemical quenching. A response to excess light energy. Plant Physiol 125:1558–1566

Müller-Moulé P, Conklin PL, Niyogi KK (2002) Ascorbate deficiency can limit violaxanthin de-epoxidase activity in vivo. Plant Physiol 128:970–977

Müller-Moulé P, Golan T, Niyogi KK (2004) Ascorbate-deficient mutants of Arabidopsis grow in high light despite chronic photooxidative stress. Plant Physiol 134:1163–1172

Müller-Moulé P, Havaux M, Niyogi KK (2003) Zeaxanthin deficiency enhances the high light sensitivity of an ascorbate-deficient mutant of Arabidopsis. Plant Physiol 133:748–760

Navrot N, Collin V, Gualberto J, Gelhaye E, Hirasawa M, Rey P, Knaff DB, Issakidis E, Jacquot J-P, Rouhier N (2006) Plant glutathione peroxidases are functional peroxiredoxins distributed in several subcellular compartments and regulated during biotic and abiotic stresses. Plant Physiol 142:1364–1379

Neill SJ, Desikan R, Clarke A, Hurst RD, Hancock JT (2002) Hydrogen peroxide and nitric oxide as signalling molecules in plants. J Exp Bot 53:1237–1247

Neubauer C, Yamamoto HY (1994) Membrane barriers and Mehler-peroxidase reaction limit the ascorbate availability for violaxanthin de-epoxidase activity in intact chloroplasts. Photosynth Res 39:137–147

Nieto-Sotclo J, Ho T-HD (1986) ElTect of heal shock on the metabolism of glutathione in maize roots. Plant Physiol 82:1031–1015

Niewiadomska E, Borland AM (2008) Crassulacean acid metabolism: a cause or consequence of oxidative stress in planta? In: Lüttge UE, Beyschlag W, Murata J (eds): Progress in botany 69. Springer-Verlag, Berlin, Heidelberg 247 (ISBN: 978-3-540-72953-2)

Niyogi KK (2000) Safety valves for photosynthesis. Current Opinion Plant Biol 3:455–460

Niyogi KK, Shih C, Chow WS, Pogson BJ, DellaPenna D, Björkman O (2001) Photoprotection in a zeaxanthin- and lutein-deficient double mutant of Arabidopsis. Photosynth Res 67:139–145.

Noctor G (2006) Metabolic signalling in defence and stress: the central roles of soluble redox couples. Plant Cell Environ 29:409–425

Noctor G, Arisi ACM, Jouanin L, Kunert KJ, Rennenberg H, Foyer CH (1998) Glutathione: biosynthesis, metabolism and relationship to stress tolerance explored in transformed plants. J Exp Bot 49:623–647

Noctor G, Foyer CH (1998) Ascorbate and glutathione: keeping active oxygen under control. Annu Rev Plant Physiol Plant Mol Biol 49:249–279

Noctor G, Foyer CH (2005) Redox homeostasis and antioxidant signaling: a metabolic interface between stress perception and physiological responses. Plant Cell 17:1866–1875

Noctor G, Gomez L, Vanacker H, Foyer CH (2002) Interactions between biosynthesis, compartmentation and transport in the control of glutathione homeostasis and signalling. J Exp Bot 53:1283–1304

Noctor G, Queval G, Gakiere B (2006) NAD(P) synthesis and pyridine nucleotide cycling in plants and their potential importance in stress conditions. J Exp Bot 57:1603–1620

Nunes-Nesi A, Carrari F, Gibon Y, Sulpice R, Lytovchenko A, Fisahn J, Graham J, Ratcliffe RG, Sweetlove LJ, Fernie AR (2007) Deficiency of mitochondrial fumarase activity in tomato plants impairs photosynthesis via an effect on stomatal function. Plant J 50:1093–1106

Nunes-Nesi A, Carrari F, Lytovchenko A, Smith AM, Loureiro ME, Ratcliffe RG, Sweetlove LJ, Fernie AR (2005) Enhanced photosynthetic performance and growth as a consequence of decreasing mitochondrial malate dehydrogenase activity in transgenic tomato plants. Plant Physiol 137:611–622

Nunes-Nesi A, Sweetlove LJ, Fernie AR (2007) The influence of the tricarboxylic acid cycle on photosynthetic metabolism of the illuminated leaf. Physiol Plant 129:45–56

Nunes-Nesi A, Sulpice R, Gibon Y, Fernie AR (2008) The enigmatic contribution of mitochondrial function in Photosynthesis. J Exp Bot 59:1675–1684

Ohashi Y, Matsuoka M (1987) Induction and secretion of pathogenesis-related proteins by salicylate or plant hormones in tobacco suspension cultures. Plant Cell Physiol 28:573–580

Olaizola M, La Roche J, Kolber Z, Falkowski P (1994) Non-photochemical fluorescence quenching and the diadinoxanthin cycle in a marine diatom. Photosynth Res 41:357–370

Ort DR, Baker NR (2002) A photoprotective role for O_2 as an alternative electron sink in photosynthesis? Curr Opin Plant Biol 5:193–198

Packer JE, Slater TF, Willson RL (1979) Direct observation of a free radical interaction between vitamin E and vitamin C Nature (London) 278:737–738

Padh H (1990) Cellular functions of ascorbic acid Biochem Cell Biol 68:1166–1173

Pallanca JE, Smirnoff N (2000) The control of ascorbic acid synthesis and turnover in pea seedlings. J Exp Bot 51:669–674

Panchuk II, Volkov RA, Schöffl F (2002) Heat stress- and heat shock transcription factor-dependent expression and activity of ascorbate peroxidase in Arabidopsis. Plant Physiol 129:838–853

Papadakis AK, Paschalidis KA, Roubelakis-Angelakis KA (2005a) Biosynthesis profile and endogenous titers of polyamines differ in totipotent and recalcitrant plant protoplasts. Physiol Plant 125:10–20

Papadakis AK, Roubelakis-Angelakis KA (2005b) Polyamines inhibit NADPH oxidase-mediated superoxides generation and putrescine prevents programmed cell death syndrome induced by the polyamine oxidase generated hydrogen peroxide. Planta 220:826–837

Parisy V, Poinssot B, Owsianowski L, Buchala A, Glazebrook J, Mauch F (2007) Identification of PAD2 as a -glutamylcysteine synthetase highlights the importance of glutathione in disease resistance of Arabidopsis. Plant J 49:159–172

Paschalidis KA, Roubelakis-Angelakis KA (2005a) Spatial and temporal distribution of polyamine levels and polyamine anabolism in different organs/tissues of the tobacco plant: correlations with age, cell division/expansion, and differentiation. Plant Physiol 138:142–152

Paschalidis KA, Roubelakis-Angelakis KA (2005b) Sites and regulation of polyamine catabolism in the tobacco plant: Correlations with cell division/expansion, cell-cycle progression, and vascular development. Plant Physiol 138:2174–2184

Pasqualini S, Batini P, Ederli L, Porceddu A, Piccioni C, De Marchis F, Antonielli M (2001) Effects of short-term ozone fumigation on tobacco plants: response of the scavenging system and expression of the glutathione reductase. Plant Cell Environ 24:245–252

Pasternak M, Lim B, Wirtz M, Hell R, Cobbett CS, Meyer AJ (2008) Restricting glutathione biosynthesis to the cytosol is sufficient for normal plant development. Plant J 53:999–1012

Pastori GM, Foyer CH (2002) Common components, networks and pathways of cross tolerance to stress. The central role of 'redox' and abscisic acid-mediated controls. Plant Physiol 129:460–468

Pastori GM, Kiddle G, Antoniw J, Bernard S, Veljovic-Jovanovic S, Verrier PJ, Noctor G, Foyer CH (2003) Leaf vitamin C contents modulate plant defense transcripts and regulate genes controlling development through hormone signaling. Plant Cell 15:1212–1226

Pastori GM, Mullineaux PM, Foyer CH (2000) Posttranscriptional regulation prevents accumulation of glutathione reductase protein and activity in the bundle sheath cells of maize. Plant Physiol 122:667–675
Paul MJ, Foyer CH (2001) Sink regulation of photosynthesis. J Exp Bot 52:1383–1400
Pavet V, Olmos E, Kiddle G, Mowla S, Kumar S, Antoniw J, Alvarez ME, Foyer CH (2005) Ascorbic acid deficiency activates cell death and disease resistance responses in Arabidopsis. Plant Physiol 139:1291–1303
Pei ZM, Murata Y, Benning G, Thomine S, Klusener B, Allen GJ, Grill E, Schroeder JI (2000) Calcium channels activated by hydrogen peroxide mediate abscisic acid signalling in guard cells. Nature 406:731–734
Pfannschmidt T, Allen JF, Oelmuller R (2001) Principles of redox control in photosynthesis gene expression. Physiol Plant 112:1–9
Pfannschmidt T (2003) Chloroplast redox signals: how photosynthesis controls its own genes. Trends Plant Sci 8:33–41
Pfannschmidt T, Liege K (2005) Redox regulation and modification of proteins controlling chloroplast gene expression. Antioxid Redox Signal 7:607–618
Pfannschmidt T, Bräutigam K, Wagner R, Dietzel L, Schröter Y, Steiner S, Nykytenko A (2009) Potential regulation of gene expression in photosynthetic cells by redox and energy state: approaches towards better understanding. Ann Bot 103:599–607
Pfanz H, Dietz KJ (1987) A fluorescence method for the determination of the apoplastic proton concentration in intact leaf tissues. J Plant Physiol 129:41–48
Pifanelli P, Zhou F, Casais C, Orme J, Jarosch B, Schaffrath U, Collins NC, Panstruga R, Schulze-Lefert P (2002) The barley MOL modulator of defense and cell death is responsive to biotic and abiotic stimuli. Plant Physiol 120:1076–1085
Pignocchi C, Fletcher JM, Barnes JD, Foyer CH (2003) The function of ascorbate oxidase (AO) in tobacco (*Nicotiana tabacum* L.). Plant Physiol 132:1631–1641
Pignocchi C, Foyer CH (2003) Apoplastic ascorbate metabolism and its role in the regulation of cell signaling. Curr Opin Plant Biol 6:379–389
Pignocchi C, Kiddle G, Hernández I, Asensi A, Taybi T, Barnes JD, Foyer CH (2006) Apoplast redox state modulates gene transcription leading to modified hormone signalling and defence in tobacco. Plant Physiol 141:423–435
Piippo M, Allahverdiyeva Y, Paakkarinen V, Suoranta UM, Battchikova N, Aro EM (2006) Chloroplast-mediated regulation of nuclear genes in Arabidopsis thaliana in the absence of light stress. Physiol Genomics 25:142–152
Pineau B, Layoune O, Danon A, De Paepe R (2008) L-galactono-1,4-lactone dehydrogenase is required for the accumulation of plant respiratory complex I. J Biol Chem 283: 32500–32505
de Pinto MC, De Gara L (2004) Changes in the ascorbate metabolism of apoplastic and symplastic spaces are associated with cell differentiation. J Exp Bot 55:2559–2569
de Pinto MC, Francis D, De Gara L (1999) The redox state of the ascorbate-dehydroascorbate pair as a specific sensor of cell division in tobacco BY-2 cells. Protoplasma 209:90–97
de Pinto MC, Tommasi F, De Gara L (2000) Enzymes of the ascorbate biosynthesis and ascorbate–glutathione cycle in cultured cells of tobacco Bright Yellow 2. Plant Physiol Biochem 38:541–550
Plaxton WC, Podestá FE (2006) The functional organization and control of plant respiration. Crit Rev Plant Sci 25:59–198
Potters G, Horemans N, Bellone S, Caubergs J, Trost P, Guisez Y, Asard H (2004) Dehydroascorbate influences the plant cell cycle through a glutathione-independent reduction mechanism. Plant Physiol 134:1479–1487
Potters G, Horemans N, Caubergs RJ, Asard H (2000) Ascorbate and dehydroascorbate influence cell cycle progression in a tobacco cell suspension. Plant Physiol 124:17–20
Prasad TK, Anderson MD, Martin BA, Stewart CR (1994) Evidence for chilling-induced oxidative stress in maize seedlings and a regulatory role for hydrogen peroxide. Plant Cell 6:65–74

Preger V, Scagliarini S, Pupillo P, Trost P (2005) Identification of an ascorbate-dependent cytochrome b of the tonoplast membrane sharing biochemical features with members of the cytochrome b561 family. Planta 220:365–375

Quan LJ, Zhang B, Shi WW, Li HY (2008) Hydrogen peroxide in plants: a versatile molecule of the reactive oxygen species network. J Integr Plant Biol 50:2–16

Queval G, Hager J, Gakière B, Noctor G (2008) Why are literature data for H2O2 contents so variable? A discussion of potential difficulties in the quantitative assay of leaf extracts. J Exp Bot 59:135–146

Queval G, Issakidis-Bourguet E, Hoeberichts FA, Vandorpe M, Gakière B, Vanacker H, Miginiac-Maslow M, Van Breusegem F, Noctor G (2007) Conditional oxidative stress responses in the Arabidopsis photorespiratory mutant cat2 demonstrate that redox state is a key modulator of daylength-dependent gene expression and define photoperiod as a crucial factor in the regulation of H_2O_2-induced cell death. Plant J 52:640–657

Queval G, Thominet D, Vanacker H, Miginiac-Maslow M, Gakiere B, Noctor G (2009) H_2O_2-activated up-regulation of glutathione in Arabidopsis involves induction of genes encoding enzymes involved in cysteine synthesis in the chloroplast. Mol Plant 2:344–356

Quinones MA, Zeiger E (1994) A putative role of the xanthophyll, zeaxanthin, in blue light photoreception of corn coleoptiles. Science 264:558–561

Rabinowitch HD, Budowski P, Kedar N (1975) Carotenoids and epoxide cycles in mature-green tomatoes. Planta 122:91–97

Rautenkranz AAF, Li L, Machler F, Martinoia E, Oertli JJ (1994) Transport of ascorbic and dehydroascorbic acid across protoplasts and vacuole membranes isolated from barley (*Hordeum vulgare L. cv. Gerbel*) leaves. Plant Physiol 106:187–193

Rausch T, Gromes R, Liedschulle V, Müller I, Bogs J, Galovic V, Wachter A (2007) Novel insight into the regulation of GSH biosynthesis in higher plants. Plant Biol 9:565–572

Rausch T, Wachter A (2005) Sulfur metabolism: a versatile platform for launching defence operations. Trends Plant Sci 10:503–509

Rennenberg H (1980) Glutathione metabolism and possible biological roles in higher plants. Phytochemistry 21:2771–2781

Rennenberg H, Herschbach C, Haberer K, Kopriva S (2007) Sulfur metabolism in plants: are trees different? Plant Biol 9:620–637

Rentel MC, Knight MR (2004) Oxidative stress-induced calcium signaling in Arabidopsis. Plant Physiol 135:1471–1479

Retsky KL, Freeman MW, Frei B (1993) Ascorbic acid oxidation product(s) protect human low density lipoprotein against atherogenic modification. J Biol Chem 268:1304–1309

del Río LA, Corpas FJ, Sandalio LM, Palma JM, Gómez M, Barroso JB (2002) Reactive oxygen species, antioxidant systems and nitric oxide in peroxisomes. J Exp Bot 53:1255–1272

Rizhsky L, Hallak-Herr E, Van Breusegem F, Rachmilevitch S, Barr JE, Rodermel S, Inzé D, Mittler R (2002) Double antisense plants lacking ascorbate peroxidase and catalase are less sensitive to oxidative stress than single antisense plants lacking ascorbate peroxidase or catalase. Plant J 32:329–342

Robinson JM, Bunce JA (2000) Influence of drought-induced water stress on soybean and spinach leaf ascorbate-dehydroascorbate level and redox status. Int J Plant Sci 161:271–279

Rodriguez AA, Grunberg KA, Taleisnik EL (2002) Reactive oxygen species in the elongation zone of maize leaves are necessary for leaf extension. Plant Physiol 129:1627–1632

Rodríguez-Serrano M, Romero-Puertas MC, Pazmiño DM, Testillano PS, Risueño MC, del Río LA, Sandalio LM (2009) Cellular response of pea plants to cadmium toxicity: cross talk between reactive oxygen species, nitric oxide, and calcium. Plant Physiol 150: 229–243

Rossel JB, Walter PB, Hendrickson L, Chow WS, Poole A, Mullineaux PM, Pogson BJ (2006) A mutation affecting ASCORBATE PEROXIDASE 2 gene expression reveals a link between responses to high light and drought tolerance. Plant Cell Environ 29:269–281

Rossel JB, Wilson IW, Pogson BJ (2002) Global changes in gene expression in response to high light in Arabidopsis. Plant Physiol 130:1109–1120

Rouhier N, Lemaire SD, Jacquot JP (2008) The role of glutathione in photosynthetic organisms: emerging functions for glutaredoxins and glutathionylation. Annu Rev Plant Biol 59:143–166

Ruiz JM, Blumwald E (2002) Salinity-induced glutathione synthesis in *Brassica napus*. Planta 214:965–969

Saito K (2004) Sulfur assimilatory metabolism, the long and smelling road. Plant Physiol 136:2443–2450

Sakuta M (2000) Transcriptional control of chalcone synthase by environmental stimuli. J Plant Res 113:327–333

Sanmartin M, Drogouti PD, Lyons T, Barnes J, Kanellis AK (2003) Over-expression of ascorbate oxidase in the apoplast of transgenic tobacco results in altered ascorbate and glutathione redox states and increased sensitivity to ozone. Planta 216:918–928

Sandermann H Jr, Ernst D, Heller W, Langebartels C (1998) Ozone: an abiotic elicitor of plant defense reactions. Trends Plant Sci 3:47–50

Sapozhnikov DI, Krasovskaya TA, Maevskaya AN (1957) Change in the interrelationship of the basic carotenoids of the plastids of green leaves under the action of light. Dokl Akad Nauk USSR 113:465–467

Sarry JE, Montillet JL, Sauvaire Y, Havaux M (1994) The protective function of the xanthophyll cycle in photosynthesis. FEBS Lett 353:147–150

Sasaki-Sekimoto Y, Taki N, Obayashi T, Aono M, Matsumoto F, Sakurai N, Suzuki H, Hirai MY, Noji M, Saito K, Masuda T, Takamiya K, Shibata D, Ohta H (2005) Coordinated activation of metabolic pathways for antioxidants and defence compounds by jasmonates and their roles in stress tolerance in Arabidopsis. Plant J 44:653–668

Sattelmacher B (2001) The apoplast and its significance for plant mineral nutrition. New Phytol 149:167–192

Schafer FQ, Buettner GR (2001) Redox environment of the cell as viewed through the redox state of the glutathione disulfide/glutathione couple. Free Radic Biol Med 30:1191–1212

Schäfer HJ, Haag-Kerwer A, Rausch T (1998) cDNA cloning and expression analysis of genes encoding GSH synthesis in roots of the heavy metal accumulator *Brassica juncea* L: evidence for Cd induction of a putative mitochondrial -glutamylcysteine synthetase isoform. Plant Mol Biol 37:87–97

Scheibe R, Backhausen JE, Emmerlich V, Holtgrefe S (2005) Strategies to maintain redox homeostasis during photosynthesis under changing conditions. J Exp Bot 56:1481–1489

Schlaeppi K, Bodenhausen N, Buchala A, Mauch F, Reymond P (2008) The glutathione-deficient mutant pad2-1 accumulates lower amounts of glucosinolates and is more susceptible to the insect herbivore *Spodoptera littoralis*. Plant J 55:774–786

Scheerer U, Haensch R, Mendel RR, Kopriva S, Rennenberg H, Herschbach C (2010) Sulphur flux through the sulphate assimilation pathway is differently controlled by adenosine 5'-phosphosulphate reductase under stress and in transgenic poplar plants overexpressing g-ECS, SO, or APR. J Exp Bot 61:609–622

Schopfer P, Liszkai A, Bechtold M, Frahry G, Wagner A (2002) Evidence that hydroxyl radicals mediate auxin-induced extension growth. Planta 214:821–828

Schöner S, Krause GH (1990) Protective systems against active oxygen species in spinach. Response to cold-acclimation in excess light. Planta 180:383–389

Schupp R, Rennenberg H (1988) Diurnal changes in the glutathione concentration of spruceneedles (*Picea abies* L.). Plant Sci 57:113–117

Schützendübel A, Polle A (2002) Plant responses to abiotic stresses: heavy metal-induced oxidative stress and protection by mycorrhization. J Exp Bot 53:1351–1365

Sen CK (2000) Cellular thiols and redox-regulated signal transduction. Curr Top Cell Regul 36:1–30

Shalata A, Neumann PM (2001) Exogenous ascorbic acid (vitamin C) increases resistance to salt stress and reduces lipid peroxidation. J Exp Bot 52:2207–2211

Shao HB, Chu LY, Lu ZH, Kang CM (2008) Primary antioxidant free radical scavenging and redox signaling pathways in higher plant cells. Int J Biol Sci 4:8–14

Siefermann-Harms D (1985) Carotenoids in photosynthesis. I. Location in photosynthetic membranes and light-harvesting function. Biochim Biophys Acta 811:325–335

Singla-Pareek SL, Reddy MK, Sopory SK (2003) Genetic engineering of the glyoxalase pathway in tobacco leads to enhanced salinity tolerance. Proc Natl Acad Sci USA 100:14672–14677

Smirnoff N (1996) The function and metabolism of ascorbic acid in plants. Ann Bot 78:661–669

Smirnoff N (2000) Ascorbic acid: metabolism and functions of a multi-facetted molecule. Curr Opin Plant Biol 3:229–235

Smirnoff N, Conklin PL, Loewus FA (2001) Biosynthesis of ascorbic acid in plants: a renaissance. Annu Rev Plant Physiol Plant Mol Biol 52:437–467

Smirnoff N, Dowdle J, Ishikawa T (2007) The role of VTC2 in vitamin C biosynthesis in *Arabidopsis thaliana*. Comp Biochem Physiol A 146:S250

Smirnoff N, Running JA, Gatzek S (2004) Ascorbate biosynthesis: a diversity of pathways. In: Asard H, May JM, Smirnoff N (eds) Vitamin C. Functions and biochemistry in animals and plants. BIOS Scientific Publishers, Oxford, pp 7–29

Smirnoff N, Wheeler GL (2000) Ascorbic acid in plants: biosynthesis and function. Crit Rev Biochem Mol Biol 35:291–314

Smith IK, Pole A, Rennenberg H (1990) Glutathione. In: Alscher RG, Cumming J (eds) Stress responses in plants: adaptation and acclimation mechanisms. Wiley-Liss, New York,pp 201–217

Smith PK, Krohn RI, Hermanson GT, Mallia AK, Gartner FH, Provenzano MD, Fujimoto EK, Goeke NM, Olson BJ, Klenk DC (1985) Measurement of protein using bicinchoninic acid. Anal Biochem 150:76–85

Srivastava A, Zeiger E (1995) The inhibitor of zeaxanthin formation, dithiothreitol, inhibits blue-light stimulated stomatal opening in *Vicia faba*. Planta 196:445–449

Stohs SJ, Bagchi D (1995) Oxidative mechanisms in the toxicity of metal ions. Free Radic Biol Med 18:321–336

Sweetlove L, Fait A, Nunes-Nesi A, Williams T, Fernie A (2007) The Mitochondrion: An Integration Point of Cellular Metabolism and Signalling Critical Reviews in Plant Sciences 26:17–43

Sweetlove LJ, Heazlewood JL, Herald V, Holtzapffel R, Day DA, Leaver CJ, Millar AH (2002) The impact of oxidative stress on Arabidopsis mitochondria. Plant J 32:891–904

Szalai G, Kellős T, Galiba G, Kocsy G (2009) Glutathione as an antioxidant and regulatory molecule in plants under abiotic stress conditions. J Plant Growth Regul 28:66–80

Ślesak I, Karpinska B, Surówka E, Miszalski Z, Karpinski S (2003) Redox changes in the chloroplast and hydrogen peroxide are essential for regulation of C3-CAM transition and photooxidative stress responses in the facultative CAM plant Mesembryanthemum crystallinum L. Plant Cell Physiol 44:573–581

Ślesak I, Libik M, Karpinska B, Karpinski S, Miszalski Z (2007) The role of hydrogen peroxide in regulation of plant metabolism and cellular signalling in response to environmental stresses. Acta Biochim Pol 54:39–50

Tabata K, Takaoka T, Esaka M (2002) Gene expression of ascorbic acid related enzymes in tobacco. Phytochem 61:631–635

Takahama U (1993) Redox state of ascorbic acid in the apoplast of stems of Kalanchoë daigremontiana. Physiol Plant 89:791–798

Takahama U (1994) Changes induced by abscisic acid and light in the redox state of ascorbate in the apoplast of epicotyls of Vigna angularis. Plant Cell Physiol 35:975–978

Tamaoki M, Mukai F, Asai N, Nakajima N, Kubo A, Aono M, Saji H (2003) Light-controlled expression of a gene encoding l-galactono-γ-lactone dehydrogenase which affects ascorbate pool size in *Arabidopsis thaliana*. Plant Sci 164:1111–1117

Tambussi EA, Bartoli CG, Beltrano J, Guiamet JJ, Araus JL (2000) Oxidative damage to thylakoid proteins in water-stressed leaves of wheat (*Triticum aestivum*). Physiol Plant 108:398–404

Tardy F, Havaux M (1997) Thylakoid membrane fluidity and thermostability during the operation of the xanthophyll cycle in higher-plant chloroplasts. Biochim Biophys Acta 1330:179–193

Tausz M, Bytnerowicz A, Arbaugh MJ, Wonisch A, Grill D (2001) Multivariate patterns of biochemical responses of *Pinus ponderosa* trees at field plots in the San Bernardino Mountains, southern California. Tree Physiol 21:329–336

Tausz M, Šircelj H, Grill D (2004) The glutathione system as a stress marker in plant ecophysiology: is a stress-response concept valid? J Exp Bot 55:1955–1962

Thordal-Christensen H, Zhang Z, Wei YD, Collinge DB (1997) Subcellular localization of H_2O_2 in plants: H_2O_2 accumulation in papillae and hypersensitive response during the barley-powdery mildew interaction. Plant J 11:1187–1194

Townsend DM (2007) S-glutathionylation: indicator of cell stress and regulator of the unfolded protein response. Mol Interv 7:313–324

Turcsányi E, Lyons T, Plöchl M, Barnes J (2000) Does ascorbate in the mesophyll cell walls form the first line of defence against ozone? Testing the concept using broad bean (*Vicia faba* L.). J Exp Bot 51:901–910

Ullmann P, Gondet L, Potier S, Bach TJ (1996) Cloning of *Arabidopsis thaliana* glutathione synthetase (GSH2) by functional complementation of a yeast gsh2 mutant. Eur J Biochem 236: 662–669

Urbańczyk-Wochniak E, Usadel B, Thimm O, Munes-Nesi A, Carrari F, Davy M, Blasing O, Kowalczyk M, Weicht D, Polinceusz A, Meyer S, Stitt M, Fernie AR (2006) Conversion of MapMan to allow the analysis of transcript data from Solanaceous species: effects of genetic and environmental alterations in energy metabolism in the leaf. Plant Mol Biol 60:773–792

Vanacker H, Carver TLW, Foyer CH (2000) Early H_2O_2 accumulation in mesophyll cells leads to induction of glutathione during the hyper-sensitive response in the barley–powdery mildew interaction. Plant Physiol 123:1289–1300

Vanacker H, Carver TLW, Foyer CH (1998) Pathogen-induced changes in the antioxidant status of the apoplast in barley leaves. Plant Physiol 117:1103–1114

Vanacker H, Harbinson J, Ruisch J, Carver TLW, Foyer CH (1998b) Antioxidant defences of the apoplast. Protoplasma 205:129–140

Veljovic-Jovanovic SD, Pignocchi C, Noctor G, Foyer CH (2001) Low ascorbic acid in the vtc-1 mutant of Arabidopsis is associated with decreased growth and intracellular redistribution of the antioxidant system. Plant Physiol 127:426–435

Venis MA, Napier RM (1997) Auxin perception and signal transduction. In: Basel AP (ed) Signal transduction in plants. Birkhäuser Verlag Basel, Switzerland, pp 45–63

Vernoux T, Wilson RC, Seeley KA, Reichheld JP, Muroy S, Brown S, Maughan SC, Cobbett CS, Van Montagu M, Inzé D, May MJ, Sung ZR (2000) The ROOT MERISTEMLESS1/CADMIUM SENSITIVE2 gene defines a glutathione dependent pathway involved in initiation and maintenance of cell division during postembryonic root development. Plant Cell 12:97–110

Vieira Dos Santos C, Laugier E, Tarrago L, Massot V, Issakidis-Bourguet E, Rouhier N, Rey P (2007) Specificity of thioredoxins and glutaredoxins as electron donors to two distinct classes of Arabidopsis plastidial methionine sulfoxide reductases B. FEBS Lett 581:4371–4376

Vieira Dos Santos C, Rey P (2006) Plant thioredoxins are key actors in the oxidative stress response. Trends Plant Sci 11:329–334

Wachter A, Rausch T (2005) Regulation of glutathione (GSH) synthesis in plants: Novel insight from Arabidopsis. FAL Agricultural Research 283:149–155

Wachter A, Wolf S, Steininger H, Bogs J, Rausch T (2005) Differential targeting of GSH1 and GSH2 is achieved by multiple transcription initiations: implications for the compartmentation of glutathione biosynthesis in the Brassicaceae. Plant J 41:15–30

Wagner GJ (1981) Vacuolar deposition of ascorbate-derived oxalic acid in barley. Plant Physiol 67:591–593

Wheeler GL, Jones MA, Smirnoff N (1998) The biosynthetic pathway of vitamin C in higher plants. Nature 393:365–369

Willekens H, Chamnongpol S, Davey M, Schraudner M, Langebartels C, Van Montagu M, Inzé D, Van Camp W (1997) Catalase is a sink for H_2O_2 and is indispensable for stress defence in C3 plants. EMBO J 16:4806–4816

Wingate VPM, Lawton MA, Lamb ChJ (1988) Glutathione causes a massive and selective induction of plant defense genes. Plant Physiol 87:206–210

Wingsle G, Karpinski S, Hällgren JE (1999) Low temperature, high light stress and antioxidant defence mechanisms in higher plants. Phyton 39:253–268

Wojtaszek P (1997) Oxidative burst: an early plant response to pathogen. Biochem J 322:681–692

Wolucka BA, Van Montagu M (2007) The VTC2 cycle and de novo biosynthesis pathways for vitamin C in plants: an opinion. Phytochemistry 68:2602–2613

Wormuth D, Heiber I, Shaikali J, Kandlbinder A, Baier M, Dietz KJ (2007) Redox regulation and antioxidative defence in Arabidopsis leaves viewed from a systems biology perspective. J Biotechnol 129:229–248

Xiang C, Oliver DJ (1998) Glutathione metabolic genes coordinately respond to heavy metals and jasmonic acid in Arabidopsis. Plant Cell 10:1539–1550

Xiong L, Schumaker KS, Zhu JK (2002) Cell signaling during cold, drought, and salt stress. Plant Cell 14:S165–S183

Yabuta Y, Mieda T, Rapolu M, Nakamura A, Motoki T, Maruta T, Yoshimura K, Ishikawa T, Shigeoka S (2007) Light regulation of ascorbate biosynthesis is dependent on the photosynthetic electron transport chain but independent of sugars in Arabidopsis. J Exp Bot 58:2661–2671

Yadav SK, Singla-Pareek SL, Sopory SK (2008) An overview on the role of methylglyoxal and glyoxalases in plants. Drug Metabol Drug Interact 23:51–68

Yamamoto HY (1979) Biochemistry of the Violaxanthin Cycle in Higher-Plants. Pure Appl Chem 51:639–648

Yamamoto A, Bhuiyan MNH, Waditee R, Tanaka Y, Esaka M, Oba K, Jagendorf AT, Takabe T (2005) Suppressed expression of the apoplastic ascorbate oxidase gene increases salt tolerance in tobacco and Arabidopsis plants. J Exp Bot 56:1785–1796

Yamamoto HY, Higashi RM (1978) Violaxanthin de-epoxidase. Lipid composition and substrate specificity. Arch Biochem Biophys 190:514–522

Yamasaki H, Takahashi S, Heshiki R (1999) The tropical fig Ficus microcarpa L. f. cv. golden leaves lacks heat-stable dehydroascorbate reductase activity. Plant Cell Physiol 40:640–646

Yen TCH, King KL, Lee HCH, Yen SH, Wei YK (1994) Age dependent increase of mitochondrial DNA deletions together with lipid peroxides and superoxide dismutase in human liver mitochondria. Free Radical Biol Med 16:207–214

Yu O, Jez JM (2008) Nature's assembly line: biosynthesis of simple phenylpropanoids and polyketides. Plant J 54:750–762

Zechmann B, Zellnig G, Müller M (2006) Immunocytochemical localization of glutathione precursors in plant cells. J Electron Microsc 55:173–181

Zechmann B, Mauch F, Sticher L, Müller M (2008) Subcellular immunocytochemical analysis detects the highest concentrations of glutathione in mitochondria and not in plastids. J Exp Bot 59:4017–4027

Chapter 2
Ascorbate and Glutathione in Organogenesis, Regeneration and Differentiation in Plant In vitro Cultures

Jarosław Tyburski and Andrzej Tretyn

Abstract The prerequisite for shoot, root or somatic embryo formation in plant in vitro culture is the development of meristem from dedifferentiated cells of the explant tissue. Auxin and cytokinin levels and their relative ratios play a decisive role in inducing the morphogenetic pathways leading to shoot, root or somatic embryo formation in plant in vitro cultures. Exogenous auxin is required to maintain the high rate of an unorganised growth in plant cell suspension cultures. On the other hand, the proliferation of hairy root cultures is usually dependent on endogenous hormonal factors. Auxin and cytokinin execute their regulatory role by being involved in a cross-talk with numerous endogenous factors affecting cell division and differentiation. Among them, ascorbate/dehydroascorbate (ASC/DHA), glutathione/glutathione disulphide (GSH/GSSG) redox pair, H_2O_2 and other components of cellular redox systems play an important role in triggering developmental responses in plant in vitro culture. Ascorbate, glutathione and related enzymes participate in the responses to auxin/cytokinin treatments. In addition, they can even directly affect hormone metabolism in tissue. Ascorbate and glutathione have important regulatory roles in the process of cell-cycle progression within the meristems, where they participate in redox-dependent determination of proliferation and quiescence patterns. The mechanism underlying the regulatory effects of ascorbate and glutathione in cell divisions is not fully elucidated; however, it seems to be related to the regulation of nucleotide synthesis. Ascorbate levels in apoplast modulate the rate of organ elongation by increasing cell wall extensibility. Besides the effects on cell proliferation and growth, ascorbate and glutathione concentrations as well as the enzymes of their metabolism protect the in vitro cultured tissues against oxidative stress. This function is of particular importance during root regeneration and the elicitation of metabolite production by hairy root cultures, where increased levels of oxidising agents are often required to stimulate both processes. In this review, we report recent studies on the involvement of ascorbate and glutathione in the processes of regeneration and proliferation in plant tissue culture.

J. Tyburski (✉) and A. Tretyn
Department of Biotechnology, Institute of General and Molecular Biology,
Nicolaus Copernicus University, Gagarina 9, 87-100 Toruń, Poland
e-mail: tybr@umk.pl

N.A. Anjum et al. (eds.), *Ascorbate-Glutathione Pathway and Stress Tolerance in Plants*,
DOI 10.1007/978-90-481-9404-9_2, © Springer Science+Business Media B.V. 2010

Keywords Ascorbate • Glutathione • In vitro • Rooting • Shoot regeneration • Somatic embryogenesis

1 Regenerating Plants from In vitro Cultured Cells, Tissues and Organs

The capacity of plant cells to undergo inducible morphogenetic pathways is essential for various biotechnological applications of plant cell culture, such as clonal propagation and genetic transformation. Morphogenesis in plant tissue culture may occur in two ways. It may result in the formation of bipolar structures called somatic embryos, equipped with shoot and root meristem. Alternatively, organogenesis may occur leading to the formation of unipolar structures, i.e. shoots or roots. Whereas, a somatic embryo directly develops into a complete plantlet, shoot organogenesis must be followed by root organogenesis at the base of a shoot before the newly regenerated plant is transferred to ex vitro conditions (Warren 1993) (Fig. 1).

The process of somatic embryogenesis may be divided into two phases: induction and expression. During the induction phase, differentiated somatic cells of an explant, undergo de-differentiation, acquire embryogenic competence and proliferate as embryogenic cells. In the expression phase, the embryogenic cells differentiate to form somatic embryos (Namasivayam 2007). In few experimental models, somatic embryos follow a sequence of development comprising the stages of embryo formation similar to those, which are observed during zygotic embryogenesis, i.e. globular, heart shape and torpedo. Such developmental pattern occurs during embryogenesis of carrot callus cells cultured as a suspension in liquid medium. This system is one of the most controllable somatic embryogenesis processes, and is often used as a model system in studies on cell differentiation. However, usually, developmental events that give rise to a somatic embryo in culture show far more variation than the equivalent process in the ovule. Somatic embryos exhibit a much greater range of sizes and shapes than the zygotic embryos (Warren 1993).

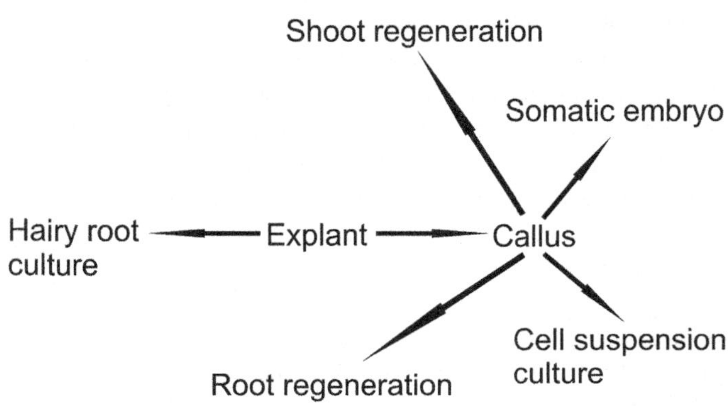

Fig. 1 Most important developmental pathways in plant tissue culture. See text for details

Somatic embryogenesis can proceed as a direct or indirect process. In direct embryogenesis, which is a relatively rare event, embryos develop directly on the surface of organised tissue such as leaf, stem, zygotic embryo, etc. Alternatively, the much more common indirect somatic embryogenesis requires an intermediary step of meristematic cluster formation preceding the embryo development. This kind of embryogenesis often occurs in cell suspensions and callus culture (Namasivayam 2007). Typically in meristematic clusters, only a few surface cells give rise to embryos; however the presence of cells that do not undergo embryogenesis is necessary within the cluster to complement cells, which are directly involved in embryo formation (Kreuger and van Holst 1993).

Induction of plant explants for somatic embryo formation generally requires a pretreatment on auxin-supplemented medium. Among auxins, the most frequently used was 2,4-diphenoxyacetic acid (2,4-D). However, other auxins like β-naphtoxyacetic (NAA) and indole butyric acid (IBA) were also used for this purpose (Razdan 2003). The formation of meristematic clusters containing embryogenically competent cells occurs during the period of auxin treatment. On the other hand, the following phase, when competent cells develop into embryos, requires the reduction or removal of auxin from the medium. This is achieved by transferring cells to a new medium with a very low level of auxin or no auxin at all (Razdan 2003). Nevertheless, auxin treatment is most frequently used to induce somatic embryogenesis. However, the effect of other plant growth regulators should not be overlooked (Jimenez 2001). For example, in the case of nucellus cultures of *Vitis*, simultaneous presence of NAA and benzylaminopurine (BAP) were inductive for embryogenesis. In some systems, somatic embryogenesis was induced by abscisic acid (ABA) (Nishiwaki et al. 2000) or even occurred on hormone-free medium (Choi et al. 1998).

Plant organogenesis in vitro is a more controllable process than embryogenesis (Warren 1993). Different morphogenetic pathways, i.e. shoot or root formation, may be induced in culture by application of the appropriate hormones in the medium. Although the exact nature of these hormonal signals may vary between species, the balance of auxin to cytokinin has been found to have a consistent effect on the type of regenerated organs. A relatively high ratio between auxin and cytokinin promotes the regeneration of roots. Whereas, shoot organogenesis is a preferred type of differentiation on culture media supplemented with high cytokinin and low auxin concentration (Skoog and Miller 1957; Christianson and Warnick 1983). The explants cells proliferate to form callus when the same concentrations of auxin and cytokinin are added to the medium. Callus cells are usually grown on solid media. Alternatively, long term cultures of friable callus may be grown on liquid media as cell suspension cultures (Zhao et al. 2008) (Fig. 1).

Root cultures may be derived without the use of exogenous growth regulators by infecting plant explants with *Agrobacterium rhizogenes*. This gram-negative soil bacterium transfers a DNA segment (T-DNA) from its large root-inducing (Ri) plasmid into the genome of the infected plant. This T-DNA carries a set of genes that encode enzymes which control auxin and cytokinin biosynthesis. The new hormonal balance induces the formation of proliferating roots, called hairy roots, that emerge at the wounding site. The hairy root phenotype is characterized by fast hormone-independent growth, increased lateral root branching and genetic stability (Guillon et al. 2006) (Fig. 1).

Temporal requirements for a specific balance of phytohormones for the organogenesis process indicate that organ regeneration is accomplished in three phases (Christianson and Warnick 1983). In the first phase, cells of the explants acquire competence, which is defined as the ability to respond to hormonal signals. The competence acquisition involves dedifferentiation of explants cells, which re-enter cell cycle. Competent cells are then canalized and determined for specific organ formation under the influence of phytohormone balance through the second phase, referred to as induction phase. Organ primordia differentiate from induced explants cells and their further development and outgrowth occurs during the third phase. These processes usually proceed independently of the exogenously supplied phytohormones (Christianson and Warnick 1983; Sugiyama 1999). Studies using *Arabidopsis* temperature sensitive mutants (*srd1*, *srd2*, *srd3*) defective for shoot and/or root redifferentiation revealed further complexity of the process of competence acquisition (Ozawa et al. 1998). It has been found that hypocotyl explants grown on callus-inducing medium first become competent in root organogenesis and then gain competence in shoot organogenesis. Therefore, dedifferentiation stage is divided into two sub-phases occurring sequentially one after another. Finally, explants cells are competent in regeneration of both roots and shoots. The transition from an incompetent state to competence in root organogenesis and from competence in root organogenesis to competence in root and shoot organogenesis requires the functions of *SRD2* and *SRD3* gene, respectively (Ozawa et al. 1998).

In spite of more than 50 years of studies, molecular and biochemical processes underlying morphogenesis in tissue culture are not fully understood. However, the growing list of genes which are known to be specifically involved in organogenesis and/or somatic embryogenesis mark significant advances in the field of elucidating the mechanism of plant regeneration in vitro (Philips 2004). It was found that the *SRD2* gene, involved in control of proliferation competence and dedifferentiation, encodes for a nuclear protein responsible for activation of snRNA transcription. At present, it remains unclear which molecular event, subsequent to the activation of snRNA transcription, is responsible for the elevation of cell proliferation competence (Ohtani and Sugiyama 2005). Re-entering the cell cycle by quiescent cells during dedifferentiation is correlated with expression of cell cycle-related genes such as cyclins and cyclin-dependent kinases (CDK). Among them, gene *cdc2At* coding PSTAIRE domain-containing CDK and *CYCD3* coding a D-type cyclin are possibly involved in committing the dedifferentiating cells to the cell cycle (Sugiyama 1999).

A developmental pathway leading to shoot organogenesis was found to be related to genes involved in cytokinin perception and signalling (Sugiyama 1999, 2000). The establishment and maintenance of shoot meristem is dependent on meristem identity genes such as *SHOOT MERISTEMLESS* (*STM*), *WUSHEL* (*WUS*) and *CLAVATA* (*CLV*). These genes function for the establishment of shoot apical meristems only after dedifferentiated cells are determined for shoot organogenesis (Philips 2004). In accordance with the essential role of auxin in root regeneration, an important role of genes engaged in auxin perception and signalling was also identified. The *ROOTING AUXIN CASCADE* (*RAC*) gene coding for Rac/Rop GTPase (Tao et al. 2002) mediating an auxin-signalling pathway is involved in an early stage of auxin perception specific to the formation of adventitious roots (Sugiyama 1999). Similar to organogenesis, the process of somatic embryogenesis involves reprogramming of gene expression patterns. Vegetative-embryonic transition is marked by changes in expression of gene coding for somatic embryogenesis receptor-like kinase (SERK), which was identified as a specific marker distinguishing individual embryo-forming cells from non-embryogenic cells in carrot suspension culture. The transcription factors *LEAFY COTYLEDON 1* (*LEC1*) and *WUSHEL* (*WUS*) are possibly involved in inducing and maintaining embryogenesis in culture (Namasivayam 2007).

2 Interaction of Ascorbate and Glutathione with Auxin and Cytokinin

A decisive role in regulating the morphogenetic pathways in plant tissue culture is attributed to auxin and cytokinin. Below, we describe possible links between these growth regulators and several aspects of ascorbate and glutathione metabolism. These interactions may possibly be important for understanding the roles of these antioxidants in differentiation and growth in plant in vitro culture.

2.1 Auxin

An apoplastic enzyme ascorbate oxidase (AOX) provides a clear link between regulatory action of auxin and ASC. Highest levels of AOX activity were detected in young and growing parts of the plant. For example, AOX activity in rapidly growing immature pumpkin fruits was 15–20 times higher than that in mature fruits, which no longer increase in size (Esaka et al. 1992). In tobacco plants, high levels of AOX transcript were detected in young and growing parts like upper leaf, upper stem and root but little or none in old tissues such as lower leaf and lower stem (Kato and Esaka 1996). Similarly, in tomato seedling roots, AOX activity was stimulated by auxin treatment in the apical part of the organ, whereas no stimulation was observed in the proximal part of root (Tyburski et al. 2008).

The relation between auxin and AOX expression and activity in pumpkin fruits was demonstrated by Esaka et al. (1992). AOX activity in pumpkin fruit tissue, cultured in vitro, strongly increased when culture medium was supplemented with synthetic auxin, 2,4-D. It was shown that the enzyme's activity is regulated by auxin at the level of gene expression. Gene expression analysis showed that AOX mRNA level also increased after transfer onto the culture medium in the presence of 2,4-D. The highest steady-state level of AOX mRNA appeared 2 days after transfer, and decreased thereafter. In the absence of 2,4-D, no increase in AOX transcript level was detected in fruit tissue (Esaka et al. 1992). AOX expression was also induced by auxin in tobacco leaves and this induction was associated with stimulation of plant growth (Pignocchi et al. 2003).

AOX is able to react with indole-3-acetic acid (IAA) as a substrate at least in vitro. It was demonstrated that AOX may effectively decrease IAA concentration in radish roots via oxidative decarboxylation forming, as a main product, oxindole-3-methanol (Kerk et al. 2000). While the principal mechanism of auxin turnover in shoots engages non-decarboxylative pathway (Woodward and Bartel 2005), the analysis of IAA metabolite in various parts of root tissue shows that decarboxylation is a main IAA-degradation pathway in roots. Oxidative decarboxylation occurred almost exclusively in the root tip with possible participation of AOX. This finding implicates an existence of a regulatory loop involving AOX, IAA and ASC in the root tip cells (Kerk et al. 2000).

Further insight into the relation between AOX and auxin was provided by studies on AOX-over-expressing tobacco plants. It was demonstrated that over-expression of AOX results in oxidation of the redox state of apoplast, which is followed by the reduction in sensitivity to exogenous auxin. While wild-type seedlings reacted to spraying with 0.5 μM NAA with enhancement of shoot growth (i.e. advanced development of cotyledon and leaves) and increase in fresh weight, AOX–over-expressing seedlings, submitted to the same treatment, accumulated 25–30% less biomass. Although lateral root proliferation following auxin treatment was observed in both, control line and AOX–over-expressing line, this effect was much more evident in wild-type seedlings. Desensitisation to auxin in AOX-over-producing plants was related to constitutive induction of auxin signalling pathway in these plants (Pignocchi et al. 2006). Auxin stimulates ROS production in apoplast and employs them as mediators in the regulation of plant growth reactions (Joo et al. 2001; Schopfer et al. 2002; Liszkay et al. 2004). Because apoplast is already oxidised in AOX-over-expressing plants, a further stimulation of ROS production by auxin has minimal effects on oxidative signalling processes leading to auxin insensitivity. Moreover, the activity of auxin-dependent MAP kinase pathway was doubled in leaves of AOX-over-expressing seedlings when compared to control wild-type seedlings. These data emphasise the role of apoplast redox-mediated changes in developmental reactions of plants to auxin treatments (Pignocchi et al. 2006).

Besides AOX, the effects of auxin treatment on the activities of other ASC metabolising enzymes were reported. Changes in the activities of AOX and Halliwel–Asada cycle enzyme activity were analysed in the proximal and distal parts of tomato seedling roots grown in vitro (Tyburski et al. 2008). Roots respond to auxin treatment with an increase in lateral root formation which occurs primarily in the proximal part of root and concurrent inhibition of root elongation, which affects root elongation zone localised in proximity of the root apex (Cleland 2004; Tanimoto 2005). It was demonstrated that a 3-day-long treatment with 1 μM IAA decreased the activities of ASC-regenerating enzymes, i.e. monodehydroascorbate reductase (MDHAR) and dehydroascorbate reductase (DHAR) within both, distal (elongating) and proximal (lateral root producing) part of the root. ASC-oxidising enzymes were stimulated by IAA. However, auxin's stimulatory effect on ascorbate peroxidise (APX) activity was restricted to proximal part of the root, whereas, AOX activity increased exclusively within apical part of the root. These changes were accompanied with an increase in the participation of DHA in total pool of ascorbate. These findings show that the effects of exogenous auxin on ascorbate metabolism depend on the zone of root which is exposed to an auxin stimulus (Tyburski et al. 2008). Generally, auxin application is usually followed by increased oxidation of the ascorbate pool due to stimulatory effects on AOX or ascorbate peroxidase (APX) activity and/or expression. This finding is consistent with an overall mode of auxin action, which triggers auxin-specific signalling pathways, oxidising the cellular environment (Joo et al. 2001; Schopfer et al. 2002).

2.2 Cytokinin

A role of endogenous cytokinin in regulation of ASC and GSH enzyme activity was studied using transgenic tobacco plants expressing supplementary *ipt* – gene under a control of the constitutive promoter for small subunit of RUBISCO (P*ssu-ipt*). Over-expression of bacterial *ipt* gene encoding isopentenyl transferase, a key enzyme in cytokinin synthesis, in plant cells led to an increase in endogenous cytokinin content in transgenic plants. When compared to non-transformed control, active forms of cytokinins increased seven times in transgenic shoots grafted onto nontransgenic rootstock and 12 times in transgenic shoots propagated in vitro. Tobacco plants over-expressing *ipt*, exhibited typical traits of cytokinin over-production syndrome, including reduced root growth, reduced apical dominance, and retarded plant and leaf senescence. Transgenic plants were also characterised by alterations in water regime, disturbance in photosynthesis and over-expression of *PATOGENESIS RELATED PROTEIN 1* (*PR1*). These findings were interpreted as symptoms of permanent stress, which is caused directly and indirectly by cytokinin over-production (Synková et al. 2004. Transgenic plants produced more H_2O_2 than control plants; however, higher activities of ascorbate/glutathione enzymes were detected in cytokinin over-producing plants as well. APX activity was two times higher in transgenic plants grown either in vitro or in vivo conditions when compared to the control. Glutathione reductase (GR) activity was also doubled in *ipt* transformants, however, this increase was observed only in in vitro cultured plants, whereas there was no difference in GR activity between control and transgenic plants grown in ex vitro conditions. In addition, catalase (CAT) activity was stimulated in transgenic plants; on the other hand, superoxide dismutase activity was halved. These findings suggest that cytokinins may regulate H_2O_2 levels by changing the activities of principal enzymes of H_2O_2 turnover (Synková et al. 2004, 2006).

Cytokinin over-expression had important consequences for redox metabolism during senescence. This process is, to a large extent, dependent on ROS accumulation in chloroplasts, which causes chlorophyll bleaching, lipid peroxidation and eventual loss of chloroplast integrity (Navabpour et al. 2003; Bhattacharjee 2005).

Control plants exhibited high activities of antioxidant enzymes in early stages of plant development, i.e. until the onset of flowering, and a decline in later stages. Contrary to the control, transgenic plants showed increase in CAT, GR and APX in later stages of development, i.e. during flowering and forming seeds. These data suggest that cytokinin control of senescence involves the regulation of the activities of aforementioned H_2O_2-processing enzymes (Synková et al. 2006).

Similar effects of exogenously applied cytokinin were demonstrated. Wheat leaf segments treated with 10^{-4} M BAP exhibit delayed senescence, which manifests itself in the retention of chlorophylls and chloroplast proteins (Zavaleta-Mancera et al. 2007). Consequently, during incubation period following exposure to BAP, chloroplasts in the cells of treated leaf segments remained intact significantly longer, whereas those of control segments exhibited various symptoms of degradation. BAP-treated leaves accumulated significantly lower levels of H_2O_2. This finding together with increasing the content of xanthophylls, at least partly, explains delayed senescence of leaves pre-treated with BAP. Decrease in H_2O_2 was due to enhanced activity in CAT and APX enzymes. During 6-day-long incubation period, in control leaves CAT activity decreased from the fourth day on and APX activity decreased on the sixth day. BAP treatment prior to incubation period efficiently prevented the decrease in the activities of both enzymes. It was concluded that the mechanism of cytokinin-dependent delay in leaf senescence involves the reduction in H_2O_2 levels due to the hormone's stimulatory effect on CAT and APX activities (Zavaleta-Mancera et al. 2007).

3 Ascorbate and Glutathione as Regulators of Cell Division in the Root Apical Meristem

There is a growing body of evidence indicating that the role of ascorbate and glutathione in plants extends beyond their intensively explored antioxidant function (De Pinto and De Gara 2004). It has been demonstrated that high ascorbate and glutathione levels are required for normal progression of the cell cycle in meristematic tissues (Liso et al. 1984, 1988; Potters et al. 2000; Vernoux et al. 2000; Jiang et al. 2003). Ascorbate is directly involved in the regulation of two processes that mediate morphogenic responses in plants: cell division and elongation. The reduced ascorbate (ASC) as well as oxidised form of this compound (dehydroascorbate, DHA) play an important role in the regulation of mitotic activity in the meristems (De Tullio et al. 1999; De Pinto et al. 2000; Potters et al. 2002). It was observed that ASC promotes cell-cycle progression in the root apical meristem by stimulating G1-S transition. If ASC is added to the cells of the root quiescent centre (QC), it induces these normally non-dividing cells to pass from G1 into the S phase (Liso et al. 1988).

A direct link between ASC/DHA and GSH/GSSG redox states and auxin-dependent regulation of mitotic activity in discrete parts of the apical root meristem has been demonstrated. The apical root meristem comprises of proximal meristem (PM), which produces the tissues of the root proper and the distal meristem (DM) that produces root cap. Between the actively dividing cells of PM and DM, the population of slowly dividing cells, called quiescent centre (QC) is localised. QC cells spend extended periods of time in G1 phase, dividing, on average, every 200 h (Jiang et al. 2006). It was demonstrated that laser ablation of QC cells leads to differentiation of the initial cells previously attached to it. Therefore, it was concluded that the function of QC is maintaining the dedifferentiated state of initial cells in adjacent meristems in a cell-contact-dependent manner (Bonke et al. 2005).

Differences in the cell division rate between the meristems and QC depend on the polar transport of auxin from shoot apex, where the hormone synthesis takes place, to the root tip. Directions of hormone transport in the root are determined by the patterns of PIN family auxin – efflux carrier protein expression (Friml 2003). Auxin is transported acropetally towards the root tip through the root stele tissue employing cells that express PIN1 auxin efflux carrier (Friml 2003). The hormone is transported to QC and columella initials where high levels of auxin accumulate (Kerk and Feldman 1995). Part of auxin is further transported from QC to columella by PIN4 protein. An auxin efflux protein PIN3, specifically expressed by columella cells is localised in lateral sides of plasma membrane that enables basipetally directed lateral auxin transport through lateral parts of the root cap and cortex/endoderm cells expressing PIN2 auxin carrier (Benkova et al. 2003; Friml 2003; Blilou et al. 2005).

It was demonstrated that auxin accumulation in QC switches the redox balance in the cells towards a more oxidised state (Jiang et al. 2003). When compared to rapidly dividing PM initials, cells of the QC are characterised by the elevated levels of DHA (×1,000), GSSG (×10), $O_2^{\cdot -}$ (×15.6) and H_2O_2 (×34). On the other hand, rapidly dividing cells of PM contain reduced forms of ASC and GSH and do not accumulate ROS. Auxin-induced oxidised intracellular environment plays a decisive role in maintaining the low cell division rate in QC (Kerk and Feldman 1995; Jiang et al. 2003). Treating roots with an inhibitor of auxin polar transport, such as NPA, results in a relocation of auxin maximum from QC to cortical and procambial region of proximal meristem. This alteration in auxin distribution was followed by mitotic activation of QC. This correlated with the development of less oxidised status in the QC and more oxidised status in the adjacent PM, to which the auxin maximum was shifted (Jiang et al. 2003).

An increase in ASC oxidation in QC is attributed to high AOX activity. Both, AOX transcript and protein accumulate very distinctly in QC, whereas proximal and distal meristem cells were characterised by much lower AOX activity. Because AOX expression is stimulated by auxin (Kato and Esaka 1999; Kerk and Feldman 1995; Liso et al. 2004), high levels of enzyme activity are maintained in QC by IAA accumulating within this structure. Consequently, high rate of ASC oxidation keeps the oxidised ascorbate redox balance (Kerk and Feldman 1995). QC cells exhibit lower abilities for regenerating ASC from DHA because of the absence of DHAR activity. However, the activity of MDHAR was similar in QC and adjacent meristem cells, which possibly prevents the ascorbate pool in QC from total oxidation. QC cells were equipped with several times lower levels of GR activities, which explains decreased abilities for recycling GSH from GSSG (Jiang et al. 2003).

Highly oxidised intracellular environment of QC seems to slow down cell division rate by reducing mitochondrial activity. QC cells are characterised by lowered mitochondrial membrane potential, which indicates the decrease in the production of ATP and NADH. Critical levels of these compounds are necessary to satisfy the G1-S checkpoint energy requirement. Therefore, reduced ATP/NADH levels in QC cells result in a decrease in the energy supply, which may cause the decrease in the cell division rate in QC (Jiang et al. 2006). Alteration in mitochondrial activity in QC may be related to inhibition of expression of some nuclear-encoded mitochondrial proteins. Among them, the reductions in tricarboxylic acid cycle enzymes have been demonstrated (Jiang et al. 2006). On the other hand, transcripts encoded by the mitochondrial genome were not decreased in QC cells when compared to adjacent meristem cells. It was also demonstrated that, in spite of constitutive oxidative stress conditions, which often result in a loss of mitochondrial membrane integrity, followed by the induction of apoptosis pathway, the overall cellular ultra-structure of the QC, including that of the mitochondria is typical of that found in unstressed plant cells. The intactness of the mitochondria, and other cell organelles of QC cells is possibly due to protective functions of residual levels of reduced ascorbate and glutathione still present within these cells (Jiang et al. 2006).

The mechanism of ASC-dependent stimulation of cell divisions is still not sufficiently explained. Some data suggest that the stimulatory effect of ASC on cell divisions in root apical meristem results from the involvement of ASC in hydroxyproline synthesis. Hydroxyproline-containing proteins play a decisive role in cell cycle regulation and ASC is a co-factor of peptidyl-prolyl hydroxylase – a hydroxyproline-synthesising enzyme (De Tullio et al. 1999). Another hypothesis indicates the role of ASC in inducing the activity of ribonucleotide reductase (RNR). The enzyme reduces ribonucleotides to deoxyribonucleotids and therefore is an essential enzyme for DNA synthesis during the S phase. ASC is supposed to be required for effective reconstitution of an iron centre of RNR by increasing the release of Fe^{2+} from intracellular stock such as ferritin (Potters et al. 2002). A complementary explanation stresses the role of AOX in oxygen management and the regulation of metabolic rate in plant cells. According to this idea, ASC oxidation catalysed by AOX as an oxygen-consuming reaction decreases the oxygen availability in the cell. This results in slowing down of the metabolism and is followed, as observed in QC cells, by a decrease in the rate of cell divisions (De Tullio et al. 2007).

Finally, ascorbate affects plant development being engaged in the synthesis of several growth regulators, such as ethylene, abscisic acids and gibberellins. It functions in these processes as a co-factor of dioxygenases – enzymes playing a decisive role in the synthesis of aforementioned hormones (Dong et al. 1992; Liu et al. 1999; Arrigoni and De Tullio, 2000; Lopéz-Carbonell et al. 2006).

4 Glutathione as a Regulatory Factor in Plant Development

It has been previously demonstrated that GSH has many functions in plants including an important role in the antioxidant system, sulphur metabolism and detoxification of xenobiotics (Noctor and Foyer 1998). In addition, similar to ASC, a regulatory role of GSH in some aspects of plant development has been reported. It was demonstrated that the timing of the development of inflorescence and flowering in the rosette plants: *Arabidopsis thaliana* (Ogawa et al. 2001) and *Eustoma grandiflorum* (Yanagida et al. 2004) is regulated by changes in the rate of GSH biosynthesis. Ogawa et al. (2001) have shown that the effects of endogenous GSH levels on flowering in *Arabidopsis* are dependent on the stage of plant development. Depleting GSH by treating plants with a specific inhibitor of GSH biosynthesis, buthionine-L- suphoxymine (BSO), promoted flowering when it was applied at the onset of transition to flowering. On the other hand, when plants were exposed to BSO from the beginning of culture, flowering was delayed. The GSH-deficient *Arabidopsis* mutant *cad2* also exhibited delayed flowering when compared to wild type plants (Ogawa et al. 2001). Later it was demonstrated that levels of endogenous GSH, decisive for the proper timing of flowering, are regulated by the availability of ATP synthesised during photosynthesis and therefore are dependent on the intensity of photosynthesis (Ogawa et al. 2004). GSH is also required for vernalization-induced flowering of rosette plant *Eustoma grandiflorum.* It was shown that vernalization was efficiently replaced by feeding with GSH, or its precursor cysteine (however not by other thiols) in promoting flower induction. On the other hand, the inductive effect of vernalizaton on bolting was suppressed by BSO, without decreasing the plant growth rate (Yanagida et al. 2004).

GSH is also involved in the regulation of the phytohormone-induced differentiation of tracheary elements (TE) in *Zinnia* and *Arabidopsis* (Henmi et al. 2001). This process was promoted by GSSG, if the substance was applied during the early period of culture, whereas the effect was completely reversed when GSSG was applied at a later period of culture. The expression of glutathione reductase (GR) was down-regulated during TE development and exogenous GSH suppressed TE formation. Over-expressing GR in *Arabidopsis* had the same effect. Because GSSG stimulated TE differentiation only in the presence of appropriate growth regulators, the authors conclude that GSSG-dependent regulation cooperates with phytohormones to induce TE differentiation (Henmi et al. 2001).

Presence of the reduced form of glutathione (GSH) in tissue was found to be necessary to maintain cell divisions in *Arabidopsis* root meristem (Sánchez-Fernández et al. 1997; Vernoux et al. 2000). GSH was localised in actively dividing initial cells while it was absent in slowly dividing cells of the quiescent centre, that suggests that growing tissues have a requirement for glutathione. Cell divisions of meristem initials were stimulated by exogenous GSH and inhibited by treatment with the inhibitor of GSH biosynthesis BSO (Sánchez-Fernández et al. 1997). Moreover, *Arabidopsis* plants homozygous for a mutation in the *ROOT MERISTEMLESS 1* (*RML1*) gene, coding for γ-glutamylcysteine synthetase (enzyme of GSH biosynthesis), accumulating only 3% of the wild type GSH level, were unable to maintain cell divisions in the root apical meristem. On the other hand, root cells of the mutant plants divided normally when fed with exogenous GSH. Cells in the root meristem of the mutant plant were arrested at the G1-S transition of the cell cycle, which resulted in inhibition of root growth. However, while post-embryonic root development was blocked due to cell division arrest, the overall organisation of cell types in the root apex was not changed when compared to wild type plants. The *rml1* phenotype can be induced by treating wild-type seedlings by lowering GSH levels with BSO. Similar to *rlm1* plants, cells in the root apical meristem of BSO-treated seedlings are arrested in G1 phase. It should be noted that the GSH requirement of cell division process is a root-specific phenomenon. The shoot meristem of *rlm1* seedlings is able to produce all the above-ground organs with timing similar to that of wild type plants (Vernoux et al. 2000).

5 Plant Cell Suspension Cultures as Model Systems in Studies on the Mechanism of Ascorbate and Glutathione Role in Cell Proliferation

The long-term cultures of plant cells and small cell aggregates named cell suspension cultures, are characterised by the highest structural and metabolic homogeneity in plant in vitro culture. Cell suspension cultures are usually derived from callus cultures and grown in liquid media under constant shaking and aeration. As quickly dividing and easily synchronizable cell populations, which follow specific growth kinetics, cell suspensions are frequently used in studies on the mechanism of cell cycle regulation (Menges and Murray 2002). Cell suspension cultures of *Arabidopsis* and tobacco cell line BY2 were used as model systems in studies on ascorbate and glutathione involvement in the regulation of cell divisions.

Timing of phases during cell suspension growth kinetics was found to be synchronized with changes in cellular ASC and GSH levels. A three- to fourfold increase in the endogenous ASC content was observed during the exponential growth phase, and a peak in ASC coincided with a peak in the mitotic index in BY2 culture (De Pinto et al. 1999). Ascorbate was also abundant in the exponential phase of *Arabidopsis* cell culture growth cycle. When the growth rate in culture declined, ascorbate levels decreased to about half its original value (Pellny et al. 2009). In addition, an increase in GSH levels was observed during a proliferation phase of the growth cycle (De Pinto et al. 1999).

Cell proliferation in tobacco cell culture was strongly stimulated by exogenous GSH, however, was not affected by GSSG (De Pinto et al. 1999, 2000). Stimulation of ASC biosynthesis in BY-2 by adding ascorbate precursor galactono-γ-lactone (GalL) to the culture medium accelerated culture growth by promoting cell division. In contrast to the reduced form of ascorbate, exogenous DHA strongly blocked cell divisions in BY2 culture (De Pinto et al. 1999). Studies using synchronized BY-2 cells revealed that addition of 1 mM DHA to cells in G1 phase induced a delay in cell cycle progression. DHA-treated cells reached the same value of mitotic index as untreated cells but several hours later. DHA added to the medium was quickly reduced after the uptake by the cells, leading to a strong increase in ASC intracellular level. Because oxidative stress induced by various environmental factors favours oxidation of apoplastic ASC to DHA, it was concluded that slowing down the cell cycle progression in the presence of high DHA levels may function as an adaptation strategy to surveillance under stress condition. (Potters et al. 2000).

Studies on BY2 cell culture revealed that the DHA effect on cell proliferation could not be reproduced by GSSG treatment (De Pinto et al. 1999). However, total depletion of cellular GSH with BSO resulted in a cell cycle delay similar to that caused by DHA treatment. Simultaneous application of both compounds completely blocked mitotic activity. On the other hand, combined addition of GSH and DHA resulted in delay similar to DHA alone. This finding suggests that the ASC/DHA pair has a specific regulatory role on cell division rather than merely acting as a general redox pair and that ascorbate and glutathione control cell cycle using independent pathways (Potters et al. 2004).

It is noteworthy that DHA is only capable of slowing down the cell cycle when added during G1 phase. Addition of DHA during G2 phase did not affect the cell cycle progression. This suggests that DHA influences cell cycle through processes that are specific for G1 or S phase. However, the precise mechanism of DHA-dependent cell cycle regulation still remains to be deciphered (Potters et al. 2004).

Changes in the activities in ASC and GSH metabolising enzymes were synchronized with ASC, GSH levels and rate of cell division in BY2 cell culture. Similar to ASC and GSH content the activities of the Halliwel–Asada cycle enzymes rise during exponential phase of the cell culture growth cycle. These finding suggests that cell division process is marked by the intensification in ascorbate and glutathione turnover. Apart from the ASC-dependent H_2O_2 scavenging that is particularly high in dividing cells, other ascorbate-consuming processes like hydroxyproline-rich protein or nucleotide synthesis are responsible for an increased ascorbate recycling (De Pinto et al. 2000).

A regulatory role for glutathione with respect to cell proliferation in *Arabidopsis* cell culture was demonstrated to be linked to pyridine nucleotide metabolism and activity of poly (ADP-ribose) polymerase (PARP) activity (Pellny et al. 2009). PARP is a nuclear enzyme which transfers ADP-ribose units from NAD^+ residues on target proteins, resulting in branched chains of ADP-ribose polymers. ADP-ribosylation is important in regulating processes such as DNA repair, modification of histone structure and chromatin remodelling (Kraus and Lis 2003). PARP activity level correlated with the rate of cell division in *Arabidopsis* cell culture; lower activity was observed on the first day after inoculation but values increased rapidly during the exponential growth phase. Subsequently, when the growth rate in culture declined, PARP activity decreased (Pellny et al. 2009). Expression patterns of the major *Arabidopsis* PARP-coding genes (*PARP1* and *PARP2*) increased progressively during the exponential phase, giving a peak when growth was the highest and declined thereafter. High *PARP1* and *PARP2* expression levels were associated with increases in the NAD + NADH pool and oxidised state of NAD/NADH redox pair. Because PARP is a NAD-cleaving enzyme, this reflects the requirement for the pyridine nucleotide substrate. Intracellular glutathione (but not ascorbate) levels increased in parallel with PARP activity and expression. The putative reason for glutathione increase during exponential phase is the necessity to buffer cellular oxidation caused by enhanced oxidation of NAD pool. At this phase of the growth cycle, glutathione is present in the cytoplasm. However, almost total cellular glutathione is recruited to the nucleus at the end of the phase of exponential growth. GSH movement between cytoplasm and nucleus may possibly have important consequences for both, redox buffering in the cytoplasm and direct regulation of gene expression. A close correlation between glutathione levels and PARP activity suggests that there is a relationship between these parameters. PARP is a redox-sensitive enzyme whose activity could be regulated by the GSH in the nucleus, via thiol-disulphide exchange mechanism or glutathionylation (Pellny et al. 2009). However, these ideas should be supported by further experiments.

6 Regulation of Cell Growth by Ascorbate

Besides the effects on cell division rate, ascorbate affects organ growth by participating in the process of cell wall stiffening, which regulates the cell expansion within the organ elongation zones (Green and Fry 2005a). An increase in activity of apoplastic AOX, detected in rapidly elongating tissues, results in the production of monodehydroascorbate (MDHA) in the plant cell wall. This compound activates plasma membrane H^+-ATPase (Gonzales-Reyes et al. 1992; Asard et al. 1995). An increase in the activity of this enzyme is followed by an acidification of apoplast that according to the acid growth theory leads to the loosening of cell wall structure and facilitates cell growth (Cosgrove 2001). Moreover, MDHA in apoplast functions as an acceptor in transmembrane electron transport. The plasma membrane redox system, involving cytochrome B, transports electrons from NADH across the plasma membrane onto MDHA, which is reduced to ASC. The intensification of this process results in plasma membrane hyperpolarisation which is followed by activation of plasma membrane H^+-ATPase that promotes cell growth as mentioned above. Simultaneously, local cytoplasm acidification resulting from NADH oxidation activates vacuolar H^+-ATPase. Consequently, an increase in cell vacuolization occurs which further enhances cell expansion (Gonzales-Reyes et al. 1992; Horemans et al. 2000). MDHA generated by AOX activity in the apoplastic spaces may undergo further spontaneous oxidation to DHA (Noctor and Foyer 1998). This compound as well as a product of its degradation – oxalate – may directly cause cell wall loosening and/or enhance cell vacuolization (Lin and Varner 1991).

Another mechanism linking apoplastic ASC with loosening of cell walls in growing segments of plant organs involves the inhibitory effect of this antioxidant on peroxidase-dependent processes of cross-linking cell wall polymers. ASC prevents oxidative formation of diferulate bridges between cell wall polysaccharides in elongating onion roots (Córdoba-Pedregosa et al. 1996; Takahama and Oniki 1992) and pine hypocotyls (Sánchez et al. 1997). ASC inhibited the polymerisation of monomers of cell wall structural protein extension by blocking isodityrosine cross-linking (Córdoba and Gonzales-Reyes 1994) and delayed cell wall lignification by reducing phenoxyl radicals, which serve as precursors in peroxidase-dependent lignin synthesis (Takahama and Oniki 1991; Padu 1999).

Ascorbate may also stimulate a pro-oxidative process within the plant cell wall, which results in local production of hydroxyl radical ($^{\cdot}OH$) (Fry 1992; Green and Fry 2005a, b). This highly reactive compound is able to cause an oxidative scission of cell wall polysaccharide chains, which is followed by loosening the cell wall structure that again promotes cell growth (Schopfer et al. 2001; Schopfer 2002). The mechanism of ASC-dependent $^{\cdot}OH$ was proposed to have a non-enzymatic nature and to require O_2 and a transition metal such as Cu^{2+}. It was demonstrated that in the presence of traces of Cu^{2+}, ASC reduces O_2 to H_2O_2, ASC also reduces Cu^{2+} to Cu^{+}. These two products: H_2O_2 and Cu^{+}, can participate in a Fenton reaction. In the course of this reaction, Cu^{2+} is regenerated with concurrent $^{\cdot}OH$ formation (Fry 1992).

Besides ascorbate itself, the components of its breakdown pathway may also act as pro-oxidative agents that stimulate the generation hydroxyl radical ($^{\cdot}OH$) in the cell wall. DHA generated by ASC oxidation may undergo an apoplast-specific degradation pathway via oxidase-dependent or non-enzymatic, conversion of DHA to 4-O-oxalyl-L-threonate. The latter compound is further converted to L-threonate and oxalate as final degradation products. This reaction is catalysed by apoplastic oxalyl esterases but may also occur non-enzymatically. During these steps additional H_2O_2 molecules are formed, which can take part in a Fenton reaction. Moreover, in many plants L-threonate is oxidated to L-threarate in a reaction, which uses two O_2 and generates two H_2O_2 molecules. In tissues possessing apoplastic oxalate oxidase, the oxalate can also yield H_2O_2 when oxidised (Green and Fry 2005a, b).

7 A Role of Ascorbate in Somatic Embryogenesis

Studies on the roles of ASC in the formation of somatic embryos were conducted using white spruce (*Picea glauca*) embryogenesis as a model system. An embryogenic tissue for the induction of somatic embryogenesis was generated from zygotic embryos (Stasolla and Yeung 1999). A well-characterised process of somatic embryogenesis in this plant generally consists of proliferation and maturation stages. These stages are characterised by different hormonal requirements. The first stage is characterised by proliferating tissue on the auxin- and cytokinin-containing medium. At this stage, the formation of filamentous embryos occurs. These structures are composed of a suspensor region subtending clusters of small, non-vacuolated cells of the embryo proper. Then, the tissue is transferred onto the ABA-supplemented maturation medium. During the culture in the presence of ABA, cell divisions occurs which leads to an increase in the size of the embryo proper. Subsequently, a well-developed shoot and root pole become visible. Finally, the embryos develop a ring of cotyledons at the shoot apical region. Mature embryos are transferred onto the germination medium where, root and shoot emergence occurs (Stasolla et al. 2002).

White spruce somatic embryo maturation and germination were marked by important changes in ASC metabolism. The differences in ASC content, ASC/DHA redox state, and the activities of ascorbate-metabolising enzymes were analysed using both embryogenic and non-embryogenic cell lines. While no differences in ASC metabolism were observed between embryogenic and non-embryogenic cell lines in the proliferation medium, after the transfer onto the maturation medium, the rate of ASC synthesis sharply increased in embryogenic lines, whereas it remained constant in the non-embryogenic line. Gradual decrease in the APX activity followed by a shift within the total ascorbate pool towards the reduced form was observed in embryogenic lines. On the other hand, the non-embryogenic line was characterised with significantly higher APX activity and oxidised state of ascorbate pool. Switching of the ASC/DHA ratio towards the reduced state and increase in ASC levels in embryogenic lines are possibly required for the progression of cell divisions within developing embryos (Stasolla and Yeung 2001).

In contrast to zygotic embryos, which undergo desiccation before subsequent germination preceded by imbibition, mature somatic embryos usually directly develop to plantlets. However, the germination frequency of somatic embryos of coniferous species, including white spruce, was found to be greatly increased by partial drying treatment (PDT) preceding transfer onto germination medium. Stimulatory effect of PDT results from an alteration in storage product deposition, decreased synthesis of ethylene and ABA, and changes in the pattern of nucleotide synthesis and utilization (Stasolla et al. 2002). PDT applied to the mature embryos was characterised by several changes in ASC metabolism. Firstly, ASC levels in the embryos as well as the activities of its redox enzymes significantly declined. Secondly, ASC/DHA ratio shifted to a more oxidised state, with equal participation of ASC and DHA in total ascorbate pool observed at the end of PDT. After PDT, white spruce embryos are transferred onto a hormone-free germination medium, where the induction of root and shoot meristem activity occurs. Upon the transfer onto the germination medium a restoration of ASC synthesis and ASC metabolism were observed. Besides the onset of ASC synthesis, a reduction of DHA accumulated during PDT occurs which contributes to an increase in the ASC level. Elevated ASC levels are required for activation of meristematic activity in germinating embryo. Moreover, an increase in ASC levels are followed by rise in APX activity necessary for detoxification of H_2O_2 generated after the recovery of oxidative metabolism (Stasolla and Yeung 2001).

The process of shoot emergence was found to be significantly affected by ASC added to the germination medium (Stasolla and Yeung 1999). An optimised concentration of ASC significantly enhanced shoot development. Moreover, ASC-treated embryos were larger and produced dark green leaves at the shoot pole. Control embryos, germinated on ascorbate-free medium were smaller than and not as green as those cultured in the presence of ASC. In contrast to shoot meristem, the root pole of the embryo did not react with an increase of root growth to ASC treatments (Stasolla and Yeung 1999). An importance of ASC for embryo meristem reactivation at germination was confirmed by a complementary approach using lycorine – an inhibitor of ASC biosynthesis. Lycorine inhibits the last step of ASC synthesis, i.e. conversion of GalL to ascorbate (Arrigoni 1994). It was demonstrated that ASC depletion following the lycorine administration to the germination medium prevents cellular divisions in the shoot apical meristem (Stasolla and Yeung 2007).

The mechanism of ASC stimulatory effect on the shoot apex development involves prevention of the peroxidase-dependent cell wall stiffening within shoot apical meristem. Experimental manipulations resulting in an increase in ASC content, i.e. supplementing germination medium with ASC or GalL, decrease the activities of ferulic acid peroxidase (FPOX) and guaiacol peroxidase (GPOX). On the other hand, activities of the aforementioned enzymes increase in lycorine-treated embryos. FPOX and GPOX are responsible for cross-linkage of cell wall compounds, which leads to cell wall stiffening and prevents cell division and cell growth. Therefore, the critical levels of endogenous ASC in the apical meristem are necessary to maintain cell wall plasticity prerequisite for both cell growth and proliferation (Stasolla and Yeung 2007).

8 A Role of Glutathione in Somatic Embryogenesis

A series of studies revealed an importance of glutathione concentration and redox state for the process of somatic embryo formation in the white spruce tissue culture. In contrast to ASC treatments, which stimulated mainly postembryonic developmental events within the shoot meristem, altering glutathione level affected all stages of somatic embryo formation and further development: cell proliferation on maintenance medium, embryo maturation and germination. GSH supplementation to the maintenance medium had a stimulatory effect on cell divisions in the embryogenic tissue resulting in higher fresh weigh increase when compared to untreated control. On the other hand, the rate of growth of GSSG-treated tissue was comparable to that of control (Belmonte et al. 2003).

The addition of exogenously supplied GSH or GSSG to the ABA-containing maturation medium affected the number of embryos formed in culture as well as their quality. An optimised GSH concentration in the medium resulted in an increase in the number of embryos formed, this increased number was, however, mostly due to the production of low quality embryos, unable to regenerate viable plants. In contrast to GSH-treatment, supplementing medium with GSSG had a marginal effect on total embryo population. However, supplementing with GSSG strongly increased embryo quality, which manifested in higher number of embryos equipped with four or more cotyledons and possessing high germination potential. It should be noted that the aforementioned effect of GSH or GSSG applications were limited to the narrow ranges of concentration. Higher level of both glutathione redox forms inhibited overall embryo-forming capacity in cultured tissue and impeded embryo development (Belmonte and Yeung 2004; Belmonte et al. 2005a).

GSSG treatment caused an increased accumulation of starch granules, lipids and protein bodies in cortical and procambial cells of GSSG-treated embryos. Storage product deposition in cells of treated embryos may be required for the acquisition of desiccation tolerance, thus enhancing the viability of embryos. Another striking feature of embryos exposed to GSSG was the proper structure of meristems, whereas meristems of untreated embryos were frequently disrupted by intercellular spaces, which led to a failure in meristem reactivation during postembryonic development (Belmonte et al. 2005a). The improved architecture of shoot apical meristems in GSSG treated embryos may be due to the reduction of ethylene production. Disorganisation of cell division patterns in meristem resulting in intercellular spaces formation was related to the accumulation of this hormone in the culture vessel. In embryos grown on GSSG-supplemented medium endogenous ethylene level did not vary throughout the culture period. On the other hand, in control cuttings ethylene production increased markedly after 10 days of culture and remained high until the end of culture period (Belmonte et al. 2005a).

Ethylene synthesis in white spruce seems to be under control of GHS/GSSG redox status. It was demonstrated that the gene encoding aminocyclopropane-1-carboxylate (ACC) oxidase, an enzyme responsible for the final step of ethylene biosynthesis is induced by GSH treatment at all the stages of somatic embryo development. A switch of the glutathione pool towards oxidised form by GSSG administration is believed to have inhibitory effect on ACC oxidase expression that consequently decreases ethylene levels (Stasolla et al. 2004). However, the down regulation of ACC oxidase transcript level imposed by GSSG application still needs to be demonstrated to prove this hypothesis.

Given that supplementing GSH to the medium favours cell proliferation during early stages of somatic embryogenesis and increases total embryo yield, whereas effect of GSSG treatment manifests itself at embryo maturation period, a protocol of sequential treatments with GSH and GSSG was developed to profit the beneficial effects of both glutathione redox forms. An optimised procedure that resulted in the highest embryo yield, containing the highest percentage of high quality embryos consisted of 7 days of GSH treatment followed by culture in the presence of GSSG during the remaining part of 40-days-long culture period (Belmonte et al. 2005a). Similar stimulation of somatic embryogenesis was achieved by replacing sequential GSH/GSSG treatment with supplementing maturation medium with optimised concentration of BSO. It was observed that the glutathione redox state turned towards a more oxidised one with 0.01 mM or higher concentration of BSO. Compared to control embryos, the GSH/GSH + GSSG ratio declined quickly in BSO-treated embryos over the course of culture period, reaching its minimal value in fully mature embryos. In contrast to BSO-treated embryos, control ones were characterised by reduced glutathione redox status throughout the culture period (Belmonte and Stasolla 2007; Stasolla et al. 2008).

Oxidation of glutathione pool in tissues grown in the presence of BSO was beneficial for the efficiency of embryogenesis and the quality of embryos formed. In contrast to control embryos, those formed in the presence of 0.01 mM BSO were characterised by proper organisation of shoot apical meristem with its sub-apical domain composed by tightly packed cells. Moreover, BSO-treated embryos contained more initial cells in root apical meristem when compared to the untreated control (Belmonte and Stasolla 2007; Stasolla et al. 2008). This finding is consistent with the present knowledge which emphasises the importance of an oxidised environment for the establishment of quiescence and cell division patterns in the apical root meristem (Jiang et al. 2003).

Since glutathione levels and redox state are coupled to ascorbate through ASC–GSH cycle, BSO treatment affected both ascorbate levels and the activities of several ascorbate enzymes. Generally, BSO treatment decreased total ascorbate levels in embryos and transiently stimulated the expression of genes coding for APX and MDHAR, which resulted in an increase in the activities of these enzymes. A stimulatory effect of BSO on the APX activity during the early phase of embryo development may have particular importance for somatic embryo development due to decreasing toxic levels of H_2O_2 generated during active embryonic growth (Stasolla et al. 2008).

9 Glutathione-Dependent Changes in Nucleotide Metabolism During Somatic Embryogenesis

Studies on nucleotide metabolism in cultures grown on GSH-supplemented media revealed that stimulatory effect of GSH treatment on cell proliferation in white spruce culture grown on maintenance medium is possibly related to alterations in purine and pyrimidine metabolism (Belmonte et al. 2003, 2005b). Both purine and pyrimidine nucleotides serve as building blocks for nucleic acid synthesis. Moreover, purine nucleotides participate in bio-energetic processes. In white spruce cells, both purine and pyrimidine nucleotides can be synthesised de novo from precursor molecules, or through the salvage mechanism, that utilises bases and nucleosides as substrates (Stasolla et al. 2003).

The effects of GSH or GSSG supplementation to the maintenance medium on pyrimidine metabolism was studied by following the metabolic fate of radiolabelled orotic acid, precursor of the de novo synthesis pathway and uridine and uracil, respective intermediates of the salvage and degradation pathway. Inclusion of either GSH or GSSG to the medium increased the levels of pyrimidine nucleoside triphosphates and nucleic acids in cultured cells. It was demonstrated that elevated GSH levels result in activation of the de novo synthesis and reduction of the degradation pathway, whereas GSSG increases the activity of salvage pathway. Compared to the control tissue, grown on GSH-free medium, tissues cultured in the presence of 0.2 mM GSH were characterised by increased production of UMP from orotic acid leading to high cellular levels of UTP and CTP. The enlargement of pyrimidine nucleotide pool observed in GSH-treated tissue was due to an increase in the activity of orotate phosphoribosyl transferase, an enzyme that converts orotic acid to orotidine-5′-monophosphate during the first step of pyrimidyne nucleotide synthesis pathway (Bellmonte et al. 2005b).

Induction of the de novo synthesis of pyrimidine nucleotides by GSH was accompanied by a reduction of the degradation pathway. Pyrimidine nucleotides undergoing degradation are converted to uracil, which in further steps is degraded to β-ureidopropionate and CO_2 (Stasolla et al. 2003). If the tissue was grown in the presence of GSH, substantially less radiolabelled uracil was converted to these degradation products. Instead, uracil was recovered into nucleotides and nucleic acids (Belmonte et al. 2005b).

In contrast to GSH, in the GSSG-treated tissue, the activities of synthesis and degaradation pathways of pyrimidine nucleotides were at control level. On the other hand, GSSG increased the efficiency of the salvage pathway by stimulating the phosphorylation of uridine to uridine monophosphate (UMP). Following further phosphorylation UMP converts to UTP, which may directly be incorporated to RNA synthesis or serve as a substrate in CTP- or UDP-glucose synthesis. The reaction of UMP synthesis from uridine may be catalysed by uridine kinase or nucleoside phosphotransferase. Given that both enzymes respond to GSSG treatment with a simultaneous rise in their activities, indicates the possible mechanism of GSSG-dependent increase in the rate of uridine salvage (Belmonte et al. 2005b).

Similar approach was applied to test the effects of GSH or GSSG supplementation on purine nucleotide metabolism in white spruce embryogenic tissue grown on maintenance medium. The activity of synthesis pathway was studied determining the rate of incorporation of purine synthesis precursor AICAR to AMP, ATP + ADP and nucleic acids. It was observed that GSH or GSSG treatments did not significantly increase the efficiency of AICAR conversion into aforementioned compounds. This finding suggests that GSH and GSSG do not affect purine synthesis pathway (Belmonte et al. 2003). On the other hand, the differential salvage activity was observed between control and GSH-treated tissue manifested in an increased incorporation of the salvage pathway intermediates, adenine and adenosine, during ATP synthesis in GSH-treated tissue. High levels of ATP during exponential growth phase of the embryogenic tissue may be required not only as a building block for nucleic acid synthesis but also as an intermediate involved in bio-energetic processes related to cell proliferation. Finding that only low levels of labelled adenine and adenosine was recovered in nucleic acids, suggests that increased ATP synthesis is the principal outcome of GSH-dependent stimulation of purine salvage pathway (Belmonte et al. 2003). The rate of the degradation of purine nucleotides was not affected by GSH or GSSG treatments. However, in GSSG-treated tissue, a substantial amount of degradation pathway intermediate – inosine – was recruited for the ATP production compared to untreated control and GSH-treated tissue (Belmonte et al. 2003).

The data reported above show that the possible mechanism underlying the stimulation of the cell proliferation in white spruce embryogenic tissue by glutathione is related to an enhancement of the production of nucleotides serving as "building blocks" for nucleic acid synthesis of substrates in bio-energetic reactions (Belmonte et al. 2003, 2005b). It was demonstrated that GSH availability regulates the activities of enzymes of nucleotide metabolism at the level of gene expression. The level of transcript coding for adenosine kinase was markedly higher in GSH-treated embryos during embryo maturation phase and germination. GSH supplementation to the medium reduced the expression of genes encoding for uracil phosphotransferase and nucleoside diphosphate kinase during the maturation phase. On the other hand, GSH stimulated the expression of uridinylate kinase in mature embryos of white spruce (Stasolla et al. 2004).

10 Glutathione-Induced Changes in Patterns of Gene Expression During Somatic Embryo Formation

The beneficial effect of GSH on embryo conversion is due to profound changes in gene expression patterns observed upon GSH administration. Microarray studies revealed that genes involved in large number of metabolic and regulatory pathways are differentially expressed between control and GSH-treated tissue. Compared with early stages of somatic embryo development, the total number of differentially expressed genes increased during embryo maturation (Stasolla et al. 2004). Many genes encoding for proteins involved in protein synthesis like ribosomal proteins, initiation and translation factors as well as late embryogenic abundant proteins were down-regulated in the presence of GSH during embryo maturation phase. The lower transcript level of genes related to protein synthesis, together with the reduced accumulation of protein bodies in embryos grown on GSH-supplemented medium is interpreted to be directly related to the switch to germination mode without PDT (Stasolla et al. 2004).

Some genes involved in carbohydrate metabolism, such as several glucanases, acetyl-CoA synthase, phosphoenolpyruvate carboxylase were repressed by the presence of GSH. On the other hand, expression levels of other group of genes involved in carbohydrate processing, represented by transketolase, fructokinase, aldolase and aconitase hydratase were elevated in GSH-treated embryos (Stasolla et al. 2004). Changes in the expression of aforementioned genes may account for an increased starch deposition in the cell of GSH-treated embryos (Stasolla et al. 2004).

An establishment of functional shoot and root apical meristem during somatic embryo maturation is dependent on expression of genes involved in meristem identity and organisation (Golz 2005). It was observed that several genes falling into that category, including the homologues of *CLAVATA 1* (*CLV 1*), *NO APICAL MERISTEM* (*NAM*) and *ARGONAUTE* (*AGO*) were substantially stimulated by GSH treatment during the late phases of embryo development (Stasolla et al. 2004). The aforementioned genes regulate division patterns within shoot apical meristem (Golz 2005). Among genes that regulate differentiation within root apical meristem, a *SCARECROW* (*SCR*) gene was found be up-regulated in mature embryos grown on GSH-supplemented medium (Stasolla et al. 2004).

Relation between expression patterns of genes involved in meristem patterning and glutathione levels was studied with regard to spruce *KNOTTED*-like homeobox (*KNOX*) gene *HBK 1*. The *HBK 1* gene is preferentially expressed in the shoot apical meristem of spruce where, similar to its *Arabidopsis* homologue *SHOOT MERISTEMELESS* (*STM*), plays decisive role in meristematic cell specification (Sundås-Larsson et al. 1998; Hjortswang et al. 2002). In situ hybridisation experiments using probes specific to *HBK1* mRNA revealed that long-term culture in the presence of GSSG extended the *HBK 1* expressing zone within an apical shoot pole. Localisation pattern of *HBK 1* was similar for both control and GSSG-treated embryos during the first 10 days of culture. *HBK 1* transcript was restricted to the apical pole of embryos. Differences in the expression pattern were visible at day 20 when *HBK 1* expressing zone was still restricted to the apical cells in control embryos, whereas was extended to the sub-apical cells in treated embryos. Moreover, less cells expressed *HBK 1* gene if meristem was disrupted by intercellular spaces. These findings raise a possibility that the improvement of the meristem organisation by GSSG is possibly related to conferring stem cell identity to larger population of cells through stimulation the *HBK 1* expression (Belmonte et al. 2005b).

A transformation approach using white spruce cells transformed with *HBK 3* gene in sense or antisense orientation revealed that representatives of *HBK* gene family control both early and late stages of somatic embryogenesis in spruce. Overexpression of *HBK 3* gene in cells of *HBK 3* – sense line promoted the initiation of embryo formation in embryogenic tissue. On the other hand, embryo formation was strongly reduced in transgenic lines where down-regulation of *HBK 3* expression was imposed due to antisense *HBK 3* transformation. Compared to the control embryos, those over-expressing *HBK 3* formed larger shoot apical meristems and exhibited higher competence for generating viable plants with no phenotypic aberrations (Belmonte et al. 2007). Further studies on *HBK 3* over- or under-expressing lines of embryogenic cells revealed an intimate relationship between *HBK 3* expression level and ASC and GSH metabolism. Both, GSH and ASC levels were significantly higher in cells over-expressing *HBK 3* especially after the onset of somatic embryo formation from embryogenic tissue induced by transfer from plant growth regulator-supplemented medium onto plant growth regulator-free medium. This increase was due to the activities of ASC and GSH regenerating enzymes. The activities of DHAR, MDHAR and GR were elevated in cells of *HBK 3* over-expressing line. These findings suggest that *HBK 3* expression might regulate the transition from cell proliferation to embryo formation in embryogenic tissue through alterations in ascorbate and glutathione metabolism (Belmonte and Stasolla 2009).

Similar to GSH/GSSG treatments, an application of BSO, resulting in a shift in the glutathione pool towards its oxidised form, induced the expression of genes involved in shoot meristem formation. Treatment of *Brassica napus* somatic embryos with 0.1 mM BSO resulted in an increase in the expression levels of *ARGONAUTE 1* (*AGO 1*) transcript during the globular and early torpedo stages of development (Stasolla et al. 2008). *AGO 1* is a protein required for both shoot meristem formation and identity, characterised by conserved *PAZ* and *PIWI* domains that are engaged in protein-protein interactions. This protein is closely related to *CLAVATA 3* – a well recognised factor engaged in meristem patterning and maintenance (Lynn et al. 1999). During the same stage of embryo development, BSO treatment increased the expression of gene coding for *CLE 27* protein (Stasolla et al. 2008), engaged in meristem size regulation (Fiers et al. 2007). During next stages of embryo formation, i.e. when embryos complete their histodifferentiation programme and switch to post-embryonic development, besides *AGO 1* and *CLE 27*, the level of *ZWILLE* transcript increased in BSO-treated embryos (Stasolla et al. 2008). *ZWILLE* gene codes for protein specifying the expression pattern of *STM*, which is also involved in the maintenance of apical shoot meristem (Moussian et al. 1998). Finally, in embryos cultured on BSO-supplemented medium, the expression of *B. napus STM* homologue increased in a late stage of embryo development (Stasolla et al. 2008). The data mentioned above, demonstrate that the imposition of an oxidised environment, effected by BSO application results in a sequential induction of a set of genes responsible for proper organisation of shoot apical meristem. This finding accounts for a beneficial effect of BSO treatment on the structure of the apical meristem.

11 Ascorbate and Glutathione Involvement in Adventitious Root Formation In vitro

Adventitious rooting consists of two stages: formation of root primordia and its subsequent outgrowth. The phase of primordia formation can be subdivided on the induction phase, when molecular and biochemical events related to dedifferentiation and competence acquirement occur, and initiation phase, characterised by organised cell divisions in the developing root primordium (Gaspard et al. 1997; Li et al. 2009a). Rooting is affected by numerous endogenous and exogenous factors, with the principal role of auxin as a chief regulator of adventitious root formation. Blocking the transport of endogenous auxin to seedling rooting zone inhibits rooting (De Klerk et al. 1999). In the regulation of adventitious rooting process, ascorbate and glutathione seem to be involved in a complex interplay between auxin and other components of cellular redox systems.

It was demonstrated that auxin stimulatory effect on rooting is mediated by NO (Pagnussat et al. 2002) and the NO-dependent signalling pathway, leading to root formation, is mediated by H_2O_2 (Liao et al. 2009). The signalling role of H_2O_2 in root formation was explored in studies on rooting of cucumber (Li et al. 2007) and Mung bean seedling cuttings (Li et al. 2009a). It was found that the culture in the presence of 20–40 mM H_2O_2 significantly increased the number of adventitious roots per explant formed by cucumber cuttings, when compared to the untreated control (Li et al. 2007). Mung bean cuttings reacted with increased rooting to H_2O_2 concentrations ranging from 1–100 mM. Cuttings underwent 8–24-h-long pulse H_2O_2 treatment followed by culture on the H_2O_2-free medium. Stimulatory effect of H_2O_2 treatment was observed both in the absence and in the presence of exogenous

IBA as an auxin source. Promoting effect of H_2O_2 on rooting decreased if a relatively higher H_2O_2 concentration and longer treatment times were applied (Li et al. 2009a). Decreasing H_2O_2 content, with diphenyloiodonium (DPI, NADPH oxidase inhibitor), catalase or ASC prevented the stimulation of rooting. Application of 4 mM ASC significantly decreased the number of regenerated roots and eliminated the stimulatory effect of H_2O_2 if these two substances were applied simultaneously (Li et al. 2009a). In contrast to H_2O_2, IBA still stimulated rooting in the presence of 4 mM ASC in Mung bean, and IAA stimulated rooting in the presence of 2 mM ASC in cucumber (Li et al. 2007, 2009a). On the other hand, ASC in concentration of 6 mM completely prevented rooting in the absence or presence of either H_2O_2 or IBA. These results suggested that H_2O_2 or IBA promotion of adventitious root formation was blocked by certain concentrations of ASC (Li et al. 2009a).

Measurements of H_2O_2 concentration in cucumber seedling explants during subsequent phases of rooting on auxin-free medium revealed that H_2O_2 peaks during the early induction phase. It may suggest that elevated H_2O_2 levels may possibly be required in early events of rooting, i.e. dedifferentiation and acquiring competence (Li et al. 2007). In Mung bean cuttings cultured on basal medium an induction phase of rooting was marked by a progressive increase in POX and APX activities. At the switch from induction to initiation phase, activities of both enzymes strongly decrease. It was demonstrated that IBA treatment transiently decreased POX and APX activities during induction phase of rooting, which may be correlated with an increase in endogenous H_2O_2 level, required for the induction of adventitious roots. Therefore, APX activity may be involved in an auxin-dependent mechanism of regulation of H_2O_2 level during adventitious rooting and the early decrease in its activity may be one mechanism by which IBA and H_2O_2 promote adventitious rooting (Li et al. 2009b). It should be noted that although the moderate oxidation stimulates rooting, H_2O_2-overproduction, resulting in severe oxidative stress inhibited root formation (Pal Singh et al. 2009).

Imin and co-workers (2007) show that both reduced (GSH) and oxidised (GSSG) form of glutathione markedly enhance the number of roots formed by callus derived from leaf explants of *Medicago truncatula* cultured on auxin-supplemented medium. In their experiments leaf segments were grown for 3 weeks on proliferation medium supplemented with NAA or NAA in combination with either GSSG or GSH. Significantly, more roots were produced by explants grown on NAA-containing media supplemented with GSH or GSSG than on media supplemented with NAA alone, which suggests that glutathione may act synergistically with exogenous auxin in the stimulation of root regeneration (Imin et al. 2007). Because in the absence of NAA, root formation was negligible (Imin et al. 2007), it would be interesting to find out if GSH or GSSG may, to some extent, replace the hormone in the stimulation of root formation. However, the authors do not include any data on effects of GSH or GSSG on root formation on NAA-free media (Imin et al. 2007).

During early stages of rooting in tomato seedling cuttings, when the root primordia are formed, total glutathione pool is characterised by higher participation of GSSG in the total glutathione pool in comparison to the later stages when the outgrowth and elongation of newly formed roots occurs (Tyburski and Tretyn 2010). Higher oxidation of glutathione pool during the formation of root primordia by tomato seedling cuttings may result from an increase in the rate of ascorbate turnover in Halliwell–Asada cycle and a strong rise in the activity of dehydroascorbate reductase at this stage of root formation. Subsequent days of rooting are characterised by a decline in the activity of dehydroascorbate reductase and other enzymes of ascorbate metabolism (Tyburski et al. 2006). Consequently, higher amounts of reduced glutathione can be accumulated in the rooting zone (Tyburski et al. 2006).

Supplementing the rooting medium with GSH increased the number of roots formed by tomato seedling cuttings grown on an auxin-free medium. However, the stimulatory effect of GSH was restricted to a narrow range of concentrations spanning from 1 or 2.5 mM GSH. Strongest stimulation of root formation occurred when plants were simultaneously treated with auxin and GSH. Treatments with GSSG did not affect root formation if cuttings were grown on the basal medium, however, an optimised GSSG concentration strongly enhanced the rooting-stimulatory effect of auxin treatment. BSO did not affect the number of roots formed by cuttings grown on BM and only slightly decreased the efficiency of rooting in the presence of IAA, which shows that depletion of GSH from rooting zones does not inhibit rooting. This finding suggests that root formation may occur in the absence of GSH but the process is stimulated by its presence (Tyburski and Tretyn 2010).

Root regeneration by tomato seedling cuttings proceeds on a highly synchronized manner with a 3-day-long period of primordia formation encompassing initiation and induction phase and subsequent phase of primordia elongation, which manifests in root emergence on the fourth day of culture. Induction and initiation phases of rooting in tomato seedling cuttings are characterised by rapid increase in the content of ASC in the explant rooting zone. The accumulation of ASC in the tissue correlates with the biosynthetic capability from GalL and with the activities of enzymes regenerating ASC from its oxidised forms (MDHAR and DHAR). Simultaneously, the sharp increase in H_2O_2 content and in the activities of APX and AOX were observed. ASC peaks on the third day of rooting, i.e. at the switch from the phase of primordia organisation to the phase of primordia elongation. With the beginning of elongation phase a dramatic decrease in ASC content was observed. In contrast to ASC, DHA levels remained constant during root formation. Similar to ASC, the H_2O_2 level, as well as, the activities of ASC-metabolising enzymes dropped at the onset of the root elongation phase. These finding suggest that the increase in the endogenous level of ASC observed at the beginning of rooting may be explained by the necessity of the regulation of H_2O_2 level in the rooting zone during the formation of root primordia. This point of view is supported by the finding that the activities of H_2O_2 scavenging enzymes: APX, POX and CAT rise simultaneously with the ASC (Tyburski et al. 2006).

Functions of ascorbate in adventitious rooting seem to extend beyond the regulation of H_2O_2 levels (Tyburski et al. 2006). It has been observed that the addition of exogenous GalL, ASC or DHA to the rooting medium affect the root formation by tomato seedling cuttings. The ascorbate precursor (GalL) as well as ASC and DHA modified the rooting response in a similar way, i.e. stimulated the formation

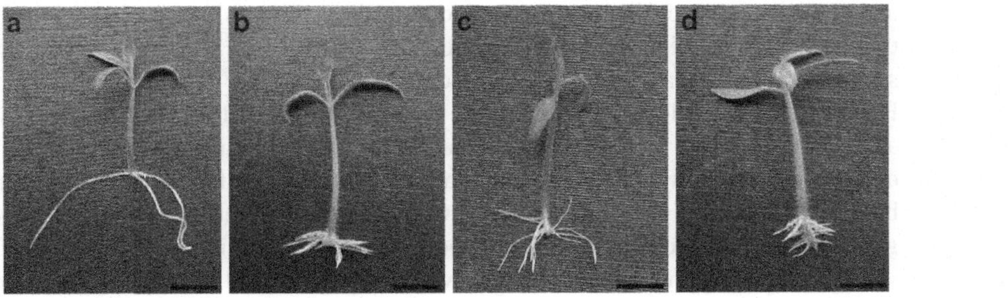

Fig. 2 The effect of ASC, DHA and IAA on rooting of tomato seedling cuttings. The cuttings were cultured 7 days on basal medium (**a**), or on the same medium supplemented with 2 mM ASC (**b**), 2 mM DHA (**c**) or 1 μM IAA 9 (**d**). Similar to exogenous auxin, ASC and DHA stimulate the formation of roots but inhibit their elongation

of roots but inhibited their elongation. Therefore, their effect was similar to that of exogenous auxin, which also increased the number of regenerated roots, simultaneously reducing their length (Tyburski et al. 2006) (Fig. 2). It is noteworthy that DHA was more effective in inducing abundant root formation than ASC and GalL, having higher stimulatory effect than auxin. Moreover, because ASC added to the medium, being an unstable molecule, is partly oxidised to DHA and GalL after being converted to ASC is subsequently oxidised to DHA. It has been found that treatments with DHA and GalL that induced more roots than ASC, resulted in a significant increase in DHA content in the rooting zones in comparison with ASC-treated cuttings. Therefore, the oxidised form of ascorbate is supposed to have a decisive role in the stimulation of rooting (Tyburski et al. 2006).

Rooting of cuttings does not occur, or is severely reduced, if auxin transport to the rooting zone is blocked by application of auxin transport inhibitors like NPA or TIBA. These treatments result in a suboptimal auxin level in explant rooting zone that block root formation (Visser et al. 1995; Ludwig-Müller et al. 2005). Several redox agents were able to restore root development in NPA or TIBA treated explants. NO and H_2O_2, which are believed to mediate in the auxin signalling pathway, were able to trigger adventitious root formation in *Tagetes erecta* seedling cuttings cultured on medium supplemented with 10 mM NPA (Liao et al. 2009). On the other hand, an inhibitory effect of 1 μM TIBA on rooting of tomato seedling cuttings was reversed by 4 mM ASC. It has been found that only ASC (but neither GalL nor DHA) was able to reverse the stimulatory effect of TIBA on root formation, although the stimulatory effect of GalL and DHA on rooting of cuttings grown on medium without TIBA was comparable to the one of ASC or even stronger. The question of why ASC added to rooting the medium is able to reverse the inhibition of rooting by TIBA while the GalL administration, which actually efficiently increases the ASC content in the root-forming tissue remains unresolved (Tyburski et al. 2006).

12 The Roles of Ascorbate, Glutathione and Related Enzymes in the Elicitation of Metabolite Synthesis in Root Cultures

Isolated root systems grown in bioreactors as suspensions on liquid media serve as important source of metabolites for commercial purposes (Flores et al. 1999; Guillon et al. 2006). It was demonstrated that changes in activities of the enzymes of ASC and GSH metabolism may influence both root development in culture as well as the rate of metabolite production, which often needs elicitation by oxidative stress-inducing factors. The production of ginsenoside by adventitious cultures of *Panax ginseng* was enhanced by increasing O_2 concentration from 20% to 40%. The same effect was achieved when roots were treated with optimised concentrations of H_2O_2. An increase in metabolite production was observed between 15 and 45 days after the exposition to O_2 started. On the other hand, treatments with 25 and 50 μM H_2O_2 resulted in increased ginsenoside levels after only 7 days after treatment. As inferred from malonyl dialdehyde (MDA) content and H_2O_2 levels in O_2-treated roots, this treatment imposed a mild oxidative stress, which seems to have a stimulatory effect on the induction of ginsenoside production. To counter the oxidative damage, antioxidant enzymes of ascorbate–glutathione cycle were induced by O_2 application. An increase in APX activity was detected after 15 days of treatment and from that day on high enzyme activity was observed during the rest of culture period with maximum after 45 days. Other enzymes, i.e. MDHAR, DHAR and GR behaved in a similar manner. Increase in their activities was coordinated with increases in the activities of other constituents of antioxidant system such as superoxide dismutase (SOD), CAT, glutathione peroxidase (GPX) and guaiacol peroxidase (POX). These finding suggest that during O_2/H_2O_2-elicited ginsenoside production, the intensity of oxidative stress is regulated in a coordinated manner by synchronized changes in the activities of antioxidant enzymes (Ali et al. 2005a).

Redox agents played a decisive role also in the regulation of root proliferation in culture. The proliferation of *Panax* isolated adventitious roots was stimulated by NO. It was demonstrated that supplementing media with NO producing producers like sodium nitroprusside (SNP) increased the number of rootlets per explant. On the other hand, decreasing NO levels with NO-scavenger 2-phenyl-4,4,5,5-tetramethylimidazoline-1-oxyl3-oxide (PTIO), prevented root proliferation (Tewari et al. 2007). NO stimulatory effect was blocked if roots were treated with DPI – an inhibitor of $O_2\cdot^-$-producing enzyme: NADPH-oxidase. This observation together with the finding that NO stimulates NADPH oxidase in ginseng roots suggest that NO stimulatory effect on root proliferation is mediated by $O_2\cdot^-$. However, along with increasing the pro-oxidant activity of NADPH oxidase, NO stimulated the activity of the set of antioxidant enzymes including SOD, CAT, POX, APX, DHAR and GR. Treatment with SNP also increased the levels of total ascorbate and the participation of ASC in the pool of total ascorbate. These effects were prevented by the application of NO scavenger. Therefore, a coordinated stimulation of pro-oxidant and antioxidant systems by NO impose a moderate oxidative stress conditions which promote root proliferation in culture (Tewari et al. 2008).

The onset of secondary metabolite production in hairy root culture often requires the treatments with elicitors; the signalling molecules of plant defence responses (Guillon et al. 2006). Elicitor treatments frequently induce oxidative stress, which sometimes is a direct factor triggering metabolite synthesis (Huerta-Heredia et al. 2009). When biotransformation process in a hairy root culture is dependent of ROS production, antioxidant activity of ASC may have an inhibitory effect, as it was shown for the tetracycline phytoremediation by hairy roots of sunflower (Gujarathi et al. 2005). However, in other experimental systems, ASC, GSH and associated enzymes play an important role in protecting cultured roots from oxidative damage during elicitation (Ali et al. 2005b). The stimulation of saponin production by hairy root cultures of *Panax ginseng* and *Panax quinqefolium* by methyl jasmonate (MeJA) was accompanied by a gradual increase in H_2O_2 content. In addition, *P. quinqefolium* root culture showed a marked increase in lipooxygenase activity upon MeJA treatment, which usually results in increased lipid peroxidation. However, the markers of lipid peroxidation declined upon elicitation in roots of both species of ginseng, which suggests that in spite of increased oxidation no severe oxidative damage occurred in treated roots. It was shown that MeJA-induced oxidative stress was mitigated by an increase in the activities of various components of the antioxidant system. Upon elicitation, ASC content and ASC/DHA ratio as well as SOD and POX activities increased in both species. On the other hand, a decrease in CAT activity was observed. APX activity increased after MeJA treatment in *P. ginseng* but not in *P. quinqefolium*. In contrast, *P. quinqefolium* was characterised by increased DHAR activity but the activity of this enzyme decreased in *P. ginseng*. These findings indicate that, the enzymes of ASC–GSH cycle play a protective role against oxidative stress during the elicitation of metabolite production in hairy root culture (Ali et al. 2005b).

13 Ascorbate and Glutathione Roles in Shoot Organogenesis In vitro

Relatively less attention, when compared with somatic embryogenesis and root formation, was paid on the role of ascorbate and glutathione in shoot differentiation in plant tissue culture. Early reports suggest that ASC may enhance shoot organogenesis in tobacco callus culture (Joy et al. 1988). Addition of 0.8 mM ASC to the shoot-forming medium increased shoot formation by 45% over the control in young callus, having a high organogenetic potential. Different concentrations of exogenous ASC (0.4–0.8 mM) were able to restore regenerative capacities of old callus tissue, characterised by reduced ability to regenerate shoots. Treatment with ASC also speeded up the shoot-forming process with the appearance of primordia as early as day 10 in culture, compared to more than 12 days in control tissue. It has also been shown that ASC treatment reversed the inhibitory effect of exogenous gibberelic acid on root formation. The stimulatory effect of ASC on shoot differentiation in tobacco callus was linked to the content of soluble sugars, which increased in tissue during ASC treatment. However, the mechanism of ascorbate involvement in shoot organogenesis in tobacco has not been explained (Joy et al. 1988).

Treatments with reduced glutathione improved the development of isolated shoot tips of apple (Nomura et al. 1998). Isolated shoot meristems were grown on basal medium supplemented with BAP for shoot development and then transferred onto the medium containing IBA for rooting the differentiated shoots. Efficiency of shoot tip culture was strongly reduced by explant browning observed within 1 day after the start of culture. Explant browning was prevented by dipping the shoot tips in 0.1 mM GSH, prior to inoculation on a solid medium. Besides inhibiting browning, dip treatment increased the number of shoot tips that developed into normal shoots (Nomura et al. 1998). Treated shoots developed into normal shoots with an efficiency of 100%. In the untreated control, rates of shoot development from untreated shoots did not exceed 60%. Promotion of normal shoot development by the dip treatment was also expressed in terms of the length of first leaves from cultured shoot tips. The dip-treated shoot tips produced longer first leaves when compared to untreated controls. Interestingly, the promotion of shoot development was observed only if pulse GSH treatment was applied before the start of culture and after that, shoot tips were grown on basal medium. If instead of GSH pulse treatment, shoot tips were cultured on medium supplemented with GSH, shoot development was not stimulated. On the other hand, cultures on medium containing GSH promoted callus proliferation at the base of shoot tips (Nomura et al. 1998).

Reducing conditions imposed by adding antioxidants (ASC, GSH and α-tocopherol) stimulated shoot organogenesis from leaf segment of gladiolus (Dutta Gupta and Datta 2003). Both, frequency of shoot organogenesis as well as the number of number of shoots per explant were increased with the addition of antioxidants at 0.5 mM concentration. However, higher concentrations were found to be inhibitory. Compared to α-tocopherol and ASC, GSH was noted to be most effective. On the other hand, H_2O_2 added to the culture medium in concentrations ranging from 0.005 to 0.15 mM strongly inhibited shoot organogenesis. Moreover, the process of shoot formation was marked by a gradual increase in the CAT and POX activities with a concurrent decrease in the SOD activity. Given that, CAT and POX are H_2O_2-consuming enzymes, whereas high SOD activity results in elevated H_2O_2 levels, it was concluded that shoot organogenesis requires a reducing environment (Dutta Gupta and Datta 2003).

In contrast to shoot organogenesis, somatic embryo formation on the gladiolus leaf explants was promoted by oxidising factors. The frequency of somatic embryogenesis and the number of embryos per responding culture were decreased by ASC, GSH and α- tocopherol in a dose-dependent manner. On the other hand, H_2O_2 at 100 μM concentration increased the frequency of somatic embryogenesis by about 18%. The process of somatic embryo formation was accompanied, in its initial phase, by an increase in SOD activity and progressive reduction in CAT and POX activities, which creates conditions favourable for H_2O_2 accumulation (Dutta Gupta and Datta 2003). Similar to gladiolus, initial phase of somatic embryo formation in the callus of *Lycium barbarum* was characterised by increase in SOD activity accompanied by decrease in POX and CAT activities, which resulted in a marked increase in intracellular H_2O_2 levels. Moreover, the treatments that increase H_2O_2 levels, such as inhibiting

CAT activity with aminotriazole or adding H_2O_2 to the culture medium promoted somatic embryogenesis. On the other hand, inhibiting SOD activity with *N,N'*-diethyldithiocarbonate caused a decrease in the frequency of somatic embryogenesis (Kairong et al. 1999). The aforementioned data suggest that shoot organogenesis and somatic embryogenesis differ in their redox requirements, being promoted by reducing or oxidising environment, respectively. However, it should be kept in mind that in some shoot regeneration systems, similar to somatic embryogenesis, oxidising processes may also have a stimulatory effect on organogenesis. This was the case for shoot organogenesis in strawberry callus, where H_2O_2 levels and SOD activity increased during early phases of shoot regeneration when meristemoid formation and vascular tissue occurs in calli grown on regeneration medium. Simultaneously, a decline in H_2O_2 scavenging enzymes; CAT and POX was observed (Tian et al. 2003). Similar to somatic embryogenesis systems (Kairong et al. 1999; Dutta Gupta and Datta 2003), shoot organogenesis percentage in strawberry was stimulated by exogenous H_2O_2 and decreased by SOD inhibitor – *N,N'*-diethyldithiocarbonate. Moreover, it was shown that types of strawberry calli, which showed high regeneration capacity, had several times higher H_2O_2 and O_2^- levels compared to calli with low-regeneration capacity (Tian et al. 2004).

14 Conclusion

The studies reported in this review demonstrate that ascorbate- and glutathione-dependent biochemical systems play important roles in many aspects of plant tissue culture, affecting growth, differentiation and metabolism. Ascorbate and glutathione participate in plant regeneration, being involved in the mechanisms regulating cell divisions in newly formed meristems and participating in hormone metabolism and signalling. Moreover, an antioxidant activity of these molecules manifested in Halliwell–Asada cycle protects in vitro cultured tissues against oxidative stress. Diverse functions of these antioxidants open vast possibilities of using them for the improvement of tissue culture and plant regeneration methods. However, further studies are required to fully exploit the properties of ascorbate and glutathione for manipulating developmental processes in plant tissue culture.

References

Ali MB, Thanh NT, Yu K-W, Hanh E-J, Paek K-Y, Lee HL (2005a) Induction in the antioxidative systems and lipid peroxidation in suspension culture roots of *Panax ginseng* induced by oxygen in bioreactors. Plant Sci 169:833–841

Ali MB, Yu K-W, Hanh E-J, Paek K-Y (2005b) Differential responses of anti-oxidants enzymes, lipooxygenase activity, ascorbate content and the production of saponins in tissue cultured root of mountain *Panax ginseng* C.A. Mayer and *Panax quinquefolium* L. in bioreactor subjected to methyl jasmonate stress. Plant Sci 169:83–92

Arrigoni O (1994) Ascorbate system in plant development. J Bioenerg Biomem 26:407–419

Arrigoni O, De Tullio MC (2000) The role of ascorbic acid in cell metabolism: between gene – directed functions and unpredictable chemical reactions. J Plant Physiol 157:481–488

Asard H, Horemans N, Caubergs RJ (1995) Involvement of ascorbic acid and b-type cytochrome in plant plasma membrane redox reactions. Protoplasma 184:36–34

Belmonte MF, Donald G, Reid DM, Yeung EC, Stasolla C (2005a) Alterations of the glutathione redox state improve apical meristem structure and somatic embryo quality in white spruce (*Picea glauca*). J Exp Bot 56:2355–2364

Belmonte MF, Stasolla C (2007) Application of DL-buthionine-[S, R]-sulfoximine deplete cellular glutathione and improve white spruce (Picea glauca) somatic embryo development. Plant Cell Rep 26:517–523

Belmonte MF, Stasolla C (2009) Altered *HBK3* expression affects glutathione and ascorbate metabolism during the early phases of Norway spruce (*Picea abies*) somatic embryogenesis. Plant Physiol Biochem 47:904–911

Belmonte MF, Stasolla C, Katahira R, Loukanina N, Yeung EC, Thorpe TA (2005b) Glutathione – induced growth of embryogenic tissue of white spruce correlates with changes in pyrimidine nucleotide metabolism. Plant Sci 168:803–812

Belmonte MF, Stasolla C, Loukanina N, Yeung EC, Thorpe TA (2003) Glutathione modulation of purine metabolism in cultured white spruce embryogenic tissue. Plant Sci 165:1377–1385

Belmonte MF, Tahir M, Schroeder D, Stasolla C (2007) Overexpression of *HBK3*, a class I KNOX homeobox gene, improves the development of Norway spruce (*Picea abies*) somatic embryos. J Exp Bot 58:2851–2861

Belmonte MF, Yeung EC (2004) The effects of reduced and oxidized glutathione on white spruce somatic embryogenesis. In Vitro Cell Dev Biol Plant 40:62–66

Benkova E, Michniewicz M, Sauer M, Teichman T, Seifertiva D, Jurgens D, Friml J (2003) Local, efflux – dependent gradients as a common module for plant organ formation. Cell 115:591–602

Bhattacharjee S (2005) Reactive oxygen species and oxidative burst: Roles in stress, senescence and signal transduction in plants. Curr Sci 89:1113–1121

Bilolu I, Xu J, Wildwater M, Willemsen V, Paponov I, Friml J, Heidstra R, Aida M, Palme K, Scheres B (2005) The PIN auxin efflux facilitator network controls growth and patterning in *Arabidopsis* roots. Nature 433:39-44

Bonke M, Tähtiharju S, Helariutta Y (2005) Lessons from the shoot apex. In: Fleming AJ (ed) Intercellular communication in plants. Blackwell Publishing, Oxford, pp 199–224

Choi YE, Yang DC, Park JC, Soh WY, Choi KT (1998) Regenerative ability of somatic single and multiple embryos from cotyledons of Korean ginseng on hormone – free medium. Plant Cell Rep 17:544–551

Christianson ML, Warnick DA (1983) Competence and determination in the process of *in vitro* shoot organogenesis. Dev Biol 95:288–293

Cleland RE (2004) Auxin and cell elongation. In: Davies PJ (ed) Plant hormones. Biosynthesis, signal transduction, action. Kluwer, Dordrecht/Boston/London, pp 204–220

Córdoba F, Gonzales-Reyes JA (1994) Ascorbate and plant cell growth. J Bioenerg Biomembr 26:399–405

Córdoba-Pedregosa MC, González-Reyes JA, Cañadillas MS, Navas P, Córdoba F (1996) Role of apoplastic and cell-wall peroxidases on the stimulation of root elongation by ascorbate. Plant Physiol 112:1119–1125

Cosgrove DJ (2001) Wall structure and wall loosening.A look backwards and forwards. Plant Physiol 125:131–134

De Pinto MC, Francis D, De Gara L (1999) The redox state of the scorbate-dehydroascorbate pair as a specific sensor of cell division in tobacco BY-2 cells. Protoplasma 209:90–97

De Pinto MC, Tommasi F, De Gara L (2000) Enzymes of the ascorbate biosynthesis and ascorbate–glutathione cycle in cultured cells of tobacco Bright Yellow 2. Plant Physiol Biochem 38:541–550

De Pinto MC, De Gara L (2004) Changes in the ascorbate metabolism of apoplastic and symplastic spaces are associated with cell differentiation. J Exp Bot 55:2559–69

De Tullio MC, Ciraci S, Liso R, Arrigoni O (2007) Ascorbic acid oxidase is dynamically regulated by light and oxygen. A tool for oxygen management in plants? J Plant Physiol 164:39–46

De Tullio MC, Paciolla C, Vecchia FD, Rascio N, D'Emerico S, De Gara L, Liso R, Arrigoni O (1999) Changes in onion root development induced by the inhibition of peptidyl-prolyl hydroxylase and influence of the ascorbate system on cell division and elongation. Planta 209:424–434

Dong JG, Fernández-Maculet JC, Yang SF (1992) Purification and characterization of 1-aminocyclopropane-1-carboxylate oxidase from apple fruit. Proc Natl Acad Sci USA 89:9789–9793

Dutta Gupta S, Datta S (2003) Antioxidant enzyme activities during *in vitro* morphogenesis of gladiolus and the effect of application of antioxidants on plant regeneration. Biol Plant 47:179–183

Esaka M, Fujisawa K, Goto M, Kisu Y (1992) Regulation of ascorbate oxidase expression by auxin and copper. Plant Physiol 100:231–237

Fiers M, Ku KL, Liu C-M (2007) CLE peptide ligands and their roles in establishing meristems. Curr Opin Plant Biol 10:39–43

Flores HE, Vivanco JM, Loyola-Vargas VM (1999) 'Radicle' biochemistry: the biology of root-specific metabolism. Trends Plant Sci 4:220–226

Friml J (2003) Auxin transport – shaping the plant. Curr Opin Plant Biol 6:7–12

Fry SC (1992) Oxidative scission of plant cell wall polysaccharides by ascorbate-induced hydroxyl radicals. Biochem J 332:507–515

Gaspard Th, Penel C, Greppin H (1997) Do rooting induction and flowering evocation involve a similar interplay between indoleacetic acid, putrescine and peroxidases? In: Greppin H, Penel C, Simon P (ed.), Travelling shot on plant development, University of Geneva, Geneve, Switzerland, pp 35–50

Golz JF (2005) Lessons from vegetative shoot apex. In: Fleming AJ (ed) Intercellular communication in plants. Blackwell, Oxford, pp 199–224

Gonzales-Reyes JA, Döring O, Navas P, Obst G, Böttger M (1992) The effect of ascorbate free radical on the energy state of the plasma membrane of onion (*Allium cepa* L.) root cells: alteration of K+ efflux by ascorbate? Biochim Biophys Acta 1098:177–183

Green MA, Fry SC (2005a) Vitamin C degradation in plants via enzymatic hydrolysis of 4-O-oxalyl-L-threonate. Nature 433:83–88

Green MA, Fry SC (2005b) Apoplastic degradation of ascorbate: novel enzymes and metabolites permeating the cell wall. Plant Biosyst 139:2–7

Guillon S, Trémouillaux-Guiller PPK, Rideau M, Gantet P (2006) Hairy root research: recent scenario and exciting prospects. Curr Opin Plant Biol 9:341–346

Gujarathi NP, Haney BJ, Park HJ, Wickramasinghe SR, Linden JC (2005) Hairy roots of *Helianhus annuus*: a model system to study phytoremediation of tetracycline and oxytetracycline. Biotechnol Prog 21:775–780

Henmi K, Tsuboi S, Demura T, Fukuda H, Iwabuchi M, Ogawa K (2001) A possible role of glutathione and glutathione disulphide in tracheary element differentiation in the cultured mesophyll cells of *Zinnia elegans*. Plant Cell Physiol 42:673–676

Hjortswang HI, Sundås-Larsson A, Bharathan G, Bozkhov PV, von Arnold S, Vahala T (2002) *KNOTTED1*-like homeobox genes of gymnosperm, Norway spruce, expressed during somatic embryogenesis. Plant Physiol Biochem 40:837–843

Horemans N, Foyer CH, Potters G, Asard H (2000) Ascorbate function and associated transport systems in plants. Plant Physiol Biochem 38:531–540

Huerta-Heredia A, Marín-Lopez R, Ponce-Noyola Cerda-García-Rojas CM, Trejo-Tapia G, Ramos-Valdivia AC (2009) Oxidative stress induces alkaloid production in Uncaria tomentosa root and cell cultures in bioreactors. Eng Life Sci 9:211–218

Imin N, Nizamidin M, Wu T, Rolfe BG (2007) Factors involved in root formation in Medicago truncatula J Exp Bot 58:439–451

Jiang K, Ballinger T, Li D, Zhang S, Feldman LJ (2006) A role for mitochondria in the establishment and maintenance of the maize root quiescent center. Plant Physiol 140:1118–1125

Jiang K, Meng YL, Feldman LJ (2003) Quiescent center formation in maize roots is associated with an auxin-regulated oxidizing environment. Development 130:1429–1438

Jimenez VM (2001) Regulation of *in vitro* somatic embryogenesis with emphasis on the role of endogenous hormones. R Bras Fisiol Veg 13:196–223

Joo JH, Bae YS, Lee JS (2001) Role of auxin-induced reactive oxygen species in root gravitropism. Plant Physiol 126:1055–1060

Joy RW, Patel KR, Thorpe TA (1988) Ascorbic acid enhancement of organogenesis in tobacco callus. Plant Cell Tiss Org Cult 13:219–228

Kairong C, Gengsheng X, Xinmin L, mGengmei X, Yafu W (1999) Effect of hydrogen peroxide on somatic embryogenesis of *Lycium barbarum* L. Plant Sci 146:9–16

Kato N, Esaka M (1996) cDNA cloning and gene expression of ascorbate oxidase in tobacco. Plant Mol Biol 30:833–837

Kato N, Esaka M (1999) Changes in ascorbate oxidase gene expression and ascorbate levels in cell division and cell elongation in tobacco cells. Physiol Plant 105:321–329

Kerk NM, Feldman LJ (1995) A biochemical model for the initiation and the maintenance of the quiescent center: implications for organization of root meristems. Development 121:2825–2833

Kerk NM, Jiang K, Feldman LJ (2000) Auxin metabolism in a root distal meristem. Plant Physiol 112:925–932

Kraus WL, Lis JT (2003) PARP goes transcription. Cell 113:677–683

Kreuger M, van Holst G-J (1993) Arabinogalactan proteins are essential in somatic embryogenesis of *Daucus carota* L. Planta 189:243–248

Li S, Xue L, Xu S, Feng H, An L (2007) Hydrogen peroxide involvement in formation and development of adventitious roots in cucumber. Plant Growth Regul 52:173–180

Li S, Xue L, Xu S, Feng H, An L (2009a) Hydrogen peroxide acts as a signal molecule in the adventitious root formation of mung bean seedlings. Environ Exp Bot 65:63–71

Li S, Xue L, Xu S, Feng H, An L (2009b) IBA-induced changes in antioxidant enzymes during adventitious rooting in mung bean seedlings: The role of H_2O_2. Environ Exp Bot 66:442–450

Liao W, Xiao H, Zhang M (2009) Role and relationship of nitric oxide and hydrogen peroxide in adventitious root development of marigold. Acta Phys Plant 31:1279–1289

Lin LS, Varner JE (1991) Expression of ascorbic acid oxidase in zucchini squash (*Cucurbita pepo* L.). Plant Physiol 96:159–165

Liso R, Calabrese G, Bitonti MB, Arrigoni O (1984) Relationship between ascorbic acid and cell division. Exp Cell Res 150:314–20

Liso R, Innocenti AM, Bitonti MB, Arrigoni O (1988) Ascorbic acid induced progression of quiescent centre cells from G1 to S phase. New Phytol 110:469–471

Liso R, De Tullio MC, Ciraci S, Balestrini R, La Rocca N, Bruno L, Chiappetta A, Bitonti MB, Bonfante P, Arrigoni O (2004) Localization of ascorbic acid, ascorbic acid oxidase, and glutathione in roots of *Cucurbita maxima* L. J Exp Bot 55:2589–2597

Liszkay A, van der Yalm E, Schopfer P (2004) Production of reactive oxygen intermediates ($O_2^{\cdot-}$, H_2O_2 and $^{\cdot}OH$) by maize roots and their role in wall loosening and elongation growth. Plant Physiol 135:3114–3123

Liu X, Shiomi S, Nakatsuka A, Kubo Y, Nakamura R, Inaba A (1999) Characterization of ethylene biosynthesis associated with ripening in banana fruit. Plant Physiol 121:1257–1266

Lopéz-Carbonell M, Munné-Bosch S, Alegre L (2006) The ascorbate-deficient *vtc-1 Arabidopsis* mutant shows altered ABA accumulation in leaves and chloroplast. J Plant Growth Regul 25:137–144

Ludwig-Müller J, Verticnik A, Town CD (2005) Analysis of indole-3-butyric acid-induced adventitious root formation on *Arabidopsis* stem segments. J Exp Bot 56:2095–2105

Lynn K, Fernandez A, Aida M, Sedbrook J, Tasaka M, Masson P, Barton MK (1999) The *PINHEAD/ZWILLE* gene acts pleiotropically in *Arabidopsis* development and has overlapping functions with the *ARGONAUTE1* gene. Development 126:469–481

Menges M, Murray JAH (2002) Synchronous *Arabidopsis* suspension cultures for analysis of cell-cycle activity. Plant J 30:203–212

Moussian B, Schoof H, Haecker A, Jürgens G, Laux T (1998) Role of *ZWILLE* gene in the regulation of central shhot meristem cell fate during *Arabidopsis* embryogenesis. EMBO J 17:1799–1899

Namasivayam P (2007) Acquisition of embryogenic competence during somatic embryogenesis. Plant Cell Tissue Organ Cult 90:1–8

Navabpour S, Morris K, Allen R, Harrison E, Mackerness S, Buchanan-Wollaston V (2003) Expression of cenescence-enhanced genes in response to oxidative stress. J Exp Bot 54:2285–2292

Nishiwaki M, Fujino K, Koda Y, Masuda K, Kikuta Y (2000) Somatic embryogenesis induced by the simple application of abscisic acid to carrot (*Daucus carota* L.) seedlings in culture. Planta 211:756–759

Noctor G, Foyer CH (1998) Ascorbate and glutathione: keeping active oxygen under control. Annu Rev Plant Physiol Plant Mol Biol 49:249–279

Nomura K, Matsumoto S, Masuda K, Inoue M (1998) Reduced glutatione promotes callus growth and shoot development in a shoot tip culture of apple root stock M26. Plant Cell Rep 17:597–600

Ogawa K, Hatano-Iwasaki A, Yanagida M, Iwabuchi M (2004) Level of Glutathione is regulated by ATP-dependent ligation of glutamate and cysteine through photosynthesis in *Arabidopsis thaliana*: mechanism of strong interaction of light intensity and flowering. Plant Cell Physiol 45:1–8

Ogawa K, Tasaka Y, Mino M, Tanaka Y, Iwabuchi M (2001) Association of glutathione with flowering in *Arabidopsis thaliana*. Plant Cell Physiol 42:524–530

Ohtani M, Sugiyama M (2005) Involvement of SRD2-mediated activation of snRNA transcription in the control of cell proliferation competence in *Arabidopsis*. Plant J 43:479–490

Ozawa S, Tasutani I, Fukuda H, Komamine A, Sugiyama M (1998) Organogenic responses in tissue culture of *srd* mutants of *Arabidopsis thaliana*. Development 125:135–142

Padu E (1999) Apoplastic peroxidases, ascorbate and lignification in relation to nitrate supply in wheat stem. J Plant Physiol 154:576–583

Pagnussat GC, Simontacchi M, Puntarulo S, Lamattina L (2002) Nitric oxide is required for root organogenesis. Plant Physiol 129:954–956

Pal Singh H, Kaur S, Batish DR, Kohli RK (2009) Caffeic acid inhibits in vitro rooting in mung bean [*Vigna radiata* (L.) Wilczek] hypocotyls by inducing oxidative stress. Plant Growth Regul 57:21–30

Pellny TK, Locato V, Vivancos PD, Markovic J, De Gara L, Pallardó FV, Foyer CH (2009) Pyridine nucleotide cycling and control of intracellular redox state in relation to poly (ADP-ribose) polymerase activity and nuclear localization of glutathione during exponential growth of *Arabidopsis* cells in culture. Mol Plant 2:442–456

Philips GC (2004) *In vitro* morphogenesis in plants – recent advances. In Vitro Cell Dev Biol Plant 40:342–345

Pignocchi C, Flether JM, Wilkinson JE, Barnes JD, Foyer CH (2003) The function of ascorbate oxidase in tobacco. Plant Physiol 132:1631–1641

Pignocchi C, Kiddle G, Hernández I, Foster SJ, Asensi A, Taybi T, Barnes J, Foyer CH (2006) Ascorbate oxidase-dependent changes in the redox state of the apoplast modulate gene transcript accumulation leading to modified hormone signaling and orchestration of defense processes in tobacco. Plant Physiol 141:423–435

Potters G, De Gara L, Asard H, Horemans N (2002) Ascorbate and glutathione: guardians of cell cycle, partners in crime? Plant Physiol Biochem 40:537–548

Potters G, Horemans N, Bellone S, Caubergs RJ, Trost P, Guisez Y, Asard H (2004) Dehydroascorbate influences the plant cell cycle through a glutathione-independent mechanism. Plant Physiol 134:1479–1487

Potters G, Horemans N, Caubergs RJ, Asard H (2000) Ascorbate and dehydroascorbate influence cell cycle progression in tobacco cell suspension. Plant Physiol 124:17–20

Razdan MK (2003) Somatic embryogenesis. In: Razdan MK (ed) Introduction to plant tissue culture, 2nd edn. Science Publishers Inc., Enfield, USA, pp 71–86

Sánchez M, Queijeiro E, Revilla G, Zarra I (1997) Changes in ascorbic acid levels in apoplastic fluid during growth of pine hypocotyls. Effect of peroxidase activities associated with cell walls. Physiol Plant 101:815–820

Sánchez-Fernández R, Fricker M, Corben LB, White NS, Sheard N, Leaver CJ, Van Montagu M, Inze D, May MJ (1997) Cell proliferation and hair tip growth in the *Arabidopsis* root are under mechanistically different forms of redox control. Proc Natl Acad Sci USA 94:2745–2750

Schopfer P (2002) Hydroxyl radical-induced cell-wall loosening in vitro and in vivo: implications for the control of elongation growth. Plant J 28:679–688

Schopfer P, Liszkay A, Bechtold M, Frahry G, Wagner A (2002) Evidence that hydroxyl radicals mediate auxin-induced extension growth. Planta 214:821–828

Schopfer P, Plachy C, Frahry G (2001) Release of reactive oxygen intermediates (superoxide radicals, hydrogen peroxide and hydroxyl radicals) and peroxidases in germinating radish seeds controlled by light, gibberellin and abscisic acid. Plant Physiol 125:1591–1602

Skoog F, Miller CO (1957) Chemical regulation of growth and organ formation in plant tissue cultures *in vitro*. Symp Soc Exp Biol 11:118–131

Stasolla C, Belmonte MF, Tahir M, Elhiti M, Khamiss K, Joosen R, Maliepaard C, Sharpe A, Gjetvaj B, Boutilier K (2008) Buthionine sulphoximine (BSO)-mediated improvement in cultured embryo quality *in vitro* entails changes in ascorbate metabolism meristem development and embryo maturation. Planta 228:255–272

Stasolla C, Belmonte MF, van Zyl L, Craig DL, Liu W, Yeung EC, Sederoff RR (2004) The effect of reduced glutathione on morphology and gene expression of white spruce (*Picea glauca*) somatic embryos. J Exp Bot 55:695–709

Stasolla C, Katahira R, Thorpe TA, Ashihara H (2003) Purine and pyrimidine nucleotide metabolism in higher plants. J Plant Physiol 160:1271–1295

Stasolla C, Kong L, Yeung EC, Thorpe T (2002) Maturation of somatic embryos in conifers: morphogenesis, physiology, biochemistry, and molecular biology. In Vitro Cell Dev Biol Plant 38:93–105

Stasolla C, Yeung EC (1999) Ascorbic acid improves conversion of white spruce somatic embryos. In Vitro Cell Dev Biol Plant 35:316–319

Stasolla C, Yeung EC (2001) Ascorbic acid metabolism during white spruce somatic embryo maturation and germination. Physiol Plant 111:196–205

Stasolla C, Yeung EC (2007) Cellular ascorbic acid regulates the activity of major peroxidases in the apical poles of germinating white spruce (*Picea glauca*) somatic embryos. Plant Physiol Biochem 45:188–198

Sugiyama M (1999) Organogenesis *in vitro*. Curr Opin Plant Biol 2:61–64

Sugiyama M (2000) Genetic analysis of plant morphogenesis *in vitro*. Int Rev Cytol 196:67–84

Sundås-Larsson A, Svenson M, Liao H, Engström P (1998) A homeobox gene with potential developmental control function in the meristem of the conifer *Picea abies*. Proc Natl Acad Sci USA 95:15118–15122

Synková H, Semorádová Š, Burketová L (2004) High content of endogenous cytokinin stimulates activity of enzymes and proteins involved in stress response in *Nicotiana tabacum*. Plan Cell Tissue Organ Cult 79:169–179

Synková H, Semorádová Š, Schnablová R, Witters E, Hušák E, Valcke R (2006) Cytokinin-induced activity of antioxidant enzymes in transgenic P*ssu-ipt* tobacco during plant ontogeny. Biol Plant 50:31–41

Takahama U, Oniki T (1992) Regulation of peroxidases-dependent oxidation of phenolics in the apoplast of spinach leaves by ascorbate. Plant Cell Physiol 33:379–387

Tanimoto E (2005) Regulation of root growth by plant hormones-roles for auxin and gibberellin. Crit Rev Plant Sci 24:249–265

Tao L, Cheung AY, Wu H (2002) Plant Rac-like GTPases are activated by auxin and mediate auxin-responsive gene expression. Plant Cell 14:2745–2760

Tewari RK, Hahn E-J, Paek K-T (2008) Function of nitric oxide and superoxide anion in the adventitious root development and antioxidant defence in *Panax ginseng*. Plant Cell Rep 27:563–573

Tewari RK, Lee SY, Hahn E-J, Paek K-T (2007) Temporal changes in the growth, saponin content and antioxidant defense in the adventitious roots of Panax ginseng subjected to nitric acid elicitation. Plant Biotech Rep 1:227–235

Tian M, Gu Q, Zhu M (2003) The involvement of hydrogen peroxide and antioxidant enzymes in the process of shoot organogenesis of strawberry callus. Plant Sci 165:701–707

Tian M, Han N, Bian HW, Zhu M (2004) The possible relationship between the regeneration capacity and reactive oxygen species in the strawberry calli. Acta Hort Sin 31:372–374

Tyburski J, Jasionowicz P, Tretyn A (2006) The effects of ascorbate on root regeneration in seedling cuttings of tomato. Plant Growth Regul 48:157–173

Tyburski J, Krzemiński Ł, Tretyn A (2008) Exogenous auxin affects ascorbate metabolism in roots of tomato seedlings. Plant Growth Regul 54:203–215

Tyburski J, Tretyn A (2010) Glutathione and glutathione disulphide affect adventitious root formation and growth in tomato seedling cuttings. Acta Phys Plant 32:411–417

Vernoux T, Wilson RC, Seeley KA, Reichheld J-P, Muroy S, Brown S, Maughan SC, Cobbett CS, Van Montagu M, Inze D, May MJ, Sung ZR (2000) The *ROOT MERISTEMLESS 1/CADMIUM SENSITIVE 2* gene defines a glutathione-dependent pathway involved in initiation and maintenance of cell division during postembryonic root development. Plant Cell 12:97–109

Visser EJW, Heijink CJ, van Hout JGM, Voesenek LACJ, Barendse GWM, Blom CWPM (1995) Regulatory role of auxin in adventitious root formation in two species of *Rumex*, differing in their sensitivity to waterlogging. Physiol Plant 93:116–122

Warren G (1993) The regeneration of plants from cultured cells and tissues. In: Stafford A, Warren G (eds) Plant cell and tissue culture. Wiley, New York, pp 82–100

Woodward AW, Bartel B (2005) Auxin: regulation, action, interaction. Ann Bot 95:707–735

Yanagida M, Mino M, Iwabuchi M, Ogawa K (2004) Reduced glutathione is a novel regulator of vernalization-induced bolting in the rosette plant *Eustoma grandiflorum*. Plant Cell Physiol 45:129–137

Zavaleta-Mancera HA, Lopez-Delgado H, Loza-Tavera H, Mora-Herrera M, Trevilla-Garcia C, Vargas-Suares M, Ougham H (2007) Cytokinin promotes catalase and ascorbate peroxidase activities and preserves the chloroplast integrity during dark-senescence. J Plant Physiol 164:1572–1582

Zhao XY, Su YH, Cheng ZJ, Zhang XS (2008) Cell fate switch during *in vitro* plant organogenesis. J Int Plant Biol 50:816–824

Chapter 3
Role of Ascorbate Peroxidase and Glutathione Reductase in Ascorbate–Glutathione Cycle and Stress Tolerance in Plants

Cai-Hong Pang and Bao-Shan Wang

Abstract Ascorbate–glutathione (AsA–GSH) cycle is an important component of the scavenging system for reactive oxygen compounds in plants. The member of this pathway involves the antioxidants: AsA, GSH and the antioxidatnt enzymes such as ascorbate peroxidase (APX), monodehydroascorbate reductase (MDAsAR), dehydroascorbate reductase (DAsAR) and glutathione reductase (GR). APX and GR are the key enzymes of the AsA–GSH cycle. APX utilizes AsA as the electron donor reducing H_2O_2 to water, and prevents the accumulation of a toxic level of H_2O_2 in photosynthetic organisms under stress conditions. GR converts oxidized glutathione (GSSG) to GSH using NAD(P)H as an electron donor. And thus a highly ratio of GSH/GSSG and AsA/DAsA is maintained at the intracellular level by this reaction during oxidative stress. In general, APX and GR activity have been shown to increase in various plant species under different stress conditions. Transgenic plants showed that APX and GR play an important role in providing resistance to oxidative stress caused by diferent stressors such as paraquat, methyl viologen, ozone, drought, heavy metals, high light, salinity, and chilling. In this review, recent progress in research on APX, GR and stress tolerance will be discussed.

Keywords Ascorbate peroxidase • Glutathione reductase • AsA-GSH cycle • Stress tolerance

C.-H. Pang
Shandong Academy of Forestry, Jinan 250014, Shandong, P.R. China

B.-S. Wang (✉)
Key Lab of Plant Stress Research, College of Life Sciences,
Shandong Normal University, Jinan 250014, Shandong, P.R. China
e-mail: bswang@sdnu.edu.cn

N.A. Anjum et al. (eds.), *Ascorbate-Glutathione Pathway and Stress Tolerance in Plants*,
DOI 10.1007/978-90-481-9404-9_3, © Springer Science+Business Media B.V. 2010

1 Properties and Functions of Ascorbate Peroxidase and Glutathione Reductase of the AsA–GSH cycle in plants

In plants, the glutathione–ascorbate cycle operates in the cytosol, mitochondria, plastids and peroxisomes (Jimenez et al. 1998; Meyer 2009). In the first step of the cycle, H_2O_2 is reduced to water by APX using AsA as the specific electron donor. At the same time AsA was oxidized to monodehydroascorbate (MDAsA). MDAsA is directly reduced to AsA by monodehydroascorbate reductase (MDAsAR) or spontaneously disproportionated to dehydroascorbate (DAsA). DAsA is reduced to AsA by dehydroascorbate reductase (DAsAR) at the expense of GSH, yielding GSSG. Finally, GSH must be regenerated using GSSG which is reduced by GR using NADPH as electron donor. Thus AsA and GSH are not consumed; the net electron flow is from NADPH to H_2O_2. We simplified the ascorbate–glutathione cycle in a plant cell in Fig. 1. The ascorbate–glutathione cycle provides a mechanism for the dissipation of the excess of reducing power in plant cell, in particular in chloroplasts under stress. This cycle play an immportant role in keeping the equilibrium between the ROS production and scavenging (Pang and Wang 2007). High levels of both ascorbate and glutathione pools are a requisite for this cycle (Anderson et al. 1983).

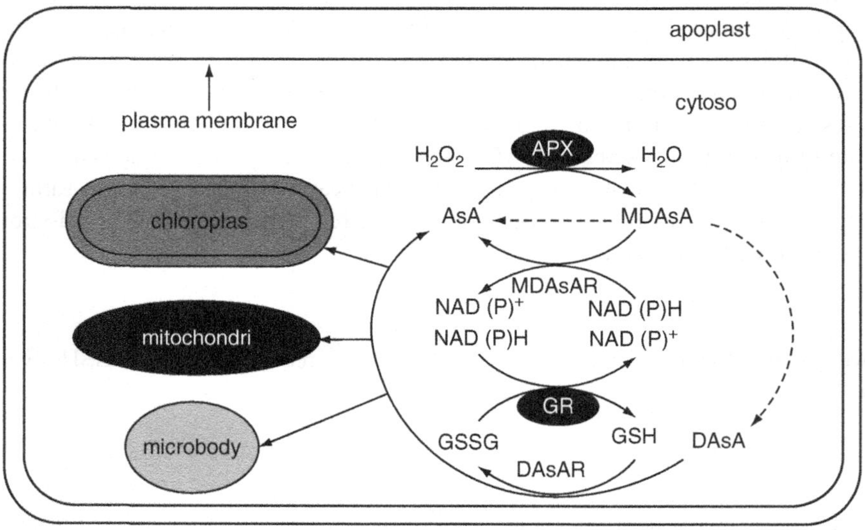

Fig. 1 Simplified scheme of AsA–GSH cycle in plant cellsThis cycle is mainly localized in the cytosol, chloroplasts, mitochondria and microbody. APX and GR play a critical role in AsA–GSH cycle. In the AsA–GSH cycle, *solid arrows* represent enzyme-mediated reactions and *dotted arrows* indicate spontaneous reactions. Abbreviations: AsA, Ascorbate; MDAsA, monodehydroascorbate; DAsA, dehydroascorbate; MDAsAR, monodehydroascorbate reductase; DAsAR, dehydroascorbate reductase; GSH, reduced glutathione; GSSG, oxidized glutathione; APX, ascorbate peroxidase; GR, glutathione reductase

2 Characteristics of Ascorbate Peroxidase

APXs (EC1.11.1.11) are enzymes that detoxify H_2O_2 using AsA as a substrate. The reaction they catalyzed is the transfer of electrons from AsA to a H_2O_2, producing dehydroascorbate and water (Raven 2000). The reaction is as follows:

$$\text{Ascorbate} + \text{Hydrogenperoxide} \rightarrow \text{Dehydroascorbate} + \text{Water}$$

APX isoenzymes have high affinity for AsA as electron donors. One of the most characteristic properties of APX, which distinguishes it from guaiacol peroxidase, Cyt *c* peroxidase and glutathione peroxidase, is its instability in the absence of AsA. When the concentration of AsA is lower than 20 μM, APX was inactivated rapidly and lost their stability (Shigeoka et al. 2002). Jespersen et al. (1997) suggested that tryptophan-175 of chloroplast APX increased their substrate specificity. So the chlAPX is more labile than cAPX in the absence of AsA. The half-life of chlAPX in ascorbate-depleted media is about 15 s, while those of cAPX isoforms are about 60 min (Miyake and Asada 1996; Yoshimura et al. 1998). The instability of chlAPX may be one reason that APX was not found in photosynthetic orgnisms for a long time.

Ascorbate peroxidase is a haem-containing enzyme whose prosthetic group is protoporphyrin. Iron plays an important role in the catalytic site. This was clearly shown by applying iron deficiency conditions in sugar beet (Zaharieva and Abadia 2003). Cyanide and azide are the inhibitors of APX (Shigeoka et al. 1980). APX is also inhibited by thiolmodifying reagents (ρ-chloromercuribenzoate), thiol and suicide inhibitors such as hydroxyurea and ρ-aminophenol (Chen and Asada 1990).

Some functional amino acid residues have been identified in the APX catalytic domain by analysis site-directed mutagenesis. Arg172 was changed to lysine, glutamine and asparagine, the variants of APX are incapable of oxidizing AsA to form Compound II. It is suggested that Arg172 of pea cAPX plays a key role in AsA utilization (Bursey and Poulos 2000). Arg172, Lys30 are the location of ascorbate binding site and the hydrogen-bonding interactions (Sharp et al. 2003). Arg-38 has a possible functional role in the control of substrate binding and orientation (Raven et al. 2001). The active site structures, including the hydrogen-bonding interactions between the proximal His-163 ligand, a buried Asp-208 residue, and a Trp-179 residue, are same to cytochrome c peroxidase (CCP). The difference between the APX and CCP is the presence of a cation binding site in APX (Patterson and Poulos 1995). Glu 112 forms a salt bridge with Lys20 and Arg24 of the opposing subunit near the axis of dyad symmetry between the subunits and is related to the dimer interface including an alteration in solvent structure (Mandelman et al. 1998a). AsA oxidation does not occur at the exposed heme edge but at an alternate binding site in the vicinity of Cys32 near Arg172 and the heme propionates (Mandelman et al. 1998b).

2.1 Isoforms and Subcelluar Localization of Acorbate Peroxidase

As shown in Fig. 1, APX isoenzymes are distributed in at least four distinct cellular compartments: cytosolic APX (cAPX), chloroplast APX (chlAPX) including stromal APX (sAPX) and thylakoid membrane-bound APX (tAPX), microbody membrane-bound APX (mAPX, including glyoxysome and peroxisome), and mitochondrial APX (mitAPX, a membrane-bound form). APX has also been found in the protozoan *Trypanosoma cruzi* (Boveris et al. 1980) and the bovine eye (Wada et al. 1998).

The subcellular localization of APX isoforms is determined by the signal peptides and transmembrane domains in the N- and C- terminal regions (Shigeoka et al. 2002; Teixeira et al. 2004). A novel pumpkin APX isoenzyme was found to be localized on membranes of microbodies (Yamaguchi et al. 1995) and spinach glyoxysomes APX was found to be located on the external side of the organelles (Ishikawa et al. 1998). The targeting signal of peroxisomal APX comprises a C-terminal transmembrane domain (TMD, rich in valine and alanine) and the followed positively charged domain containing five amino acid rsidues (Mullen and Trelease 2000). These isoforms are indirectly sorted to the peroxisomes via a subdomain of the rough endoplasmic reticulum (Lisenbee et al. 2003). chlAPX isoforms possess a transit peptide consisting of approximately 70 residues in the N-terminus rich in Ser and Thr (Shigeoka et al. 2002). At C-terminal, tAPX isoforms present a transmembrane hydrophobic domain which is responsible for spanning to the stroma-exposed thylskoid membranes in chloroplasts (Ishikawa et al. 1996). Chew et al. (2003) showed that the *Arabidopsis* stromal APX is dual-targeted to chloroplasts stroma and mitochondria. An additional methionine residue 25 residues from the initiator methionine resulted in an additional band upon translation.This is maybe the reason that the APX isoform produces an ambiguity targeting peptide in the N-terminus.

2.2 Ascorbate Peroxidase Family and its Evolution in Plants

On the basis of the amino acid sequences plant peroxidases were classified into three classes (Welinder 1992). Class I contains the intracellular peroxidase of prokaryotic origin including yeast cytochrome *c* peroxidase (CCP) (Dabrowska et al. 2007; Shigeoka et al. 2002), class II consists of secretory fungal peroxidases which include ligninases, or lignin peroxidases (LiPs), and manganesedependent peroxidases (MnPs). And class III consists of the secretory plant peroxidases. APXs belong to the class I family (Dabrowska et al. 2007; Shigeoka et al. 2002). They are found in chloroplastic organisms and putatively in cyanobacterial (Miyake et al. 1991).

The phylogenetic tree was constructed using ClustalW software based on the deduced amino acid sequences of the APX isoenzymes. It was found that the APX isoforms can be divided into four groups (two kinds of cAPX, chlAPX and mAPX) (Shigeoka et al. 2002). Diversification of cAPX and chlAPX was the first step of the APX evolution. They have the same originand and were resulted from duplication events in the basal Virdiplantae (Shigeoka et al. 2002; Passardi et al. 2007; Dabrowska et al. 2007). ChlAPX can be divided into two subclasses: tAPX and sAPX. Peroxisomal APX isoforms are only present in land plants. They probably originated from cytosolic isoforms that gained an exon encoding the membrane-bound domain and the targeting peptide (Teixeira et al. 2004). APX isoforms in higher plants show high homology (70–90%) within each group and 50–70% homology with each other (Shigeoka et al. 2002).

2.3 *Ascorbate Peroxidase-Encoding Genes in Higher Plants*

The genes encoding APXs in higher plants have been isolated and characterized. Seven APX genes were identified in the tomato genome including three cytosolic isoforms, two peroxisomal isoforms and two chloroplastic ones (Najami et al. 2008). Rice APX gene famiy is comprised of genes encoding two cytosolic, two putative peroxisomal, and four putative chloroplastic APXs (Teixeira et al. 2006). Nine genes were suggested as APX genes in *Arabidopsis thaliana* which include three cytosolic, three peroxisomal and two chloroplastic (one thylakoid-bound and one whose product is targeted to both chloroplast stroma and mitochondria) (Chew et al. 2003).

There are two or more putative cAPXs in higher plants.The pea cAPX gene (*APX1*) is the one whose structure is best known (Mittler and Zilinskas 1992). *APX1* is encoded by a single copy gene and contains nine introns which have a high AT nucleotides content. The first intron is in the 5′UTR region. Steady state *APX1* transcript level was increased by drought, heat, and application of ethephon, abscisic acid, and the superoxide-generating agent paraquat. This phenomenon is possibly in relation to the regulatory elements in the promoter of *APX1*. These include two repeats of a putative plant heat-shock element (HSE) and a reversed antiperoxidative element (ARE). In addition, two repeats of a putative xenobiotic responsive element (CACGCA) were also found at the positions: −282 and −302 and a putative GPEI enhancer was found at position 97. Recent analysis of knockout plants deficient in cytosolic APX1 (KO-Apx1) revealed that cAPX plays a key role in protecting chloroplasts under light stress conditions (Pnueli et al. 2003; Davletova et al. 2005). The cAPX isoforms are homodimers. Subunit molecular weight of rice cytosolic APX1 and APX2 is 28.5 and 34 kDa respectively (Sharma and Dubey 2004). The molecular weight of pea cAPX is 57.5 kDa and its subunit is 29.5 kDa (Mittler and Zilinskas 1991).

Ascorbate peroxidase gene, ecoding spinach chlAPX isoenzymes, consists of 13 exons and 12 introns. Exons 1 ~ 11 encode the common amino acid sequence of sAPX and tAPX. There are two splice acceptor sites in the exon 12 and 13 seperated by 14 bp nucleotides. Exon 12 contains one condon for Asp-365 before the termination codon and the entire 3′-UTR including potential polyadenylation signal (AATAA) of sAPX mRNA. The final exon consists of the sequence coding hydrophobic C-terninal region,TGA terminal codon and the entire 3′-UTR including potential polyadenylation signal (AATATA) of tAPX mRNA. By alternative splicing of the intron 11 and 12 in the 3′-termina, two kinds of mRNA are generated by one chlAPX gene. The N-termianl amino acids of sAPX are completely identical with those of tAPX, while c-terminal amino acids differ. A similar finding was also made in spinach (Ishikawa et al. 1997), pumpkin (Mano et al. 1997) tabacco and *Suaeda salsa* (data unpublished). While in *Arabidopsis thaliana* sAPX and tAPX are encoded by different genes. The nucleotide sequences of their cDNAs share only 66.1% identity (Jespersen et al. 1997). APX isoforms located in organelle are monomer with different molecular weight. tAPX with a molecular weight of 40 kDa is about 10 kDa higher than that of the sAPX from spinach chloroplasts (Miyake et al. 1993). The reason for difference in molecular mass between tAPX and sAPX is related to the transmembrane domain in the C-terminal.

The spatial expression patterns of the APX genes were studied in tomato (Najami et al. 2008). Among the cytosolic APX genes, *SlAPX1* was mainly expressed in the root, *SlAPX2* in the root, stem, and fruit while *SlAPX3* in the stem and fruit. Interestingly, the relative abundance of the three cytosolic *APX* transcripts, and especially that of *SlAPX3*, was lower in the leaves than in the root, stem, and fruit. Transcripts of *SlAPX4* were found to be accumulated mostly in the leaves and stem, and *SlAPX5* in the leaves, stem, and fruit. Of the two chloroplastic genes, *SlAPX6* was highly expressed in leaves, less in fruit, stem and root and *SlAPX7* was contrary. It suggested that the expression pattern of APX genes is different at tissue level.

3 Characteristics of Glutathione Reductase

Glutathione reductase (GR, EC1.6.4.2), which converts oxidized glutathione (GSSG) to reduced glutathione (GSH) using NADPH, is ubiquitous in living organisms. It is necessary for maintaining the high ratio of GSH/GSSG in the plant cells and accelerating the H_2O_2 scavenging pathway in plants particularly under stress conditions (Smith et al. 1989). GR catalyses following reaction:

$$\text{GSSG} + \text{NADPH} + \text{H}^+ = 2\,\text{GSH} + \text{NADP}^+$$

GR has high specificities for its substrates, although some glutathione conjugates and mixed glutathione disulphides can be reduced by GR (Gaullier et al. 1994). Most GRs from plants have a high affinity for NADPH (<10 μM). GRs can reduce GSSG by using NADH too, but the affinities for NADH is very low (Halliwell and Foyer 1978). The affinity of GR for GSSG is varied from 10 to 7,300 μM (Mullineaux and Creissen 1997). GR activity is regulated by the combined effect of pH, NADPH and GSSG. Low concentration of NADPH inactivated the activity of GR from yeast below pH 5.5 and over 7.0, while no significant inhibition was found at pH values between these two values (Pinto et al. 1984, 1985). Wheat Chloroplastic GR was inhibited by NADPH about 93% up to pH 7.5, the inhibition was little at pH 8.0–9.5. The NADPH-dependent inhibition was almost completely prevented by GSSG (Lascano et al. 1999). The Km of GR isoforms for GSSG is different. This may be one mechanism of GR in response to stress (Edwards et al. 1994).

Glutathione reductase has been purified and characterized from different sources such as bacteria, fungi, plants and human. They belong to the flavor-protein oxidoreductase family (Takeda et al. 1993). The molecular weight of plant GRs ranges from 60 to 190 kDa (Mullineaux and Creissen 1997). GRs from different organisms have different quateruary structure. Most of GR isoforms are homodimer (Wingsle 1989; Anderson et al. 1990; Edwards et al. 1990; Madamanchi et al. 1992). While in pea and maize GR is a heterodimer (Kalt-Torres et al. 1984; Mahan and Burke 1987). Takeda et al. (1993) found that Chlamydamaonas reinhardtii GR is a monomer. GR dimers may be further assembled into tetramers or even higher aggregative states. Temperature and pH determine the type of higher-order state. This may be one of the regulation mechanisms of GR activity (Mullineaux and Creissen 1997).

As a general rule, GR is composed of four domains: a FAD-binding domain, a NADPH-binding domain, GSSG binding domain and an interface domain (Karplus and Schulz 1987; Chalapathi Rao and Reddy 2008). Two arginine residues (Arg287 and Arg293) required for binding of NADPH are conserved (Scrutton et al. 1990; Tang and Webb 1994). The two cysteines of the GR for forming the redox-active disulphide bridge are separated by four amino acid residues in a highly conserved motif GGTCV (I/V) RGCVPKK (I/L) LVY GP (Creissen et al. 1991; Rouhier et al. 2006). The Cys residue responsible for the inter-subunit disulphide bond in the human enzyme is replaced with Serl64 in higher plants such as pea (Creissen et al. 1991).

3.1 Isoforms and Subcellular Localization of Glutathione Reductase

Glutathione reductase isoforms have been reported in many plant species such as tobacco (Foyer et al. 1991), spinach (Guy and Carter 1984), eastern white pine (Anderson et al. 1990), pea (Edwards et al. 1994), *Chlamydomonas reinhardtii* (Serrano and Liobell 1993) and wheat (Lascano et al. 2001). GR has been mainly localized in chloroplasts, mitochondria and the cytosol (Edwards et al. 1990; Madamanchi et al. 1992; Creissen et al. 1995; Jiménez et al. 1997; Rudhe et al. 2004). It was also found in peroxisomes (Jiménez et al. 1997; del Río et al. 2002). Generally, GR in the chloroplast has higher activities (Stevens et al. 2000) and about 80% of GR activities in leaf tissues are accounted for chloroplastic isoforms (Edwards et al. 1990).

Different signal peptides in the GR direct its targeting to different cell compartments. In the N-termina of *Arabidopsis thaliana* GR, there is a 74 amino acids leader sequence, which is rich in hydroxylated amino acids Ser and Thr, leading the protein transit into chloroplast (Kubo et al. 1993). A 52-amino acid signal peptide was found in the GR of cowpea leaves and soybean, while pea GR signal peptide was 60 amino acids at the N-terminus (Contour-Ansel et al. 2006). This region is rich in Ser (15%), Thr (15%) and Ala (10%), indicating that this may be a chloroplastidic, rather than a mitochondria1 transit peptide (von Heijne et al. 1989). GR is a matrix-associated enzyme in mitochondria and peroxisomes from pea leaves

(Jiménez et al. 1997). At the same time, pea GR was the first dual targeting protein reported. It imported into either chloroplasts or mitochondria owing to an ambiguous-targeting sequence located at the *N*-terminus (Romero-Puertas et al. 2006). But the cotargeting function parts of the signal peptide sequence could not be identified (Creissen et al. 1995). Dual targeting signals are more hydrophobic than mitochondrial or chloroplastic targeting signals (Peeters and Small 2001).

3.2 Glutathione Reductase Family and Its Evolution in Plants

Glutathione reductase is conserved between all kingdoms and was thought to be diverged from lipoamide dehydrogenase. It has been found in many eukaryotes and prokaryotes. In bacteria, yeasts and animals, only one GR gene is found. In plant genomes, there are two genes encoding GR. Surprisingly, Drosophila and Trypanosomes do not have any GR at all (Kanzok et al. 2001).

The cytosolic GR genes from *Brassica campestris* (Lee et al. 2002), *Pisum sativum* (Stevens et al. 1997), *Oryza sativa* (Kaminaka et al. 1998) and chloroplastic/mitochondrial GR genes from *Arabidopsis thaliana* (Kubo et al. 1993), *Pisum sativum* (Mullineaux et al. 1996) are reported to be a single copy. However, chloroplastic/mitochondrial GR gene from root nodule is composed of a multigene family (Tang and Webb 1994). They have similar coding regions but different 3′ untranslated regions.

Jiang et al. (1995) compared GR sequences from *Arabidopsis thaliana*, *Pisum sativum*, *Anabaena* PCC7120, *Pseudomonas aeruginosa*, *Escherichia coli* and human. GR of *Anabaena* PCC7120 exhibits higher similarity to plants GR than to those from bacteria and human. Codons which are low in G + C content are preferentially used in *Anabaena* PCC 7120 and plants.

According to the dendrogram of mature protein sequences of various plant GRs (Chen et al. 2007), it is found that the GRs in the same cellular organs have closer genetic relationship than those in the same plants. Cowpea cytosolic GR shared 89.2%, 80.2% and 74.6% identity with pea, *Arabidopsis* and rice cytosolic GR, respectively (Contour-Ansel et al. 2006). When the two subgroups are compared, the homology was lower (Rouhier et al. 2006).The homology of the GR sequences from *Arabidopsis thaliana* cytosol and chloroplast is only 55%.

3.3 Glutathione Reductase-Encoding Genes in Higher Plants

The genes of the GRs from a variety of plants are classified into two groups. One is the gene encoding cytosolic GR (Lee et al. 1998, 2002; Stevens et al. 2000) and the other accounts for organelle isoforms (Mullineaux et al. 1996; Rudhe et al. 2002; Chew et al. 2003; Contour-Ansel et al. 2006).

Two genes were identified in *Nicotiana tobaccum* (Creissen and Mullineaux 1995), *Arabidopsis thaliana* (Kubo et al. 1993), *Pisum sativum* (Creissen et al. 1991; Stevens et al. 1997) and three genes in *Oryza sativa* and *Populus trichocarpa* (Rouhier et al. 2006). The structures of cytosolic GR genes are more comprehensive than the organelle isoforms. The cytosolic GR genes of *Brassica campestris* (Lee et al. 2002) and *Oryza sativa* (Kaminaka et al. 1998) are composed of 17 exons and 16 introns while chloroplastic/mitochondrial GR genes from *Pisum sativum* (Mullineaux et al. 1996) and *Arabidopsis thaliana* (Lee et al. 2002) contain 10/9 and 11/10 exons/introns, respectively. The size of open reading frame of the genes varies from 2,744 bp of *Arabidopsis thaliana* to 5,733 bp of *Oryza sativa* (Lee et al. 2002).

Glutathione reductase from plants contain an unknown function C-terminal extension about 22–26 amino acids compared to the C-terminus of the human, *Escherichia coli*, and *Pseudomonas aeruginosa* enzymes (Creissen et al. 1991; Kubo et al. 1993). This domain may be in part responsible for the appearance of different enzyme isoforms in plants exposed to a variety of environmental stresses (Creissen and Mullineaux 1995). Furthermore, two ABA-responsive elements (ABREs) were observed in the 5′-flanking region of the RGRC2 gene. Expression of the RGRC2 gene was strongly induced by ABA treatment and drought stress, but weakly by salt stress and chilling treatment. And a parallel change was observed between the amounts of mRNA and protein of RGRC2 (Kaminaka et al. 1998).

4 Regulation of Ascorbate Peroxidase and Glutathione Reducatse Under Environmental Stress

4.1 Changes in Ascorpate Peroxidase and Glutathione Rreducatse Activity

Ascorpate peroxidase and GR play a crucial role in determining the tolerance of a plant under envirmental stress conditions such as salt, drought, high light, extremes of temperature, ozone and paraquat. However, the results about changes in APX and GR acticity of plants under stress conditions are controversial. Several studies showed that the activity of APX and GR was increased in plants under stress conditions. While others suggested that the activity was decreased. The changes of the activity may be due to the protein synthesis or by the changed kinetic properties of the enzymes (Lascano et al. 1998, 1999).

Activity of APXs and GRs is differently responsed to salinity. Under salinity stress, increased activity of APX and GR was reported in *Casuarina* (Desingh et al. 2006), pea (Hernandez et al. 2000), citrus (Gueta-Dahan et al. 1997), soybean (Khan et al. 2009), rice (Lin and Kao 2000; Vaidyanathan et al. 2003), *Cicer arietinum* (Kukreja et al. 2005), tomato (Mittova et al. 2004), wheat (Sairam et al. 2005), *Setaria italica* (Sreenivasulu et al. 2000) and cotton (Desingh and Kanagaraj 2007).

The activities of APX and GR in barley were increased in the root and shoot under the NaCl stress. But the increase was more significant and consistent in the root, indicating rapid responses of antioxidant enzymes to salt stress in barley roots (Kim et al. 2005b). The roots may be the primary site of salt stress. Maintaining ion homeostasis and redox potential is critical for the normal root growth and salinity resistance (Greenway and Munns 1980; Hasegawa et al. 2000). Salt stress markedly enhanced APX and GR activity in the shoots of salt-tolerant potato, while the activity decreased in salt-sensitive cultivar. The reduction of the APX and GR activity in shoots may be responsible for the sensitivity of the cultivar to salt stress (Aghaei et al. 2009). APX and GR activity remained unchanged in the roots of salt-tolerant maize and decreased in the salt-sensitive genotype (Azevedo-Neto et al. 2006). Higher APX and GR activities in a salt-tolerant cultivar were reported in rice (Kim et al. 2005a; Vaidyanathan et al. 2003), soybean (Khan et al. 2009), bean (Yasar et al. 2008), wheat (Sairam et al. 2005; Mandhania et al. 2006) and cotton (Gossett et al. 1994). APX and GR activity in the cholorplast of the halophyte *Suaeda salsa* leaves was distinctly increased with 200 mM NaCl treatment (Pang et al. 2005). There is a relative high APX and GR activity under environmental stress in tolerant species. This is possibly an important tolerant mechanism of plants to stress conditions.

Increased APX and GR activity during drought was reported in various plant species, including barley (Smirnoff and Colombe 1988), maize (Jiang and Zhang 2002; Pastori and Trippi 1992), wheat (Sairam et al. 1997,1998; Selote and Chopra 2006), rice (Selote and Chopra 2004; Sharma and Dubey 2005; Srivalli et al. 2003), *Tortula ruralis* (Dhindsa 1991). GR activities significantly increased under drought stress in drought tolerant *Lycopersicon peruvianum* while decreased in drought sensitive genotype. APX activity decreased under all stress treatments in both species (Ünyayar et al. 2005). Under water stress, drought tolerant genotype C 306 had higher APX and GR activity in comparison to susceptible genotype (Sairam et al. 1997, 1998). A similar finding was also detected in rice (Selote and Chopra 2004) and tomato (Ünyayar et al. 2005). In rice seedlings, the activities of GR and APX showed increases with increasing levels of drought stress (Sharma and Dubey 2005). While the activities of APX and GR decrease under drought in pea nodules (Gogorcena et al. 1995). The decline may result from a restricted supply of NAD(P)H in vivo for the ascorbate–glutathione pathway and from the Fe-catalyzed Fenton reactions of ascorbate and glutathione with activated oxygen (Gogorcena et al. 1995). In desert plant *Reaumuria soongorica* (Pall.) maxim, APX and GR activities declined in the initial stages of drought process, elevated significantly with further increasing water deficit progression and recovered after rewatering (Bai et al. 2009). In seedlings of *Helianthus annuus*, GR activity decreased under drought stress but recovered with ABA treatment (Kele and Ünyayar 2004).Under drought conditions, the activities of plastidial APX and GR were lower than those of cytosolic fractions in *Populus euramericana*. At same time the percent of inhibition of plastidial APX and GR was higher in a drought-sensitive cultivar than that of a drought-tolerant one (Edjolo et al. 2001). GR activity increased in the symplastic and apoplastic areas of the leaf. As for the petiole, GR activity increased in the apoplastic area but decreased in the symplastic area (Saruhan et al. 2009). Therefore, the changes in activity of APXs and GRs are dependent on the plant species, cultivars, organs and stress degree under drought condition.

High or low temperature also affects activities of GR and APX. Activities of APX and GR of tolerant cultivar rice were remained high, while those of a susceptible cultivar decreased under chilling (Huang and Guo 2005). In a high temperature sensitive species *Eupatorium adenophorum*, APX and GR activities were increased at low temperature and were not increased during the heat treatments. In a low temperature sensitive species *Eupatorium odoratum*, increase of the APX and GR were observed in heat-treated plants, but the antioxidant enzymes were unable to operate in cold stress (Lu et al. 2008).

Ascorpate peroxidase and GR activities of plants are also affected by heavy metal, photooxidative stress and oxidative stress. The activity of APX and GR in sunflower were significantly suppressed by Cd treatments in the first 24 h. After 48 h, these activities were strongly increased particularly at 50 and 100 μM Cd. whereas the activities of the APX were declined with 96 h treatment (Hatata and Abdel-Aal 2008). Under photooxidative stress, the activity of APX in green barley (*Hordeum vulgare*) leaves was increased, while GR axtivity did not change (Kozel and Shalygo 2009). Rice seedlings fumigated by ozone, activities of GR and APX of ascorbate–glutathione cycle were higher in sensitive rice cultivar than those in the tolerant cultivar (Lin et al. 2001).

These data suggested that the changes of APX and GR activity were not only related to different organs and plant species but also to stress conditions and/or a given period of exposure. The reasons for the changes in APX and GR activities induced by stressors are related to gene expression, enzyme mofications and signal transduction.

4.2 *Changes in Ascorbate Peroxidase and Glutathione Reducatse at Transcript Level*

Under stress conditions, the changes of the APX and GR activity is different. The changes of the activity may be due to the mRNA level and the enzyme protein amount or by the changed kinetic properties of the enzymes. Many reports showed that environmental stressors affected APX and GR transcript level. Under salt stess, the transcript level of *cAPX* and stromal *APX* was increased in the third leaves of rice seedings while transcripts of the thylakoid-bound *APX* genes were down-regulated (Kim et al. 2007). The mRNA levels of chloroplast APX in euhalophyte *Suaeda salsa* was up-regulated by NaCl treatment (data unpublished). In spinach leaves (Yoshimura et al. 2000), the transcript level of *cAPX*, *sAPX*, *tAPX* and microbody-bound *APX* were not changed in response to salinity. In rice seedings cytosolic *GR* was strongly induced by salinity (Kaminaka et al. 1998; Kim et al. 2007). Exogenously applied H_2O_2 and treatment with NaCl enhanced the expression of *GR* in rice roots (Tsaia et al. 2005). Cytosolic *GR* was strongly induced in the NaCl-tolerant pea but not in the NaCl-sensitive variety under salt stress (Hernandez et al. 2000).

The transcript level of *APXs* is regulated by drought. The steady-state transcript level of pea *cAPX* increased during drought stress but was even more dramatically enhanced following recovery from drought, and this was accompanied by a slight enhanced APX protein content and APX activity (Mittler and Zilinskas 1994). In spinach leaves (Yoshimura et al. 2000), the transcript level of *cAPX*, *sAPX*, *tAPX* and microbody-bound *APX* were not changed in response drought. In rice seedings cytosolic *GR* was strongly induced by drought (Kaminaka et al. 1998). Drought stress induced an upregulation of the expression of *cGR* in tolerant and susceptible cowpea cultivars. Whereas the progression of the drought treatment down-regulated *dtGR* expression of tolerant cowpea cultivar and stimulated *dtGR* expression in the susceptible cultivar (Contour-Ansel et al. 2006). The level of transcripts was relatively higher for cGR than for the dual isoform.

Heat or low temperature also affects the transcript level of APXs and GRs. In the pea (*Pisum sativum*), the steady state transcript levels of *APXI* was found to increase in response to heat. While the protein level of cAPX was unchanged (Mittler and Zilinskas 1992). Increase in APX activity is possibly at the posttranslational level. The expression of theAPX*1* gene from *Arabidopsis thaliana* was induced by heat treatment (Storozhenko et al. 1998). The level of the chlorplastic APX was poorly modified by heat stress (Panchuk et al. 2002). In rice seedings cytosolic *GR* was strongly induced by chilling (Kaminaka et al. 1998). Baek and Skinner (2003) demonstrated that expression level of *GR* transcripts was increased during cold acclimation in wheat plants. In the pea leaves (Romero-Puertas et al. 2006), chloroplastic *GR* (*GR1*) and cytosolic *GR* (*GR2*) were dramatically induced by low temperature. Moreover, chloroplastic *GR* (*GR1*) was repressed by high temperature.

The regulation of *APXs* and *GRs* transcript level also occurs under stressors such as high light and oxidative stress. In *Arabidopsis thaliana*, *cAPX* was induced by excess light stress and the transcript levels of *GORl* did not change markedly during the excessive light and the poststress period (Karpinski et al. 1997). The transcript level of cytosolic *APX* was significantly increased when suspension cultures of germinating rice embryos were treated with paraquat or H_2O_2 (Morita et al. 1999; Donahue et al. 1997). Paraquat also increased the cytosolic APX mRNA in pea (*Pisum* sativum) (Donahue et al. 1997), while paraquat treatment caused a decline in transcript abundance of plastidic *GR*. Exposure of plants to 0.1 ppm ozone enhanced the APX1 mRNA level within 3 h, which showed a diurnal rhythm (Kubo et al. 1995). In spinach leaves (Yoshimura et al. 2000), the transcript level of *cAPX* remarkably increased in response to high-light and methyl viologen (MV). While the sAPX, tAPX and microbody-bound APX were not changed in response to any of the stress treatments at the levels of transcript. During pathogen-induced programmed cell death in tobacco leaves, c*APX* expression was suppressed by inhibition of protein synthesis (Mittler et al. 1998). In the pea leaves (Romero-Puertas et al. 2006), chloroplastic *GR* (*GR1*) and cytosolic *GR* (*GR2*) appeared to be repressed by high light and continuous dark.

When peas (*Pisum sativum*) was treated with cold conditions (4°C) in conjunction with light, paraquat, ozone, and illumination of etiolated seedlings (greening), none of the applied stresses had any effect on steady-state levels of GR mRNA. Except for ozone fumigation, the activity of GR was increased. The K_m of GR for glutathione disulphide showed a marked decrease (Edwards et al. 1994). This indicates that GR from stressed plants has an increased affinity for glutathione disulphide. GR transcripts levels and protein did not show significant changes respect to controls under cadmium stress, while the GR activity in roots was increased (Yannarelli et al. 2007). These results are in agreement with those reported in Cd-stressed peas by Romero-Puertas et al. (2006). It suggests that post translational modification could occur. After heat, MV and H_2O_2 treatments, the activities of cAPX, sAPX, mAPX increased after a slight decline throughout the experiment. Consistent closely with sAPX activity, the expression of *sAPX* followed a similar change pattern, indicating that sAPX was regulated at the transcriptional level. In contrast, constitutive expression was observed in tAPX activity and no significant changes in tAPX activity were found throughout the experiment (Song et al. 2005).

In conclusion, it is likely that the expression patterns of the APX and GR isoenzymes are individually regulated at each cellular compartment under various environmental stress conditions. There are a number of different results about the expression of the APX and GR isoenzymes in plants. For this reason further research is necessary to be done, in particular, in the characteristics of pomoters of the APX and GR isoenzymes.

4.3 *Signal for the Regulation of Ascorbate Peroxidase and Glutathione Reducatse*

Under normal conditions, production and removal of ROS must be strictly controlled in different cellular compartments. However, the equilibrium between production and scavenging of ROS may be perturbed by abiotic stress factors such as salinity, high light, drought, low temperature, high temperature, metal and mechanical stress. As a result of these disturbances, ROS level may rapidly rise. High concentration of ROS can lead to phytotoxicity whereas relatively low levels can be used for signaling.

4.3.1 Signal Molecules of Oxidative Stress

Hydrogen peroxide (H_2O_2) has been implicated as a signal in a wide range of abiotic stress responses. In germinating rice embryos, when APX and CAT were inhibited by hydroxyurea or aminotriazole, cellular H_2O_2 content was elevated and cAPX mRNA was markedly increased without paraquat or H_2O_2 treatment. This suggests that cytosolic APX is regulated by the H_2O_2 level within cells (Morita et al. 1999).

H_2O_2 also induced the expression of APX gene in *Arabidopsis thaliana* leaves (Karpinski et al. 1999). These results indicate that H_2O_2 is involved in oxidative stress signalling, leading to the induction of APX. While Hong et al (2007), Tsaia et al (2005) and Sung et al (2009) demonstrated that H_2O_2 is not involved in the regulation of *APX* and *GR* expression. But they cannot exclude the involvement of H_2O_2 in this signaling pathway. It seems that other factors execute the function with H_2O_2 for up-regulation of APX and GR transcription under stress conditions (Sung et al. 2009). Induction of *APX2* expression has been suggested to be mediated by ABA in *Arabidopsis thaliana* (Fryer et al. 2003). Kaminaka et al (1998) suggested that the expression of the rice cytosolic GR gene is regulated via ABA-mediated signal transduction under environmental stresses.

4.3.2 Sensors of Reactive Oxygen Species-Perception

Under oxidative stress, ROS production in plant cells including extracellular and intracellular spaces can be sensed by three different mechanisms: ROS receptors, redox-sensitive transcription factors (such as NPR1 or HSFs) and direct inhibition of phosphatases by ROS (Mittler et al. 2004). The expression of the stress-sensitive cultivated tomato Sl*APX1* and Sl*APX2* decreased in response to NaCl stress (Najami et al. 2008). At*APX1* (putative co-ortholog of these tomato genes) of the glycophyte *Arabidopsis thaliana* was also not up-regulated in response to NaCl stress (Mittler et al. 2004). The cadmium-induced increase in Sl*APX2* transcript level suggests that different sensors, which probably transduce the signals along different transduction pathways, sense cadmium and NaCl stress (Najami et al. 2008).

4.3.3 Reactive Oxygen Species-Signal Transduction Cascade

Reactive oxygen species-receptors sense the ROS signal and results in the generation of Ca^{2+} signals. And then Ca^{2+} activates the protein kinase OXI1. The activation of OXI1 results in the activation of MAPK casade. MAPK casade induced or activated different transcript factors (WRKY, Zat, RAV, GRAS and Myb families) regulating the ROS-scavenging pathway (Mittler et al. 2004).

H_2O_2 may directly oxidise redox-sensitive transcription factors including HSF, NPR1 in either the cytoplasm or the nucleus (Neill et al. 2002). Activated transcription factors would subsequently interact with cognate H_2O_2-response elements and modulate gene expression such as *APX* and *GR*. Alternatively, redox-sensitive transcription factors activated MAPK casade. ROS is sensed via modification of thiol groups in certain proteins either. The inactivation of *Arabidopsis* protein tyrosine phosphatases (AtPTP1) by H_2O_2 may be mediated by the oxidation state of the active-site cysteine. AtPTP1 can inactivate the *Arabidopsis* MPK6 (Gupta and Luan 2003). When oxidative stress occurred, AtPTP1 was inactivated by H_2O_2. Then the MAPK casade was activated and transcription factors were regulated. Inhibition of phosphatases by ROS might result in the activation of kinases of OXI1 or MAPK casade.

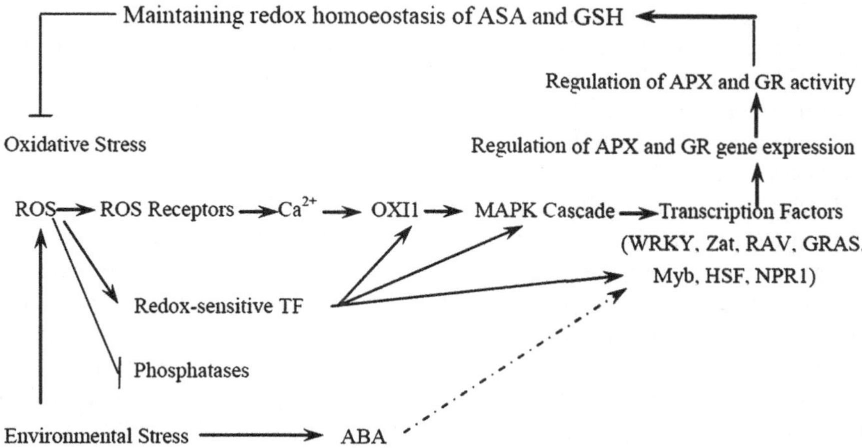

Fig. 2 A possible reactive oxygen species-signal transduction pathway of ascorbate peroxidase and glutathione reducatse regulation in plant cells under environmental stress conditionReactive oxygen species can be sensed by three different mechanisms: ROS receptors, redox-sensitive transcription factors and direct inhibition of phosphatases. Detection ROS by receptors results in the generation of Ca^{2+} signals. And then Ca^{2+} activates the protein kinase OXI1. The activation of OXI1 results in the activation of MAPK casade. MAPK casade induced or activated different transcript factors regulating the APX and GR. H_2O_2 may directly oxidise redox-sensitive transcription factors. Activated transcription factors would subsequently interact with cognate H_2O_2-response elements and modulate gene expression such as *APX* and *GR*. Alternatively, redox-sensitive transcription factors activated MAPK casade. Inhibition of phosphatases by ROS might result in the activation of kinases of OXI1 or MAPK casade. ABA might be involved in the signal transduction pathway. Abbreviations: TF, transcription factors; HSF, heat shock factors; Zat, zinc-finger protein; NPR1, non-expressor of PR1; OXI1, serine/threonine protein kinase

Thus, it can be seen that transcription factors and the MAPK casade are key components in ROS signal transduction. A possible signal transduction pathway of regulation of APX and GR activity under environmental stress is proposed (Fig. 2). The expression of different transcription factors is enhanced by ROS and this includes members of the WRKY, Zat, RAV, GRAS and Myb families (Mittler et al. 2004). Large-scale transcriptome analyses coupled with proteomic and metabolomic analyses of plants perturbed at the levels of individual or multiple components of the environmental stress induced ROS network will be essential for future studies.

5 Perspectives

Biochemical and molecular studies have shown that APX and GR are important enzymes for scavenging the ROS, which are continuously generated in different compartments during environmental stresses. The changes of APX and GR activity were not only related to different organs and plant species but also to stress conditions and/or a given period of exposure. APX and GR activity are relative higher in tolerant species than those in sensitive species under environmental stress. This is possibly an important tolerant mechanism of plants to stress conditions. Further experiment should be done in the realtionship between plant stress tolerance and APX and GR activity, in particular, ragulation mechanism of APX and GR activity under stress.

During environmental stress, the changes of APX and GR activities may be due to the protein synthesis or isoform population. It is important to characterize which isoform is crucial for a particular stress and how the expression of that particular isoform is regulated. At the same time the cooperation mechanism of different isoforms at different compartments for scavenging the ROS is also required to study.

At the present stage, the signal transduction pathway that regulates the expression of APX and GR has been reported in some plants under stress conditions. But there are still many uncertainties in understanding of the signal transduction under stress conditions such as ROS sensors and components of the signal transduction. This needs more direct proof.

Acknowledgements We are grateful for financial support from the NSFC (National Natural Science Research Foundation of China, project No. 30870138) and the HI-Tech Research and Development (863) Program of China (2007AA091701).

References

Aghaei K, Ehsanpour AA, Komatsu S (2009) Potato responds to salt stress by increased activity of antioxidant enzymes. J Integ Plant Biol 51(12):1095–1103

Anderson JV, Hess JL, Chevone BI (1990) Purification, characterization and immunological properties for two isoforms of glutathione reductase from eastern white pine needles. Plant Physiol 94:1402–9

Anderson JW, Foyer CH, Walker DA (1983) Light dependent reduction of hydrogen peroxide by intact spinach chloroplasts. Biochim Biophys Acta 724:69–74

Azevedo-Neto AD, Ptisco JT, Eneas-Filho J, Abreu CEB, Gomez-Filho E (2006) Effect of salt stress on antioxidative enzymes and lipid peroxidation in leaves and roots of salt-tolerant and salt-sensitive maize genotypes. Environ Exp Bot 56:87–94

Baek KH, Skinner DZ (2003) Alteration of antioxidant enzyme gene expression during cold acclimation of near-isogenic wheat lines. Plant Sci 165:1221–1227

Bai J, Gong CM, Chen K, Kang HM, Wang G (2009) Examination of antioxidative system's responses in the different phases of drought stress and during recovery in desert plant *Reaumuria soongorica* (Pall.) maxim. J Plant Biol 52:417–425

Boveris A, Sies H, Martino EE, Docampo R, Turrens JF, Stoppani AO (1980) Deficient metabolic utilization of hydrogen peroxide in *Trypanosoma cruzi*. Biochem J 188:643–648

Bursey EH, Poulos TL (2000) Two substrate binding sites in ascorbate peroxidase: the role of arginine 172. Biochemistry 39:7374–7379

Chalapathi Rao ASV, Reddy AR (2008) Glutathione reductase: a putative redox regulatory system in plant cells. In: Khan NA, Singh S, Umar S (eds) Sulfur assimilation and abiotic stress in plants. Springer, Berlin, Heidelberg, pp 111–147

Chen GX, Asada K (1990) Hydroxyurea and p-aminophenol are the suicide inhibitors of ascorbate peroxidase. J Biol Chem 265:2775–2781

Chen YP, Xing LP, Wu GJ, Wang HZ, Wang XE, Cao AZ, Chen PD (2007) Plastidial glutathione reductase from *Haynaldia villosa* is an enhancer of powdery mildew resistance in wheat (*Triticum aestivum*). Plant Cell Physiol 48:1702–1712
Chew O, Whelan J, Millar AH (2003) Molecular definition of the ascorbate-glutathione cycle in *Arabidopsis* mitochondria reveals dual targeting of antioxidant defenses in plants. J Biol Chem 278:46869–46877
Contour-Ansel D, Torres-Franklin ML, De Carvalho MHC, D'Arcy-Lameta A, Zuily-Fodil Y (2006) Glutathione reductase in leaves of cowpea: cloning of two cDNAs, expression and enzymatic activity under progressive drought stress, desiccation and abscisic acid treatment. Ann Bot 98:1279–1287
Creissen GP, Reynolds H, Xuem Y, Mullineaux PM (1995) Simultaneous targeting of pea glutathione reductase and of a bacterial fusion protein to chloroplasts and mitochondria in transgenic tobacco. Plant J 8:167–175
Creissen GP, Edwards EA, Enard C, Wellburn A, Mullineaux PM (1991) Molecular characterization of glutathione reductase cDNAs from pea (*Pisum sativum* L.). Plant J 2:129–31
Creissen GP, Mullineaux PM (1995) Cloning and characterization of glutathione reductase cDNAs and identification of two genes encoding the tobacco enzyme. Planta 197:422–5
Dabrowska G, Katai A, Goc A, Szechynska-Hebda M, Skrzypek E (2007) Characteristics of the plant ascorbate peroxidase family. Acta Biol Cracov Ser Bot 49:7–17
Davletova S, Rizhsky L, Liang H, Zhong SQ, Oliver DJ, Coutu J, Shulaev V, Schlauch K, Mittlera R (2005) Cytosolic ascorbate peroxidase 1 is a central component of the reactive oxygen gene network of *Arabidopsis*. Plant Cell 17:268–281
del Río LA, Corpas FJ, Sandalio LM, Palma JM, Gómez M, Barroso JB (2002) Reactive oxygen species, antioxidant systems and nitric oxide in peroxisomes. J Exp Bot 53:1255–1272
Desingh R, Jutur PP, Reddy AR (2006) Salinity stress-induced changes in photosynthesis and antioxidative systems in three casuarina species. J Plant Biol 33:1–2
Desingh R, Kanagaraj G (2007) Influence of salinity stress on photosynthesis and antioxidative systems in two cotton varieties. Gen Appl Plant Physiol 33:221–234
Dhindsa RS (1991) Drought stress, enzymes of glutathione metabolism, oxidative injury and protein synthesis in *Tortula ruralis*. Plant Physiol 95:648–5
Donahue JL, Okpodu CM, Cramer CL, Grabau EA, Alscher RG (1997) Responses of antioxidants to paraquat in pea leaves. Plant Physiol 113:249–257
Edjolo A, Laffray D, Guerrier G (2001) The ascorbate-glutathione cycle in the cytosolic and chloroplastic fractions of drought-tolerant and drought-sensitive poplars. J Plant Physiol 158:1511–1517
Edwards EA, Enard C, Creissen GP, Mullineaux PM (1994) Synthesis and properties of glutathione reductase in stressed peas. Planta 192:137–143
Edwards EA, Rawsthorne S, Mullineaux PM (1990) Subcellular distribution of multiple forms of glutathione reductase in leaves of pea (*Pisum sativum* L.). Planta 180:278–284
Khan F, Siddiqi TO, Mahmooduzzafar AA (2009) Morphological changes and antioxidant defence systems in soybean genotypes as affected by salt stress. J Plant Interact 4:295–306
Foyer CH, Lelandais M, Galap C, Kunert KJ (1991) Effects of elevated cytosolic glutathione reductase activity on the cellular glutathione pool and photosynthesis in leaves under normal and stress conditions. Plant Physiol 97:863–872
Fryer MJ, Ball L, Oxborough K, Karpinski S, Mullineaux PM, Baker NR (2003) Control of Ascorbate peroxidase 2 expression by hydrogen peroxide and leaf water status during excess light stress reveals a functional organisation of *Arabidopsis* leaves. Plant J 33:691–705
Gaullier JM, Lafontant P, Valla A, Bazin M, Giraud M, Santus R (1994) Glutathione peroxidase and glutathione reductase activities towards glutathione-derived antioxidants. Biochem Biophys Res Commun 203:1668–1647
Gogorcena Y, Iturbe-Ormaetxe I, Escuredo PR, Becana M (1995) Antioxidant defenses against activated oxygen in pea nodules subjected to water stress. Plant Physiol 108:753–9

Gossett DR, Millhollon EP, Rucas MC, Banks SW, Marney MM (1994) The effects of NaCl on antioxidant enzyme activities in callus tissue of salt-tolerant and salt-sensitive cotton cultivars (*Gossypium hirsutum* L.). Plant Cell Rep 13:498–503
Greenway H, Munns H (1980) Mechanisms of salt tolerance in nonhalophytes. Annu Rev Plant Physiol 31:149–190
Gueta-Dahan Y, Yaniv Z, Zilinkas BA, Ben-Hayyim G (1997) Salt and oxidative stress: similar and specific responses and their relation to salt tolerance in citrus. Planta 203:460–469
Gupta R, Luan S (2003) Redox control of protein tyrosine phosphatases and mitogen-activated protein kinases in plants. Plant Physiol 132:1149–1152
Guy CL, Carter JV (1984) Characterization of partially purified glutathione reductase from cold-hardened and non hardened spinach leaf tissue. Cryobiology 21:454–464
Halliwell B, Foyer CH (1978) Properties and physiological function of a glutathione reductase purified from spinach leaves by affinity chromatography. Planta 139:9–17
Hasegawa PM, Bressan RA, Zhu JK, Bohnert H (2000) Plant cellular and molecular responses to high salinity. Annu Rev Plant Mol Biol 51:463–499
Hatata MM, Abdel-Aal EA (2008) Oxidative stress and antioxidant defense mechanisms in response to cadmium treatments. Am-Euras J Agric Environ Sci 4:655–669
Huang M, Guo Z (2005) Responses of antioxidative system to chilling stress in two rice cultivars differing in sensitivity. Biol Plant 49:81–84
Hernandez JA, Jimenez A, Mullineaux PM, Sevilla F (2000) Tolerance of pea (*Pisum sativum* L.) to long term salt stress is associated with induction of antioxidant defenses. Plant Cell Environ 23:853–862
Hong CY, Hsu YT, Tsai YC, Kao CH (2007) Expression of ASCORBATE PEROXIDASE 8 in roots of rice (*Oryza sativa* L.) seedlings in response to NaCl. J Exp Bot 58:3273–3283
Ishikawa T, Yoshimura K, Sakai K, Tamoi M, Takeda T, Shigeoka S (1998) Molecular characterization and physiological role of a glyoxysome-bound ascorbate peroxidase from spinach. Plant Cell Physiol 39:23–34
Ishikawa T, Yoshimura K, Takeda T, Shigeoka S, Sakai K (1996) cDNAs encoding spinach stromal and thylakoidbound ascorbate peroxidase, differing in the presence or absence of their 3′-coding regions. FEBS Lett 384:289–293
Ishikawa T, Yoshimura K, Tamoi T, Takeda T, Shigeoka S (1997) Alternative mRNA splicing of 3′-terminal exons generates ascorbate peroxidase isoenzymes in spinach (*Spinacia oleracea*) chloroplasts. Biochem J 328:795–800
Jespersen HM, Kjaersgård IV, Ostergaard L, Welinder KG (1997) From sequence analysis of three novel ascorbate peroxidases from *Arabidopsis thaliana* to structure, function and evolution of seven types of ascorbate peroxidase. Biochem J 326:305–310
Jiang FY, Hellman U, Sroga GE, Bergman B, Mannervik B (1995) Cloning sequencing and regulation of the glutathione reductase gene from the cyanobacterium *Anabaena* PCC 7120. J Biol Chem 270:22882–22889
Jiang M, Zhang J (2002) Water stress-induced abscisic acid accumulation triggers the increased generation of reactive oxygen species and up-regulates the activities of antioxidant enzymes in maize leaves. J Exp Bot 53:2401–2410
Jiménez A, Hernández JA, del Río LA, Sevilla F (1997) Evidence for the presence of the ascorbate-glutathione cycle in mitochondria and peroxisomes of pea leaves. Plant Physiol 114:175–284
Jimenez A, Hernandez JA, Pastori G, del Rio LA, Sevilla F (1998) "Role of the ascorbate-glutathione cycle of mitochondria and peroxisomes in the senescence of pea leaves. Plant Physiol 118:1327–1335
Kalt-Torres W, Burke JJ, Anderson JM (1984) Chloroplast glutathione reductase: purification and properties. Physiol Plant 61:271–278
Kaminaka H, Morita S, Nakajima M, Masmura T, Tanaka K (1998) Gene cloning and expression of cytosolic glutathione reductase in rice (*Oryza sativa* L.). Plant Cell Physiol 39:1269–1280
Kanzok SM, Fechner A, Bauer H, Ulschmid JK, Müller HM, Botella-Munoz J, Schneuwly S, Schirmer R, Becker K (2001) Substitution of the thioredoxin system for glutathione reductase in *Drosophila melanogaster*. Science 291:643–646

Karpinski S, Escobar C, Karpinska B, Creissen G, Mullineaux PM (1997) Photosynthetic electron transport regulates the expression of cytosolic ascorbate peroxidase genes in *Arabidopsis* during excess light stress. Plant Cell 9:627–640

Karpinski S, Reynolds H, Karpinska B, Wingsle G, Creissen G, Mullineaux P (1999) Systemic signaling and acclimation in response to excitation energy in Arabidopsis. Science 284:654–657

Karplus PA, Schulz GE (1987) Refined structure of glutathione reductase at 1.54 A resolution. J Mol Biol 195:701–729

Kim DW, Rakwal R, Agrawal GK, Jung YH, Shibato J (2005a) A hydroponic rice seedling culture model system for investigating proteome of salt stress in rice leaf. Electrophoresis 26:4521–4539

Kim SY, Lim JH, Park MR, Kim YJ, Park TI, Seo YW, Choi KG, Yun SJ (2005b) Enhanced antioxidant enzymes are associated with reduced hydrogen peroxide in barley roots under saline stress. J Biochem Mol Biol 38:218–224

Kim DW, Shibato J, Agrawa GK, Fujihara S, Iwahashi H, Kim DH, Shim IS, Rakwal R (2007) Gene transcription in the leaves of rice undergoing salt-induced morphological changes (*Oryza sativa* L.). Mol Cells 24:45–59

Kele Y, Ünyayar S (2004) Responses of antioxidant defense system of *Helianthus annuus* to abscisic acid treatment under drought and waterlogging. Acta Physiol Plant 26:149–156

Kozel NV, Shalygo NV (2009) Barley leaf antioxidant system under photooxidative stress induced by Rose Bengal. Russ J Plant Physl 56:316–322

Kubo A, Saji H, Tanaka K, Kondo N (1995) Expression of Arabidopsis cytosolic ascorbate peroxidase gene in response to ozone or sulfur dioxide. Plant Mol Biol 29:479–489

Kubo A, Sano T, Saji H, Tanaka K, Kondo N, Tanaka K (1993) Primary structure and properties of glutathione reductase from *Arabidopsis thaliana*. Plant Cell Physiol 34:1259–1266

Kukreja S, Nandwal AS, Kumar N, Sharma SK, Sharma SK, Unvi V, Sharma PK (2005) Plant water status, H_2O_2 scavenging enzymes, ethylene evolution and membrane integrity of *Cicer arietinum* roots as affected by salinity. Biol Plant 49:305–308

Lascano HR, Gomez LD, Casano LM, Trippi VS (1998) Changes in glutathione reductase activity and protein content in wheat leaves and chloroplasts exposed to photooxidative stress. Plant Physiol Biochem 36:321–329

Lascano HR, Gomez LD, Casano LM, Trippi VS (1999) Wheat chloroplastic glutathione reductase activity is regulated by the combined effect of pH, NADPH and GSSG. Plant Cell Physiol 40:683–90

Lascano HR, Casano LM, Melchiorre MN, Trippi VS (2001) Biochemical and molecular characterisation of wheat chloroplastic glutathione reductase. Biol Plant 44:509–516

Lee H, Jo J, Son D (1998) Molecular cloning and characterization of the gene encoding glutathione reductase in *Brassica campestris*. Biochim Biophys Acta 1395:309–314

Lee H, Won SH, Lee BH, Park HD, Chung WI, Jo J (2002) Genomic cloning and characterization of glutathione reductase gene from *Brassica campestris* var. Pekinensis. Mol Cells 13:245–251

Lin CC, Kao CH (2000) Effect of NaCl stress on H_2O_2 metabolism in rice leaves. Plant Growth Regul 30:151–155

Lin DI, Lur HS, Chu C (2001) Effects of abscisic acid on ozone tolerance of rice (*Oryza sativa* L.) seedlings. Plant Growth Regul 35:295–300

Lisenbee CS, Heinze M, Trelease RN (2003) Peroxisomal ascorbate peroxidase resides within a subdomain of rough endoplasmic reticulum in wild-type *Arabidopsis* cells. Plant Physiol 132:870–882

Lu P, Sang WG, Ma KP (2008) Differential responses of the activities of antioxidant enzymes to thermal stresses between two invasive eupatorium species in China. J Integr Plant Biol 50:393–401

Madamanchi NR, Anderson JV, Alscher RG, Cramer CL, Hess JL (1992) Purification of multiple forms of glutathione reductase from pea (*Pisum sativum* L.) seedlings and enzyme levels in ozone-fumigated pea leaves. Plant Physiol 100:138–145

Mahan JR, Burke JJ (1987) Purification and characterization of glutathione reductase from corn mesophyll chloroplasts. Physiol Plant 71:352–358

Mandelman D, Jamal J, Poulos TL (1998a) Identification of two electron-transfer sites in ascorbate peroxidase using chemical modification, enzyme kinetics, and crystallography. Biochemistry 37:17610–17617

Mandelman D, Li HY, Poulos TL, Schwarz FP (1998b) The role of quaternary interactions on the stability and activity of ascorbate peroxidase. Protein Sci 7:2089–2098

Mandhania s, Madan S, Sawhney V (2006) Antioxidant defense mechanism under salt stress in wheat seedlings. Biol Plant 50:227–231

Mano S, Yamaguchi K, Hayashi M, Nishimura M (1997) Stromal and thylakoid-bound ascorbate peroxidases are produced by alternative splicing in pumpkin. FEBS Lett 413:21–26

Meyer A (2009) The integration of glutathione homeostasis and redox signaling. J Plant Physiol 165:1390–403

Mittler R, Feng X, Cohen M (1998) Post-transcriptional suppression of cytosolic ascorbate peroxidase expression during pathogen-induced programmed cell death in tobacco. Plant Cell 10:461–473

Mittler R, Zilinskas BA (1991) Purification and characterization of pea cytosolic ascorbate peroxidase. Plant Physiol 97:962–968

Mittler R, Zilinskas BA (1992) Molecular cloning and characterization of a gene encoding pea cytosolic ascorbate peroxidase. J Biol Chem 267:21802–21807

Mittler R, Zilinskas BA (1994) Regulation of pea cytosolic ascorbate peroxidase and other antioxidant enzymes during the progression of drought stress and following recovery from drought. Plant J 5:397–405

Mittler R, Vanderauwera S, Gollery M, Van Breusegem F (2004) The reactive oxygen gene network of plants. Trends Plant Sci 9:490–498

Mittova V, Guy M, Tal M, Volokita M (2004) Salinity up-regulates the antioxidative system in root mitochondria and peroxisomes of the wild salt-tolerant tomato species *Lycopersicon pennellii*. J Exp Bot 55:1105–1113

Miyake C, Asada K (1996) Inactivation mechanism of ascorbate proxidase at low concentrations of ascorbate; hydrogen peroxide decomposes compound I of ascorbate peroxidase. Plant Cell Physiol 37:423–430

Miyake C, Cao WH, Asada K (1993) Purification and molecular properties of the thylakoid-bound ascorbate peroxidase in spinach chloroplasts. Plant Cell Physiol 34:881–889

Miyake C, Michihata F, Asada K (1991) Scavenging of hydrogen peroxide in prokaryotic and eukaryotic algae: acquisition of ascorbate peroxidase during the evolution of cyanobacteria. Plant Cell Physiol 32:33–43

Morita S, Kaminaka H, Masumura T, Tanaka K (1999) Induction of rice cytosolic ascorbate peroxidase mRNA by oxidative stress; the involvement of hydrogen peroxide in oxidative stress signalling. Plant Cell Physiol 40:417–422

Mullen RT, Trelease RN (2000) The sorting signals for peroxisomal membrane-bound ascorbate peroxidase are within its C-terminal tail. J Biol Chem 275:16337–16344

Mullineaux PM, Creissen GP (1997) Glutathione reductase: regulation and role in oxidative stress. In: Scandalios JG (ed) Oxidative stress and the molecular biology of antioxidants. Cold Spring Harbor Laboratory Press, New York, pp 667–713

Mullineaux PM, Enard C, Hellens R, Creissen G (1996) Characterization of a glutathione reductase gene and its genetic locus from pea (*Pisum sativum* L.). Planta 200:186–194

Najami N, Janda T, Barriah W, Kayam G, Tal M, Guy M, Volokita M (2008) Ascorbate peroxidase gene family in tomato: its identification and characterization. Mol Genet Genomics 279:171–182

Neill S, Desikan R, Hancock J (2002) Hydrogen peroxide signalling. Curr Opin Plant Biol 5:388–395

Panchuk II, Volkov RA, Schöffl F (2002) Heat stress- and heat shock transcription factor-dependent expression and activity of ascorbate peroxidase in *Arabidopsis*. Plant Physiol 129:838–853

Pang CH, Zhang SJ, Gong ZZ, Wang BS (2005) NaCl treatment markedly enhances H_2O_2-scavenging system in leaves of halophyte *Suaeda salsa*. Physiol Plant 125:490–499

Pang CH, Wang BS (2007) Oxidative stress and salt tolerance. In: Luettge U, Beyschlag W, Murata J (eds) Progress in botany, vol 69. Springer-Verlag, Berlin Heidelberg, pp 231–245

Passardi F, Bakalovic N, Teixeira FK, Margis-Pinheiro M, Penel C, Dunand C (2007) Prokaryotic origins of the non-animal peroxidase superfamily and organelle-mediated transmission to eukaryotes. Genomics 89:567–579

Pastori GM, Trippi VS (1992) Oxidative stress induces high rate of glutathione reductase synthesis in a drought resistant maize strain. Plant Cell Physiol 33:957–961

Patterson WR, Poulos TL (1995) Crystal structure of recombinant pea cytosolic ascorbate peroxidase. Biochemistry 34:4331–4341

Peeters N, Small I (2001) Dual targeting to mitochondria and chloroplasts. Biochim Biophys Acta 1541:54–63

Pinto MC, Mata AM, Lopez-Barea J (1984) Reversible inactivation of *Saccharomyces cerevisiae* glutathione reductase under reducing conditions. Arch Biochem Biophys 228:1–12

Pinto MC, Mata AM, Lopez-Barea J (1985) The redox interconversion mechanism of *Saccharomyces cerevisiae* glutathione reductase. Eur J Biochem 15:275–278

Pnueli L, Liang H, Rozenberg M, Mittler R (2003) Growth suppression, altered stomatal responses, and augmented induction of heat shock protein in cytosolic ascobate peroxidase (Apx1) deficient *Arabidopsis* plants. Plant J 34:187–203

Raven EL (2000) Peroxidase-catalyzed oxidation of ascorbate. Structural, spectroscopic and mechanistic correlations in ascorbate peroxidase. Subcell Biochem 35:317–349

Raven EL, Çelik A, Cullis PM, Sangar R, Sutcliffe MJ (2001) Engineering the active site of ascorbate peroxidase. Biochem Soc Trans 29:105–111

Romero-Puertas MC, Corpas FJ, Sandalio LM, Leterrier M, Rodríguez-Serrano M, del Río LA, Palma JM (2006) Glutathione reductase from pea leaves: response to abiotic stress and characterization of the peroxisomal isozyme. New Phytol 170:43–52

Rouhier N, Couturier J, Jacquot JP (2006) Genome-wide analysis of plant glutaredoxin systems. J Exp Bot 57:1685–1696

Rudhe C, Chew O, Whelan J, Glaser E (2002) A novel in vitro system for simultaneous import of precursor proteins into mitochondria and chloroplasts. Plant J 30:213–220

Rudhe C, Clifton R, Chew O, Zeman K, Richter S, Lamppa G, Whelan J, Glaser E (2004) Processing of the dual targeted precursor protein of glutathione reductase in mitochondria and chloroplasts. J Mol Biol 343:639–647

Sairam RK, Deshmukh PS, Saxena DC (1998) Role of sntioxidant systems in wheat genotypes tolerance to water stress. Biol Plant 41:387–394

Sairam RK, Shukla DS, Saxena DC (1997) Stress-induced injury and antioxidant enzymes in relation to drought tolerance in wheat genotypes. Biol Plant 40:357–364

Sairam RK, Srivastava GC, Agarwal S, Meena RC (2005) Differences in antioxidant activity in response to salinity stress in tolerant and susceptible wheat genotypes. Biol Plant 49:85–91

Saruhan N, Terzi R, Saglam A, Kadioglu A (2009) The relationship between leaf rolling and ascorbate-glutathione cycle enzymes in apoplastic and symplastic areas of *Ctenanthe setosa* subjected to drought stress. Biol Res 42:315–326

Scrutton NS, Berry A, Perham RN (1990) Redesign of the coenzyme specificity of a dehydrogenase by protein engineering. Nature 343:38–43

Selote DS, Chopra RK (2004) Drought-induced spikelet sterility is associated with an inefficient antioxidant defense in rice panicles. Physiol Plant 121:462–471

Selote DS, Chopra RK (2006) Drought acclimation confers oxidative stress tolerance by inducing co-ordinated antioxidant defense at cellular and subcellular level in leaves of wheat seedlings. Physiol Plant 127:494–506

Serrano A, Liobell A (1993) Occurrence of two isoforms of glutathione reductase in the green alga *Chlamydomonas reinhardtii*. Planta 190:199–205

Sharma P, Dubey RS (2004) Ascorbate peroxidase from rice seedlings: properties of enzyme isoforms, effects of stresses and protective roles of osmolytes. Plant Sci 67:541–550

Sharma P, Dubey RS (2005) Drought induces oxidative stress and enhances the activities of antioxidant enzymes in growing rice seedlings. Plant Growth Regul 46:209–221

Sharp KH, Mewies M, Moody PCE, Raven EL (2003) Crystal structure of the ascorbate peroxidase–ascorbate complex. Nat Struct Biol 10:303–307

Shigeoka S, Ishikawa T, Tamoi M, Miyagawa Y, Takeda T, Yabuta Y, Yoshimura K (2002) Regulation and function of ascorbate peroxidase isoenzymes. J Exp Bot 53:1305–1319

Shigeoka S, Nakano Y, Kitaoka S (1980) Purification and some properties of L-ascorbic acid-specific peroxidase in *Euglena gracillis*. Arch Biochem Biophys 20:1221–1227

Smirnoff N, Colombe SV (1988) Drought influences the activity of enzymes of the chloroplast hydrogen peroxide scavenging system. J Exp Bot 39:1097–1108

Smith IK, Vierheller TL, Thorne CA (1989) Properties and functions of glutathione reductase in plants. Physiol Plant 77:449–456

Song XS, Hu WH, Mao WH, Ogweno JO, Zhou YH, Yu JQ (2005) Response of ascorbate peroxidase isoenzymes and ascorbate regeneration system to abiotic stresses in *Cucumis sativus* L. Plant Physiol Biochem 43:1082–1088

Sreenivasulu N, Grimma B, Wobusa U, Weschkea W (2000) Differential response of antioxidant compounds to salinity stress in salt-tolerant and salt-sensitive seedlings of foxtail millet (*Setaria italica*). Physiol Plant 109:435–442

Srivalli B, Sharma G, Chopra RK (2003) Antioxidative defense system in an upland rice cultivar subjected to increasing intensity of water stress followed by recovery. Physiol Plant 119:503–512

Stevens RG, Creissen GP, Mullineaux PM (1997) Cloning and characterization of a cytosolic glutathione reductase cDNA from pea (*Pisum sativum* L.) and its expression in response to stress. Plant Mol Biol 35:641–654

Stevens RG, Creissen GP, Mullineaux PM (2000) Characterization of pea cytosolic glutathione reductase expressed in transgenic tobacco. Planta 211:537–545

Storozhenko S, Pauw PD, Montague MV, Inze D, Kushnir S (1998) The heat-shock element is a functional component of the Arabidopsis APX1 gene promotor. Plant Physiol 118:1005–1014

Sung MS, Hsu YT, Hsu YT, Wu TM, Lee TM (2009) Hypersalinity and hydrogen peroxide upregulation of gene expression of antioxidant enzymes in *Ulva fasciata* against oxidative stress. Mar Biotechnol 11:199–209

Takeda T, Ishikawa T, Shigeoka S, Hirayama O, Mitsunaga T (1993) Purification and characterization of glutathione reductase from *Chlamydomonas reinhardtii*. J Gen Microbiol 139:2233–2238

Tang X, Webb MA (1994) Soybean root nodule cDNA encoding glutathione reductase. Plant Physiol 104:1081–1082

Teixeira FK, Menezes-Benavente L, Galvao VC, Margis R, Margis-Pinheiro M (2006) Rice ascorbate peroxidase gene family encodes functionally diverse isoforms localized in different subcellular compartments. Planta 224:300–314

Teixeira FK, Menezes-Benavente L, Margis R, Margis-Pinheiro M (2004) Analysis of the molecular evolutionary history of the ascorbate peroxidase gene family: inferences from the rice genome. J Mol Evol 59:761–770

Tsaia YC, Hongb CY, Liua LF, Kao CH (2005) Expression of ascorbate peroxidase and glutathione reductase in roots of rice seedlings in response to NaCl and H_2O_2. J Plant Physiol 162:291–299

Ünyayar S, Kele Y, Çekiç FÖ (2005) The antioxidative response of two tomato species with different drought tolerances as a result of drought and cadmium stress combinations. Plant Soil Environ 51:57–64

Vaidyanathan H, Sivakumar P, Chakrabarty R, Thomas G (2003) Scavenging of reactive oxygen species in NaCl-stressed rice (*Oryza sativa* L.) differential responses in salt tolerant and sensitive varieties. Plant Sci 165:1411–1418

von Heijne G, Steppuhn J, Herrmann RG (1989) Domain structure of mitochondrial and chloroplast targeting peptides. Eur J Biochem 180:535–545

Wada N, Kinoshita S, Matsuo M, Amako K, Miyake C, Asada K (1998) Purification and molecular properties of ascorbate peroxidase from bovine eye. Biochem Biophys Res Commun 242:256–261

Welinder KG (1992) Superfamily of plant, fungal and bacterial peroxidases. Curr Opin Struct Biol 2:388–393

Wingsle G (1989) Purification and characterization of glutathione reductase from Scots pine needles. Physiol Plant 76:24–30

Yamaguchi K, Nishimura M, Mori H (1995) A novel isoenzyme of ascorbate peroxidase localized on glyoxysomal and Leaf peroxisomal membranes in pumpkin. Plant Cell Physiol 36:1157–1162

Yannarelli GG, Fernández-Alvarez AJ, Santa-Cruz DM, Tomaro ML (2007) Glutathione reductase activity and isoforms in leaves and roots of wheat plants subjected to cadmium stress. Phytochemistry 68:505–512

Yasar F, Ellialtioglu S, Yildiz K (2008) Effect of salt stress on antioxidant defense systems, lipid peroxidation, and chlorophyll content in green bean. Russ J Plant Physiol 55:782–786

Yoshimura K, Ishikawa T, Nakamura Y, Tamoi M, Takeda T, Tada T, Nishimura K, Shigeoka S (1998) Comparative study on recombinant chloroplastic and cytosolic ascorbate peroxidase isozymes of spinach. Arch Biochem Biophys 363:66–63

Yoshimura K, Yabuta Y, Ishikawa T, Shigeoka S (2000) Expression of spinach ascorbate peroxidase isoenzymes in response to oxidative stresses. Plant Physiol 123:223–233

Zaharieva TB, Abadia J (2003) Iron deficiency enhances the levels of ascorbate, glutathione, and related enzymes in sugar beet roots. Protoplasma 211:269–275

Chapter 4
The Ascorbate–Gluathione Cycle and Related Redox Signals in Plant–Pathogen Interactions

Elżbieta Kuźniak

Abstract Reactive oxygen species (ROS) in plants are known to accumulate during biotic stress and different cellular compartments respond to them by a distinct antioxidant repertoire. The ascorbate–glutathione (AsA–GSH) cycle serves as the main antioxidant pathway in plant cells linking the protection against ROS to redox-regulated plant defence. This chapter makes a selective overview on the compartment-specific role of the AsA–GSH cycle constituents in the process of sensing and transducing signals and in regulation of nuclear gene expression in response to pathogens. The focus of this review is the specific interplay between ascorbate and glutathione, the two major cellular redox determinants. The potent biological significance of the effects mediated by ascorbate and glutathione in plant–pathogen interactions has been considered with reference to the lifestyle of the invading pathogen and to the local versus systemic defence.

Keywords Ascorbate-glutathione cycle • Compartmentation of stress responses • Reactive oxygen species • Plant-pathogen interactions • Redox signalling

1 Introduction

The oxidative stress related to the imbalance between the production of reactive oxygen species (ROS) such as superoxide anion radical ($O_2^{\bullet -}$), hydrogen peroxide (H_2O_2) and hydroxyl radical ($OH^\bullet$), and the capacity of the antioxidant system is a widespread phenomenon commonly observed in plants exposed to most, if not all, environmental stresses (Bolwell et al. 2002; Dat et al. 2003; Apel and Hirt 2004; Gechev et al. 2006). Traditionally, the term "oxidative stress" denoted a harmful process with potentially deleterious effects on plant metabolism that have to be

E. Kuźniak (✉)
Department of Plant Physiology and Biochemistry, University of Łódź, 90-237 Łódź, Banacha 12/16, Poland
e-mail: elkuz@biol.uni.lodz.pl

N.A. Anjum et al. (eds.), *Ascorbate-Glutathione Pathway and Stress Tolerance in Plants*,
DOI 10.1007/978-90-481-9404-9_4, © Springer Science+Business Media B.V. 2010

avoided by mobilization of the antioxidant defence. However, recent studies have substantially extended our understanding of the role of ROS in plant biology, suggesting that they are key regulators of biological processes. It is now generally accepted that the effects of ROS result from direct or indirect responses to sensing systems involving antioxidants, rather than from oxidative damage to biomolecules per se.

Under stress, controlling the toxicity of ROS, simultaneously enabling them to act as signalling molecules, requires a complex regulatory network including ROS sensing, producing and scavenging systems. The interplay between ROS production and scavenging determines the steady-state level of ROS in cells, as well as the ROS signature, i.e. the duration, localization, and amplitude of ROS signals conditioning stress responses (Mahalingam and Fedoroff 2003; Miller et al. 2008). The processing of ROS involves modifications of redox components, and the adjustment in cellular redox state has been shown integral to the mechanism by which plants respond to the constantly fluctuating environment (Foyer and Noctor 2005). In plant cells, the reductive detoxification of ROS is strongly dependent on ascorbate and glutathione being the two main hydrophilic antioxidants and redox buffers (Noctor and Foyer 1998). However, ascorbate and glutathione are multifunctional compounds with functions extending far beyond the antioxidative system (De Tulio and Arrigoni 2004; Meyer 2008; Noctor 2006).

Much evidence points to a pivotal role of the ascorbate–glutathione (AsA–GSH) cycle that scavenges ROS, and particularly H_2O_2, in activating a successful defence strategy under environmental stress (Noctor and Foyer 1998). The AsA–GSH cycle involves successive oxidations and reductions of ascorbate and glutathione catalysed by the enzymes constituting the cycle, namely ascorbate peroxidase (APX, EC 1.11.1.11), monodehydroascorbate reductase (MDHAR, EC 1.6.5.4), dehydroascorbate reductase (DHAR, EC 1.8.5.1) and glutathione reductase (GR) (Noctor 2006; Polle 2001). It operates in all cellular compartments in which ROS detoxification is needed, i.e. in apoplast, cytosol, chloroplasts, mitochondria and peroxisomes (Potters et al. 2002; Noctor and Foyer 1998). Apart from its direct antioxidant role, the AsA–GSH cycle functions for ROS sensing and signalling. Modulation of the appropriate defense response under biotic stress can be achieved by changes in the absolute concentrations of the total, reduced and oxidized forms of ascorbate and glutathione as well as their redox ratios (Foyer and Noctor 2005; Kuźniak and Skłodowska 2005a). Owing to the action of the AsA–GSH cycle the information-rich metabolic pools of ascorbate and glutathione are tightly linked by redox flux and function. However, each of them may have specific roles in redox regulation of the defence response, as they have different redox potentials. Moreover, compensatory mechanisms can be involved in regulation of the ascorbate- and glutathione-mediated defence (May et al. 1996; De Tulio and Arrigoni 2004; Foyer and Noctor 2005). In general, the interplay between ROS, ascorbate and glutathione, mediated by the AsA–GSH cycle, linking the antioxidant function with the redox-dependent defence reactions, is likely a key part in the signalling mechanism by which plants perceive and respond to the environmental stresses. As sessile organisms plants cannot escape from the stressful environment. This implies that to maintain the successful growth and propagation during their life span, plants must be able to continuously adjust their metabolism to the fluctuating external conditions. In view of metabolic regulations, redox signalling is recognized as one of the major ways used by plants to sense and respond to the environmental cues from early evolution onwards.

According to the concept that redox regulation of metabolic processes is a universal feature of life, a plant cell could be viewed as a unit composed of compartments with specific sets of antioxidants, differing in their antioxidant buffering capacities and redox status. The inherent ascorbate- and glutathione-mediated redox homeostasis in individual compartments is maintained at different levels complementary to their metabolic requirements and differentially affected by the oxidative stress of biotic origin (Kuźniak and Skłodowska 2005a). On pathogen attack, the interactions between ROS and the AA–GSH cycle could generate compartment-specific redox signals interfering with other signalling pathways to achieve a complex rearrangement of the primary metabolism from photo-assimilatory pathways into an emergency survival strategy (Scharte et al. 2005).

2 Role of Apoplast Oxidative Metabolism in Plant–Pathogen Interactions

The accumulation of ROS, especially $O_2^{\cdot -}$ and H_2O_2, has been routinely recorded in infected plants and is considered one of the most characteristic early signs of plant–pathogen interactions (Wojtaszek 1997; Hückelhoven et al. 1999; Pitzschke et al. 2006; Shetty et al. 2008). The overproduction of ROS under biotic stress, known as the oxidative burst, is generally attributed to enzymatic mechanisms, namely NADPH-oxidase, amine oxidase and peroxidases located mainly extracellularly (Doke 1983; Thordal-Christensen et al. 1997; Wojtaszek 1997). However, the mechanisms for ROS generation have been shown to vary depending on the plant–pathogen interaction (Bolwell 1999). The localized extracellular H_2O_2 generation has been shown to be necessary and sufficient to account for fungal penetration failure (Mellersh et al. 2002; Asselbergh et al. 2007). Rapid infection-induced production of ROS in the apoplast could also inhibit pathogen penetration by cell wall lignification, cross-linking of cell wall polymers (Almagro et al. 2009) as well as by the production of antimicrobial phytoalexins (Colvile and Smirnoff 2008). In addition ROS have the potential to be directly toxic to pathogens (Lu and Higgins 1999) and to induce the expression of defense genes (Levine et al. 1994; Desikan et al. 2001). Thus, most attention has been focused on the apoplast contributing to the inducible defence responses following biotic stress as a site where oxidants are produced and perceived.

Taking into account the broad spectrum of defence functions attributed to ascorbate and glutathione (Noctor and Foyer 1998; Arrigoni and De Tulio 2002; Noctor 2006), their apoplastic pools appear to play the main role in maintaining the defensive potential of this cellular compartment. Ascorbate is the major redox active compound in the apoplast and it has been shown to be crucial for the stress response when stress factors, like pathogen, are perceived firstly by the apoplast. In general, the apoplastic ascorbate pool has been calculated to be from 5% to 10% of the total cellular pool and it is set at a comparatively oxidized level (Noctor and Foyer 1998; Veljovic-Jovanovic et al. 2002). In this compartment, ascorbate is oxidized either by ROS or by ascorbate oxidase (AO) or ascorbate peroxidase (APX, Patykowski and Urbanek 2003; Pignocchi and Foyer 2003). There is no evidence that NAD(P)H, the AsA–GSH cycle driving force for ascorbate regeneration, is found in apoplast although some enzymes of the cycle are present (Vanacker et al. 1998a, b). Thus, the ascorbate regeneration process takes place at the expense of the reducing power of the cytosol. It is performed by the action of reduced ascorbate/dehydroascorbate (AsA/DHA) transporter that is linked to the cytosolic AsA–GSH cycle, or either by a membrane-bound cytochrome b or MDHAR (Horemans et al. 2000). This mechanism requires collaboration of different cellular compartments. Moreover, antioxidant buffering capacity could depend on the rate of ascorbate degradation that has been shown to be located in the apoplast (Green and Fry 2005). Considerable attention has been focused on the apoplastic ascorbate pool during biotic stress, as AsA negatively regulates the peroxidase-dependent oxidation of phenolics by the reduction of the oxidized phenolic intermediates (Takahama and Oniki 1992). This mechanism hinders the cell wall strengthening processes like lignification known to be of key importance in the plant cell attempt to stop pathogen spread (Fig. 1). As lignification of the cell wall only occurs in the absence of AsA, the apoplastic ascorbate pool must be oxidized to allow this stress response to proceed. It was shown that in the resistant tomato cultivar the level of reduced ascorbate in the apoplast decreased significantly after infection with *Botrytis cinerea* and this corresponded to the higher activity of peroxidase assayed with a substrate indicative of lignification (Patykowski and Urbanek 2003, 2005). Importance of the cell wall strengthening processes mediated by the apoplastic class III peroxidases for plant resistance to infection has been identified for different pathosystems (Lamb and Dixon 1997; Almagro et al. 2009). Similarly, the ascorbate-deficient (*vtc*) *Arabidopsis thaliana* mutants with increased resistance to infection by virulent pathogens (Barth et al. 2004; Pavet et al. 2005) exhibit elevated cell wall peroxidase activity, increased transcript levels of pathogenesis-related (PR) genes and they accumulated phytoalexin camalexin (Colvile and Smirnoff 2008). Moreover, the results of Pastori et al. (2003) demonstrated that 12.9% of the 171 genes differentially expressed in *vtc* mutants of *A. thaliana*, when compared with the wild type, were involved in plant defence. Thus, apoplastic concentration of ascorbate can control the rate of oxidation of lignin monomers or other phenolics which is catalyzed by cell wall-bound peroxidases (Takahama 2004). Moreover, it has been revealed that ascorbate availability may control the biosynthesis/signalling of ethylene as well as salicylic and abscisic acids (Pastori et al. 2003; Pignocchi and Foyer 2003). The evidence outlined above suggests that the redox state of ascorbate is important in modulating plant hormone-dependent signalling pathways controlling defence responses. In addition, changes in the ascorbate pool may affect the age-related resistance/susceptibility to pathogens which is typically seen as an increase/decreasae in the resistance of plants with development (Barth et al. 2004; Pavet et al. 2005). Ascorbate deficiency in *vtc1* and *vtc2* mutants stimulated the innate immune responses via activation of localized programmed cell death (PCD) and systemic acquired resistance (SAR) and the onset of SAR was linked to the cessation of expansion growth (Pavet et al. 2005).

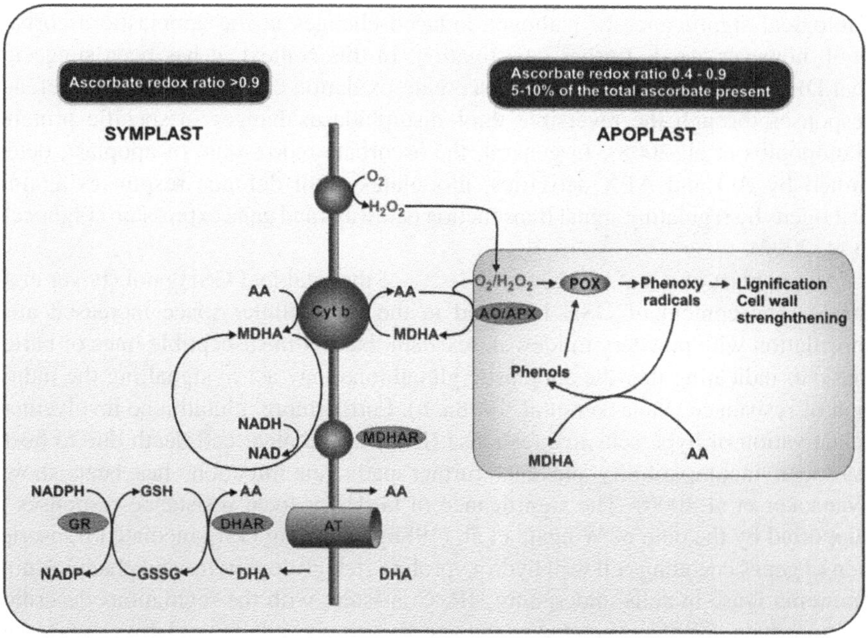

Fig. 1 The apoplastic ascorbate/phenolic/peroxidase system. Redox interactions across the plasma membrane including the possible presence of AA/DHA transporters and DHA/AA exchange carriers as well as the AA–GSH cycle constituents are indicated. The apoplastic ascorbate can be oxidized by the action of ascorbate peroxidase (APX) or ascorbate oxidase (AO). After production of monodehydroascorbate (MDHA) in apoplast, AA is either directly re-reduced by a b-type cytochrome-mediated electron transfer (cyt b) or by the action of cytosolic monodehydroascorbate reductase (MDHAR), or it disproportionates to dehydroascorbate (DHA). The extracellular DHA is transported by an AA/DHA carrier (AT) to the cytosol where it is re-reduced by dehydroascorbate reductase (DHAR) using reduced glutathione (GSH) as an electron donor. Glutathione reductase (GR) mediates the reduction of oxidized glutathione (GSSG) to GSH by using NADPH as an electron donor. Reduced ascorbate (AA) negatively regulates the peroxidase (POX)-dependent oxidation of phenolics by the reduction of the oxidized phenolic intermediates (depicted on the *grey panel*). This mechanism hinders lignification and other cell wall strengthening processes known to be of key importance in local defence against biotrophic pathogens (Based on Horemans et al. 2000; Takahama 2004)

The results suggest that ascorbate may be one of the components that influence the inverse relationship between growth and defence that is frequently observed in plants (Heil 1999). However, tobacco plants overexpressing the cell wall AO in the apoplast, and exhibiting an increased H_2O_2 content, display increased susceptibility to *B. cinerea* infection when compared with wild-type plants (Fotopoulos et al. 2006, 2008). Thus, plants with lower ascorbate redox state in the leaf apoplast are impaired in their defence response to infection. In general, these results are in line with the concept that plants can sense shifts in both the amount and redox state of ascorbate and they point to a potentially crucial role of apoplast-located ascorbate in the first line of plant defence against external agents promoting oxidative stress.

Biological significance of pathogen-induced changes in the apoplastic ascorbate pool, however, needs further investigation. In this context, it has been suggested that DHA resulting from apoplast ascorbate oxidation can modulate plant defence responses through the reversible thiol–disulphide exchanges in specific proteins (Fotopoulos et al. 2008). In general, the ascorbate redox state of apoplast, determined by AO and APX activities, modulates plant defence responses against pathogens by regulating signal transduction pathways and gene expression (Pignocchi et al. 2006).

An apoplast of a leaf cell contains 1–2% of the total leaf GSH pool (Foyer et al. 2001). The content of GSH localized in the extracellular space increased after inoculation with powdery mildew in resistant, but not in susceptible lines of barley and oat, indicating that the apoplastic glutathione may act in signalling the induction of resistance (Vanacker et al. 1998a, b). Furthermore, glutathione involvement in activation of hypersensitive response (HR), where plant cell death due to host–pathogen incompatibility prevents further pathogen infection, has been shown (Vanacker et al 2000). The significance of GSH for local resistance responses is supported by the data of Wingate et al. (1988) indicating GSH-mediated transcription of genes encoding cell wall hydroxyproline-rich glycoproteins and phenylalanine ammonia lyase in cells undergoing HR. Consistent with the regulations described for ascorbate, GSSG generated in the apoplast as a result of oxidative stress must be returned to the cytosol for re-reduction to GSH and plasma membrane GSH and GSSG transport has an important function in apoplast-mediated protection against biotic stress.

In general, the concentrations of ascorbate and glutathione and their redox ratios are important factors in apoplastic metabolism and plant defence responses. The ascorbate and glutathione homeostasis in the apoplast is easily perturbed, as this compartment lacks systems for the regeneration of reduced forms of these antioxidants. Taking into consideration low proportions of the total leaf ascorbate and glutathione present in the leaf apoplast and its low buffering capacity, the specific plasma membrane transport systems, linking the apoplastic ascorbate and glutathione redox couples to their cytoplasmic counterparts, appear to play a decisive role in maintaining the AsA–GSH cycle-dependent defensive capabilities of this compartment (Fig. 1). The role that apoplast plays in the plant–pathogen interactions could acquire significance as the low redox buffering capacity allowing oxidative signals to accumulate has been suggested to facilitate redox signalling (Foyer and Noctor 2005).

With respect to the function of ROS in pathogenesis, plant defence mechanisms against necrotrophic pathogens, which kill host tissues, are considered to differ from those that are effective against biotrophic pathogens, which grow on living hosts. It has been suggested that H_2O_2 from the oxidative burst plays a dual role in plant–pathogen interactions. H_2O_2 acts as an active defence compound against biotrophic pathogens as it accumulates at the sites of attempted penetration and assists the development of HR that inhibits pathogen growth by surrounding it with an extremely oxidative environment (Bestwick et al. 1997; Hückelhoven et al. 1999). However, for the necrotrophic pathogens the accumulation of H_2O_2 has been postulated to be beneficial for infection, leading to enhanced colonization of the plant tissue (Govrin and Levine 2000). Necrotrophic fungi, e.g. *B. cinerea* and *Sclerotinia sclerotiorum* are apparently not deterred by the HR of the host, and stimulation of the HR facilitates host infection (Mayer et al. 2001; Govrin et al. 2006). This effect could be prevented by ROS inhibitors or scavengers (Govrin and Levine 2000). It seems likely that the plants maintaining high antioxidant potential could be compromised in their ability to sustain the attack of a biotrophic pathogen. This relationship has been questioned in other studies (Małolepsza and Urbanek 2002; Unger et al. 2005; Walz et al. 2008); Dong et al. (2008) were not able to confirm the accelerated and enlarged lesion formation in *S. sclerotiorum*-infected transgenic plants constitutively overexpressing the oxalate oxidase gene. However, there is no doubt that necrotrophs can successfully cope with the HR-related defence mechanisms evolved to defend against biotrophic pathogens, or even use them to their advantage (Spoel et al. 2007). In this respect, dead cells resulting from HR provide resistance to biotrophs, but also provide potential entry points for the invading necrotrophs in the natural environment (Kliebenstein and Rowe 2008). Moreover, it has been demonstrated that contrary to local tissues where a successful infection by a biotroph reduces resistance to a necrotroph, in systemic tissues trade-off between biotroph and necrotroph resistance was negligible (Spoel et al. 2007). Moreover the success of the HR is related to the unbalance of the AsA–GSH redox state (de Pinto et al. 2002) and transient suppression of the enzymatic antioxidant defence as transgenic plants with suppressed activities of catalase and APX are hyperresponsive to pathogens (Mittler et al. 1999). Thus the physiological significance of the AsA–GSH cycle-related changes in pathogenesis should be interpreted with respect to the lifestyle of an invading pathogen and the local versus systemic defence response. Thereby, the physiological significance of the AsA–GSH cycle may be difficult to estimate under natural environmental conditions, in plants simultaneously coping with multiple pathogens with different lifestyles.

Both a pathogen and a plant host deploy similar oxidant and antioxidant systems and at the interface there must be a fine balance between them dependent on the relative sensitivity of each partner to ROS and other products of oxidation reactions generated (Lu and Higgins 1999; Mayer et al. 2001). Moreover, the mechanisms by which plant cells regulate the level of ROS, and especially of H_2O_2, could be crucial for the outcome of plant–pathogen interaction because the role that H_2O_2 plays in plant defence depends on its concentration (Mittler 2002). Low content of H_2O_2 induces protective antioxidant mechanisms and activates signalling pathways triggering a systemic response (Solomon et al. 1999). Moderate to high H_2O_2 levels can be directly toxic to pathogens (Peng and Kuć 1992) or can trigger PCD in the form of HR (Tenhaken et al. 1995; Mittler et al. 1999). In this context the prooxidant/antioxidant balance dependent on the AsA–GSH cycle activity may represent a mechanism conditioning the defence strategy of the infected plant. However, it is noteworthy that the measurements of endogenous H_2O_2 concentration are technically challenging, and the literature data for H_2O_2 content in leaves show great variability ranging from 0.05 to 5 μmol g^{-1} FW (Queval et al. 2008). These relatively high values are inconsistent with the fact that the photosynthetic CO_2 fixation as well as the tricarboxylic acid cycle can be inhibited by H_2O_2 at the concentration of 10 μmol g^{-1} FW (Veljovic-Jovanovic et al. 2002). This implies that the leaf H_2O_2 content can be overestimated (see Veljovic-Jovanovic et al. 2002 for further discussion). If this is the case, the physiological significance of these data should be interpreted with extreme caution. With respect to the methodological difficulties in accurately measuring H_2O_2 in the whole leaf extracts (Queval et al. 2008) it is unclear to what extent these assays provide useful information on tissue redox state determined by the AsA–GSH cycle-related mechanism, especially when its compartment-specific role is taken into consideration.

3 Significant Role of Compartment-Specific Responses of the Ascorbate–Glutathione Cycle in the Regulation of Plant Defence Against Biotic Stress

Emerging evidence points to the role of the subcellular compartmentation in the antioxidant defence (Kuźniak and Skłodowska 1999, 2001; Foyer and Noctor 2005). Unfortunately, the response of the AsA–GSH cycle under biotic stress has been studied almost exclusively at the whole cell level and in the apoplast (El-Zahaby et al. 1995; Vanacker et al. 1998a; Patykowski and Urbanek 2003). In plants, chloroplasts, mitochondria and peroxisomes have been proposed to be the main sources of ROS under "normal" conditions. Simultaneously, these organelles are equipped with effective antioxidant systems, including the AsA–GSH cycle (Jiménez et al. 1997; Noctor and Foyer 1998; del Río et al. 2002; Kuźniak and Skłodowska 2005b). Although in plant–pathogen interactions ROS are generated mainly extracellularly, the antioxidants located in cytosol, chloroplasts, mitochondria and peroxisomes could be also affected. In tomato plants, the successful *B. cinerea* infection was found to overcome the protective antioxidant function of the AsA–GSH cycle in chloroplasts, mitochondria and peroxisomes, and mitochondria and peroxisomes underwent the most pronounced oxidative changes (Kuźniak and Skłodowska 2005a). The general shift of the cellular redox equilibrium toward the oxidative state was manifested by the decrease in the concentrations and redox ratios of ascorbate and glutathione as well as lowered activity of MDHAR, DHAR and GR, the enzymes responsible for ascorbate and glutathione recycling. Thus, success of the pathogen was related to the decline of the antioxidant capacity at the subcellular level. In respect of the timing and intensity of changes in the AsA–GSH cycle activity, the organelles were differentially affected by *B. cinerea* infection. However, a progressive decrease in glutathione content was found in all cellular compartments (Kuźniak and Skłodowska 2005a).

The need for coordinating antioxidant defence in different cellular compartments during plant reactions to pathogens has been also discussed with respect to its contribution to the spatial and temporal tuning of ROS levels and their signalling properties (Dutilleul et al. 2003; Torres et al. 2006; Kuźniak et al. 2009; Song et al. 2009). The AsA–GSH cycle elements, operating in all organelles where the generation of ROS occurs, form an integrated network with extensive crosstalk that could be triggered by ROS and redox signals (Vranová et al. 2002). Several links between chloroplastic production of ROS and plant stress responses, including basal resistance, have been provided by studies on the lesion mimic mutants that spontaneously display necrotic lesions on leaves (Lorrain et al. 2003; Mateo et al. 2004). The formation of these lesions, that are very similar to those resulting from HR, are often accompanied by PR gene expression. There are results indicating that the regulation of pathogen-triggered cell death requires both chloroplastic and mitochondrial signals and communication between these organelles (Yao and Greenberg 2006). It appears that in the context of the mitochondrial and chloroplastic AsA–GSH cycle, this coordination occurs by dual targeting of the cycle enzymes rather than retrograde signalling (Creissen et al. 1995; Chew et al. 2003). Dual targeting of APX, MDHAR, and GR, but not DHAR gene products, to mitochondria and chloroplasts has been reported in *A. thaliana* (Obara et al. 2002; Chew et al. 2003). Thus, the fact that several key antioxidative enzymes are dual targeted to chloroplasts and mitochondria may be significant in PCD regulation. Coordination of chloroplastic and mitochondrial mechanisms with defence as a general response to pathogens is an interesting point emerging from these studies, although the underlying details can be quite variable depending on the specific plant–pathogen system and on the environmental conditions.

Although there are considerable data showing interference between the ascorbate and glutathione systems in plant response to stress (Conklin et al. 1996; Leipner et al. 2000) it is possible that the redox state of glutathione could be to some extent independent of ascorbate (Noctor 2006). Taking into account the relative redox potentials of ascorbate and glutathione redox pairs, GSSG can exist in the presence of the high ascorbate concentration, thus it should not interfere with redox regulations via the thioredoxin (TRX) system and other thiol–disulphide interconversions. However, under severe stress when the oxidation of ascorbate increases substantially, the ascorbate–glutathione cycle is activated resulting in increased GSSG formation.

Perturbation of the redox processes in plants infected with *B. cinerea* is suggested to be an attack strategy of this necrotrophic pathogen (Von Tiedemann 1997; Kuźniak and Skłodowska 2005a). However, the infection-induced changes vary depending on the host plant. In *A. thaliana* leaves massive depletion of ascorbate and glutathione and no increases in lipid peroxidation products were observed whereas *B. cinerea*-infected leaves of *Phaseolus vulgaris* showed significant increases in lipid peroxidation products concomitant with slight decline in the content of non-enzymatic antioxidants (Muckenschnabel et al. 2001, 2002). Therefore, the results obtained for *A. thaliana*, a model plant, should be extrapolated to other plants with extreme caution.

The H_2O_2-decomposing function of the AsA–GSH cycle is associated with APX, the unique enzyme found mainly in plants and algae that possesses a very high affinity to H_2O_2 (Nakano and Asada 1981; Asada 1999; Raven 2003).The function of APX depends on the availability of the AsA pool, and on GSH with respect to the activity of DHAR, the GSH-dependent enzyme that could limit the regeneration of AsA from DHA. It has been reported that in plants strong suppression of APX activity may coincide with the induction of cell death (de Pinto et al. 2002), which in the context of defence strategy to biotic stress is regarded the most effective response against biotrophic pathogens. Inhibition of APX activity may occur at various stages and may depend on numerous external factors, e.g. the TMV induced PCD in tobacco leaves was caused by post-transcriptional suppression of cytosolic APX, which was correlated with an increase in intracellular H_2O_2 content and promoted cell death (Mittler et al. 1998).

Gene expression studies in plants subjected to stress reveal that among nine members of *A. thaliana* APX gene family expressed in cytosol, chloroplasts, mitochondria and peroxisomes, the cytosolic APX isozymes are the most stress-responsive (Mittler and Poulos 2005). Davletova et al. (2005) demonstrated that in the absence of the cytosolic H_2O_2-scavenging enzyme APX1, the entire chloroplastic H_2O_2-scavenging system of *A. thaliana* collapsed, H_2O_2 accumulated and protein oxidation occurred. Transgenic tobacco plants with suppressed APX1 expression are hyper-responsive to pathogen attack and they activate programmed cell death in response to low amounts of pathogens that do not trigger a cell death response in the control plants (Mittler et al. 1999). These studies pointed to the importance of ROS scavenging in cytosol that mediates the export and import of ascorbate and glutathione as well as their recycling in order to maintain large redox gradients between cellular compartments (Kuźniak and Skłodowska 2005a). Regulations of the cytosolic redox state could be key determinants for the plant response to biotic stress, when a massive influx of DHA and GSSG resulting from the pathogen-induced oxidative burst in the apoplast occurs. In *B. cinerea*-infected tomato leaves, the redox state of the cytoplasm was maintained on a significantly higher level than in other cellular compartments. Moreover, the pronounced redox equilibrium decrease, indicating that in the infected leaves a high burden was put on the cytosolic antioxidant system, overlapped the initiation of disease progression concomitant with the breakdown of the antioxidant plant defence (Kuźniak and Skłodowska 2005a). The ROS/redox signals transmitted from different compartments are buffered by the cytosolic antioxidants which can limit their strength and spread (Baier and Dietz 2005). Whether they can pass the cytosol without loss of function and reach the nucleus to control gene expression depends to some extent on the capacity of the cytosolic AsA–GSH cycle.

The AsA and GSH are maintained in their reduced forms by a set of four enzymes regenerating AsA and GSH, namely MDHAR, DHAR, GR and glutaredoxins whose activities rely on NADPH (Mittler and Poulos 2005). Thus, for ascorbate and glutathione to act as redox couples an adequate supply of NADPH is necessary. In plants, NADPH is supplied mainly via the oxidative pentose phosphate pathway (OPP). However, information on the co-regulation of the OPP and AsA–GSH cycle under stress is surprisingly scarce although increases in the activity of glucose-6-phosphate dehydrogenase, the main OPP enzyme, have been observed in response to fungal elicitors (Kombrink and Hahlbrock 1986). In contrast, 6-aminonicotinamide, an inhibitor of OPP, diminished the PR1 gene expression in *Arabidopsis* treated with SAR inducer (Mou et al. 2003).

Furthermore, compartmentalization of both ROS production and activation of antioxidants could contribute to fine-tuning of ROS levels and their signalling properties (Torres et al. 2006). Although the primary oxidative burst following pathogen recognition occurs in the apoplast, ROS produced in other organelles with oxidative metabolism or characterized by rapid electron flux, e.g. chloroplasts, mitochondria and peroxisomes, may also have functions in defence. The complex regulatory network involved in managing ROS and determining the intensity, duration and localization of the ROS/redox signals is beyond the scope of this chapter. It has been covered by several recent reviews (Laloi et al. 2004; Foyer and Noctor 2009; Miller et al. 2008).

Rapid adaptive flux in the cellular redox metabolism is a prerequisite to counteract the pathogen-imposed oxidative stress. Although there are some reports on the involvement of AsA–GSH cycle enzymes in biotic stress responses, there is a dearth of literature data with respect to the position of the AsA–GSH cycle in the overall defence network. The perturbations in glutathione and ascorbate pools as well as their redox ratios could be either beneficial for plant defence against pathogens when an adequate response is induced, or they could reflect inability to withstand the stress impact. Glutathione is acknowledged as the most important thiol antioxidant in the plant–pathogen interactions. The increase is GSH pool is generally regarded as a protective response against pathogens because a positive relationship between GSH concentration/glutathione pool homeostasis and resistance to pathogens was found in several studies (Gullner et al. 1999; Vanacker et al. 2000; Kuźniak and Skłodowska 2005a). The resistant tomato cultivar infected with *Pseudomonas syringae* was able to maintain homeostasis of the glutathione pool whereas a sustained accumulation of GSSG and glutathione redox ratio decreased, in addition to GSH depletion, were characteristic features of the susceptible interaction (Kuźniak and Skłodowska 2004). However, the response of an *A. thaliana* mutant with a 70% deficit in the content of glutathione to bacterial and fungal pathogens was unaltered (May et al. 1996). This effect was attributed to the considerable increase in AsA level found in this mutant under biotic stress conditions suggesting that a compensatory mechanism was induced to afford adequate protection in the absence of glutathione. Similar compensatory regulations, superior to changes in individual elements of the AsA–GSH cycle, have been reported for tomato-*P. syringae* interaction. In this pathosystem compensation of GSH depletion by AsA increase to maintain the overall redox balance under stress has been observed in the susceptible tomato cultivar whereas in the resistant one the redox homeostasis was maintained throughout *P. syringae* attack and no significant changes in the ascorbate pool occurred (Kuźniak and Skłodowska 2004). These studies as well as numerous data from transgenic plants with altered antioxidant capacity (Kuźniak 2002) pointed further to the interdependence between different pathways of protection as a mechanism which helps plants to withstand the pathogen attack.

Dehydroascorbate reductase is commonly recognized as an enzyme that links ascorbate and glutathione in the AsA–GSH cycle and DHA-reducing proteins have been purified from different plant sources (Hossain and Asada 1984; Dipierro and

Borracino 1991; Hou and Lin 1997). Substrate affinity is an important parameter in assessing the relative contribution of these different enzymes to DHA reduction. However, except DHAR isolated from spinach (Hossain and Asada 1984), these proteins exhibit low affinity to DHA (K_m for DHA in the range of 0.26–0.50 mM, Yang et al. 2009) and at present their relative contribution to DHA reduction is suggested less significant than previously postulated (Polle 2001). DHA is readily reduced non-enzymatically to ascorbate by thiols such as GSH (Polle 2001). Moreover, some of the DHA-reducing proteins have been shown to fulfil completely different functions (Urano et al. 2000). Plant glutathione transferases (GST), most intensively studied in relation to their role in xenobiotic detoxification by means of conjugation with GSH, have been reported to contribute to ascorbate redox homeostasis acting as potent DHAR (Dixon et al. 2002). In tomato responding to *P. syringae* infection a positive relationship between GST activity and DHA content was observed in the resistant cultivar only, providing evidence that GST possessing a DHA-reducing activity contributed to tomato resistance to bacterial speck disease (Kuźniak and Skłodowska 2004).

4 Glutathione- and Ascorbate-Related Redox Regulations of Defence Reactions to Pathogens

Redox signalling is regarded as a fundamental form of signal transmission in all living organisms. Efficient redox signalling, however, can be obtained only when the system is stabilized by a redox buffer. The cellular redox environment is formed by a set of various linked redox couples (Potters et al. 2010). Glutathione, owing to its chemistry and low milimolar concentration in plant tissues is considered as a key element in signalling cascades transducing information from the external environment to the specific cellular targets (Mullineaux and Rausch 2005; Noctor 2006). Accumulating evidence suggests that changes in the glutathione pool reset the cellular redox homeostasis to ensure an adequate response under stress (Tausz et al. 2004; Foyer and Noctor 2005). Unlike ascorbate, the redox potential of glutathione is a function of both GSH/GSSG ratio and the concentration of GSH. The antioxidant functions of GSH consume two molecules of GSH per one molecule of GSSG formed ($[GSH]^2/[GSSG]$), and according to the Nernst equation the glutathione redox state is a second order function of GSH concentration (Schafer and Buettner 2001). Thus as the redox state of glutathione is concentration-dependent, change in the content of GSH even with constans GSH/GSSG ratio could alter the cellular redox state. Unfortunately, this aspect of the glutathione-mediated redox regulation of plant defence through modifications of metabolic enzymes or changes of gene expression is often neglected.

The redox state of a given cellular compartment is routinely characterized by the reduction/oxidation ratio of a specific redox compartment ? metabolite, and this is recognized as an indicator of the overall redox state. Taking into account the numerous redox active compounds creating the cellular redox state, the validity of this approach seems to be limited and depends on the choice of the representative redox compound. On this point, it is interesting to note that DHA rarely accumulates in plant cells (Arrigoni and De Tulio 2002). Moreover, the redox states of ascorbate and glutathione can be differentially regulated (Kuźniak and Skłodowska 2005a). In this context, determination of the physiological significance of pathogen-induced redox changes in the ascorbate and glutathione pools remains a challenging research task.

The GSH/GSSG redox couple in plants appears to be important in regulating the expression of defense genes via redox-sensitive mediators. To date, one of the best-known examples of how the glutathione-related redox signalling pathways may work, inducing alterations in plant gene expression, is the activation of NPR1 protein (Nonexpressor of Pathogenesis-Related protein 1, also called Non-Inducible Immunity1, NIM1). NPR1 was identified as a redox-sensitive transcription factor in *Arabidopsis*. The conformation of NPR1 protein is redox sensitive (Fig. 2). The reduction of NPR1 preceding defence gene induction requires an increase in GSH content and a concomitant shift in the cellular redox environment toward the reducing conditions (Mou et al. 2003; Fobert and Despres 2005). Under these conditions, NPR1 is reduced from an inactive oligomeric complex localized in the cytosol to an active monomeric state through the reduction of intermolecular disulfide bonds. Monomeric NPR1 is then translocated into the nucleus where it interacts with transcription factors of the TGA class, such as TGA1 and TGA2 (Mou et al. 2003; Pieterse and Van Loon 2004). NPR1 is an intrinsic component of the salicylic acid (SA)-triggered SAR, which confers long-lasting broad-spectrum resistance. The redox dependence of the NPR1 pathway implies that biotic stimuli that perturb the cellular redox state can up-regulate defence genes via the NPR1 pathway (Mou et al. 2003). Interestingly the kinetics of the changes in the glutathione pool and the GSH/GSSG ratio required for in vitro NPR1 reduction were similar to those evoked after SAR induction (Mou et al. 2003). Glutathione has been suggested to be the main antioxidant providing the redox environment for activation of transcriptional regulators such as NPR1. However, ascorbate and the antioxidative enzymes undergoing dynamic changes under biotic stress can serve as other redox mediators modulating the effect of glutathione (Fig. 2).

Maintaining the appropriate glutathione equilibrium in the time course of pathogenesis could be crucial for the effectiveness of plant response to infection. The observed prooxidant changes at the initial infection stage seem to be beneficial to apoplast-based defense and putative oxidative signaling (Mittler et al. 2004). However, a reduced environment is critical for the outcome of plant–pathogen interaction and glutathione-related systems constitute a central mechanism regulating, via redox-sensitive mediators, the transcription of resistance genes (Fobert and Despres 2005). In accordance with this concept, the progressive shift of the glutathione- and ascorbate-related redox balance towards the oxidative state, manifested by a massive decline in concentrations and redox ratios of the antioxidants as well as by the insufficient activity of GR and DHAR needed for their regeneration, found in the tomato-*B. cinerea* interaction described earlier, seems to prevent a successful resistant response (Kuźniak and Skłodowska 2005a).

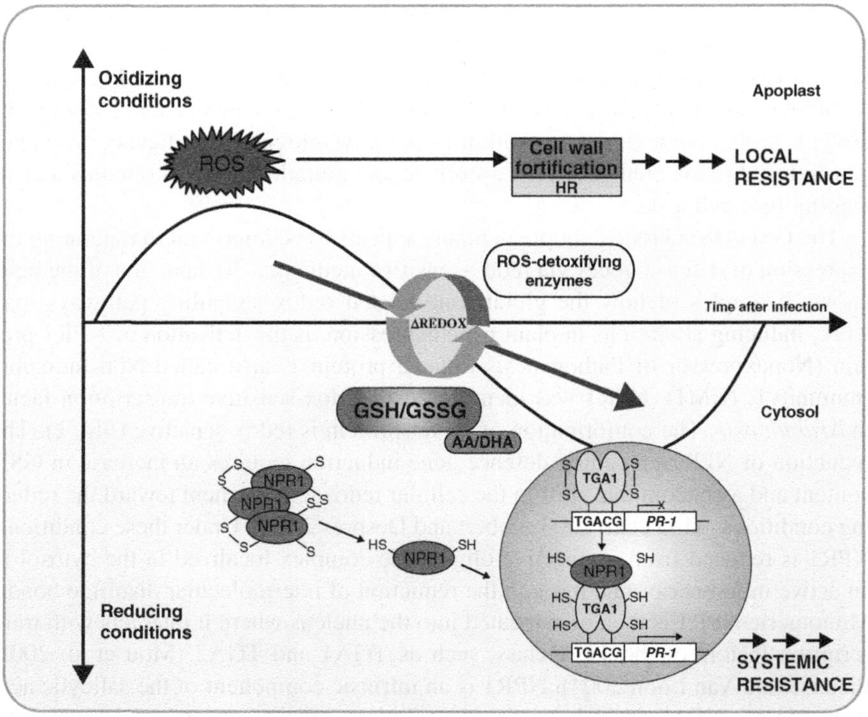

Fig. 2 The schematic representation of glutathione- and ascorbate-mediated redox regulation of resistance implicating both an oxidizing and reducing phase in controlling defence responses. ROS and the redox changes can modify the redox state of transcription factors, their localization and interaction with other partners. Upon attack of a pathogen, reactive oxygen species (ROS) generated in the apoplast create a highly oxidizing environment favoring lignification, cell wall fortification and hypersensitive response (HR) which determine local resistance to pathogens. Under these oxidative conditions the NPR1 protein (Nonexpressor of Pathogenesis-Related protein 1) and transcription factor from the TGA family (TGA1) are oxidized and inactive. NPR1 is present in the cytosol as an oligomer with subunits aggregated via disulphide bonds. The TGA1 transcription factor bound to the promoter of salicylic acid-responsive gene (TGACG) is not competent to activate systemic acquired resistance (SAR) genes, exemplified as PR1. Mobilization of enzymatic and non-enzymatic antioxidant mechanisms results in a shift of the cellular redox environment towards reducing conditions (Δ redox). This enables the reduction of cysteine residues in NPR1 and TGA1, possibly by thioredoxins or glutaredoxins. The reduced monomeric form of NPR1 accumulates in the nucleus and interacts with the reduced form of TGA1 which binds specifically to regulatory sequences and activates SAR-related genes. The reduced/oxidized glutathione redox couple (GSH/GSSG) appears to be crucial in regulating the PR gene expression via the redox-activated NPR1 pathway. Reduction of NPR1 prior to gene induction requires an increase in GSH concentration and a concomitant over-reduction of the cellular redox environment. Ascorbate (AA/DHA) can modulate this effect (Modified from Fobert and Despres 2005; Mou et al. 2003)

Apart from its role in regulating SAR in the nucleus, a novel cytosolic function of NPR1 in cross-communication between SA- and jasmonic acid (JA)-dependent defence signalling pathways has been identified. The JA-triggered gene expression has been suggested to be negatively regulated through SA-activated NPR1 in the cytosol (reviewed by Pieterse and van Loon 2004). Taking into account the role of JA-dependent signalling in defence, the suppression of JA-inducible gene expression, exemplifying the cross-talk between the SA- and JA-dependent pathways, can modulate the plant resistance response to necrotrophic pathogens.

At present the redox-dependent thiol/disulphide transition extends beyond the well characterized TRX-mediated regulation of the Calvin cycle enzymes (Lemaire et al 2007) and is considered to be one of the most important modifications affecting many cell protein functions (Buchanan and Balmer 2005). Because H_2O_2 is a mild oxidant that can oxidize thiol groups, it has been speculated that H_2O_2 generated under stress could be also sensed via modification of thiol residues in target proteins (Pitzschke et al 2006). Taking into account that among different ROS generated under biotic stress H_2O_2 is the predominant one (Wojtaszek 1997), and that H_2O_2 is the most stable of the ROS freely crossing plant membranes through specific aquaporins (Bienert et al. 2007) the postulated regulatory effects of H_2O_2 in plant resistance response could be mediated by thiol/disulphide exchanges.

The thiol-mediated signalling could be further enhanced by glutathionylation, i.e. the posttranslational protein modification consisting of the formation of a mixed disulfide bond between a cysteine residue and GSH. Glutathionylation is favoured under oxidative conditions. In vitro studies have suggested that ROS promote glutathionylation most effectively than GSSG alone (Rouhier et al. 2008). In addition to its postulated role in protecting protein thiols from irreversible inactivation, this mechanism can also affect, both positively and negatively, the activation of target proteins via thiol-disulphide exchange (Rouhier et al. 2008). In *A. thaliana* glutathione transferase possessing DHAR activity has been shown to be glutathionylated in the presence of GSSG. This reaction was suggested to be an intermediary step in the GSH-dependent reduction of DHA (Rouhier et al. 2008). Glutaredoxins (GRXs), thiol oxidoreductases reducing mixed disulfide between a protein and glutathione or a disulfide bridge, have been suggested to be implicated in the response to oxidative stress through regeneration of thiol-dependent antioxidant enzymes involved in peroxide reduction, namely thiol peroxidases (Rouhier et al. 2008). The first true GRX target protein was identified as a specific subclass of thiol-peroxidases, called type II peroxiredoxin. The best documented function for GRXs, by far, is their involvement in the stress response, in particular through the regeneration of the Prx II subclass (reviewed by Rouhier 2010). Moreover the participation of some GRXs in response to pathogens by interacting and presumably regulating certain TGA transcription factors has been elucidated (Ndamukong et al. 2007). This finding expanded the redox regulatory network controlling plant defence responses as the authors postulate that GRX which acts as a TGA-interacting protein is a negative effector of JA-inducible gene expression. Thus GRX represents a potential regulatory component of the SA/JA antagonism decisive for the outcome of plant–pathogen interactions (Ndamukong et al. 2007).

In general, the scenario of the involvement of the AsA/DHA and GSH/GSSG redox couples in the cellular redox network provides for the possibility that they can undergo distinct physiological activation corresponding to their midpoint potentials.

According to the model proposed by Dietz (2008) the redox transmitters such as plant TRXs and GRXs, in collaboration with redox sensors such as peroxiredoxins and glutathione peroxidase, distribute the electrons to the redox target proteins, e.g. enzymes and transcription factors. It is hypothesized that ROS drain electrons from the redox network mostly through the redox sensors (Dietz 2008). The ascorbate system, tightly linked to that of glutathione through the ascorbate–glutathione cycle, may modulate the redox signal strength acting alongside or downstream of the ROS accumulation (Foyer and Noctor 2005; Pavet et al. 2005).

Under biotic stress the AsA–GSH cycle fulfils functions that go far beyond the ROS-scavenging activity per se. According the conceptual model presented here, the AsA–GSH cycle activity can modulate the ROS signal in different cellular compartments and determine its intensity, duration and localization, i.e. the ROS signature (Mittler et al. 2004) that can determine in turn the type and intensity of the response to a particular pathogen. In addition, ascorbate and glutathione act as main cellular redox buffers and the related redox signals have roles in the orchestration of local and systemic defence reactions.

5 Concluding Remarks

The AsA–GSH cycle, initially discovered in chloroplasts in the context of photosynthetic regulations, is now known to occur throughout the plant cell acting as ROS-scavenging and signalling pathway. The studies of the protective role of AsA–GSH pathway in plants deal primarily with abiotic stress responses. This review attempts to present a broad picture of how the AsA- and GSH-mediated system works and interacts with signalling networks complementary to this antioxidant system under biotic stress. There is extensive body of evidence indicating that the AsA–GSH cycle, linking the antioxidant activity to redox signalling and regulation of defence gene expression, plays a pivotal role in the strategy of plant cell response to infection. Another point of interest will be to study the biological significance of all the AsA- and GSH-mediated functions in plant–pathogen interactions, as this field is still in its infancy. This chapter describes spatial and temporal heterogenecity of the AsA–GSH cycle response in plants subjected to infection. Only selected aspects of the present knowledge have been summarized from the perspective of the local and systemic defence.

References

Almagro L, Gómez Ros LV, Belchi-Navarro S, Bru R, Ros Barceló A, Pedreño MA (2009) Class III peroxidases in plant defence reactions. J Exp Bot 60:377–390

Apel K, Hirt H (2004) Reactive oxygen species: metabolism, oxidative stress, and signal transduction. Annu Rev Plant Biol 55:373–399

Arrigoni O, De Tulio MC (2002) Ascorbic acid: much more than just an antioxidant. Biochim Biophys Acta 1569:1–9

Asada K (1999) The water–water cycle in chloroplasts: scavenging of active oxygen and dissipation of excess photons. Annu Rev Plant Physiol Plant Mol Biol 50:601–639

Asselbergh B, Curvers K, França SC, Audenaert K, Vuylsteke M, Van Breusegem F, Höfte M (2007) Resistance to *Botrytis cinerea* in *sitiens*, an abscisic acid-deficient tomato mutant, involves timely production of hydrogen peroxide and cell wall modifications in the epidermis. Plant Physiol 144:1863–1877

Baier M, Dietz K-J (2005) Chloroplasts as source and target of cellular redox regulation: a discussion on chloroplast redox signals in the context of plant physiology. J Exp Bot 56:1449–1462

Barth C, Moeder W, Klessig DF, Conklin PL (2004) The timing of senescence and response to pathogens is altered in the ascorbate-deficient *Arabidopsis* mutant *vitamin c-1*. Plant Physiol 134:1784–1792

Bestwick CS, Brown IR, Bennet MH, Mansfield JW (1997) Localization of hydrogen peroxide accumulation during the hypersensitive reaction of lettuce cells to *Pseudomonas syringe* pv *phaseolicola*. Plant Cell 9:209–221

Bienert GP, Møller ALB, Kristiansen KA, Schulz A, Møller IM, Schjoerring JK, Jahn TP (2007) Specific aquaporins facilitate the diffusion of hydrogen peroxide across membranes. J Biol Chem 282:1182–1192

Bolwell GP (1999) Role of active oxygen species and NO in plant defense responses. Curr Opin Plant Biol 2:287–294

Bolwell GP, Bindschedler LV, Blee KA, Butt VS, Davies DR (2002) The apoplastic oxidative burst in response to biotic stress in plants: a three-component system. J Exp Bot 53:1367–1376

Buchanan BB, Balmer Y (2005) Redox regulation: a broadening horizon. Annu Rev Plant Biol 56:187–220

Chew O, Whelan J, Millar AH (2003) Molecular definition of the ascorbate-glutathione cycle in *Arabidopsis* mitochondria reveals dual targeting of antioxidant defenses in plants. J Biol Chem 278:46869–46877

Colvile L, Smirnoff N (2008) Antioxidant status, peroxidase activity, and PR protein transcript levels in ascorbate-deficient *Arabidopsis thaliana vtc* mutants. J Exp Bot 59:1–12

Conklin PL, Williams EH, Last RL (1996) Environmental stress sensitivity of an ascorbic acid-deficient *Arabidopsis* mutant. Proc Natl Acad Sci USA 93:9970–9974

Creissen G, Reynolds H, Xue Y, Mullineaux P (1995) Simultaneous targeting of pea glutathione reductase and of a bacterial fusion protein to chloroplasts and mitochondria in transgenic tobacco. Plant J 8:167–175

Dat JF, Pellinen R, Beeckman T, Van De Cotte B, Langebartels C, Kangasjärvi J, Inzé D, Van Breusegem F (2003) Changes in hydrogen peroxide homeostasis trigger an active cell death process in tobacco. Plant J 33:621–632

Davletova S, Rizhsky L, Liang H, Shengqiang Z, Oliver DJ, Coutu J, Shulaev V, Schlauch K, Mittler R (2005) Cytosolic ascorbate peroxidase 1 is a central component of the reactive oxygen gene network of *Arabidopsis*. Plant Cell 17:268–281

de Pinto MC, Tommasi F, De Gara L (2002) Changes in the antioxidant systems as part of the signaling pathway responsible for the programmed cell death activated by nitric oxide and reactive oxygen species in tobacco bright-yellow 2 cells. Plant Physiol 130:1–11

De Tulio MC, Arrigoni O (2004) Hopes, disillusions and more hopes form vitamin C. Cell Mol Life Sci 61:209–219

del Río LA, Corpas FJ, Palma JM, Sandalio LM, Palma JM, Gómez M, Barroso JB (2002) Reactive oxygen species, antioxidant systems and nitric oxide in peroxisomes. J Exp Bot 53:1255–1272

Desikan R, Mackerness SAH, Hancock JT, Neill SJ (2001) Regulation of the *Arabidopsis* transcriptome by oxidative stress. Plant Physiol 127:159–172

Dietz K-J (2008) Redox signal integration: from stimulus to networks and genes. Physiol Plant 133:459–468

Dipierro S, Borracino G (1991) Dehydroascorbate reductase from potato tubers. Phytochemistry 30:427–429
Dixon D, Davis PBG, Edwards R (2002) Functional divergence in the glutathione transferase superfamily in plants. J Biol Chem 277:30859–30869
Doke N (1983) Involvement of superoxide anion generation in the hypersensitive response of potato-tuber tissues to infection with an incompatible race of *Phytophthora infestans* and to the hyphal wall components. Physiol Plant Pathol 23:345–357
Dong X, Ji R, Guo X, Foster SJ, Chen H, Dong C, Liu Y, Hu Q, Liu S (2008) Expressing a gene encoding wheat oxalate oxidase enhances resistance to *Sclerotinia sclerotiorum* in oilseed rape (*Brassica napus*). Planta 228:331–340
Dutilleul C, Garmier M, Noctor G, Mathieu C, Chétrit P, Foyer CH, De Paepe R (2003) Leaf mitochondria modulate whole cell redox homeostasis, set antioxidant capacity and determine stress resistance through altered signaling and diurnal regulation. Plant Cell 15:1212–1226
El-Zahaby HM, Gullner G, Király Z (1995) Effects of powdery mildew infection of barley on the ascorbate-glutathione cycle and other antioxidants in different host-pathogen interactions. Phytopathology 85:1225–1230
Fobert PR, Despres C (2005) Redox control of systemic acquired resistance. Curr Opin Plant Biol 8:378–382
Fotopoulos V, De Tullio MC, Barnes J, Kanellis AK (2008) Altered stomatal dynamics in ascorbate oxidase over-expressing tobacco plants suggest a role for dehydroascorbate signalling. J Exp Bot 59:729–737
Fotopoulos V, Sanmartin M, Kanellis AK (2006) Effect of ascorbate oxidase over-expression on ascorbate recycling gene expression in response to agents imposing oxidative stress. J Exp Bot 57:3933–3943
Foyer CH, Noctor G (2005) Oxidant and antioxidant signalling in plants: a re-evaluation of the concept of oxidative stress in a physiological context. Plant Cell Environ 28:1056–1071
Foyer CH, Noctor G (2009) Redox regulation in photosynthetic organisms: signaling, acclimation, and practical implications. Antiox Redox Signal 11:862–905
Foyer CH, Theodoulou FL, Delrot S (2001) The functions of inter- and intracellular glutathione transport systems in plants. Trends Plant Sci 6:486–492
Gechev TS, Van Breusegem F, Stone JM, Denev I, Laloi C (2006) Reactive oxygen species as signals that modulate plant stress responses and programmed cell death. BioEssays 28:1091–1101
Govrin EM, Levine A (2000) The hypersensitive response facilitates plant infection by the necrotrophic pathogen *Botrytis cinerea*. Curr Biol 10:751–757
Govrin EM, Rachmilevitch S, Tiwari BS, Solomon M, Levin A (2006) An elicitor from *Botrytis cinerea* induces the hypersensitive response in *Arabidopsis thaliana* and other plants and promotes the gray mold disease. Phytopathology 96:299–307
Green MA, Fry SC (2005) Degradation of vitamin C in plant cells via enzymic hydrolysis of 4-*O*-oxalyl-1-threonate. Nature 433:83–87
Gullner G, Tóbiás I, Fodor J, Kömives T (1999) Elevation of glutathione level and activation of glutathione-related enzymes affect virus infection in tobacco. Free Radic Res 31:55–161
Heil M (1999) Systemic acquired resistance: available information and open ecological questions. J Ecol 87:341–346
Horemans N, Foyer CH, Asard H (2000) Transport and action of ascorbate at the plant plasma membrane. Trends Plant Sci 5:263–267
Hossain MA, Asada K (1984) Purification of dehydroascorbate reductase from spinach and its characterisation as a thiol enzyme. Plant Cell Physiol 25:85–92
Hou WC, Lin YH (1997) Dehydroascorbate reductase and monodehydroascorbate reductase activities of trypsin inhibitors, the major sweet potato (*Ipomoea batatas* (L.) Lam) root storage protein. Plant Sci 128:151–158
Hückelhoven R, Fodor J, Preis C, Kogel K-H (1999) Hypersensitive cell death and papilla formation in barley attacked by the powdery mildew fungus are associated with hydrogen peroxide but not with salicylic acid accumulation. Plant Physiol 119:1251–1260

Jiménez A, Hernández J, del Rio L, Sevilla F (1997) Evidence for the presence of the ascorbate-glutathione cycle in mitochondria and peroxisomes of pea leaves. Plant Physiol 114:275–84
Kliebenstein DJ, Rowe HC (2008) Ecological costs of biotrophic versus necrotrophic pathogen resistance, the hypersensitive response and signal transduction. Plant Sci 174:551–556
Kombrink E, Hahlbrock K (1986) Responses of cultured parsley cells to elicitors from phytopathogenic fungi. Plant Physiol 81:216–221
Kuźniak E (2002) Transgenic plants: an insight into oxidative stress tolerance mechanisms. Acta Physiol Plant 24:97–113
Kuźniak E, Niewiadomska E, Miszalski Z, Karpiński S (2009) The role of chloroplasts and redox status in holistic regulation of stress responses in plants. In: Maksymiec W (ed) Compartmentation of responses to stresses in higher plants, true or false. Transworld Research Network, Kerala, India, pp 163–192
Kuźniak E, Skłodowska M (1999) The effect of *Botrytis cinerea* infection on ascorbate-glutathione cycle in tomato leaves. Plant Sci 148:69–76
Kuźniak E, Skłodowska M (2001) Ascorbate, glutathione and related enzymes in chloroplasts of tomato leaves infected by *Botrytis cinerea*. Plant Sci 160:723–731
Kuźniak E, Skłodowska M (2004) Differential implication of glutathione, glutathione-metabolizing enzymes and ascorbate in tomato resistance to *Pseudomonas syringae*. J Phytopathol 152:529–536
Kuźniak E, Skłodowska M (2005a) Compartment-specific role of the ascorbate-glutathione cycle in the response of tomato leaf cells to *Botrytis cinerea*. J Exp Bot 56:921–933
Kuźniak E, Skłodowska M (2005b) Fungal pathogen-induced changes in the antioxidant systems of leaf peroxisomes from infected tomato plants. Planta 222:192–200
Laloi C, Apel H, Danon A (2004) Reactive oxygen signalling: the latest news. Curr Opin Plant Biol 7:323–328
Lamb C, Dixon RA (1997) The oxidative burst in plant disease resistance. Annu Rev Plant Physiol Plant Mol Biol 48:251–275
Leipner J, Stamp P, Fracheboud Y (2000) Artificially increased ascorbate content affects zeaxanthin formation but not thermal energy dissipation or degradation of antioxidants during cold-induced photooxidative stress in maize leaves. Planta 210:964–969
Lemaire SD, Michelet L, Zaffagnini M, Massot V, Issakidis-Bourguet E (2007) Thioredoxins in chloroplasts. Curr Genet 51:343–365
Levine A, Tenhaken R, Dixon R, Lamb C (1994) H_2O_2 from the oxidative burst orchestrates the plant hypersensitive disease resistance response. Cell 79:583–593
Lorrain S, Vailleau F, Balagué C, Roby D (2003) Lesion mimic mutants: keys of deciphering cell death and defense pathways in plants? Trends Plant Sci 8:263–271
Lu H, Higgins VJ (1999) The effect of hydrogen peroxide on the viability of tomato cells and the fungal pathogen *Cladosporium fulvum*. Physiol Mol Plant Pathol 54:131–143
Mahalingam R, Fedoroff N (2003) Stress response, cell death and signalling: the many faces of reactive oxygen species. Physiol Plant 119:56–68
Małolepsza U, Urbanek H (2002) o-Hydroxyethylorutin-mediated enhancement of tomato resistance to *Botrytis cinerea* depends on a burst of reactive oxygen species. J Phytopathol 150:616–624
Mateo A, Mühlenbock P, Rusterucci C, Chang CC-C, Miszalski Z, Karpinska B, Parker JE, Mullineaux PM, Karpinski S (2004) *LESION SIMULATING DISEASE 1* is required for acclimation to conditions that promote excess excitation energy. Plant Physiol 136:2818–2830
May MJ, Hammond-Kosack KE, Jones JDG (1996) Involvement of reactive oxygen species, glutathione metabolism, and lipid peroxidation in the Cf-gene-dependent defense response of tomato cotyledons induced by race-specific elicitors of *Cladosporium fulvum*. Plant Physiol 110:1367–1379
Mayer AM, Staples RC, Gil-ad NL (2001) Mechanisms of survival of necrotrophic fungal plant pathogens in hosts expressing the hypersensitive response. Phytochemistry 58:33–41
Mellersh DG, Foulds IV, Higgins VJ, Heath MC (2002) H_2O_2 plays different roles in determining penetration failure in three diverse plant-fungal interactions. Plant J 29:257–268

Meyer AJ (2008) The integration of glutathione homeostasis and redox signalling. J Plant Physiol 165:1390–1403

Miller G, Coutu J, Shulaev V, Mittler R (2008) Reactive oxygen signalling in plants. Annu Plant Rev 33:189–201

Mittler R (2002) Oxidative stress, antioxidants and stress tolerance. Trends Plant Sci 7:405–410

Mittler R, Feng X, Cohen M (1998) Post-transcriptional suppression of cytosolic ascorbate peroxidase expression during pathogen-induced programmed cell death in tobacco. Plant Cell 10:461–474

Mittler R, Herr EH, Orvar BL, Van Camp W, Willekens H, Inzé D, Ellis BE (1999) Transgenic tobacco plants with reduced capability to detoxify reactive oxygen intermediates are hyperresponsive to pathogen infection. Proc Natl Acad Sci 96:14165–14170

Mittler R, Poulos TL (2005) Ascorbate peroxidase. In: Smirnoff N (ed) Antioxidants and reactive oxygen species in plants. Blackwell Publishing, Oxford, pp 87–100

Mittler R, Vanderauwera S, Gollery M, Van Breusegem F (2004) Reactive oxygen gene network of plants. Trends Plant Sci 9:490–498

Mou Z, Fan W, Dong X (2003) Inducers of plant systemic acquired resistance regulate the NPR1 function through redox changes. Cell 113:935–944

Muckenschnabel I, Goodman BA, Williamson B, Lyon GD, Deighton N (2002) Infection of leaves of *Arabidopsis thaliana* by *Botrytis cinerea*: changes in ascorbic acid, free radicals and lipid peroxidation products. J Exp Bot 53:207–214

Muckenschnabel I, Williamson B, Goodman BA, Lyon GD, Stewart D, Deighton N (2001) Markers of oxidative stress associated with soft rots in French beans (*Phaseolus vulgaris*) infected by *Botrytis cinerea*. Planta 212:376–381

Mullineaux PM, Rausch T (2005) Glutathione, photosynthesis and the redox regulation of stress-responsive gene expression. Photosynth Res 86:459–474

Nakano Y, Asada K (1981) Hydrogen peroxide is scavenged by ascorbate-specific peroxidase in spinach chloroplasts. Plant Cell Physiol 22:867–880

Ndamukong I, Abdallat AA, Thurow C, Fode B, Zander M, Weigel R, Gatz C (2007) SA-inducible *Arabidopsis* glutaredoxin interacts with TGA factors and suppresses JA-responsive PDF1.2 transcription. Plant J 50:128–139

Noctor G (2006) Metabolic signalling in defence and stress: the central roles of soluble redox couples. Plant Cell Environ 29:409–425

Noctor G, Foyer CH (1998) Ascorbate and glutathione: keeping active oxygen under control. Annu Rev Plant Physiol Plant Mol Biol 49:249–279

Obara K, Sumi K, Fukuda H (2002) The use of multiple transcription starts causes the dual targeting of *Arabidopsis* putative monodehydroascorbate reductase to both mitochondria and chloroplasts. Plant Cell Physiol 43:697–705

Pastori GM, Kiddle G, Antoniw J, Bernard S, Veljovic-Jovanovic S, Verrier PJ, Noctor G, Foyer CH (2003) Leaf vitamin C contents modulate plant defense transcripts and regulate genes controlling development through hormone signaling. Plant Cell 15:1212–1226

Patykowski J, Urbanek H (2003) Activity of enzymes related to H_2O_2 generation and metabolism in leaf apoplastic fraction of tomato leaves infected with *Botrytis cinerea*. J Phytopathol 151:153–161

Patykowski J, Urbanek H (2005) The activity of antioxidative enzymes, contents of H_2O_2 and of ascorbate in tomato leaves of cultivars more or less sensitive to in fection with *Botrytis* cinerea. Acta Physiol Plant 27:193–203

Pavet V, Olmos E, Kiddle G, Mowla S, Kumar S, Antoniw J, Alvarez ME, Foyer CH (2005) Ascorbic acid deficiency activates cell death and disease resistance responses in Arabidopsis. Plant Physiol 139:1291–1303

Peng M, Kuć J (1992) Peroxidase-generated hydrogen peroxide as a source of antifungal activity *in vitro* and on tobacco leaf discs. Phytopathology 82:696–699

Pieterse CMJ, Van Loon LC (2004) NPR1: the spider in the web of induced resistance signalling pathways. Curr Opin Plant Biol 7:456–464

Pignocchi C, Foyer CH (2003) Apoplastic ascorbate metabolism and its role in the regulation of cell signalling. Curr Opin Plant Biol 6:379–389

Pignocchi C, Kiddle G, Hernández I, Foster SJ, Asensi A, Taybi T, Barnes J, Foyer CH (2006) Ascorbate oxidase-dependent changes in the redox state of the apoplast modulate gene transcript accumulation leading to modified hormone signaling and orchestration of defense processes in tobacco. Plant Physiol 141:423–435

Pitzschke A, Forzani C, Hirt H (2006) Reactive oxygen species signaling in plants. Antioxid Redox Signal 8:1757–1764

Polle A (2001) Dissecting the superoxide dismutase-ascorbate-glutathione-pathway in chloroplasts by metabolic modeling. Computer simulations as a step towards flux analysis. Plant Physiol 126:445–462

Potters G, De Gara L, Asard H, Horemans N (2002) Ascorbate and glutathione: guardians of the cell cycle, partners in crime? Plant Physiol Biochem 40:537–548

Potters G, Horemans N, Jansen MAK (2010) The cellular redox state in plant stress biology – a charging concept. Plant Physiol Biochem. doi:10.1016/j.plaphy.2009.12.007

Queval G, Hager J, Gakière B, Noctor G (2008) Why are literature data for H_2O_2 contents so variable? A discussion of potential difficulties in the quantitative assay of leaf extracts. J Exp Bot 59:135–146

Raven EL (2003) Understanding functional diversity and substrate specificity in haem peroxidases: what can we learn from ascorbate peroxidase? Nat Prod Rep 20:367–381

Rouhier N (2010) Plant glutaredoxins: pivotal players in redox biology and iron-sulphur centre assembly. New Phytol. doi:10.1111/j.1469-8137.2009.03146.x

Rouhier N, Lemaire SD, Jacquot J-P (2008) The role of glutathione in photosynthetic organisms: emerging functions for glutaredoxins and glutathionylation. Annu Rev Plant Biol 59:143–66

Schafer FQ, Buettner GR (2001) Redox state of the cell as viewed through the glutathione disulfide/glutathione couple. Free Radic Biol Med 30:1191–1212

Scharte J, Schön H, Weis E (2005) Photosynthesis and carbohydrate metabolism in tobacco leaves during an incompatible interaction with *Phytophthora nicotianae*. Plant Cell Environ 28:1421–1435

Shetty NP, Jørgensen HJL, Jensen JD, Collinge DB, Shetty HS (2008) Roles of reactive oxygen species in interactions between plants and pathogens. Eur J Plant Pathol 121:267–280

Solomon M, Belenghi B, Delledonne M, Menachem E, Levine A (1999) The involvement of cysteine protease inhibitor genes in the regulation of programmed cell death in plants. Plant Cell 11:431–443

Song X-S, Wang Y-J, Mao W-H, Shi K, Zhou Y-H, Nogués S, Yu J-Q (2009) Effects of cucumber mosaic virus infection on electron transport and antioxidant system in chloroplasts and mitochondria of cucumber and tomato leaves. Physiol Plant 135:246–257

Spoel SH, Johnson JS, Dong X (2007) Regulation of tradeoffs between plant defenses against pathogens with different lifestyles. Proc Natl Acad Sci USA 104:18842–18847

Takahama U (2004) Oxidation of vacuolar and apoplastic phenolic substrates by peroxidase: physiological significance of the oxidation reactions. Phytochem Rev 3:207–219

Takahama U, Oniki T (1992) Regulation of peroxidasedependent oxidation of phenolics in the apoplast of spinach leaves by ascorbate. Plant Cell Physiol 33:379–387

Tausz M, Šircelj H, Grill D (2004) The glutathione system as a stress marker in plant ecophysiology: is a stress-response concept valid? J Exp Bot 55:1955–1962

Tenhaken R, Levine A, Brisson LF, Dixon RA, Lamb C (1995) Function of the oxidative burst in hypersensitive disease resistance. Proc Natl Acad Sci 92:4158–4163

Thordal-Christensen H, Zhang Z, Yangdou W, Collinge DB (1997) Subcellular localization of H_2O_2 in plants. H_2O_2 accumulation in papillae and hypersensitive response during the barley-powdery mildew interaction. Plant J 11:1187–1194

Torres MA, Jones JDG, Dangl JL (2006) Reactive oxygen species signaling in response to pathogens. Plant Physiol 141:373–378

Unger CH, Kleta S, Jandl G, Tiedemann A (2005) Suppression of the defence-related oxidative burst in bean leaf tissue and bean suspension cells by the necrotrophic pathogen *Botrytis cinerea*. J Phytopathol 153:15–26

Urano J, Nakagawa T, Maki Y, Masumura T, Tanaka K, Murata N, Ushimaru T (2000) Molecular cloning and characterization of a rice dehydroascorbate reductase. FEBS Lett 466:107–111

Vanacker H, Carver TLW, Foyer CH (1998a) Pathogen-induced changes in the antioxidant status of the apoplast in barley leaves. Plant Physiol 117:1103–1114

Vanacker H, Carver TLW, Foyer CH (2000) Early H_2O_2 accumulation in mesophyll cells leads to induction of glutathione during the hypersensitive response in the barley-powdery mildew interaction. Plant Physiol 123:1289–300

Vanacker H, Harbinson J, Ruisch J, Carver TLW, Foyer CH (1998b) Antioxidant defences of the apoplast. Protoplasma 205:129–140

Veljovic-Jovanovic S, Noctor G, Foyer CH (2002) Are leaf hydrogen peroxide concentrations commonly overestimated? The potential influence of artefactual interference by tissue phenolics and ascorbate. Plant Physiol Biochem 40:501–507

Von Tiedemann A (1997) Evidence for a primary role of oxygen species in induction of host cell death during infection of bean leaves with *Botrytis cinerea*. Physiol Mol Plant Pathol 50:151–166

Vranová E, Inzé D, Van Breusegem F (2002) Signal transduction during oxidative stress. J Exp Bot 53:1227–1236

Walz A, Zingen-Sell I, Theisen S, Kortekamp A (2008) Reactive oxygen intermediates and oxalic acid in the pathogenesis of the necrotrophic fungus *Sclerotinia sclerotiorum*. Eur J Plant Pathol 120:317–330

Wingate VPM, Lawton MA, Lamb CJ (1988) Glutathione causes a massive and selective induction of plant defense genes. Plant Physiol 87:206–210

Wojtaszek P (1997) Oxidative burst: an early plant response to pathogen infection. Biochem J 332:681–692

Yang H-L, Zhao Y-R, Wang C-L, Yang Z-L, Zeng Q-Y, Lu H (2009) Molecular characterization of a dehydroascorbate reductase from *Pinus bungeana*. J Integr Plant Biol. doi:10.1111/j.1744-7909.2009.00848.x

Yao N, Greenberg JT (2006) Arabidopsis ACCELERATED CELL DEATH2 modulates programmed cell death. Plant Cell 18:397–411

Chapter 5
Regulation of the Ascorbate–Glutathione Cycle in Plants Under Drought Stress

Adriano Sofo, Nunzia Cicco, Margherita Paraggio, and Antonio Scopa

Abstract Acclimation of plants to drought is often associated with increased levels of reactive oxygen species (ROS), such as superoxide anion ($O_2{\cdot}^-$), hydrogen peroxide (H_2O_2), hydroxyl radical (HO·) and singlet oxygen ($^1O_2$), which are toxic for the cells. ROS are by-products of aerobic metabolism, and their production is enhanced during drought conditions through the disruption of electron transport system and oxidizing metabolic activities occurring in chloroplasts, mitochondria and microbodies. Under non-stressful conditions, ROS are efficiently eliminated by non-enzymatic and enzymatic antioxidants, whereas during drought conditions the production of ROS exceeds the capacity of the antioxidative systems to remove them, causing oxidative stress. The non-enzymatic antioxidant system includes ascorbate and glutathione, located both within the cell and in the apoplast. They are two constituents of the antioxidative ascorbate–glutathione cycle which detoxify H_2O_2 in the chloroplasts. Ascorbate (AsA) is a major primary antioxidant compound synthesized on the inner membrane of the mitochondria which reacts chemically with $^1O_2$, $O_2{\cdot}^-$, HO· and thiyl radical, and acts as the natural substrate of many plant peroxidases. Moreover, AsA is involved in other functions such as plant growth, gene regulation, modulation of some enzymes, and redox regulation of membrane-bound antioxidant compounds. Glutathione (GSH) is a tripeptide synthesized in the cytosol and chloroplasts which scavenges $^1O_2$ and H_2O_2, and it is oxidized to glutathione disulfide (GSSG) when acts as an antioxidant and redox regulator. GSH is the substrate of glutathione S-transferases, which have a protective role in the detoxification of xenobiotics, and dehydroascorbate reductase (DHAR). Finally, GSH is a precursor of phytochelatins, which regulate cellular heavy metals levels, and is involved in gene expression. This review, based on the most significant studies published in

A. Sofo (✉) and A. Scopa
Dipartimento di Scienze dei Sistemi Colturali, Forestali e dell'Ambiente,
Università degli Studi della Basilicata, Via dell'Ateneo Lucano 10, 85100 Potenza, Italy
e-mail: adriano.sofo@unibas.it

N. Cicco and M. Paraggio
Istituto di Metodologie per l'Analisi Ambientale, Consiglio Nazionale delle Ricerche,
C. da S. Loia, Zona Industriale, 85050 Tito Scalo (PZ), Italy

N.A. Anjum et al. (eds.), *Ascorbate-Glutathione Pathway and Stress Tolerance in Plants*,
DOI 10.1007/978-90-481-9404-9_5, © Springer Science+Business Media B.V. 2010

the last decade, focuses on the changes of antioxidant enzyme activities (ascorbate peroxidase, APX; monodehydroascorbate reductase, MDHAR; dehydroascorbate reductase, DHAR; glutathione reductase, GR), and of the levels of some compounds involved in the ascorbate–glutathione cycle (ascorbate and glutathione pools , H_2O_2 and α-tocopherol) in plants grown under water shortage.

Keywords Antioxidant enzymes • Ascorbate-glutathione cycle • Ascorbate peroxidase • Dehydroascorbate reductase • Drought stress • Glutathione reductase • Oxidative stress • Water deficit

1 Introduction

Drought stress is one of the main environmental factors limiting plant growth and yield worldwide, and it is the most prevalent cause of crop yield loss but also the most difficult to tackle because of the strong link between transpiration and photosynthesis (Smirnoff 1998; Posch and Bennett 2009). Acclimation of plants to drought is often associated with increased levels of reactive oxygen species (ROS), such as superoxide anion ($O_2{\cdot}^-$), hydrogen peroxide (H_2O_2), hydroxyl radical (HO·) and singlet oxygen ($^1O_2$), which are toxic for the cells (Smirnoff 1993; Chaves et al. 2003). ROS are by-products of aerobic metabolism and their production is enhanced during drought conditions through the disruption of electron transport system, and oxidizing metabolic activities occurring in chloroplasts, mitochondria and microbodies (Asada 1999; Van Breusegem et al. 2001).

Under non-stressful conditions, ROS are efficiently eliminated by non-enzymatic and enzymatic antioxidants (Fig. 1), whereas during drought conditions the production of ROS exceeds the capacity of the antioxidative systems to remove them, causing oxidative stress (Smirnoff 1998; Morales et al. 2006).

The non-enzymatic antioxidant system includes ascorbate and glutathione, located both within the cell and in the apoplast (Horemans et al. 2000; Foyer et al. 2001). They are two constituents of the antioxidative ascorbate–glutathione cycle which detoxify hydrogen peroxide (H_2O_2) in the chloroplasts (Asada 1999) (Figs. 1 and 2). Ascorbate (AsA) is a major primary antioxidant compound synthesized on the inner membrane of the mitochondria which reacts chemically with $^1O_2$, $O_2{\cdot}^-$, HO· and thiyl radical (Noctor and Foyer 1998; Asada 1999), and acts as the natural substrate of many plant peroxidases (Mehlhorn et al. 1996). One of the important functions of AsA is the protection against oxidative damage of plant cells through the scavenging of H_2O_2 mediated by ascorbate peroxidase (APX) which has a higher affinity for H_2O_2 than catalase (CAT) or peroxidase isoforms (Srivalli et al. 2003; Mittler and Poulos 2005). In bright light, or when low temperatures and drought limit CO_2 fixation, the excess excitation energy is dissipated in the light harvesting antennae as heat by zeaxanthin, that is formed by successive de-epoxidation of the xanthophyll cycle pigments violaxanthin and antheroxanthin. The deepoxidase, which is bound to the lumen side of the thylakoid membrane, is

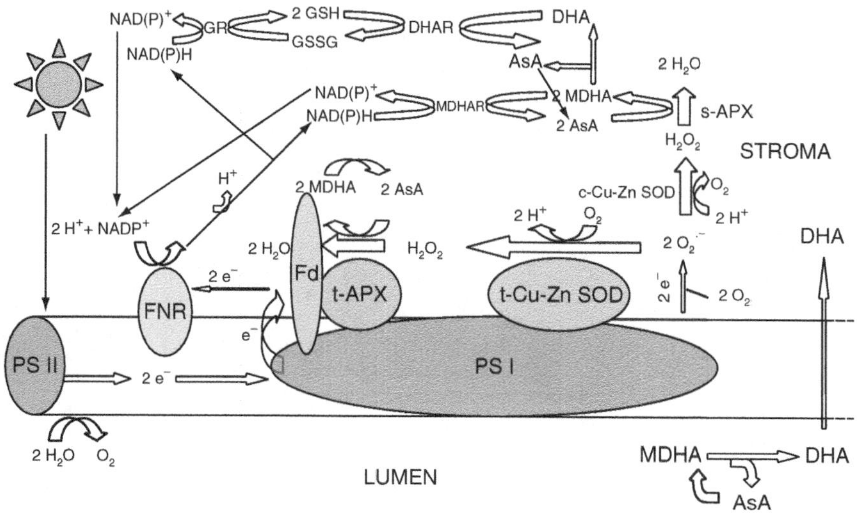

Fig. 1 Antioxidant system of plant chloroplasts.The thylakoidal antioxidant system includes Cu–Zn–superoxide dismutase (t–Cu/Zn–SOD), present on the thylakoidal surfaces (in many plant species, t–Cu–Zn–SOD is substituted by t–Fe–SOD), thylakoidal ascorbate peroxidase (t–APX) and ferredoxin (Fd). Fd reduces monodehydroascorbate (MDHA) directly to ascorbate (AsA). The stromatic antioxidant system is composed by stromatic Cu–Zn–superoxide dismutase (t–Cu/Zn–SOD), stromatic APX (s-APX), monodehydroascorbate reductase (MDHAR), dehydroascorbate reductase (DHAR) and glutathione reductase (GR). NAD(P)H is used for the reduction of monodehydroascorbate (MDHA), whereas dehydroascorbate (DHA) is photo-generated by ferredoxin-NADP$^+$-oxidoreductase (FNR). MDHA is also produced in chloroplast lumen by violaxantin de-epoxidase or when AsA releases electron to the two photosystems (PS I or PS II). MDHA is rapidly transformed in AsA and DHA. This latter enters the lumen by thylakoidal membranes and is reduced to AsA

dependent on AsA as a cofactor (Smirnoff 2005). Moreover, AsA is involved in other functions such as plant growth, gene regulation, modulation of some enzymes, and redox regulation of membrane-bound antioxidant compounds (Horemans et al. 2000; Foyer et al. 2001). Glutathione (GSH), one of the major redox buffers in most aerobic cells, is a tripeptide synthesized in the cytosol and the chloroplast which scavenges $^1O_2$ and H_2O_2, and is oxidized to glutathione disulfide (GSSG) when acts as an antioxidant and as a regulator of redox status and gene expression (Briviba et al 1997; Smirnoff 1998; Foyer et al. 2001). Furthermore, GSH is the substrate of glutathione S-transferases, which have a protective role in the detoxification of xenobiotics, phospholipid hydroperoxide glutathione peroxidase, that use glutathione to reduce H_2O_2 and lipid hydroperoxides, and dehydroascorbate reductase (DHAR), a key enzyme of the ascorbate–glutathione cycle (Foyer et al. 2001; Yang et al. 2006). Finally, GSH is a precursor of phytochelatins, which regulate cellular heavy metals levels, and is involved in gene expression (Noctor and Foyer 1998). In addition to ascorbate and glutathione, α-tocopherol (α-toc, vitamin E) found in leaf chloroplasts takes part to the ascorbate–glutathione cycle as it deactivates photosynthesis-derived ROS and prevents the propagation of lipid peroxidation by scavenging lipid peroxyl radicals in thylakoid membranes (Munné-Bosch et al. 2001).

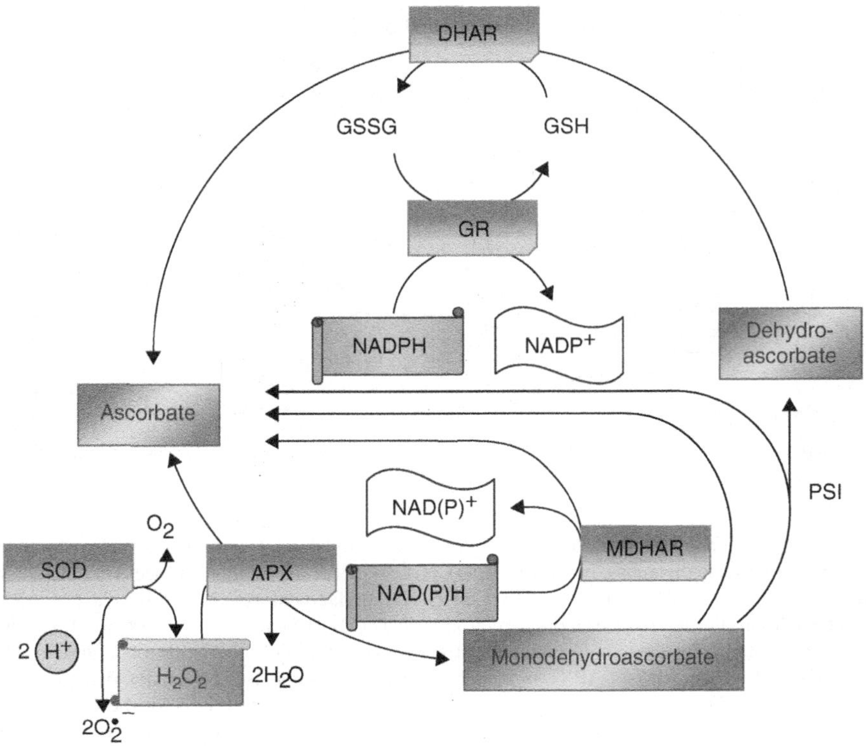

Fig. 2 The ascorbate-glutathione cycle in plants. Ascorbate peroxidase (APX), dehydroascorbate reductase (DHAR), monodehydroascorbate reductase (MDHAR), and glutathione reductase (GR)

Successively, the scavenging of lipid peroxyl radicals results in the formation of tocopheroxyl radicals, which can be recycled back to α-tocopherol by ascorbate.

The enzymatic antioxidant system, that operates both in the chloroplasts and in cytosol, includes the enzymes of the ascorbate–glutathione cycle: ascorbate peroxidase (APX, EC 1.11.1.11), monodehydroascorbate reductase (MDHAR, EC 1.6.5.4), dehydroascorbate reductase (DHAR, EC 1.8.5.1) and glutathione reductase (GR, EC 1.6.4.2) (Figs. 1 and 2). Activities of enzymes in the ascorbate glutathione cycle are increased by drought and low temperature suggesting a requirement for increased activity of the cycle under these conditions (Smirnoff 1996). Furthermore, the mRNAs corresponding to the genes of antioxidant enzymes are induced by drought stress (Reddy et al. 2004b). Ascorbate peroxidise isozymes, able to scavenge the H_2O_2 produced by SOD using ascorbate as the electron donor, are generally located in chloroplasts, but microsomal, peroxisomal and membrane-bound forms, as well as soluble cytosolic and apoplastic isozymes, also exist (Quan et al. 2008). Moreover, APX can scavenge H_2O_2 that is inaccessible for CAT because of their higher affinity for H_2O_2 and presence in different subcellular locations (Van Breusegem et al. 2001). Monodehydroascorbate (MDHA), a free radical intermediate produced by APX catalysis, can disproportionate spontaneously to AsA and dehydroascorbate (DHA) or be enzymatically reduced to AsA by MDHAR, a FAD enzyme with an high specificity to MDHA which uses NAD(P)H as a reductant (Smirnoff 2005). DHAR is a monomeric thiol enzyme that reduces DHA to AsA using GSH as an electron donor, with the consequent production of GSSG (Foyer et al. 2005). DHAR has been frequently implied as a biochemical indicator of oxidative stress in plant metabolism (Vadassery et al. 2009) but a characterization of DHAR has remained elusive because of rapid loss of enzyme activity. The isoforms of GR are flavoenzymes with a redox cystine residue in their active sites which maintain the intracellular glutathione pool in the reduced status, catalysing the NADPH-dependent reduction of GSSH to GSH (Foyer et al. 2005). Morell et al. (1997) had tried to demonstrate that the regeneration of AsA is not coupled to a glutathione-dependent DHAR, and that GR is not directly involved in the regeneration of AsA but Foyer and Mullineaux (1998) and many successive works definitively proved that both DHAR and GR have a key role against oxidative stress.

Excessive levels of ROS damage cellular structures and macromolecules, causing photoinhibition of photosynthetic apparatus, but also activate multiple defence responses, thus having also a positive role (Van Breusegem et al. 2001; Vranová et al. 2002; Foyer and Noctor 2003; Laloi et al. 2004) (Fig. 3). This dualism

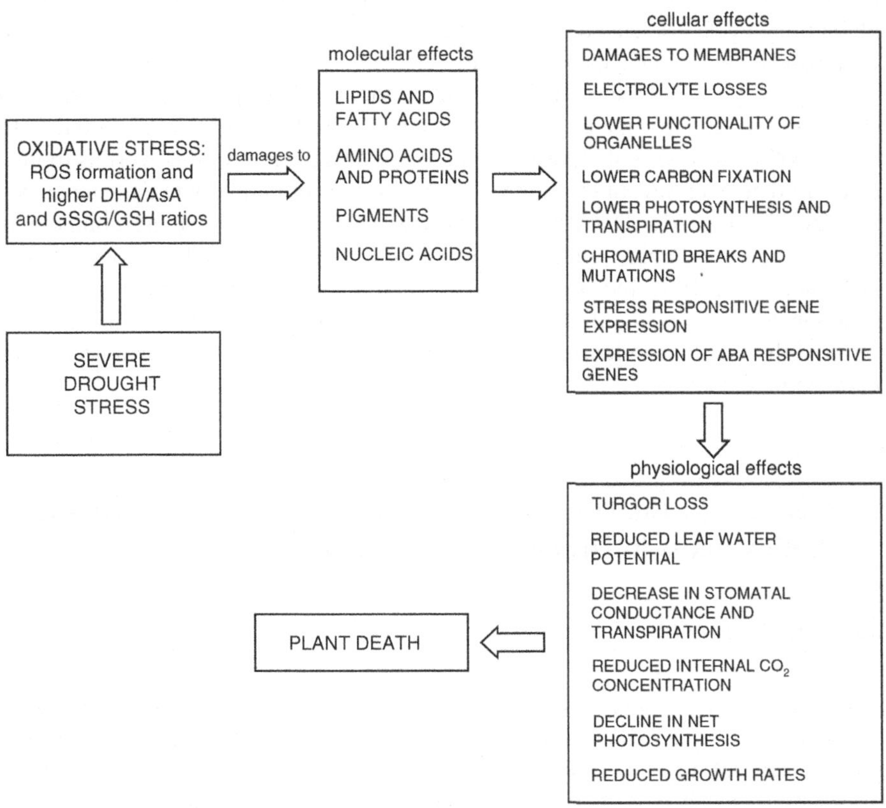

Fig. 3 Molecular and cellular effects of drought-mediated oxidative stress

can be obtained only when cellular levels of ROS are tightly controlled at both the production and consumption levels (Van Breusegem et al. 2001; Quan et al. 2008). Foyer and Noctor (2005) highlighted the crucial role of ROS as second messengers in signal transduction cascades in processes as diverse as mitosis, tropisms and cell death. In particular, the presence of H_2O_2 in the apoplast is toxic for pathogens, is involved in gene transcription and systemic acquired resistance, and slows down the spread of invading organisms by cell death round the infection and a rapid local cross-linking of the cell wall (Horemans et al. 2000; Smirnoff 2000). Other two major low molecular weight antioxidants, such as ascorbate and glutathione determine the specificity of the transduced signal in cells, and are also themselves signal-transducing molecules that can either signal independently or further transmit ROS signals (Foyer and Noctor 2005). For all these reasons, in contrast to this pejorative or negative term, implying a state to be avoided, the presence of ROS in cellular apparatus would be more usefully described as 'oxidative signalling', that is, an important and critical function associated with the mechanisms by which plant cells sense the environment and make appropriate adjustments to gene expression, metabolism and physiology.

The response to water deficit of plant species is a well documented process but relatively few studies highlighted the importance of the enzymes of ascorbate–glutathione cycle associated to drought tolerance and/or resistance, and not much is known about the linkages between drought and the components of the ascorbate–glutathione cycle in some economically important C_3 plant species (e.g., fruit trees) (Scebba et al. 2001; Lima et al. 2002; Chai et al. 2005; Pinheiro et al. 2004; Sofo et al. 2005b; Guerfel et al. 2009). For these reasons, the aim of this work is to give an up-to-date overview of the studies on the changes of antioxidant enzyme activities (APX, MDHAR, DHAR and GR), and of the levels of some compounds involved in the ascorbate–glutathione cycle (ascorbate and glutathione pools, AsA/DHA and GSH/GSSG redox couples, H_2O_2 and α-toc) in plants grown under water shortage. Some of the significant changes in enzymatic and non-enzymatic antioxidants of the ascorbate–glutathione cycle in drought-stressed plants have been summarized in Table 1.

2 Changes in Enzyme Activities and Pools of Non-enzymatic Antioxidants in Drought-Stressed Plants

2.1 *Tree Species*

Plants are sessile organisms and their only alternative to a rapidly changing environment is a fast adaptation to the abiotic and biotic stresses. This concept is particularly valid for the physiological and biochemical responses (adaptation, avoidance, resistance or tolerance) against water deficit, among which there are the

Table 1 Changes in enzymatic and non-enzymatic antioxidants of the ascorbate–glutathione cycle in drought-stressed plants

No.	Species	Level of drought stress	Antioxidant	Tissue	Change in response of drought stress	Reference
1	*Acer saccharinum*	RWC from 55% to 28%	AsA	Embryo	From 25 to 33 µmol g^{-1} DW	Pukacka and Ratajczak 2006
			DHA		From 45 to 52 µmol g^{-1} DW	
			AsA/DHA		From 4.5 to 2.5	
			GSH		From 0 to 1,100 mmol g^{-1} DW	
			GSSG		From 600 to 1,450 mmol g^{-1} DW	
			GSH/GSSG		From 7.8 to 3.6	
			H_2O_2	Seeds	From 1.04 to 2.58 µg g^{-1} DW	
			$O_2^{\cdot -}$		From 0.71 to 1.27 ΔA_{530} g^{-1} DW	
2	*Allium schoenoprasum*	RWC from 70.7% to 53.2%	APX	Leaves	From 1.60 to 2.06 units mg^{-1} protein	Egert and Tevini 2002
3	*Anoda cristata*	6 days of water withholding	APX	Leaves	From 1.1 to 1.3 units mg^{-1} protein	Ratnayaka et al. 2003
4	*Arabidopsis thaliana*	Leaves Ψ_w from −0.65 up to −2.54 MPa	GR	Young leaves	From 0.21 to 0.24 µmol min^{-1} mg^{-1}	Jung 2004
				Mature leaves	From 0.21 to 0.34 µmol min^{-1} mg^{-1}	
5	*Arbutus unedo*	RWC from 83% to 53%	t-Asc	Leaves	From 18 to 30 µmol g DW^{-1}	Munné-Bosch and Peñuelas 2004
			DHA/t-Asc		From 0.09 to 0.14	
6	*Bupleurum chinense*	RWC from 93.02% to 45.78%	APX	Roots	From 19257 to 33262 µmol Vc g^{-1} FW h^{-1}	Zhu et al. 2009
			AsA		From 19.43 to 43.32 µg g^{-1} FW	
			H_2O_2		From 0.35 to 0.95 µmol g^{-1} FW	
7	*Capparis ovata*	RWC from 83.13% to 76.29%	APX	Leaves	From 0.2 to 3.7 units mg^{-1} protein	Ozkur et al. 2009
			GR		From 0.18 to 0.55 units mg^{-1} protein	

(continued)

Table 1 (continued)

No.	Species	Level of drought stress	Antioxidant	Tissue	Change in response of drought stress	Reference
8	*Catharanthus roseus* 'Rosea'	20 days interval drought	APX	Leaves	From 39 to 44 units mg^{-1}	Jaleel et al. 2008a, b
				Roots	From 28 to 32 units mg^{-1}	
			AsA	Roots	From 7 to 9 mg g^{-1} FW	
			GSH		From 8 to 9 mg g^{-1} FW	
			α-toc		From 13 to 15 mg g^{-1} FW	
	Catharanthus roseus 'Alba'	20 days interval drought	APX	Leaves	From 35 to 38 units mg^{-1}	
				Roots	From 20 to 24 units mg^{-1}	
			AsA	Roots	From 11 to 13 mg g^{-1} FW	
			GSH		From 9 to 11 mg g^{-1} FW	
			α-toc		From 12 to 14 mg g^{-1} FW	
9	*Cistus clusii*	RWC from 82% to 64%	t-Asc	Leaves	From 38 to 94 μmol dm^{-1} leaf surface	Hernández et al. 2004
			α-toc		From 120 to 400 nmol dm^{-1}	
10	*Coffea canephora*	Predawn $\Psi_w = -3.0$ MPa	APX	Leaves	From 0.3–0.5 to 0.5–0.7 μmol AsA min^{-1} mg^{-1} protein	Pinheiro et al. 2004
			DHAR		From 0.020–0.025 to 0.025–0.030 μmol NADPH min^{-1} mg^{-1} protein	
			t-Asc		From 28 to 30 μmol g^{-1} DW	
11	*Coffea canephora*	Predawn $\Psi_w = -3.0$ MPa	APX	Leaves	From 0.31–0.42 to 0.83–1.34 units mg^{-1} protein	Lima et al. 2002

12	*Cucumis sativus*	PEG solution 10% (w/v) for 3 days	APX	Leaves	From 125 to 170 units g^{-1} DW	Liu et al. 2009
			MDHAR		From 2,150 to 2,650 units g^{-1} DW	
			DHAR		From 1,000 to 1,100 units g^{-1} DW	
			GR		From 200 to 300 units g^{-1} DW	
			AsA		From 1,500 to 1,700 mg g^{-1} DW	
			AsA/DHA		From 3 to 4	
			GSH		From 200 to 650 μg g^{-1} DW	
			GSH/GSSG		From 20 to 40	
			H_2O_2		From 30 to 85 μmol g^{-1} DW	
			$O_2^{\cdot -}$		From 40 to 90 nmol min^{-1} g^{-1} DW	
13	*Eucalyptus globules* – clone 'ST5'	Predawn $\Psi_w = -2.43$ MPa	APX	Roots	From 0.22 to 0.95 mmol H_2O_2 g^{-1} DW min^{-1}	Shvaleva et al. 2005
			GR	Leaves	From 8.2 to 3.9 mmol H_2O_2 g^{-1} DW min^{-1}	
				Roots	From 0 to 8.2 H_2O_2 g^{-1} DW min^{-1}	
14	*Eucalyptus globules* – clone 'CN5'	Predawn $\Psi_w = -1.71$ MPa	APX	Leaves	From 27 to 7 mmol H_2O_2 g^{-1} DW min^{-1}	
				Roots	From 0.19 to 1.30 mmol H_2O_2 g^{-1} DW min^{-1}	
			GR	Leaves	From 15 to 2 mmol H_2O_2 g^{-1} DW min^{-1}	
				Roots	From 0 to 3.4 H_2O_2 g^{-1} DW min^{-1}	
15	*Fagus sylvatica*	SWC from 30% to 15%	t-Asc	Leaves	From about 6 to 12 μmol g^{-1} DW	Haberer et al. 2008
			α-toc		From 100 to 150 nmol g^{-1}	
			AsA		From 0.54–0.94 to 2.56–3.18 mg g^{-1} FW	
			MDHA		From 0.25–0.39 to 1.12–1.48 mg g^{-1} FW	

(continued)

Table 1 (continued)

No.	Species	Level of drought stress	Antioxidant	Tissue	Change in response of drought stress	Reference
16	*Glycine max*	Predawn $\Psi_w = -1.0$ MPa + chilling	APX	Leaves	From 10.7 to 27.8 µmol min^{-1} mg^{-1} protein	Riekert van Heerden and Krüger 2002
			GR		From 1.9 to 4.0 µmol min^{-1} mg^{-1} protein	
17	*Gossypium hirsutum*	6 days of water withholding	APX	Leaves	From 0.8 to 1.7 units mg^{-1} protein	Ratnayaka et al. 2003
18	*Helianthus anuus*	RWC from 90% to 40%	APX	Leaves	From 110 to 130 nmol min^{-1} mg^{-1} protein	Zhang and Kirkham 1996a, b
			DHAR		From 5 to 27 nmol min^{-1} mg^{-1} protein	
			GR		From 23 to 35 nmol min^{-1} mg^{-1} protein	
			AsA		From 5 to 1 mg g^{-1} DW	
			DHA		From 0.0 to 0.2 mg g^{-1} DW	
			GSH		From 2.2 to 0.0 mg g^{-1} DW	
			GSSG		From 0.00 to 0.38 mg g^{-1} DW	
19	*Laurus azorica*	RWC from 95% to 50–55%	APX	Leaves	From 1.26 to 7.08 µmol m^{-2} s^{-1}	Sánchez-Díaz et al. 2007
20	*Licopersicum esculentum*	RWC from 95% to 65%	APX	Leaves	From 2.2 to 5.9 units mg^{-1} protein	Nasibi and Kalantari 2009
			GR		From 0.100 to 0.075 units mg^{-1} protein	
			AsA		From 4.84 to 1.79 mg g^{-1} DW	
			DHA		From 0.95 to 2.37 mg g^{-1} DW	
			t-Glu		From 0.19 to 0.05 mmol g^{-1} DW	
			H_2O_2		From 21.05 to 127.88 µmol g^{-1} DW	
21	*Ligustrum vulgare*	Predawn $\Psi_w = -0.8$ MPa	APX	Leaves	From about 8 to 14 mol min^{-1} g FW^{-1}	Guidi et al. 2008
22	*Malus domestica*	Predawn $\Psi_w = -2.0$ MPa	AsA	Leaves	From 5.49 to 7.44 mg g^{-1} DW	Šircely et al. 2005, 2007
			GSSG/t-Glu		From 8.9 to 39.2	

23	*Medicago sativa*	PEG solution 35% (w/v) during germination	APX	Shoots	From about 2 to 8 units mg^{-1} protein	Wang et al. 2009
				Roots	From about 3 to 6 units mg^{-1} protein	
			H_2O_2	Shoots	From about 0.3 to 0.5 mol g^{-1} FW	
24	*Morus alba*	$\Psi_w = -2.50$ MPa	APX	Leaves	From 450–700 to 900–1,500 μmol mg^{-1} chl min^{-1}	Ramachandra Reddy et al. 2004b
			MDHAR		From 180–200 to 420–870 μmol mg^{-1} chl min^{-1}	
			GR		From 160–300 to 280–450 μmol mg^{-1} chl min^{-1}	
25	*Musa* AAA 'Berangan'	LWC from 93% to 72–75%	APX	Leaves	From 46.28 to 67.64 nmol AsA s^{-1} mg^{-1} protein	Chai et al. 2005
			GR		From 0.65 to 1.01 nmol NADPH s^{-1} mg^{-1} protein	
26	*Musa* AA 'Mas'	LWC from 93% to 72–75 %	APX	Leaves	From 43.73 to 44.91 nmol AsA s^{-1} mg^{-1} protein	
			GR		From 1.26 to 1.93 nmol NADPH s^{-1} mg^{-1} protein	
27	*Nicotiana tabacum*	4 weeks of a water-deficit cycle	GR	Leaves	From 500–600 to 100–300 units g^{-1} protein	Synková and Valcke 2001
28	*Olea europaea* 'Coratina'	Predawn $\Psi_w = -5.73$ MPa	APX	Leaves	From 3.88 to 13.77 units mg^{-1} DW	Sofo et al. 2005a
				Roots	From 0.23–0.36 to 0.34–0.51 units mg^{-1} DW	
29	*Olea europaea* 'Chemlali'	RWC from 95% to 40%	APX	Leaves	From about 0.3 to 1.4 units mg^{-1} protein	Ennajeh et al. 2009
30	*Olea europaea* 'Meski'	RWC from 95% to 40%	APX	Leaves	From about 0.3 to 1.8 units mg^{-1} protein	

(continued)

Table 1 (continued)

No.	Species	Level of drought stress	Antioxidant	Tissue	Change in response of drought stress	Reference
31	*Olea europaea* 'Chemlali'	Leaves Ψ_w at 9:30 h = –4.10 MPa	APX	Leaves	From 23 to 34 nmol AsA mg^{-1} $proteins^{-1}$	Guerfel et al. 2009
			GR		From 0.3 to 0.7 nmol NADPH mg^{-1} $proteins^{-1}$	
32	*Olea europaea* 'Chétoui'	Leaves Ψ_w at 9:30 h = –5.51 MPa	GR	Leaves	From 0.6 to 1.0 nmol NADPH mg^{-1} $proteins^{-1}$	
33	*Oryza sativa*	Leaves Ψ_w at –3.24 MPa at 10.30 h	APX	Leaves	From 0.18 to 0.60 μmol AsA min^{-1} mg^{-1} protein	Srivalli et al. 2003
			GR		From 0.023 to 0.078 umol NADPH min^{-1} mg^{-1} protein	
			AaA/DHA		From 6.57 to 1.00	
			AsA		From 10.18 to 5.24 μmol g DW^{-1}	
34	*Oryza sativa* 'Xiangnuo no. 1' and 'Zimanuo'	PEG-6000 solution 23% (w/v) for 7 days	APX	Leaves	From about 11 to about 6 units g^{-1} DW	Guo et al. 2006
			AsA		From about 22 to 11–13 μmol g^{-1} DW	
			GSH		GSH from about 11 to 4–5 μmol g^{-1} DW	
			H_2O_2		From about 2.0 to about 4.0–6.0 μmol g^{-1} DW	
35	*Oryza sativa* 'Xiangzhongxian no. 2' and 'IR50'	PEG-6000 solution 23% (w/v) for 7 days	APX	Leaves	From about 11 to about 13 units g^{-1} DW	
			AsA		From about 22 to 24 μmol g^{-1} DW	
			GSH		GSH from about 11 to 12 μmol g^{-1} DW	
			H_2O_2		From about 1.5 to about 3.0 μmol g^{-1} DW	
36	*Picea asperata*	RWC from 76.3–87.7% to 65.9–66.1%	APX	Leaves	From 6.61–6.89 to 9.30–10.60 units g^{-1} FW	Duan et al. 2005

37	*Picea asperata*	Field capacity from 100% to 30%	APX	Leaves	From 0.30 to 0.71 μmol min^{-1} mg^{-1} protein	Yang et al. 2008
			GR		From 0.39 to 1.14 μmol min^{-1} mg^{-1} protein	
			t-Asc		From 1.44 to 1.61 mg g^{-1} FW	
			H_2O_2		From 8.50 to 15.70 μmol g^{-1} FW	
			GR		From 53.8 to 156.4 μmol m^{-2} s^{-1}	
			t-Asc/(t-Asc + DHA)		From 0.6 to 0.7	
38	*Pinus canariensis*	Ψ_w at noon = –0.44 MPa	GSSG/t-Glu	Leaves	From 10% to 20%	Tausz et al. 2001
39	*Pisum sativum*	Predawn Ψ_w = –1.0 MPa	APX	Leaves	From 0.13 to 0.48 μmol ascorbate min^{-1} mg^{-1} protein	Zabalza et al. 2008
			GR		From 0.063 to 0.068 μmol NADH min^{-1} mg^{-1} protein	
40	*Poa pratensis*	RWC from 95% to 68%	APX	Leaves	From 900 to 1,500 nmol min^{-1} mg^{-1} protein	Bian and Jiang 2009
			MDHAR		From 65 to 150 nmol min^{-1} mg^{-1} protein	
				Roots	From 260 to 360 nmol min^{-1} mg^{-1} protein	
			DHAR	Leaves	From 60 to 180 nmol min^{-1} mg^{-1} protein	
				Roots	From 275 to 150 nmol min^{-1} mg^{-1} protein	
			GR		From 75 to 110 nmol min^{-1} mg^{-1} protein	
			H_2O_2	Leaves	From 2.2 to 3.5 μmol g^{-1} FW	
				Roots	From 1.1 to 2.0 μmol g^{-1} FW	

(continued)

Table 1 (continued)

No.	Species	Level of drought stress	Antioxidant	Tissue	Change in response of drought stress	Reference
41	*Poncirus trifoliata*	RWC from 75% to 55%	APX	Roots	From 3.98 to 5.63 units mg^{-1} FW	Wu et al. 2006
			GR	Leaves	From 8.81 to 9.52 units mg^{-1} FW	
			t-Asc		From 9.32 to 6.04 μmol g^{-1} FW	
			GSH		From 2.30 to 1.43 μmol g^{-1} FW	
			H_2O_2		From 105.94 to 139.02 μmol g^{-1} FW	
42	*Populus kangdingensis*	Field capacity from 100% to 50%	APX	Leaves	From 1.27 to 1.73 μmol H_2O_2 min^{-1} g^{-1} FW	Ren et al. 2007
	Populus cathayana	Field capacity from 100% to 50%	APX	Leaves	From 0.30 to 0.41 μmol H_2O_2 min^{-1} g^{-1} FW	
43	*Populus euramericana*	$\Delta\Psi_w = -0.224$ MPa	c-APX	Leaves	From 140 to 195 nmol ascorbate oxidized min^{-1} mg^{-1} protein	Edjolo et al. 2001
44	*Populus przewalskii*	Field capacity from 100% to 25%	APX	Leaves	From 8 to 27 μmol AsA min^{-1} mg^{-1} protein	Lei et al. 2006
			GR		From 0.6 to 2.1 μmol NADH min^{-1} mg^{-1} protein	
			AsA		From about 350 to about 900 μg g^{-1} DW	

45	*Prunus spp.*	Predawn Ψ_w = −3.30 MPa	APX	Leaves	From 0.1–0.7 to 1.6–2.3 units mg^{-1} protein	Sofo et al. 2005b
			MDHAR		From 50–130 to 220–550 units mg^{-1} protein	
			DHAR		From 45–60 to 120–170 units mg^{-1} protein	
			GR		From 45–55 to 60–200 units mg^{-1} protein	
			AsA		From about 0.06 to 0.12–0.17 μmol g^{-1} FW	
			DHA		From about 0.50 to 1.25–1.80 μmol g^{-1} FW	
			GSH		From 0.18–0.28 to 0.23–0.46 μmol g^{-1} FW	
			H_2O_2		From about 0.025 to 0.125–0.150 μmol g^{-1} FW	
46	*Salvia officinalis*	RWC from 67% in June to 32% in August	α-toc	Leaves	From 24 to 14 μg g DW^{-1}	Munné-Bosch et al. 2001
47	*Solanum tuberosum*	RWC from 90% to 70%	GSSG/t-Glu	Leaves	From 19 to 30	Broin et al. 2000
48	*Sorghum bicolor*	RWC from 90% to 50%	APX	Leaves	From 100 to 280 nmol min^{-1} mg^{-1} protein	Zhang and Kirkkham 1996a, b
			MDHAR		From 10 to 50 nmol min^{-1} mg^{-1} protein	
			DHAR		From 1 to 5 nmol min^{-1} mg^{-1} protein	
			GR		From 14 to 45 nmol min^{-1} mg^{-1} protein	
			AsA		From 2 to 4 mg g^{-1} DW	
			DHA		From 0.3 to 0.5 mg g^{-1} DW	
			GSH		From 1.8 to 2.5 mg g^{-1} DW	
			GSSG		From 0.35 to 0.18 mg g^{-1} DW	

(continued)

Table 1 (continued)

No.	Species	Level of drought stress	Antioxidant	Tissue	Change in response of drought stress	Reference
49	*Triticum aestivum*	8 days of water withholding	APX	Leaves	From 160–230 to 210–370 μmol AsA min^{-1} g^{-1} FW	Sairam and Saxena 2000
			GR		From 0.7–2.2 to 1.8–3.8 μmol A_{412} min^{-1} g^{-1} FW	
50	*Triticum aestivum*	RWC from 82–91% to 61–68%	GR	Leaves	From 2.0–5.5 to 8.5–10.5 ΔA_{412} min^{-1} mg^{-1} protein	Sairam and Srivastava 2001
			t-Asc		From 27–54 to 16–27 μmol g^{-1} DW	
			H_2O_2		From 2.3–3.4 to 2.6–3.7 μmol g^{-1} DW	
51	*Triticum aestivum*	PEG-6000 solution 10% (w/v) for 10 days	APX	Leaves	From 0.5 to 1.2 units mg^{-1} protein	Qiu et al. 2008
			t-Asc		From 8.3 to 6.7 mg g^{-1} FW	
			GSSG		From 75 to 70 μg g^{-1} FW	
			H_2O_2		From 37 to 32 μmol g^{-1} FW	
52	*Triticum aestivum*	Predawn $\Psi_w = -1.49$ MPa	GR	Leaves	From 75 to 63 nmol mg^{-1} protein min^{-1}	Gong et al. 2005
			H_2O_2		From 5.3 to 7.0 μmol g^{-1} DW	
53	*Triticum aestivum*	$\Psi_w = -0.5$ MPa	APX	Leaves	From 15 to 30 μmol g^{-1} DW	Tian and Lei 2007
			H_2O_2		From 0.25 to 0.35 μmol H_2O_2 mg^{-1} protein min^{-1}	
54	*Triticum aestivum*	Predawn $\Psi_w = -1.5$ MPa	APX	Leaves	From 0.10 to 0.15 μmol AsA s^{-1} g^{-1} FW	Nayyar and Gupta 2006
			DHAR		From 0.4 to 0.7 μmol min^{-1} g^{-1} FW	
			GR		From 0.5 to 0.7 μmol min^{-1} g^{-1} FW	
			t-Asc		From 8 to 12 μmol g^{-1} DW	
			GSH		From 300 to 420 μmol g^{-1} DW	
			H_2O_2		From 10 to 35 μmol g^{-1} DW	
			APX	Roots	From 0.13 to 0.22 μmol AsA s^{-1} g^{-1} FW	
			DHAR		From 0.5 to 0.7 μmol min^{-1} g^{-1} FW	
			GR		From 0.7 to 0.9 μmol min^{-1} g^{-1} FW	

55	*Triticum aestivum*	PEG-6000 solution 10% (w/v) for 10 days + laser pre-treatment	APX	Leaves	From 0.5 to 1.2 Units mg^{-1} protein	Qiu et al. (2008)
56	*Triticum aestivum* 'Moti'	Leaves Ψ_w from −1.40 to −1.65 up to −2.40 MPa (acclimated)	H_2O_2	Leaves	From about 2.0–2.5 to 4.0 µmol g^{-1} DW	Khanna-Chopra and Selote 2007
		Leaves Ψ_w from −1.40 and then directly −2.40 MPa (non-acclimated)	H_2O_2	Leaves	From about 2.0–2.5 to 4.5 µmol g^{-1} DW	
	Triticum aestivum 'C306'	Leaves Ψ_w from −1.40 to −1.65 up to −2.40 MPa (acclimated)	MDHAR	Leaves	From 25 to 40 µmol g^{-1} DW	
			DHAR		From 20 to 25 µmol g^{-1} DW	
			H_2O_2		From about 2.0–2.5 to 3.0 µmol g^{-1} DW	
		Leaves Ψ_w from −1.40 and then directly −2.40 MPa (non-acclimated)	APX	Leaves	From 190 to 280 µmol g^{-1} DW	
			GR		From 35 to 70 µmol g^{-1} DW	
			H_2O_2		From about 2.0–2.5 to 4.5 µmol g^{-1} DW	
57	*Triticum aestivum* and *Triticum durum*	6 days of water withholding	GR	Leaves	From 2–5 to 4–8 µmol $NADPH_2$ min^{-1}	Keleş and Öncel 2002
			α-toc		From 600–800 to 900–1,200 µg g^{-1} FW	
			t-Asc		From 9 to 15 µmol g^{-1} DW	
			GSH		From 380 to 450 µmol g^{-1} DW	
			H_2O_2		From 12 to 43 µmol g^{-1} DW	
58	*Trifolium repens*	RWC from 80% to 60%	APX	Leaves	From about 20 to 60 µg g^{-1} DM min^{-1}	Bermejo et al. 2006
			GR		From about 20 to 50 µg g^{-1} DM min^{-1}	
59	*Vaccinium myrtillus*	Environmental drought stress	GR	Leaves	from 0.8–1.3 pkat g^{-1} DW in June to 0.2–0.4 pkat g^{-1} DW in December	Tahkokorpi et al. 2007

(continued)

Table 1 (continued)

No.	Species	Level of drought stress	Antioxidant	Tissue	Change in response of drought stress	Reference
60	*Zea mays*	Predawn $\Psi_w = -1.5$ MPa	APX	Leaves	From 0.14 to 0.23 µmol AsA s^{-1} g^{-1} FW	Nayyar and Gupta 2006
			DHAR		From 0.4 to 0.6 µmol min^{-1} g^{-1} FW	
			GR		From 0.6 to 0.8 µmol min^{-1} g^{-1} FW	
			t-Asc		From 9 to 15 µmol g^{-1} DW	
			GSH		From 280 to 450 µmol g^{-1} DW	
			H_2O_2		From 8 to 26 µmol g^{-1} DW	
			APX	Roots	From 0.13 to 0.32 µmol AsA s^{-1} g^{-1} FW	
			DHAR		From 0.5 to 1.6 µmol min^{-1} g^{-1} FW	
			GR		From 0.7 to 1.5 µmol min^{-1} g^{-1} FW	
			t-Asc		From 9 to 15 µmol g^{-1} DW	
			GSH		From 380 to 530 µmol g^{-1} DW	
			H_2O_2		From 12 to 31 µmol g^{-1} DW	
61	*Zea mays*	RWC from 95% to 70%	AsA	Leaves	From 260 to 0 µmol m^{-2}	Aroca et al. 2003
			t-Glu		From 2 to 5–6 µmol m^{-2}	
			H_2O_2		From about 50 to 90–13 µmol m^{-2}	
62	*Zea mays*	PEG-6000 solution 20% (w/v) for 2 days	APX	Leaves	From 47 to 55 nmol min^{-1} mg^{-1} protein	Rapala-Kozik et al. 2008
			GR		From 36 to 78 nmol min^{-1} mg^{-1} protein	
			α-toc		From 1.3 to 2.0 µmol g DW^{-1}	

Abbreviations α-toc, α-tocopherol; APX, ascorbate peroxidase; AsA, reduced ascorbate; c-APX, cytosolic ascorbate peroxidase; DHA, dehydroascorbate; DHAR, dehydroascorbate reductase; DW, dry weight; FW, fresh weight; GR, glutathione reductase; GSH, reduced glutathione; GSSG, oxidized glutathione; MDHAR, monodehydroascorbate reductase; t-Asc, total ascorbate; t-Glu, total glutathione antioxidant defenses (Foyer et al. 2005; Smirnoff 2005; Morales et al 2006). Trees carry on the same processes as other seed plants, but their larger size, slower maturation, and much longer life accentuate their susceptibility to drought-mediated oxidative stress in comparison to smaller plants having a shorter life span (Pallardy 2008). For this reason, the antioxidant response of woody plants is of key importance and will be herein discussed in detail.

Among tree species, poplar (*Populus* spp.) is a model plant for its economic importance and relative short life cycle. Therefore, poplar genome was entirely sequenced and the antioxidative responses to abiotic stresses were studied in detail. Cuttings of *Populus kangdingensis* and *P. cathayana* originating from high and low altitudes in south-west China, respectively, were used to determine the effect of drought and enhanced UV-B radiation [daily UV-B supplementation = 4.4 kJ m^{-2} day^{-1} (UV-B_{BE})] and their combination on plant growth and physiological traits in a greenhouse during one growing season (Ren et al. 2007). In both species, cuttings grown under drought conditions exhibited reduced growth. Drought and enhanced UV-B radiation, separately or together, significantly reduced plant growth, and increased APX activity. As higher APX activity was observed in *P. kangdingensis* when compared to *P. cathayana*, an interesting adaptative effect was observed by the authors: *P. kangdingensis*, originating from high altitude exhibited greater tolerance to drought and enhanced UV-B radiation than did *P. cathayana* originating from lower altitude. Lei et al. (2006) found that in a dry climate-adapted population of *Populus przewalskii* Maximowicz exposed to three different watering regimes, drought significantly induced the entire set of antioxidative systems including the increase of AsA content and the activities of APX, and GR. Poplar trees under drought stress were choosen to determine the presence and the activities of cytosolic and plastidial forms of some enzymes of the ascorbate–glutathione cycle, in order to test the ROS-scavenging system in this species. For this purpose, Edjolo et al. (2001) determined APX and GR activities in a drought-tolerant *Populus euramericana* clone (Dorskamp). Because ROS were mainly generated in illuminated chloroplasts, cytosolic and chloroplastic APX and GR were followed in seedlings exposed for 12 h to control or 100 mmol L^{-1} mannitol. Whatever the treatment, the activities of plastidial APX and GR were lower than those of cytosolic fractions (140–200 and 10–60 nmol ascorbate oxidized min^{-1} mg^{-1} protein for APX in cytosol and chloroplasts, respectively; 10–20 and 5–7 nmol NADPH oxidized min^{-1} mg^{-1} protein for GR for APX in cytosol and chloroplasts, respectively). Mannitol treatment significantly increased cytosolic APX activity. The direct linear plot of 1/V against 1/S (where V is the velocity and S is the substrate concentration in AsA, GSSG, and NADPH) was used to estimate the apparent K_m values of APX and GR. In stressed plants, the apparent K_m value for AsA decreased for both APX isoforms (this indicates an increased affinity for AsA in both cell compartments), while K_m for GSSG and NADPH increased for GR isoforms, so demonstrating the different behaviors of the two enzymes observed in cytosolic and chloroplastic subcellular compartments.

The distinction of oxidative stress levels in different of tree species, and in particular in roots, is quite rare to find in the past and recent literature. This is likely due to the difficulty of sampling metabolically active roots of the trees (usually fine roots, with a diameter <1 mm) and maintaining uniform conditions in the soils, and to the presence of interfering substances for enzyme isolation and enzyme activities determination. In a work on *Eucalyptus globulus* Labill., Shvaleva et al. (2005) compared the metabolic responses of leaves and roots of two clones differing in drought sensitivity to a slowly 7-week imposed water deficit. In addition to the general decrease in growth caused by water deficit, a decrease in osmotic potential was observed at severe drought stress. In both clones, these decreases were greater in roots than in leaves, consistent with the observed increases in concentrations of soluble sugars and proline in these organs. Leaf GR activity in both clones was significantly ($P < 0.001$) decreased by drought stress. The water-stress treatment also decreased APX activity, but only in CN5. In roots, the effects of drought stress on antioxidant enzymatic activities were more marked than in leaves. The activity of APX and GR in roots increased significantly ($P < 0.001$) in both the clones in response to drought stress. Some of the molecular and cellular effects of drought-mediated oxidative stress have been outlined in the Fig. 3.

The enzymatic antioxidative response of tree species having a high degree of tolerance against drought, such as sclerophyllous Mediterranean species or xerophytic boreal conifers, is particularly interesting. Duan et al. (2005) examined the responses of photosynthetic gas exchange, chlorophyll fluorescence, activities of antioxidant enzymes, and lipid membrane peroxidation of two contrasting *Picea asperata* Mast. populations to 30% of full sunlight (shade conditions obtained by neutral shade clothes; photosynthetically active radiation, PAR = 300 μmol m^{-2} s^{-1}), and full sunlight (PAR = 900 μmol m^{-2} s^{-1}) were investigated under well-watered and drought conditions. The two contrasting populations came from the wet and dry climate regions in China, respectively. For both populations tested, drought resulted in lower needle relative water content (RWC), gas exchange and photosynthetic efficiency, SOD and APX activities as well as malondialdehyde (MDA) levels and electrolyte leakage in sun plants, whereas these changes were not significant in shade plants. In particular, in the population from the wet climate, APX activity in sun plants increased less than in the population from the dry climate. No significant differences were observed between control and drought-stressed shaded plants in both the populations. This physiological and biochemical response is in accordance to the data of Sofo et al. (2004, 2005b, 2009), that found in olive and *Prunus* spp exposed to the synergic effects of drought and shade. Also in this case, it seems that the lower expression of the enzymatic antioxidant system in shade plants may be due to a reduced need of ROS removal. On the contrary, in sun plants, higher enzyme activities are required for a better protection against a more pronounced oxidative stress. Recently, Yang et al. (2008) also pointed out that *Picea asperata* plants under high light condition and drought significantly increased biomass partitioning to roots, and increased the foliar levels of H_2O_2, total ascorbate content (t-Asc), and APX and GR activities. Sánchez-Díaz et al. (2007) studied photoprotection and antioxidative protection in the three major species of the Canarian laurel forest (*Laurus azorica* (Seub.) Franco, *Persea indica* (L.) K. Spreng and *Myrica faya* Aiton). Trees were exposed to drought under controlled conditions by withholding water until leaf RWC reached 50–55%. Drought reduced photosynthetic rate, and the effect was associated with decreased quantum yield of photosystem II (PSII), electron transport, and increased non-photochemical quenching in *L. azorica* and *M. faya*. Drought-treated trees of *L. azorica* had the highest de-epoxidation state of the xanthophyll cycle and the highest zeaxanthin concentration, suggesting that this species had more effective photoprotective mechanisms than *M. faya* and *P. indica*. Moreover, β-carotene remained unaltered in *L. azorica* trees during drought, suggesting that the chloroplasts of this species are better protected against oxidative stress than those of *M. faya* and *P. indica*. Increased antioxidation by leaf APX, SOD and GR in *L. azorica* removed ROS generated during drought treatment. Although *M. faya* was able to increase its energy dissipation rate by forming zeaxanthin, and thus increasing the de-epoxidation state of the xanthophyll cycle, it did not respond to drought-induced oxidative stress with the result that β-carotene degradation occurred. *Persea indica* did not activate an energy dissipation mechanism in response to drought treatment, hence formation of ROS was likely high in the drought-treated trees. In general, *L. azorica* appeared to be most resistant, and *P. indica* most sensitive to photoinhibition and oxidative stress during drought.

In the last years, most of the research on antioxidants was focused on the genetic amelioration of fruit tree cultures particularly subjected to environmental changes. One of these is coffee tree, a species with a low genetic diversity, and for this reason very susceptible to drought or pathogens. Pinheiro et al. (2004) submitted four clones of *Coffea canephora* (Robusta coffee), representing drought-tolerant (14 and 120) and drought-sensitive (46 and 109A) genotypes, to a slowly imposed water deficit, until predawn leaf water potential (ψ_w) approximately –3.0 MPa was reached. Drought-tolerant clones were better able to maintain their leaf water status than drought-sensitive clones after withholding irrigation. Regardless of the clones investigated, the net carbon assimilation rate decreased under drought stress. Drought triggered increases in SOD, APX, CAT and POD, and also in DHAR, t-Asc and ascorbate redox state [t-Asc/(t-Asc + DHA)]. Activity of MDHAR was not induced in drought-stressed plants. In another paper (Lima et al. 2002), the effects of water deficit on photochemical parameters and activities of SOD, CAT and APX, as well as cellular damages, were investigated in two clones of *Coffea canephora* differing in drought tolerance. After 6 days without irrigation, predawn ψ_w fell to –3.0 MPa, and this was accompanied by the suppression of net photosynthesis in both clones. Under drought conditions, activities of SOD, CAT and APX increased to a greater extent in the drought-tolerant clone (0.42–1.34 units mg^{-1} protein) than in the drought-sensitive one (0.31–0.83 units mg^{-1} protein). This seemed to be matched with higher protection against oxidative stress, as judged from the lower levels of lipid peroxidation and electrolyte leakage in the drought-tolerant clone. Whereas Pinheiro et al. (2004) did not observe a general link between protection against oxidative stress with differences in clonal tolerance to drought, Lima et al. (2002), concluded that the ability to increase the antioxidant system activity in order to limit cellular damages might be an important attribute linked to the drought tolerance in *C. canephora*. Another crop with a low level of heterozigosis is banana tree. Chai et al. (2005) investigated oxidative injury and antioxidant responses in two banana genotypes (*Musa* AAA 'Berangan' and *Musa* AA 'Mas') subjected to a drought stress period for 14 days induced by polyethylene glycol (PEG). PEG treatment resulted in oxidative injury, as expressed in increased lipid peroxidation and reduced membrane stability index, in both cultivars; however, greater oxidative injury was detected in 'Mas'. Under PEG treatment, leaf CAT activity and GR activity were enhanced in both cultivars. Leaf APX activity was enhanced in 'Berangan' under drought stress, but was unaffected in 'Mas'.

Olive tree (*Olea europaea*) is one of the most typical and economically important tree culture species belonging to the Mediterranean basin, where water shortage occurs with regularity, often lasting throughout the spring–summer period. This evergreen sclerophyllous tree shows a high degree of drought tolerance, a parsimonious consumption of soil water and a higher ratio of transpiration rate to leaf surface area in comparison with other fruit tree species in both ideal, and in water shortage conditions (Sofo et al. 2009). Olive tree is able to resist drought stress by lowering the water content and water potentials of its tissues, establishing a high water potential gradient between leaves and roots, stopping shoot growth, and reducing transpirative and photosynthesis-related processes (Sofo et al. 2004, 2005a; Ennajeh et al. 2009; Guerfel et al. 2009). For all these reasons, the antioxidant response of olive tree to drought stress has been well documented. The first studies on olive tree's antioxidant enzymes were carried out by Sofo et al. (2004, 2005a) on olive Italian cv. 'Coratina'. They pointed out that olive tree is able to up-regulate the enzymatic antioxidant system when subjected to drought stress (very low predawn ψ_w up to −5.37 MPa), and in particular APX showed marked and significant increases in leaves, medium roots, with a diameter between 1 and 5 mm (from 0.23 to 0.34 units mg^{-1} dry weight [DW]), and in fine roots, with a diameter <1 mm (from 0.36 to 0.51 units mg^{-1} DW). After reaching the maximum level of drought stress, the same plants were subjected to a rewatering treatment for 30 days, under both environmental irradiance and semi-shade conditions of about 60% of PAR (Sofo et al. 2004). In fact, the water recovery after a period of drought is a normal condition after the dry season in the Mediterranean regions. The activity of APX decreased during the rewatering period in both leaves and roots and these decrements were faster in plants rewatered in semi-shade conditions than in plants under environmental light. A similar behavior has been also found in olive (cv. Coratina) during a drought stress period (predawn ψ_w up to −5.7 MPa), in which APX activity and H_2O_2 content were higher in fully irradiated plants than in plants under 67% of PAR (Fig. 4).

The authors concluded that the lower APX expression in shaded plants with respect to non-shaded ones may be due to a reduced need of activated oxygen species removal, whereas in non-shaded plants, higher APX activity is required for a better protection against a more pronounced oxidative stress. Similar results have been found by Ennajeh et al. (2009) in two Tunisian olive cultivars, 'Chemlali' and 'Meski'. In these cultivars, an increase of leaf APX activity with decreasing leaf RWC. In the Tunisian olive cvs. 'Chemlali' and 'Chétoui' experiencing 30 days without irrigation (Guerfel et al. 2009), GR activity increased in both the cultivars,

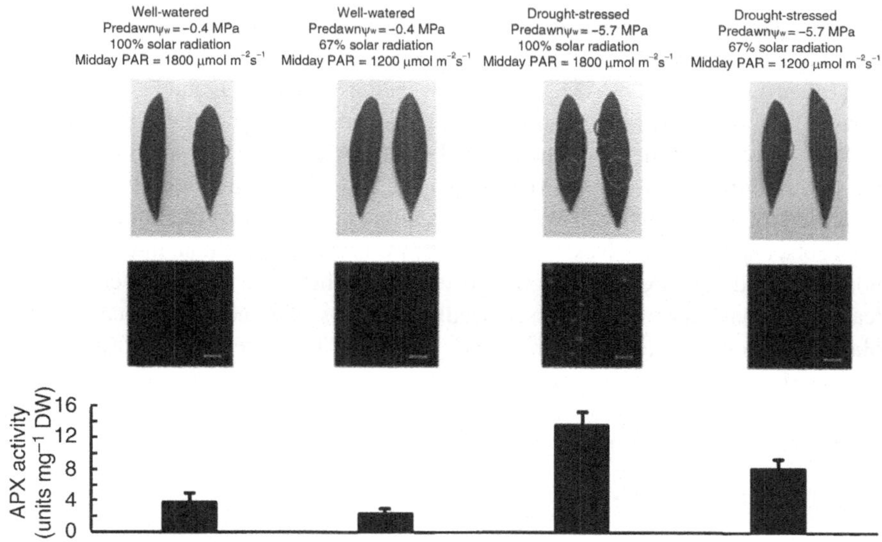

Fig. 4 Oxidative damages (in the *red circles*), H_2O_2 generation (white bar = 10 μm) and APX activity (± standard error) in olive (cv. 'Coratina') epidermal fragments of well-watered and drought-stressed plants subjected to 100% and 67% solar radiation. The levels of H_2O_2 were monitored using the H_2O_2-sensitive fluorescent probe H_2DCFDA (Sigma, MO, USA) as described by Desikan et al. (2004). The images show that the drought-stressed and fully-irradiated olive plants were found to be highly sensitive to the oxidative stress caused by H_2O_2, whereas the drought-stressed and semi-shaded plants were more tolerant to oxidative stress damage. Little visible oxidative damage was observed in well-watered fully-irradiated plants and no damage was observed in well-watered semi-shaded plants (A. Sofo)

whereas APX activity increased only in 'Chemlali'. This result is important as it indicates that enzyme activity regulation can be cultivar-specific. The different values of APX activity found in olive leaves, fine roots and medium roots by Sofo et al. (2004, 2005a) confirm their different functions: leaf tissues showed more pronounced changes (about a fourfold increase), due to the synergic effect of high irradiance levels and loss of cellular water; fine roots were more sensitive to drought stress and its consequent effects, while medium roots maintained a prolonged functionality and presented less reactivity, even at severe drought stress.

It is know that mycorrhiza infection of roots gives benefits to plants. Arbuscular mycorrhizal (AM) symbiosis can positively affect the water relations of many plants and this effect is often more pronounced in plants grown under drought stress than under well-watered conditions. The mechanisms by which AM act in plants are not completely known but there is evidence that AM symbiosis might increase the drought tolerance of plants by promoting antioxidant enzymes (Wu et al. 2006). The effect of the AM fungus, *Glomus versiforme*, on growth and reactive oxygen metabolism of trifoliate orange (*Poncirus trifoliata*) seedlings was studied in potted plants under both well-watered and drought stressed conditions (Wu et al. 2006). Drought stress significantly decreased root colonization. Shoot dry weight, plant height and stem diameter were higher in AM than in non-AM seedlings regardless of the water status. Inoculation with *G. versiforme* increased root dry weight and leaf number per plant of well-watered seedlings. There was less MDA concentration in leaves and roots of AM seedlings, as well as lower H_2O_2 and $O_2{\cdot}^-$ concentrations in AM roots. AM inoculation did not affect the H_2O_2 and $O_2{\cdot}^-$ concentrations of WW leaves. Whether drought-stressed or not, AM symbiosis notably increased the GR activity of leaves and APX activity of roots. AM infection also markedly increased the APX activity of WS leaves. Soluble proteins and glutathione in leaves and roots, and t-Asc in leaves were higher in well-watered AM than in well-watered non-AM seedlings. AM infection also enhanced the ASC and GSH contents of leaves and roots in drought-stressed seedlings. Cross-tolerance might occur in AM plants and could be enhanced by AM symbiosis. These results suggest that the increased concentrations of antioxidant enzymes and non-enzymatic antioxidants found in AM plants may serve to protect the organism against oxidative damage, enhancing drought tolerance.

2.2 Shrub Species

Plants that occur in semi-arid and arid regions are usually physiologically and biochemically adapted to cope with high levels of solar irradiation and drought, but also possess morphological adaptation (e.g., schlerophyllia, presence of trichomes and waxes on leaf surface, deep root systems, high root/shoot, prevalence of lignified tissues). A good number of species are characterized by a shrub habitus with a high degree of resilience after having experienced environmental stresses. The studies on the regulation of antioxidant enzymes under different levels of light are particularly important in these species, as in mutable semi-arid and arid environments, and light can be an additional abiotic stresses in addition to drought.

Guidi et al. (2008) studied the interactive effects of drought stress and solar irradiance on physiological and biochemical traits in *Ligustrum vulgare*, with special emphasis on antioxidant enzymes and flavonoids. Measured were carried out in plants growing in 12% (shade) or 100% (sun) sunlight, and supplied with 100% or 40% of daily evapotranspiration-demand over a 4-week period. The mild drought stress treatment caused predawn ψ_w and RWC to decline on average by –0.2 MPa and 4.5%, respectively. In response to the drought stress treatment, photosynthetic rates decreased more in sun plants than in shade plants, likely because of declines in PS II photochemistry, coupled with significant reductions in stomatal conductance. Antioxidant enzymatic activities, which were much greater in sun leaves than in shade leaves under well-watered conditions, increased (in particular the enzymatic activities associated with H_2O_2 removal, such as CAT and APX) in response to drought stress only in shade leaves. The authors suggest that assimilated carbon in sun plants was used largely to support an effective antioxidant system capable of countering water-stress-induced oxidative damage, an example of cross tolerance. Another hypothesis is that in shade plants, carbon was also diverted from growth to counter oxidative damage driven by the mild drought stress treatment. Both findings are consistent with the nearly exclusive distribution of *L. vulgare* in well-watered, partially shaded Mediterranean areas. The effects of separately or simultaneously induced dark chilling and drought stress were evaluated in *Glycine max* 'Merrill' cv. (Riekert van Heerden and Krüger 2002). For the induced drought treatment of 9 days, plants were maintained at normal growth temperatures without irrigation. For the simultaneously induced dark chilling and drought stress treatment, plants were dark chilled (incubated at 8°C during a dark period) without irrigation. All treatments caused similar decreases in predawn ψ_w, but resulted in distinct physiological and biochemical effects on photosynthesis.

Caper (*Capparis ovata* Desf.) is a xerophyte perennial shrub and drought resistant plant which is well adapted to Mediterranean Ecosystem. In a recent study, Ozkur et al. (2009) investigated the plant growth, RWC, chlorophyll fluorescence (F_V/F_M), lipid peroxidation (evaluated by the levels of TBA-reactive substances content) as parameters indicative of oxidative stress in relation to the tolerance to PEG drought stress in *C. ovata* seedlings. Total activity of antioxidative enzymes SOD, APX, POD, CAT, and GR were investigated in *C. ovata* seedlings under PEG mediated drought. For induction of drought stress, the 35 days seedlings were subjected to PEG 6000 of osmotic potential −0.81 MPa for 14 days. Lipid peroxidation increased in PEG stressed seedlings as compared to non-stressed seedlings of *C. ovata* during the experimental period. With regard to vegetative growth, PEG treatment caused decrease in shoot fresh and dry weights, RWC and photosynthetic efficiency but the decline in PEG-treated plants was more prominent on day 14; furthermore, both APX and GR activities increased under the drought period.

Inoculation of autochthonous drought tolerant fungal strains could be an important strategy that assured the greatest tolerance drought stress contributing to a better plant growth under drought. Marulanda et al. (2007) compared the effectiveness of four arbuscular mycorrhizal (AM) fungal isolates (two autochthonous drought-tolerant *Glomus* spp., and two allochthonous drought-sensitive strains) on a drought-adapted plant (*Lavandula spica*) growing under drought conditions. The autochthonous AM fungal strains produced a higher lavender biomass, specially the root biomass, and a more efficient N and K absorption than with the inoculation of similar allochthonous strains under drought conditions. The autochthonous strains of *Glomus intraradices* and *Glomus mosseae* increased root growth by 35% and 100%, respectively, when compared to similar allochthonous strains. These effects were concomitant with an increase in water content and a decline in t-Asc, GSH and H_2O_2. The low cell accumulation of ascorbate and glutathione in plants colonized by autochthonous AM fungal strains is an indication of high drought tolerance. Non-significant differences between antioxidant activities such as GR (GR activity was about 0.15–0.20 µmol NADPH oxidized g^{-1} fresh weight [FW] min^{-1}), CAT and SOD in colonized plants were found. Thus, these results do not allow the generalization that GR, CAT and SOD were correlated with the symbiotic efficiency of these AM fungi on lavender drought tolerance.

2.3 Cereals

Among herbacesous plants, cereals were the most studied for their use as food and economic importance. It seems that drought tolerance in cereals is mainly due to higher membrane stability, chlorophyll and carotenoid contents, lower lipid peroxidation, and higher antioxidant enzyme activity (Nayyar and Gupta 2006; Qiu et al. 2008). The degree of oxidative stress and antioxidant activity seems to be closely associated with the tolerance/susceptibility of a genotype to drought stress.

The role of plant antioxidant systems in drought stress tolerance was studied in three contrasting wheat genotypes (*Triticum aestivum*) (Sairam and Saxena 2000). Drought stress, imposed for 8 days at different stages after anthesis, resulted in an increase in lipid peroxidation, and a decrease in membrane stability, chlorophyll and carotenoid contents. The antioxidant enzymes APX, GR and non-specific peroxidase also increased significantly under drought stress. Genotype PBW 175, which had highest APX, GR and POD activity, showed low lipid peroxidation, and high chlorophyll and carotenoids content under drought stress, while the susceptible genotype WH 542 exhibited the lowest antioxidant enzyme activity, membrane stability and contents of chlorophyll and carotenoids and the highest lipid peroxidation.

Another experiment was conducted on five wheat (*Triticum aestivum*) cultivars, 'C 306', 'PBW 175' (tolerant to drought stress), 'DL 153–2' (moderately tolerant to drought stress), 'HD 2428' and 'HD 2329' (susceptible to drought stress, and so recommended for well-watered conditions), under pot culture conditions. The effect of 7 days of drought stress, starting at 17 days after anthesis, on oxidative injury and antioxidant activity was evaluated (Sairam and Srivastava 2001). In this study, drought stress significantly decreased RWC, t-Asc content and membrane stability, and increased H_2O_2 and MDA content, a measure of lipid peroxidation, and activities of antioxidant enzymes in all genotypes. Drought stress tolerant genotypes 'C 306' and 'PBW 175', closely followed by 'DL 153–2', were superior to 'HD 2428' and 'HD 2329' in maintaining high RWC, t-Asc content and membrane stability, and lower H_2O_2 content and lipid peroxidation (in terms of MDA) under drought stress. The highest activities of GR and CAT under drought stress were observed in 'C 306', 'PBW 175' and 'DL 153–2', and the lowest activities in 'HD 2428' and 'HD 2329' at all the stages. It is apparent that drought stress induces an increase in H_2O_2 content, and consequently lipid peroxidation and membrane injury, that in turn indicate a reduced membrane stability.

Keleş and Öncel (2002) investigated the effects of environmental stress combinations on soluble metabolic compounds in *Triticum aestivum* (cvs. 'Bezostaya-1', 'Seri-82' and 'Kıraç-66'), and *Triticum durum* Desf. (cvs. 'Kiziltan-91', 'Kunduru 414–44' and 'Ç. 1252'). The seedlings were grown at normal (24/16°C), low (LT, 5/−5°C) and high (HT, 40/30°C) temperature conditions, and then exposed to drought stress. Seedlings responses to cross interactions between temperature and drought stresses were investigated. Root and shoot elongation significantly decreased under drought and salt stresses. The content of α-toc significantly increased under drought stress but this increase was inhibited under HT stress, while CAT activity decreased especially in *T. durum* genotypes, and GR activity increased under drought.

The metabolic reasons associated with differential sensitivity of C_3 and C_4 plant species to drought stress are not well understood. In the deep and important study of Nayyar and Gupta (2006), 15-day-old wheat (*Triticum aestivum*) and maize (*Zea mays*) plants, representatives of C_3 and C_4 plants, respectively, were subjected to a drought stress (predawn $\psi_w = -1.5$MPa) induced by PEG-6000 for 7 days under controlled conditions (Nayyar and Gupta 2006). Both the roots and leaves of these species were evaluated for oxidative damage and antioxidants along with stress injury (as electrolyte leakage), water content and abscisic acid (ABA). While at mild stress, both the plant species did not vary significantly from each other for stress injury, moderate and high stress levels caused considerably more damage to wheat as compared to maize. The oxidative damage in terms of MDA and H_2O_2 content was markedly higher in wheat as compared to maize at moderate and high stress levels. Relatively, maize had significantly higher content of non-enzymatic (ascorbate and glutathione) and enzymatic antioxidants (APX, DHAR and GR, especially in its leaves). In contrast, wheat possessed more activity of CAT in roots as well as leaves in comparison to maize. Thus, leaves of both the species experienced more damage than roots, likely because more subjected to the environmental stresses. These findings suggested that differential sensitivity of C_3 and C_4 plants to drought stress appear to be partially governed by their ability to counter oxidative stress, involving ascorbate and glutathione. The levels of t-Asc/GSH appear to have greater involvement in regulating this response. Manipulation of endogenous expression of antioxidants through genetic means might elevate the defense ability of these plant species, especially of C_3 plants to drought stress. The results of Nayyar and Gupta (2006) showed that increased activity of antioxidants in leaves may be more important for stress tolerance than in roots.

In order to determine the role of laser in drought stress resistance of spring wheat (*Triticum aestivum*), Qiu et al. (2008) exposed seed embryos to CO_2 laser radiation for 5 min, and treated 12-day-old seedlings with 10% (w/v) PEG-6000 solution for 10 days. This is the first investigation reporting the use of CO_2 laser pretreatment to enhance drought stress resistance of spring wheat. Changes in the concentration of MDA, H_2O_2, t-Asc, GSH, GSSG, carotenoid, zeaxanthin, the production rate of $O_2{\cdot}^-$, the activities of APX, POD, CAT, SOD and GR, glutathione peroxidase, glutathione-S-transferase, and the growth parameters of seedlings (plant height, leaf area and dry weight) were measured to test the effects of laser pretreatment. The results showed that suitable laser pretreatment of embryos enhanced drought stress resistance in wheat seedlings by decreasing the concentration of MDA and H_2O_2, GSSG, the production rate of $O_2{\cdot}^-$, leaf area and increasing the activities of APX, glutathione peroxidase, glutathione-S-transferase and POD and t-Asc, carotenoid and zeaxanthin concentration. It is suggested that those changes in MDA, O_2^-, H_2O_2, anti-oxidative enzymes, and anti-oxidative compounds were responsible for the increase in drought stress resistance observed in the experiments. The results also showed that the laser had a long-term positive physiological effect on the growth of drought stress seedlings.

During life cycle, wheat crop may experience water deficit cycles that induce oxidative stress. An interesting study of Khanna-Chopra and Selote (2007) was conducted to evaluate the role of oxidative stress management in the leaves of two wheat (*Triticum aestivum*) cultivars, 'C306' (drought-resistant) and 'Moti' (drought-susceptible), when subjected directly to severe drought stress (non-acclimated plants) or to drought stress cycles of increasing intensity (ψ_w from control at −1.40 MPa to −1.63/−1.67 MPa at mild stress to −1.93/−2.13 MPa at severe stress; instead, directly to −2.70/−1.97 MPa in severe-stressed non-acclimated plants) with an intermittent rewatering (drought-acclimated plants). Mild drought stress during vegetative growth enabled 'C306' to acclimatize better than 'Moti' during subsequent drought stress of severe nature during post-anthesis period. The drought-acclimated 'C306' leaves maintained favorable water relations and lower membrane injury due to low H_2O_2 accumulation than non-acclimated 'C306' plants during severe drought stress. This is due to systematic increase in the activity of APX and POD, and maintenance of ascorbate and glutathione redox pool by efficient functioning of GR enzyme in the drought-acclimated 'C306' plants. MDHAR and DHAR activities increased with increasing drought stress, but only in drought-acclimated plants. In contrast, both acclimated as well as non-acclimated 'Moti' plants exhibited loss in turgor potential, high H_2O_2 levels and poor antioxidant enzyme response leading to enhanced membrane damage during severe drought stress conditions. Generally, total AsA content and AsA/DHA ratio decreased in both stress treatments for both the cultivars, and a similar trend was observed for total glutathione (t-Glu), GSH and GSH/GSSG. It is interesting to note that a drastic drought stress appears to cause more changes than a slower one, and that non-acclimated plants are more subjected to drought stress oxidative damages. Failure in the induction of APX and ascorbate–glutathione cycle enzymes makes the chloroplast susceptible to oxidative stress in non-acclimated plants. Non-acclimated plants protected the leaf mitochondria from oxidative stress by up-regulating SOD, APX, and GR activities. Rewatering led to rapid enhancement in all the antioxidant defense components in non-acclimated plants, which suggested that the excess levels of H_2O_2 during severe drought stress conditions might have inhibited or down-regulated the antioxidant enzymes.

Aroca et al. (2003) studied the photosynthetic performance and protective mechanisms against oxidative stress in two maize (*Zea mays*) genotypes differing in chilling sensitivity ('Z7', tolerant and 'Penjalinan', sensitive), subjected to 5°C for 5 days, with or without a drought pretreatment. In 'Penjalinan' plants, the drought pretreatment decreased the symptoms of chilling injury, estimated as necrotic leaf area and maximum quantum yield of PS II. Furthermore, drought pretreatment diminished the level of lipid peroxidation caused by chilling in 'Penjalinan' plants. After 1 day of recovery from chilling the 'Z7' and drought-pretreated 'Penjalinan' plants showed higher net photosynthesis rates than the non-drought-pretreated 'Penjalinan' plants, thereby decreasing the probability of generating ROS. The greater net photosynthesis was correlated with the greater NADP-malate dehydrogenase activity. No differences in either the de-epoxidation state of the xanthophyll cycle or the antioxidant enzyme activities (APX and GR) were found among the drought and drought-chilled groups of plants. However, a drastic decrease in AsA content was observed in chilled 'Penjalinan' plants without drought pretreatment (from 260 to 0 μmol m^{-2}). As the authors found an increase of H_2O_2 content after drought pretreatment, they suggested its involvement as a signal in the drought-enhanced chilling tolerance of maize. Recently, Rapala-Kozik et al. (2008) used *Zea mays* seedlings as a model system to analyze for any relation between the plant response to abiotic stress, and the properties of thiamine biosynthesis and activation. Conditions of drought were induced by PEG-6000 at 20% for 2 days, and increases in the activities of APX and GR were found. In an experiment on rice (*Oryza sativa*, cv. 'Tulsi'), Srivalli et al. (2003) subjected plants to three cycles of drought stress of increasing stress intensity. Each stress cycle was terminated by rewatering the plants for a 48-h period. The response of the antioxidant metabolites ascorbate and glutathione was analyzed in terms of activity and isozyme pattern for each cycle of stress and recovery. It was observed that drought stress caused increases in APX and GR activities, and there was a better management of toxic H_2O_2 levels. These increases were parallel to a lower AaA/DHA and to a decrease in AsA. Guo et al. (2006) investigated about the responses of antioxidative defense systems to chilling and drought stresses in four cultivars of rice (*Oryza sativa*) differing in sensitivity, two of them ('Xiangnuo no. 1' and 'Zimanuo') are tolerant to chilling but sensitive to drought, and the other two ('Xiangzhongxian no. 2' and 'IR50') are tolerant to drought but sensitive to chilling. The seedlings of rice were transferred into growth chamber for 5 days at 8°C as chilling treatment, or at 28°C as control, or at 28°C but cultured in 23% PEG-6000 solution as drought stress treatment. Under drought stress, the elevated levels of electrolyte leakage, and the contents of H_2O_2 and MDA in 'Xiangzhongxian no. 2' and 'IR50' are lower than those in 'Xiangnuo no. 1' and 'Zimanuo'. Activities of SOD, CAT, APX, and the contents t-Asc and GSH were measured during the stress treatments. APX activity, AsA and GHS showed a slight increase until 3 days after drought stress in the two drought-tolerant cultivars, or after chilling stress in the two chilling-tolerant cultivars. On the other hand, activities of antioxidant enzymes and contents of antioxidants were greatly decreased in the drought-sensitive cultivars after drought stress, and in the chilling-sensitive cultivars after chilling stress. The results indicated that tolerance to drought or chilling in rice is well associated with the enhanced capacity of antioxidative system under drought or chilling condition, and that the sensitivity of rice to drought or chilling is linear correlated to the decreased capacity of antioxidative system.

2.4 *Other Herbaceous Species*

The investigations on enzymatic and non-enzymatic antioxidants of the ascorbate–glutathione cycle in drought-stressed herbaceous plants in the last 10 years (excluding cereals) have been mainly focused on species having a known antioxidant effect and with a strong degree of drought tolerance (such as *Euphorbia, Vaccinium,*

Allium, Bupleurum, Catharanthus, Anoda spp.). A good number of studies have also regarded grasses used as cover crops in cultivated fields (*Trifolium, Medicago* spp.) or 'model' plants (*Arabidopsis thaliana*). This studies highlight the importance of antioxidant enzymes in the plant growth in different phenological stages under drought conditions. In the perennial weed leafy spurge (*Euphorbia esula*), GR specific activity increased almost 60% during drought, whereas APX activity only showed a transient increase after 3 days of drought (Davis and Swanson 2001). Tahkokorpi et al. (2007) studied the influence of drought-related winter stress on growth, oxidative responses, and possible after-effects in bilberry (*Vaccinium myrtillus*) in field conditions favorable for desiccation in winter. Bilberry plants were subjected to low temperatures, drought and irradiance, grown under snow (control plants), under a grey plate without snow, under a transparent plate without snow, and under either grey or transparent plates in the absence of snow. Tissue water content, GR activity as well as soluble proteins decreased as frost hardening approached in the three treatments (from 0.8–1.3 pkat g^{-1} DW in June to 0.2–0.4 pkat g^{-1} DW in December). Bilberry plants activate their metabolism in early spring due to increasing temperature, in agreement with the rising of tissue water content and increments in anthocyanins, photosynthetic pigments, GR activity and soluble proteins in March. GR activity increase was lower in the plants that wintered without a snow cover, which points to delayed activation of metabolism, resulting from multiple stresses (low temperatures, drought, light) that acted simultaneously. In another study (Egert and Tevini 2002), drought caused by withholding water led to significant reductions in the relative water content (−17.5%) of chives (*Allium schoenoprasum*) leaves, a significant rise in the osmolarity of the leaf sap (+18.9%) and a loss of leaf transpiration (leaf diffusion resistance >20 s cm^{-1}). While it did not affect specific POD activity, drought increased the specific activity of APX significantly by almost 29%, and reduced the specific lipoxygenase activity significantly by 60%. Recently, Zhu et al. (2009) investigated on roots of *Bupleurum* spp., one of the most popular ingredients in many oriental medicinal preparations. Potted *Bupleurum chinense* DC seedlings were subjected to progressive drought stress by withholding irrigation for 9 days, followed by a rewatering phase, and the changes in antioxidant system, H_2O_2 and $O_2{\cdot}^-$ contents were investigated. Additionally, the antioxidant activity of root extracts was evaluated. The results showed that *B. chinense* root appeared highly resistant to water deficit. Increasing levels of drought stress were accompanied by enhanced $O_2{\cdot}^-$ content and SOD, CAT and APX activity until severe drought stress.

Interference and drought are two major causes of limited productivity in plants. Since weeds reduce the amount of water available to the crop, drought stress can alter crop performance under crop-weed interference differently from that under weed-free conditions. Among weeds commonly found in cotton, spurred anoda is also a member of the Malvaceae family with a growth pattern similar to cotton. For this reason, the influence of plant interference and mild drought stress on gas exchange and oxidative stress was investigated (Ratnayaka et al. 2003) on the cotton species (*Gossypium hirsutum* cv. 'Delta Pine 5415'), and on spurred anoda (*Anoda cristata*). An individual cotton plant was grown alone, or three cotton plants were planted per pot for intraspecific interference studies. For interspecific interference, a single cotton plant was grown with two plants of spurred anoda. Without interference, cotton and spurred anoda had similar net photosynthesis but different pigment profiles. Stomatal conductance and transpiration rate were greater in spurred anoda than cotton. Net photosynthesis and biomass in cotton were reduced more by spurred anoda interference than by intraspecific interference. With interference, the xanthophyll cycle conversion state and α-toc increased in cotton, but remained unchanged in spurred anoda. The activities of CAT, APX and GR were not influenced by plant interference. Mild drought increased APX activity both in cotton and spurred anoda. Upon drought recovery, drought-induced APX activity was still higher in cotton if compared to well-watered control (1.0 and 0.6 units mg^{-1} protein, respectively), and GR activity compared with well-watered plants was higher in previously drought-stressed cotton (430 and 300 units mg^{-1} protein, respectively), and spurred anoda plants (540 and 480 units mg^{-1} protein, respectively). The authors concluded that the greater impact of spurred anoda interference than intraspecific interference on cotton biomass was due mainly to reduced carbon gain in cotton.

The variations in antioxidant potentials and indole alkaloid content were studied by Jaleel et al. (2008a, b), in two varieties (Rosea and Alba) of *Catharanthus roseus*, an important plant used in traditional as well as modern medicine, exposed to water deficit stress (20 days interval drought starting from 30 days after sowing). The antioxidant and alkaloid profiles were estimated from root, stem, leaf, flowers and pods. The antioxidant potentials were examined in terms of level of non-enzymatic antioxidants and activities of antioxidant enzymes. The non-enzymatic antioxidant molecules studied were ascorbate (AsA), α-toc and GSH, whereas the estimated antioxidant enzymes were SOD, APX, CAT, POD and polyphenol oxidase (PPO). The antioxidant concentrations and activities of antioxidant enzymes were high under water deficit stress in all parts of the plants. The very high concentration of AsA in the root system is surprising and indicates the very high antioxidant ability in this species.

Bermejo et al. (2006) evaluated the changes in water status in two *Trifolium repens* cv. 'Regal' biotypes, tolerant and sensitive to ozone, subjected to a to a short-term drought. The evolution of both soil and plant water status along with leaf gas exchange parameters, leaf APX and GR enzymatic activities for an equal number of clones of each biotype under drought and control treatments was registered for 5 days. The results obtained show that when the short-term drought is imposed, symptoms of oxidative stress and a negative impact on most of the measured parameters are displayed at an earlier stage in the sensitive biotype, proving it to be more sensitive to low water availability. Therefore, it is suggested that these differences might be linked with the existing variability in ozone sensitivity between the biotypes and that, at the same time, they could result in a poorer performance of this biomonitoring system in field assays under conditions of high evapotranspiratory demand, such as those registered during summer time in Mediterranean areas. To understand the adaptability of alfalfa (*Medicago sativa*) to environmental stresses, Wang et al. (2009) have recently analyzed the activity of several antioxidant enzymes, including SOD, POD, APX and CAT, in alfalfa shoots and roots subjected to salt and drought stresses during germination. The germination rate of six alfalfa cultivars was comparatively studied under a 200 mM NaCl or a 35% PEG treatment. 'Alfalfa Xinmu No. 1' and 'Northstar' varieties were selected as stress-tolerant and -sensitive cultivars, respectively, and were used for further characterization. After NaCl or PEG treatment, 'Xinmu No. 1' showed enhanced seedling growth, compared with 'Northstar'. 'Xinmu No. 1' also exhibited low levels of H_2O_2 production and lipid peroxidation, compared with 'Northstar'. In addition, 'Xinmu No. 1' showed higher enzymatic activity of SOD, APX, CAT, and POD in its shoots and roots than 'Northstar'. These results seem to indicate that 'Xinmu No. 1' cultivar's tolerance to salt or drought stresses during germination is associated with enhanced activity of antioxidant enzymes.

In an interesting research on *Arabidopsis* (Jung 2004), young and mature leaves of 4-week-old plants were exposed to drought stress from −0.65 up to −2.54 MPa of ψ_w by withholding water supply for 7 days. The drought-induced increase in non-enzymatic antioxidants in young and mature leaves, enzymatic antioxidants including CAT, POD, SOD and GR substantially increased only in drought-stressed mature leaves. Plants recovered rapidly 24 h after resupplying water, as indicated by the values of ψ_w, photosynthetic efficiency and pigment contents, however, the activities of POD, SOD and GR remained high. Young leaves were rather not affected in drought responses. The fact that drought-stressed mature leaves suffer more stress than drought-stressed young leaves suggests that developmental stages of leaves might contribute to the differential prevention of oxidative damage in plants exposed to drought.

2.5 Non-enzymatic Antioxidant Pools

The studies reported in this paragraph deal with the changes in the level of non-enzymatic antioxidants (ascorbate and glutathione in their reduced or oxidized status, and α-toc) that act within the ascorbate–glutathione cycle. From a general analysis of the most significant studies published in the last decade on this topic, it emerges that there was found no incontestable and definitive indication that a particular degree of metabolic tolerance to drought is related to the absolute level of some non-enzymatic antioxidants. This is in contrast to the fact that the action of drought stress on antioxidant enzyme activities seems to be clear and direct (see the paragraphs above, in which the up-regulation of the enzymes appeared to be ubiquitous among species).

More interestingly, the increase of the oxidized forms of ascorbate (DHA) and glutathione (GSSG) pools are often symptoms of the increased enzymatic activity of APX, DHAR, MDHAR, and GR in response to lower water content. High GSH/GSSG ratios (10:1) are required for the metabolism of the cytosol and chloroplast, and the high content of GSH and GSH-synthesizing enzymes in *Arabidopsis* cells indicate a specific role in drought stress tolerance (Foyer et al 2001). Posch and

Bennett (2009) found that in *Allocasuarina luehmannii* seedlings severe drought stress (predawn $\psi_w = -6.0$ MPa) decreased stomatal conductance and CO_2 assimilation rate to 5% and 15% of the control values, respectively; while t-Asc and t-Glu concentrations remained unaffected by drought treatments, and ascorbate became more oxidized (DHA). In leaves of potato plants subjected to drought (70% RWC), glutathione oxidation ratio increased by about 58%, showing an increase in GSSG/t-Glu ratio from 19% to 30% (Broin et al. 2000). Šircelj et al. (2005, 2007) investigated the influence of gradual water deprivation on potted apple trees (*Malus domestica* Borkh.) up to predawn $\psi_w = -2.0$ MPa. Mild drought (predawn $\psi_w = -0.4$ MPa) did not significantly affect the chosen stress indicators. Moderate drought (predawn $\psi_w = -1.3$ MPa) increased the concentrations of ascorbate and GSSG/t-Glu ratio, indicating the adaptation to oxidative stress in apple trees. Severe drought (predawn $\psi_w = -2.0$ MPa) negatively affected vitality of apple trees, and caused decreases in ascorbate together with the increase in GSSG concentration, indicated severe damage due to oxidative stress. A mild drought is a potential oxidative stressor due to the production of ROS in illuminated chloroplasts which lack CO_2 due to stomatal closure (Morales et al. 2006). Photoprotective pigments (e.g. xanthophyll cycle) may avoid this situation through light energy dissipation, and antioxidants such as ascorbate, tocopherols, and glutathione, may detoxify ROS (Smirnoff 2005). Tausz et al. (2001) subjected potted *Pinus canariensis* seedlings to mild drought by withholding irrigation for one week. This treatment induced a reduction in maximum stomatal conductance (50 mmol m^{-2} s^{-1}) compared to irrigated controls (130 mmol m^{-2} s^{-1}). Concentrations of ascorbate, glutathione, chlorophyll, and the xanthophyll cycle carotenoids were minimal in the evening (under low light) compared to light-saturated conditions. These short-term changes were not affected by drought but the glutathione pool was more oxidized in needles of non-irrigated trees, whereas the redox state of ascorbate remained stable.

Although ascorbate is a high-abundance metabolite, relatively little is known about the factors controlling its accumulation in leaves. To address this issue, Bartoli et al. (2005) examined the role of L-galactono-1,4-lactone dehydrogenase (GalLDH), the enzyme which catalyses the last step of this pathway, in the control of ascorbate content under optimal and stress conditions (drought was imposed on 3-week-old plants by withholding watering until the pots reached a soil ψ_w of −1.5 MPa). In a range of species, no clear relationship between ascorbate content and leaf GalLDH protein and activity was found under optimal growth conditions. To explore the effect of drought stress on GalLDH activity and protein content, wheat (*Triticum aestivum*) was selected for detailed analysis, using two cultivars that differ in their constitutive AsA level. Neither leaf DHA content nor activities of AsA regenerating enzymes were modified by drought. Although drought caused a substantial increase in GalLDH protein and activity in the low AsA in cv. 'CM', this treatment had no effect on these parameters in cv. 'BCH'. Notably, leaf AsA content was unaffected by drought. These results suggest that GalLDH protein and activity cannot be used as an indicator for changes in the capacity for ascorbate biosynthesis, and that ascorbate biosynthesis is constrained by other factors under stress.

The role of α-toc in plants under drought stress is of primary importance within the ascorbate–glutathione cycle (see introduction). In fact, α-tocopherol plays an important role in protecting chloroplastic membranes from the deleterious effects of lipid peroxy radicals and singlet oxygen. It is usually recycled back by ascorbate or GSH following oxidation by lipid peroxy radicals. However, it can be irreversibly converted to the corresponding quinine and quinone epoxides after reacting with singlet oxygen. In a interesting investigation, Munné-Bosch et al. (2001) have studied that the endogenous α-toc levels in a drought-recovery cycle in leaves of sage (*Salvia officinalis* subs. *officinalis*), a drought-susceptible Mediterranean species. The relative leaf water content of the sage plants fell markedly when exposed to drought during the summer. As the drought progressed (RWC from 67% in June to 32% in August), α-toc levels decreased progressively with drought. Therefore, the leaves contained smaller pools of antioxidant defences to counteract oxygen toxicity during the drought, and this explains, among other biochemical and structural features, the susceptibility of this species to stress. These results suggest that it is the complete set of antioxidants, and not a single mechanism, that is responsible for avoiding drought-induced damage in plants. In another study, Munné-Bosch and Peñuelas (2004) investigated the endogenous concentrations of xanthophyll cycle pigments, α-toc, and reduced and oxidized ascorbate in 2-year-old strawberry tree (*Arbutus unedo*) plants exposed to a combination of water deficit, high light, and high temperatures. In the same plants under severe stress, α-toc levels markedly increased, zeaxanthin concentrations increased by 75%, ascorbate increased from 18 to 30 μmol g DW^{-1} and its redox state shifted towards its oxidized form, and chlorophylls, lutein and β-carotene decreased by 63%, 61% and 75%, respectively. To gain insight into the role of flavonoids in the antioxidant defense system of *Cistus clusii* Dunal, Hernández et al. (2004) evaluated drought-induced changes in flavonoids in leaves and compared the response of these compounds with that of ascorbate, α-toc and carotenoids. Total ascorbate (t-Asc) and α-toc concentrations increased to a similar extent in response to a 50-days drought period, even if the kinetics of the drought-induced increases differed. Haberer et al. (2008) quantified ascorbate, glutathione, and α-toc in fine roots of mature *Fagus sylvatica* under free-air canopy ozone (O_3) exposure (twice ambient O_3 concentration, 2 × $[O_3]$) during two growing seasons that differed in the extent of summer drought (exceptional drought year 2003, average year 2004). This design allowed authors to test whether O_3 exposure or drought, or both, affected root antioxidants during the growing season. In both years, root ascorbate and α-toc showed a similar relationship with volumetric soil water content (SWC): t-Asc concentrations on a root dry mass basis increased when SWC dropped from 25% to 20%, whereas α-toc increased at SWC values below 20%. Root glutathione showed no relationship with SWC or differences between the dry and the average year. The results were inconclusive as to whether shoot–root translocation of glutathione or glutathione production in the roots was diminished. Phloem glutathione concentrations in the canopy remained constant, but reduced transport velocity in the phloem and, as a consequence, reduced mass flow of glutathione cannot be ruled out.

3 Effect of Drought Stress on the Whole Ascorbate–Glutathione Cycle

The degree of drought stress at which the up-regulation of the ascorbate–glutathione cycle increases is extremely variable among several plant species, and even between two cultivars of the same species (Reddy et al. 2004b). The level of response depends on the species, the development, and the metabolic state of the plant, as well as the duration and intensity of the stress. The distribution and the level of activity of the enzymes of the ascorbate–glutathione cycle are also known to be differently distributed among all photosynthetic cells in higher plants. In maize leaves, GR and DHAR are exclusively localized in mesophyll cells, APX is mainly localized in mesophyll and bundle sheath cells, and MDHAR is approximately equally distributed between mesophyll and bundle sheath cells (Foyer et al. 2001). Most of the studies on C_4 plants indicate that oxidative damage under stressful conditions is not uniformly distributed between mesophyll and bundle sheath cells of C_4 plants, suggesting that it is restricted to bundle sheath tissue (Nayyar and Gupta 2006).

The whole ascorbate–glutathione system was studied during desiccation of recalcitrant seeds of the silver maple (*Acer saccharinum*) (Pukacka and Ratajczak 2006). The desiccated seeds (seeds were dried at 21°C and 35–40% RH for 14 days) gradually lost their germination capacity and this was strongly correlated with an increase in electrolyte leakage from seeds. Simultaneously the increase of $O_2^{\cdot -}$ and H_2O_2 production in seeds was observed. The results indicate that remarkable changes in the concentrations and redox status of ascorbate and glutathione occur in embryo axes and cotyledons. At the first stages of desiccation, up to a level of 43% of moisture content, AsA and GSH content in embryo axes increased. The enzymes of the ascorbate–glutathione pathway: APX, MDHAR, DHAR and GR increased their activity during desiccation, but mainly in embryonic axes.

Lu et al. (2007) investigated the synergic effects of high temperature, low temperature, and water deficiency on the activities of seven antioxidant enzymes in crofton weed (*Eupatorium adenophorum* Spreng.), an invasive weed in southwestern China. The changes in activities of SOD, CAT, POD, APX, GR, MDHAR, and DHAR were evaluated to determine physiological aspects of the crofton weed that might render the plant vulnerable to environmental stress. In the drought-treated plants, APX activity increased to 202% of the control whereas GR activity increased to maximum levels and was more than 23% greater than the controls. The activity of MDHAR decreased significantly ($P < 0.01$) in the heat and drought treatments, whereas DHAR activity increased from day 0 to 4, reaching the highest levels (210% greater than the controls) on day 4.

Reddy et al. (2004a) determined foliar ascorbate content, and antioxidant enzyme activities in five mulberry (*Morus alba*) cultivars subjected to drought stress by withholding water until the ψ_w reached −2.50 MPa. The activities of antioxidant enzymes, which include SOD, CAT, APX, POD, GR and MDHAR were significantly high in the water-stressed leaves of all the cultivars. The contents of AsA and MDHA in the leaf extracts of all the five cultivars of mulberry also showed variations in response to water deficit. In order to determine if the enzymes of the ascorbate–glutathione cycle were differently affected by drought in C_3 and C_4 plants, Zhang and Kirkkham (1996a, b)) grew *Sorghum bicolor* (C_3) and *Helianthus anuus* (C_4) under dry conditions in a growth chamber. Levels of leaf enzymatic antioxidants (APX, CAT, POD, DHAR, MDHAR, GR and SOD), non-enzymatic antioxidants (ascorbate, glutathione and carotenoids) and stress parameters (chlorophyll and MDA) were determined under watered and drought conditions. Under watered conditions, inherent levels of antioxidants were not consistently higher or lower in sorghum than in sunflower. In response to drought, levels of antioxidants decrease or remained unchanged depending on crop, duration of drought and kind of antioxidants. Duration of drought was divided arbitrarily into three stages. At a late stage of drought (watering had stopped for 7–8 days) when leaf RWC had markedly decreased in sorghum and sunflower, the parameters studied resulted affected by drought. Because of the differential effect of drought, levels of antioxidants were not consistently higher or lower in sorghum than in sunflower under drought. These results show that, under both drought and watered conditions, sorghum does not have consistently higher or lower antioxidant levels than sunflower, and that antioxidant responses to drought differ in C_3 and C_4 plants.

Sofo et al. (2005b) studied the activities of APX, MDHAR, DHAR and GR, as well as the levels of ascorbate pool, glutathione pool and H_2O_2 in plants of four interspecific hybrids of *Prunus* spp. rootstocks subjected to water deficit and shade conditions. The genus *Prunus* comprises more than 400 species adapted to temperate areas and cultivated in Europe. In particular, commercial stone fruit crops, such as peach, plum, almond, apricot, and cherry tree are usually grafted plants with a lower part, the rootstock and an upper grafted part, which is the genotype of the commercial variety. Rootstocks have different genetic background compared to the commercial varieties, and can be used to confer various traits, such as drought stress resistance. After 70 days of water shortage, plants were subjected to a rewatering treatment. During water recovery, leaves fully exposed to sunlight and leaves in shade conditions of about 30% of environmental irradiance were sampled. After 70 days without irrigation, mean predawn ψ_w of all the hybrids fell from −0.34 to −3.30 MPa, and marked decreases in net photosynthesis and transpiration occurred. Generally, APX, MDHAR, DHAR and GR activities were up-regulated during the drought phase and successively down-regulated during the rewatering phase. Furthermore, enzyme activities in shaded leaves were lower than those found in non-shaded leaves. The levels of AsA, DHA, GSH and H_2O_2 were directly related to the increase of drought stress and subsequently decreased during rewatering. In the first 13 days of the drought phase, the ratios of AsA to DHA were similar to those of well-watered plants, whereas, at high level of water deficit, ASA/DHA ratio decreased and subsequently showed an increase in the last days of the drought period. Generally, the ratios of GSH/GSSG increased during high levels of water deficit. The results highlighted the capacity of *Prunus* hybrids to withstand drought conditions by regulating the ascorbate–glutathione cycle. The results obtained in this investigation, underlining the important role of some antioxidant enzymes and compounds in protecting cellular apparatus during water deficit conditions, may be useful for the selection for drought resistance in *Prunus* rootstocks material. This could lead to the characterization of different genotypes with this important characteristic. The same conclusions of Sofo et al. (2005b) were drawn from an unpublished study on various species of almond (*Prunus* spp.) during drought stress and subsequent rewatering (K. Sorkheh et al., 2010, personal communication).

4 Transcriptional Regulation of Genes Encoding Antioxidant Enzymes in Drought-Stressed Plants

In drought-stressed plants, the induction of the proteins by drought stress is strictly regulated (Fig. 5). If compared to the studies on enzyme activities, the investigations on the regulation of the transcription of antioxidant enzymes genes under drought stress are very scarce. Furthermore, the most of them are only qualitative and not quantitative, as these researches have often been conducted by standard PCR-based methods. Recently, quantitative differences in transcription levels in plants under drought conditions or between drought-tolerant and drought-sensitive have been identified by Real Time-PCR-based methods and microarray techniques (Guo et al. 2009; Vadassery et al. 2009). The gene expression of the enzymes of ascorbate–glutathione cycle profile is affected by the intensity and duration of drought stress. For example, the levels of expression of the *APX2* gene is dependent on drought stress degree and light intensity (Tamaoki et al. 2004).

Besides genes encoding for APX, one of the most studied enzymes is glutathione reductase, that play a key role in the restoration of the post-stress redox state of the cytosolic glutathione pool (Foyer et al. 2005). Two isoform of GR cDNA have been cloned and sequenced from pea (*Pisum sativum* cv. 'Birte') (Stevens et al. 1997). The cytosolic GR cDNA (*GOR2*) is significantly different at the DNA level from the chloroplastidial/mitochondrial GR (*GOR1*). *GOR2* maps to linkage group 6 on the pea genome map and it seems likely that this is the only locus for this gene. In contrast to *GOR1*, transcript levels of *GOR2* increase in the rewatering (post-stress) by about ten- and threefold, respectively. Key regulators and signalling components involved in high light-mediated oxidative stress may lead to cross protection against drought, as many high light-regulated genes are also induced under drought conditions. Molecular analyses of *Arabidopsis thaliana*, ecotype Columbia, have revealed a number of genes whose expression changes in response to high light, including the H_2O_2 scavenger *APX2*, and have provided evidence for common steps in drought and high light stress response pathways (Rossel et al. 2006). The authors described the drought tolerant mutant *alx8*, which has constitutively higher *APX2* expression and higher levels of foliar ABA than wild type. In fact, exogenous ABA increased *APX2* expression and the *APX2* promoter contains ABA response elements. The *alx8* mutant exhibits improved water-use efficiency and the up-regulaton of a number of drought-tolerance genes, including *APX2*.

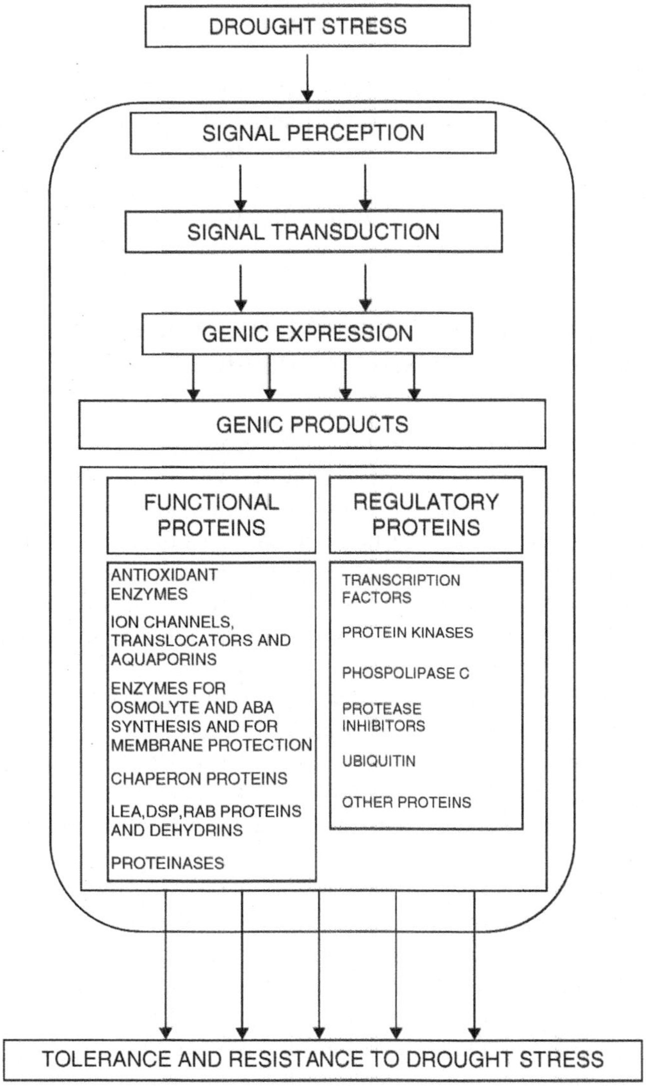

Fig. 5 Functions of the proteins induced by drought stress. The induced proteins are divided in two groups: functional proteins involved in stress tolerance and stress cellular adaption, and regulatory proteins with a role in genic expression and signal transduction in response to drought

The study of Jin et al. (2006) suggests that the regulation of APX at the transcript level may be involved in the response to water deficit stress. Cut rose (*Rosa hybrida*) cv. 'Samantha' flowers were pretreated for 12 h with 6 mM ascorbate (AsA), 5 mM β-aminophenol, or water (control) prior to exposing to water deficit stress for 24 h, and then were placed into water for recovery and vase life. Vase life, flower development, ψ_w, MDA content, and SOD and APX activities were then determined until end of vase life. Water deficit stress reduced vase life and inhibited flower development.

AsA pretreatment alleviated deterioration, while β-aminophenol pretreatment increased the deterioration. AsA pretreatment also decreased MDA content, and increased SOD and APX activities, but the opposite effects were found for the β-aminophenol pretreatment. A cDNA encoding cytosolic APX was isolated from petals, and named *Rh-APX1*. Gene expression in control petals increased in the first 9 h, then decreased until the end of water deficit stress; it recovered when water was resupplied, and peaked again on the third day after placing flowers in water. Compared with the control, the gene expression was enhanced substantially by AsA pretreatment throughout water deficit stress, water recovery, and throughout vase life. In contrast, the expression was inhibited by β-aminophenol. The changing patterns of *Rh-APX1* gene expression paralleled those of APX activity.

The objective of the recent study of Bian and Jiang (2009) was to investigate accumulation of ROS, antioxidant enzyme activities, and gene expression patterns of antioxidant enzymes of Kentucky bluegrass (*Poa pratensis*) under drought stress and recovery. Grass (cv. Midnight II) was subjected to soil drying for 5 days and then rewatered for 1 day in growth chamber. Drought stress increased $O_2{\cdot}^-$ production of leaves and H_2O_2 content of the leaves and roots. Recovery enhanced leaf $O_2{\cdot}^-$ production and root H_2O_2 content. Drought stress increased the leaf activities of APX, MDHAR and DHAR, and the root activities of GR and MDHAR, while reducing the root activities of DHAR. The increased leaf activities of APX, MDHAR, DHAR and GR, and the root activity of APX and MDHAR were also maintained after rewatering. For the leaves, the expression of DHAR was down-regulated by drought stress but recovered to control level after rewatering, while the expressions of GR and MDHAR were up-regulated and maintained high transcript levels also after water recovery. For the roots, the expressions of cytosolic APX, GR, and DHAR were down-regulated under drought stress but recovered except for GR and DHAR, while MDHAR expression was up-regulated. Antioxidant enzymes and their gene expressions may be differentially or cooperatively involved in the defense mechanisms in the leaves and roots of Kentucky bluegrass exposed to drought stress and recovery. MDHAR and DHAR are the two main enzymes that maintain ascorbate in its reduced state. *MDAR2* (At3g09940) and *DHAR5* (At1g19570) expression is up-regulated in the roots and shoots of *Arabidopsis* seedlings co-cultivated with the root-colonizing endophytic fungus *Piriformospora indica*, or that were exposed to a cell wall extract or a culture filtrate from the fungus (Vadassery et al. 2009). In fact, growth and seed production were not promoted by *Piriformospora indica* in *mdar2* (SALK_0776335C) and *dhar5* (SALK_029966C) T-DNA insertion lines, while colonized wild-type plants were larger and produced more seeds compared to the uncolonized controls. After 3 weeks of drought stress, growth and seed production were reduced in *Piriformospora indica*-colonized plants compared to the uncolonized controls, and the roots of the drought-stressed insertion lines were colonized more heavily by the fungus than were wild-type plants. Upregulation of the message for the antimicrobial PDF1.2 protein in drought-stressed insertion lines indicated that MDAR2 and DHAR5 are crucial for producing sufficient ascorbate to maintain the interaction between *Piriformospora indica* and *Arabidopsis* in a mutualistic state.

5 Proteomic Studies on Antioxidant Enzymes in Drought-Stressed Plants

Proteomic analysis provides a broad view of plant responses to stress at the level of proteins. In recent years, this approach has increased in sensitivity and power as a result of improvements in two-dimensional polyacrylamide gel electrophoresis (2DE), protein detection and quantification, fingerprinting and partial sequencing of proteins by mass spectrometry (MS), bioinformatics, and methods for gene isolation (Fig. 6). Salekdeh et al. (2002) detected more than 2,000 proteins from drought-stressed and well-watered leaves of rice (line CT9993-5-10-1-M, upland japonica). Among these proteins, the three most marked changes were seen with actin depolymerizing factor, a homologue of the S-like ribonucleases and the chloroplastic glutathione-dependent DHAR. The data showed an increase of about 60% in the abundance of DHAR in drought-stressed plants. Desert plants are exposed to a combination of environmental stress conditions, including low water availability, extreme temperature fluctuations, high irradiance and nutrient deprivation. Studying desert plants within their natural habitat may therefore reveal novel mechanisms and strategies that enable plants to resist stressful conditions. Mittler et al. (2001) studied the acclimation of *Retama raetam*, an evergreen stem-assimilating desert

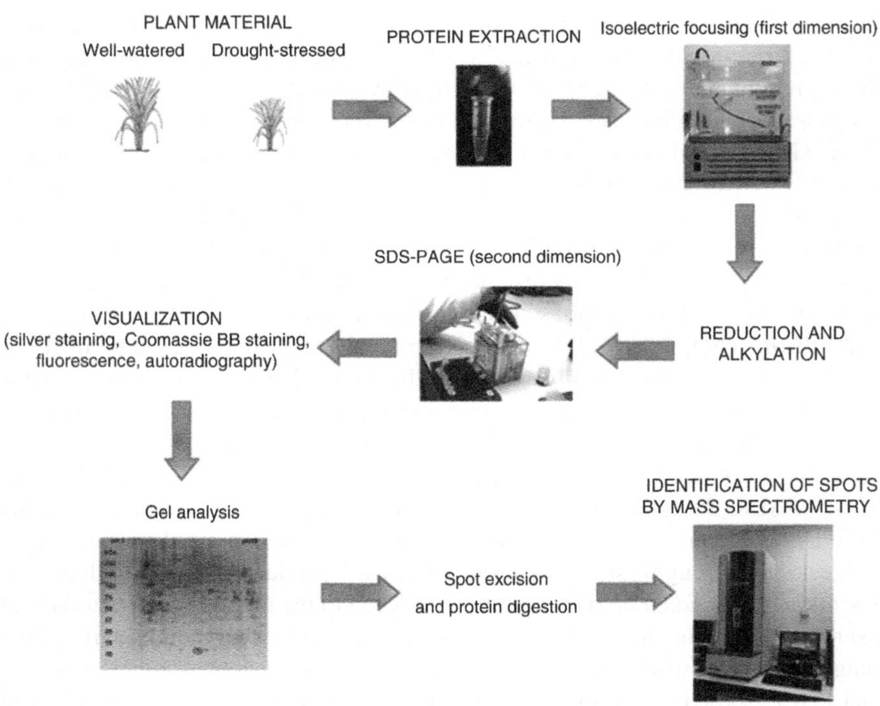

Fig. 6 Scheme for a differential protein display utilizing 2D gel electrophoresis coupled with mass spectrometry

plant, to growth within an arid dune ecosystem. *Retama raetam* contained two different populations of stems: those of the upper canopy, exposed to direct sunlight, and those of the lower canopy, protected from direct sunlight. During the dry season, stems of the upper canopy contained a very low level of a number of essential proteins, including the large and small subunits of rubisco, APX, and the D1 subunit of the reaction centre of photosystem II. Upon rewatering, as well as following the first rainfall of the season, these 'photosynthetically suppressed' stems recovered and accumulated essential proteins within 6 ± 24 h. In contrast, stems of the lower canopy contained the essential proteins throughout the dry season. The authors suggest that *R. raetam* uses an acclimation strategy of 'partial plant dormancy' in order to survive the dry season. 'Dormancy', as evident by the post-transcriptional suppression of gene expression, as well as the suppression of photosynthesis, was induced specifically in stems of the upper canopy which protect the lower canopy by shading.

In order to investigate the unique contribution of individual wine grape (*Vitis vinifera*) berry tissues and water-deficit to wine quality traits, Grimplet et al. (2009) carried out an investigation on tissue-specific differences in protein and selected metabolites, using pericarp (skin and pulp) and seeds of berries from vines grown under well-watered and water-deficit stress conditions. Water-deficit stress altered the abundance of approximately 7% of pericarp proteins, but had little effect on seed protein expression. About half of the 32 metabolites surveyed showed tissue-specific differences in abundance with water-deficit stress affecting the accumulation of seven of these compounds. In particular, the skin displayed a notable increase in the relative abundance of cytosolic APX.

6 Studies on Drought-Tolerant Transgenic Plants

In one of the first experiments on transgenic plants designed to analyze the potential of ROS scavenging system of chloroplasts during drought stress, Shikanai et al. (1998) introduced *Escherichia coli* CAT into tobacco chloroplasts. Photosynthesis of transgenic plants was tolerant to high irradiance (1,600 μmol m^{-2} s^{-1}) under drought conditions, while the wild plants suffered severe damage in photosynthesis under the same conditions. Irrespective of responses to the stress, chloroplastic APX was completely inactivated both in the transgenic and wild-type plants (from about 60 to 0 nmol mg^{-1} protein min^{-1}, and from about 70 to 0 nmol mg^{-1} protein min^{-1} in wild and transgenic plants, respectively). Considering the enzyme stability under oxidative stress conditions, however, the authors concluded that CAT is much superior to APX to confer stress-resistance. These preliminary findings were contrary to the established idea that the APX-mediated antioxidative system protects chloroplasts from oxidative stress but the most of the studies of the last decade pointed out that APX plays an important role in the metabolism of H_2O_2 in higher plants. Recently, Bhatnagar-Mathur et al. (2009) investigated on transgenic plants of peanut over-expressing the *AtDREB1A* transgene, driven by a stress-inducible promoter (*Atrd29A*) when exposed to progressive drought stress conditions. This authors found APX and GR activities were higher in the transgenic plants than in the untransformed counterparts without the promoter. Moreover, in the transgenic plants, the antioxidative machinery in plants over-expressing the *AtDREB1A* transcription factor under water-limiting conditions indirectly caused improved gas exchange and water use efficiency. Li et al. (2009) studied the effect of over-expressing a *Populus* peroxisomal APX (*PpAPX*) gene under the control of the cauliflower mosaic virus 35S promoter or the rd29 promoter in transgenic tobacco. High levels of *PpAPX* expression were observed in 35S-PpAPX transgenic plants, with a 50% increase in APX activity. The constitutive expression of *PpAPX* in the tobacco exhibited no morphological abnormalities, while significantly increased root growth was observed in transgenic plants, when compared to control plants. APX activity was nearly 80% higher in the leaves of transgenic plants in response to drought or salt stresses. Moreover, the transgenic tobacco also showed significantly improved drought resistance and salt tolerance at the vegetative stage. RNA blot analysis indicated that the *PpAPX* transcript level was very low under normal growing conditions in rd29Ap–PpAPX plants, but significantly increased under drought stress. The authors concluded that that *PpAPX* does not play a significant role under normal growing conditions, but did ameliorate oxidative injury under abiotic stress. Bhatnagar-Mathur et al. (2009) recently investigated on transgenic plants of peanut over-expressing the *AtDREB1A* transgene, driven by a stress-inducible promoter (*Atrd29A*) when exposed to progressive drought stress conditions. These authors found APX and GR activities were higher in the transgenic plants than in the untransformed counterparts without the promoter. Moreover, in the transgenic plants, the antioxidative machinery in plants over-expressing the *AtDREB1A* transcription factor under water-limiting conditions indirectly caused improved gas exchange and water use efficiency.

The effects of re-watering after drought stress and the capacity of plants to resume well after a mild drought have rarely been studied in experiments on transgenic plants. In two interesting works, two cDNAs of the enzyme GR encoding a dual-targeted isoform (dtGR) and a cytosolic isoform (cGR), were cloned from leaves of common bean (*Phaseolus vulgaris*) (accession DQ459505 for dtGR and DQ459504 for cGR) (Contour-Ansel et al. 2006; Torres-Franklin et al. 2008). Moderate drought stress (ψ_w measured at 10 a.m. = −1.5 MPa) followed by re-watering was applied to common bean cultivars, one tolerant to drought ('IPA'), the other susceptible ('Carioca') and to cowpea (*Vigna unguiculata* Walp) cultivars, one tolerant to drought (EPACE-1), and the other susceptible (1183). The results showed that mRNA levels were much higher for *cGR* than for *dtGR* in all cases. Moderate drought stress induced an up-regulation of the expression of *cGR* in the susceptible cultivars. On the contrary, *dtGR* expression decreased. In the tolerant cowpea EPACE-1, GR gene expression remained stable under drought. Total GR activity in bean leaves was 18, 15 and 32 nmol min^{-1} mg^{-1} protein in control, drought-stressed and re-hydrated 'IPA' plants; and 21, 13 and 17 nmol min^{-1} mg^{-1} protein in control, drought-stressed and re-hydrated 'Carioca' plants. The higher

GR activity can be so associated to the higher tolerance of Carioca plants against drought. It is important to note that during recovery from drought, an up-regulation of the two GR isoforms expression occurred, with a peak at 6–10 h after re-hydration. Plant response to re-watering was very rapid: except for *dtGR* in IPA, expression of both GR isoforms was enhanced as soon as 6 h re-hydration for all cultivars. This suggests that moderate drought stress may lead to a hardening process and acclimation tolerance to a subsequent more severe drought. Glutathione has a primary importance for acting also as a substrate of enzymes that take no part in the ascorbate–glutathione cycle. For example, glutathione peroxidase (GPX)-like proteins (GPX-1 and GPX-2) of *Synechocystis* PCC 6803 (S. PCC 6803) reduce unsaturated fatty acid hydroperoxides using NADPH. Gaber et al. (2006) used transgenic *Arabidopsis* plants overexpressing S. PCC 6803 GPX-2 in the cytosol (AcGPX2) or chloroplasts (ApGPX2). Both the transgenic lines showed enhanced tolerance to oxidative damage caused by treatment with H_2O_2 (10 mM), Fe ions (200 mM) or methylviologen (50 mM) and drought for 12 days. The data reported by the authors indicated that the expression of S. PCC 6803 GPX-2 contributes to the reduction in unsaturated fatty acid hydroperoxides using NADPH in situ under stress conditions in the transgenic plants.

Sometimes, a transgenic approach on genes that indirectly affect the expression of enzymatic antioxidant system can be successful, as in the case of the work of Synková and Valcke (2001). The fact that P*ssu-ipt* tobacco, despite a permanent water deficit, can maintain almost unaffected photosynthesis suggested to the authors that some efficient protecting mechanisms exist. The response of antioxidant enzymes to cyclic drought was studied in control non-transformed tobacco (*Nicotiana tabacum* cv. 'Petit Havana SRI') and two types of transgenic P*ssu-ipt* tobacco (grafted on wild rootstock and poorly rooted progeny of Fl generation) grown under different conditions of irradiation (greenhouse, referred as high light, versus growth chamber, referred as low light). Transgenic P*ssu-ipt* tobacco contains the *ipt* gene, encoding for the isopentenyl transferase, a key enzyme of the citokinins biosynthesis. Drought stress cycles (a period of 4 weeks of a water-deficit cycle, i.e., 3 days withholding irrigation followed by re-watering) started with plants at two contrasting developmental stages, i.e., at the stage of vegetative growth (young), and at the onset of flowering (old). Drought reduced the growth of 'SRI' plants compared with transgenic ones, particularly, when treatment started in earlier stage of plant development. Relative leaf water content was significantly lower (below 70%) in all transgenic grafts and plants compared with the wild type, irrespective of age, drought, and growth conditions. The response of antioxidant enzymes was significantly dependent on plant type and plant age; nevertheless, growth conditions and drought stress also affected enzyme activities. Contrary to non-transgenic tobacco, where significant changes of GR activity were found between control and drought-stressed plants grown in a greenhouse, both transgenic types exhibited unchanged activities throughout plant stress treatment. In contrast to non-transgenic and P*ssu-ipt* rooted plants, peroxidase activities (APX, POD and syringaldazine peroxidase) in older P*ssu-ipt* grafts were up to four times higher, irrespective of growth and stress, nevertheless, the effect seemed to be age-dependent. The differences observed in activities of enzymes of intermediary metabolism (i.e., malic enzyme and glucose-6-phosphate dehydrogenase) revealed that transgenic grafts probably compensated differently for a decrease of ATP and NADPH than control and transgenic rooted plants under stress.

In a recent research, Zhang et al. (2009) demonstrated that the adaptive responses of plants to drought stress is due to a better antioxidant response. For the experiment, they used a transgenic tobacco line Over-expressing the 9-*cis*-epoxycarotenoid dioxygenase gene (*SgNCED1*), with increased ABA content, and tolerance to drought and salt stresses. H_2O_2 and nitric oxide (NO) contents were enhanced in guard cells and mesophyll cells of the transgenic plants, accompanied with increased transcripts and activities of antioxidant enzymes including SOD, CAT, APX and GR. The abundance of H_2O_2 and NO levels, and of the transcripts and activities of antioxidant enzymes in the transgenic plants was blocked by pre-treatments with inhibitor of ABA biosynthesis, scavengers of H_2O_2 and NO, and inhibitors of NADPH oxidase and NO synthase. The elevated production of NO in the transgenic plants was blocked by scavenger of H_2O_2 and inhibitors of NADPH oxidase, whereas H_2O_2 level was not affected by scavenger of NO and inhibitor of NOS-like, indicating that H_2O_2 is essential for the elevated production of NO. The results demonstrate that the increased drought and salt tolerance in the transgenic plants is associated with ABA-induced production of H_2O_2 via NADPH oxidase and NO via NOS-like, which sequentially induce transcripts and activities of SOD, CAT, APX and GR.

7 Some External Applications on Plants for a Better Drought Tolerance

The positive effects of pesticides, exogenous ascorbate, sodium nitroprusside, salicylic acid, silicon or UV-B radiation on drought tolerance of plants were often associated with the increase of antioxidant defense abilities, therefore alleviating oxidative damage of cellular functional molecules induced by over produced ROS under drought and maintaining many physiological processes of stressed plants (Gong et al. 2005).

In order to examine whether paraquat modifies the functioning of antioxidants and oxidative stress levels in drought-stressed plants, Liu et al. (2009) carried out an experiment with *Cucumis sativus* (cv. 'Yuexiu no. 3') grown hydroponically. Drought stress, increased the contents of $O_2{\cdot}^-$ and H_2O_2 in cucumber leaves, while pretreatment of paraquat combined with drought increased them in a lower extent. Drought stress and paraquat application both increased the activities of antioxidants such as SOD, CAT, POD, APX, DHAR, MDHAR, GR, GSH and AsA. Furthermore, the combined effect of paraquat application and drought stress resulted in the highest activities of enzymatic and non-enzymatic antioxidants. In conclusion, paraquat is able to moderate the activities of scavenging system enzymes, and to influence oxidative stress intensity under drought stress induced by PEG. Jaleel et al. (2007) conducted a pot culture experiment to estimate the drought stress mitigating effect of ketoconazole (KCZ), a fungicide cum plant growth regulator, in *Catharanthus roseus* plants. The plants under pot culture were subjected to drought stress (10, 15 and 20 days) and drought stress with KCZ from 30 days after sowing and regular irrigation was kept as control. Antioxidant contents and activities of antioxidant enzymes were estimated from roots and leaves of both control and treated plants. Individual and combined drought stress and KCZ treatments increased ascorbate and α-toc contents, SOD, APX, CAT and PPO activities when compared to control. Unfortunately, the authors did not report the values of the control plants, so it is impossible to do a precise comparison. Another investigation was carried out to find out the extent of changes occurred in groundnut (*Arachis hypogaea* cv. 'ICG 221') in response to paclobutrazol (PBZ) treatment under water deficit (Sankar et al. 2007). Individual treatment with PBZ (10 mg l^{-1}) and drought stress showed an increase in foliar t-Asc and α-toc contents, and GSH, SOD, APX and CAT activities. PBZ with drought stressed plants determined higher levels of antioxidant and scavenging enzymes than drought alone. In the study of Manivannan et al. (2007), a pot culture experiment was conducted to estimate the ameliorating effect of propiconazole (PCZ) on drought stress in cowpea (*Vigna unguiculata*) plants. From 30 days after sowing, the plants were subjected to 9 days of drought stress, and to drought stress with 15 mg l^{-1} PCZ. The plants were separated into root, stem and leaf for estimating the antioxidant contents and activities of antioxidant enzymes. Individual and combined drought stress and PCZ treatments increased t-Asc and α-toc contents, SOD, APX, CAT and PPO activities when compared to control. The PCZ treatment mitigated the adverse effects of drought stress by increasing the antioxidant potentials and thereby paved the way for overcoming drought stress in *V. unguiculata* plants. A pot-culture experiment was conducted to estimate the ameliorating effect of triadimefon (TDM) on drought stress in sunflower plants (*Helianthus annuus*) (Manivannan et al. 2008). Triazole compounds such as TDM, hexaconazole, uniconazole and paclobutrazol, etc., are widely used as fungicides, and they also possess varying degrees of plant-growth regulating properties, mediated by their interference with the isoprenoid pathway and subsequent shift in the balance of important plant hormones, including GA, ABA and cytokinins. The plants were subjected to 3, 6, and 9 days of drought stress and drought stress with TDM at 15 mg l^{-1}. The plant samples were collected and separated into root, stem and leaf for estimating the activities of antioxidant enzymes. Individual and combined drought stress, and TDM treatments increased APX activity when compared to control plants. From the results of this investigation, it can be concluded that the application of TDM caused a partial amelioration of the adverse effects of drought stress by its influence on antioxidant potentials of *H. annuus* plants.

Although ascorbate has been firmly associated with antioxidant response, recent studies have suggested that the functions of ascorbate are related primarily to developmental processes (Smirnoff 2005). Nitrogen fixation in legumes is dramatically inhibited by drought stresses, and this reduction is often associated with oxidative damage. Zabalza et al. (2008) examined the hypothesis that ascorbate is involved in alleviating the oxidative damage to nodules caused by an increase in ROS under drought stress. The hypothesis was tested by supplying 5 mM ascorbate to pea plants (*Pisum sativum*) experiencing moderate drought stress (predawn $\psi_w = -1.0$ MPa). A supply of exogenous ascorbate increased the nodule t-Asc, whereas the levels of AsA, DHA, GSH, GSSG, GSH + GSSG were not affected. AsA application did not significantly modulate the response to drought stress of APX, whereas modulated that of GR. The effects of water-deficit stress and foliar application of ascorbate were also studied in leaves of *Zea mays* (Dolatabadian et al. 2009). Foliar application was performed by backpack sprayer and four concentrations of AsA (0, 50, 100 and 150 ppm) were applied. The activity of some antioxidant enzymes was clearly increased by water-deficit stress (soil ψ_w of −1.3 MPa). Foliar application of ascorbate reduced stress-induced and antioxidative enzymes activities. It seems that, ascorbate application helps the plants for better resistance under the stress by inactivation and scavenging of free radicals. The authors concluded that ascorbate treatment reduced the damaging action of drought and decreased enzyme activity due to ROS scavenging; thereupon it may be effective for the improvement of stressed plants in arid and semi-arid regions.

In a recent research of Nasibi and Kalantari (2009) on tomato, sodium nitroprusside (SNP) was used as NO donor in control and drought-stressed plants, and the role of NO in reduction of oxidative damages was investigated. The authors observed that SNP pretreatment (plants were sprayed with SNP 100 μM) prevented drought-induced decrease in RWC, membrane stability index, increase in lipid peroxidation and lipoxygenase activity, and increase in H_2O_2 content. However, pretreatment of plants with SNP and phenyl 4,4,5,5-tetramethylimidazoline-1-oxyl-3-oxide (a NO scavenger) (plants were sprayed with PTIO 200 μM) reversed the protective effects of SNP suggesting that protective effect by SNP is attributable to NO release. In addition, the relationship between these defense mechanisms and activity of antioxidant enzymes were checked. Results showed that in drought-stressed plants APX and CAT activities were elevated over the well-watered controls, while GR decreased under drought condition. The activity of APX and GR increased under SNP pretreatment and it seems that under this condition APX had a key role of detoxification of ROS in tomato plants. This result corresponded well with AsA, DHA and total acid-soluble thiols content. Therefore, reduction of drought-induced oxidative damages by NO in tomato leaves is most likely mediated through either NO ability to scavenge active oxygen species or stimulation of antioxidant enzyme such as APX. Exogenous salicylic acid has been also shown to confer tolerance against biotic and abiotic stresses. In the work of Horváth et al. (2007), the ability of salicylic acid to increase abiotic stress tolerance was demonstrated: it improved the drought tolerance of the winter wheat (*Triticum aestivum*) cv. 'Cheyenne'. The induction of drought tolerance in Cheyenne was correlated with an increase in APX activity (from 0.04 ΔA mg^{-1} protein in non-treated plants to almost 0.06 ΔA mg^{-1} protein in treated plants).

Drought-induced changes in oxidative damage to photosynthetic pigments, proteins and lipids, some enzyme activities and photosynthesis were investigated in wheat (*Triticum aestivum*) plants grown in pots applied with or without silicon under drought stress (Gong et al. 2005). Three treatments were prepared: well-watered plants, drought-stressed plants and drought-stressed plants + silicon (2.11 mmol of sodium silicate kg^{-1} soil). The results showed that application of silicon improved the water status of drought stressed plants. Compared with the non-silicon treatment, application of silicon increased the foliar activities of SOD, CAT, and GR, the fatty acid unsaturation of lipids, and the contents of photosynthetic pigments and soluble proteins as well as total thiols under drought, whereas the content of H_2O_2 and oxidative stress of proteins were decreased by applying silicon compared with those of non-silicon treatments under drought. The activities of POD and APX showed no significant difference between silicon treated and untreated plants. In addition, application of silicon also increased the net CO_2 assimilation rate of wheat leaves under drought. Physiological and biochemical responses of wheat seedlings to drought, UV-B radiation, and combined stress were investigated by Tian and Lei (2007). Oxidative damage caused by mild drought (ψ_w of −0.5 MPa), UV-B (3.5 kJ m^{-2} UV-B radiation), and combined stresses retarded seedling growth by 26.5%, 29.1%, and 55.9%, respectively. The activities of SOD, POD, and APX increased under drought, UV-B, and the combination of stresses, while CAT activity decreased under the combined stress as compared to the control. The combination of drought and UV-B caused more severe damage to wheat seedlings than stress factors applied separately. Thus, the combined application of drought and UV-B had more strong adverse effects on wheat seedlings. The addition of 0.2 mM sodium nitroprusside (SNP) enhanced wheat seedling growth under drought, UV-B, and combined stress, likely, due to decreasing the accumulation of H_2O_2 and lipid peroxidation as well as activating the antioxidant enzymes.

8 Conclusions and Perspectives

The last decade of research on enzymatic and non-enzymatic antioxidants within the ascorbate–glutathione cycle have revealed that photoprotection against drought-mediated oxidative stress is as complex and intricately regulated. A conspicuous increase in the studies on antioxidant enzymes and their trasnscripts, and on transgenic plants with enhanced antioxidant defenses is expected in the next years. It is not easy to predict if this new knowledge will lead to the creation of varieties with enhanced drought tolerance, as antioxidation does not readily lend itself to improvement via single-gene transgenic up-regulation (Logan 2005).

One of the most studied antioxidant enzyme will probably be ascorbate peroxidase, as it is a unique class I peroxidase found mainly in photosynthetic algae and plants, with a high structural homology to yeast cytochrome c peroxidase (Mittler and Poulos 2005), and uses ascorbic acid as its preferred reducing substrate. Most of the studies herein reported have demonstrated that APX, due to its presence in every compartment of the plant cell and its highly regulated expression (transcriptional and posttranscriptional) is a key defense enzyme involved in the removal of H_2O_2 in plants. In addition, the studies discussed show that ascorbate and glutathione are highly abundant metabolites in plants and that they have many diverse and important functions, such as the H_2O_2-mediated signal transduction cascades. Finally, the intra-cellular distribution of the enzymes of the ascorbate–glutathione cycle is still not completely clear, and more experimentation is required to determine whether the chloroplast and cytosol enzymes fulfill different roles. Further studies are required to elucidate how the function of the different genes/enzymes of the ascorbate–glutathione cycle are coordinated and whether or not they have additional functions in plants.

Acknowledgement We are grateful to Prof. Cristos Xiloyannis for his important suggestions and scientific support.

References

Aroca R, Irigoyen JJ, Sánchez-Díaz M (2003) Drought enhances maize chilling tolerance. II. Photosynthetic traits and protective mechanisms against oxidative stress. Physiol Plant 117:540–549

Asada K (1999) The water-water cycle in chloroplasts: scavenging of active oxygens and dissipation of excess photons. Annu Rev Plant Physiol Plant Mol Biol 50:601–639

Bartoli CG, Guiamet JJ, Kiddle G, Pastori GM, Di Cagno R, Theodoulou FL, Foyer CH (2005) Ascorbate content of wheat leaves is not determined by maximal L-galactono-1, 4-lactone dehydrogenase (GalLDH) activity under drought stress. Plant Cell Environ 28:1073–1081

Bermejo R, Irigoyen JJ, Santamaría JM (2006) Short-term drought response of two white clover clones, sensitive and tolerant to O_3. Physiol Plant 127:658–669

Bhatnagar-Mathur P, Devi MJ, Vadez V, Sharma KK (2009) Differential antioxidative responses in transgenic peanut bear no relationship to their superior transpiration efficiency under drought stress. J Plant Physiol 166:1207–1217

Bian S, Jiang Y (2009) Reactive oxygen species, antioxidant enzyme activities and gene expression patterns in leaves and roots of Kentucky bluegrass in response to drought stress and recover. Sci Hort 120:264–270

Briviba K, Klotz LO, Sies H (1997) Toxic and signalling effects of photochemically or chemically generated singlet oxygen in biological systems. Biol Chem 378:1259–65

Broin M, Cuiné S, Peltier G, Rey P (2000) Involvement of CDSP 32, a drought-induced thioredoxin, in the response to oxidative stress in potato plants. FEBS Lett 467:245–248

Chai TT, Fadzillah NM, Kusnan M, Mahmood M (2005) Water stress-induced oxidative damage and antioxidant responses in micropropagated banana plantlets. Biol Plant 49:153–156

Chaves MM, Maroco JP, Pereira JS (2003) Understanding plant responses to drought-from genes to the whole plant. Funct Plant Biol 30:239–264

Contour-Ansel D, Torres-Franklin ML, Cruz De Carvalho MH, D'Arcy-Lameta A, Zuily-Fodil Y (2006) Glutathione reductase in leaves of cowpea: cloning of two cdnas, expression and enzymatic activity under progressive drought stress, desiccation and abscisic acid treatment. Ann Bot 98:1279–1287

Davis DG, Swanson HR (2001) Activity of stress-related enzymes in the perennial weed leafy spurge (*Euphorbia esula* L.). Environ Exp Bot 46:95–108

Desikan R, Cheung K, Clarke A, Goulding S, Sagi M, Fluhr R, Rock C, Hancock JT, Neill SJ (2004) Hydrogen peroxide is a common signal for darkness- and ABA-induced stomatal closure in *Pisum sativum*. Funct Plant Biol 31:913–920

Dolatabadian A, Sanavy SAMM, Sharifi M (2009) Alleviation of water deficit stress effects by foliar application of ascorbic acid on *Zea mays* L. J Agron Crop Sci 195:347–355
Duan B, Yanwei L, Chunying Y, Olavi J, Chunyang L (2005) Physiological responses to drought and shade in two contrasting *Picea asperata* populations. Physiol Plant 124:476–484
Edjolo A, Laffray D, Guerrier G (2001) The ascorbate-glutathione cycle in the cytosolic and chloroplastic fractions of drought-tolerant and drought-sensitive poplars. J Plant Physiol 158:1511–1517
Egert M, Tevini M (2002) Influence of drought on some physiological parameters symptomatic for oxidative stress in leaves of chives (*Allium schoenoprasum*). Environ Exp Bot 48:43–49
Ennajeh M, Vadel AM, Khemira H (2009) Osmoregulation and osmoprotection in the leaf cells of two olive cultivars subjected to severe water deficit. Acta Physiol Plant 31:711–721
Foyer CH, Gomez LD, Van Heerdern PDR (2005) Glutathione. In: Smirnoff N (ed) Antioxidants and reactive oxygen species in plants. Blackwell Publishing, Oxford, UK, pp 53–66
Foyer CH, Mullineaux PM (1998) The presence of dehydroascorbate and dehydroascorbate reductase in plant tissues. FEBS Lett 425:528–529
Foyer CH, Noctor G (2003) Redox sensing and signalling associated with reactive oxygen in chloroplasts, peroxisomes and mitochondria. Physiol Plant 119:355–364
Foyer CH, Noctor G (2005) Oxidant and antioxidant signalling in plants: a re-evaluation of the concept of oxidative stress in a physiological context. Plant Cell Environ 28:1056–1071
Foyer CH, Theodoulou FL, Delrot S (2001) The functions of inter and intracellular glutathione transport systems in plants. Trends Plant Sci 6:486–492
Gaber A, Yoshimura K, Yamamoto T, Yabuta Y, Takeda T, Miyasaka H, Nakano Y, Shigeoka S (2006) Glutathione peroxidase-like protein of *Synechocystis* PCC 6803 confers tolerance to oxidative and environmental stresses in transgenic *Arabidopsis*. Physiol Plant 128:251–262
Gong H, Zhu X, Chen K, Wang S, Zhang C (2005) Silicon alleviates oxidative damage of wheat plants in pots under drought. Plant Sci 169:313–321
Grimplet J, Wheatley MD, Jouira HB, Deluc LG, Cramer GR, Cushman JC (2009) Proteomic and selected metabolite analysis of grape berry tissues under well-watered and water-deficit stress conditions. Proteomics 9:2503–2528
Guerfel M, Ouni Y, Boujnah D, Zarrouk M (2009) Photosynthesis parameters and activities of enzymes of oxidative stress in two young 'Chemlali' and 'Chetoui' olive trees under water deficit. Photosynthetica 47:340–346
Guidi L, Degl'Innocenti E, Remorini D, Massai R, Tattini M (2008) Interactions of water stress and solar irradiance on the physiology and biochemistry of *Ligustrum vulgare*. Tree Physiol 28:873–883
Guo P, Baum M, Grando S, Ceccarelli S, Bai G, Li R, von Korff M, Varshney RK, Graner A, Valkoun J (2009) Differentially expressed genes between drought-tolerant and drought-sensitive barley genotypes in response to drought stress during the reproductive stage. J Exp Bot 60:3531–3544
Guo Z, Ou W, Lu S, Zhong Q (2006) Differential responses of antioxidative system to chilling and drought in four rice cultivars differing in sensitivity. Plant Physiol Biochem 44:828–836
Haberer K, Herbinger K, Alexou M, Rennenberg H, Tausz M (2008) Effects of drought and canopy ozone exposure on antioxidants in fine roots of mature European beech (*Fagus sylvatica*). Tree Physiol 28:713–719
Hernández I, Alegre L, Munné-Bosch S (2004) Drought-induced changes in flavonoids and other low molecular weight antioxidants in *Cistus clusii* grown under Mediterranean field conditions. Tree Physiol 24:1303–1311
Horemans N, Foyer CH, Asard H (2000) Transport and action of ascorbate at the plant plasma membrane. Trends Plant Sci 5:263–267
Horváth E, Pál M, Szalai G, Páldi E, Janda T (2007) Exogenous 4-hydroxybenzoic acid and salicylic acid modulate the effect of short-term drought and freezing stress on wheat plants. Biol Plant 51:480–487

Jaleel CA, Manivannan P, Sankar B, Kishorekumar A, Gopi R, Somasundaram R, Panneerselvam R (2007) Induction of drought stress tolerance by ketoconazole in *Catharanthus roseus* is mediated by enhanced antioxidant potentials and secondary metabolite accumulation. Colloid Surf B 60:201–206

Jaleel CA, Gopi R, Manivannan P, Gomathinayagam M, Sridharan R, Panneerselvam R (2008a) Antioxidant potential and indole alkaloid profile variations with water deficits along different parts of two varieties of *Catharanthus roseus*. Colloid Surf B 62:312–318

Jaleel CA, Sankar B, Murali PV, Gomathinayagam M, Lakshmanan GMA, Panneerselvam R (2008b) Water deficit stress effects on reactive oxygen metabolism in *Catharanthus roseus*; impacts on ajmalicine accumulation. Colloid Surf B 62:105–111

Jin J, Shan N, Ma N, Bai J, Gao J (2006) Regulation of ascorbate peroxidase at the transcript level is involved in tolerance to postharvest water deficit stress in the cut rose (*Rosa hybrida* L.) cv. Samantha. Postharvest Biol Tec 40:236–243

Jung S (2004) Variation in antioxidant metabolism of young and mature leaves of *Arabidopsis thaliana* subjected to drought. Plant Sci 166:459–466

Kele Y, Öncel I (2002) Response of antioxidative defence system to temperature and water stress combinations in wheat seedlings. Plant Sci 163:783–790

Khanna-Chopra R, Selote DS (2007) Acclimation to drought stress generates oxidative stress tolerance in drought-resistant than -susceptible wheat cultivar under field conditions. Environ Exp Bot 60:276–283

Laloi C, Apel K, Danon A (2004) Reactive oxygen signalling: the latest news. Curr Opin Plant Biol 7:323–328

Lei Y, Yin C, Li C (2006) Differences in some morphological, physiological, and biochemical responses to drought stress in two contrasting populations of *Populus przewalskii*. Physiol Plant 127:182–191

Li YJ, Hai RL, Du XH, Jiang XN, Lu H (2009) Over-expression of a Populus peroxisomal ascorbate peroxidase (PpAPX) gene in tobacco plants enhances stress tolerance. Plant Breed 128:404–410

Lima ALS, DaMatta FM, Pinheiro HA, Totola MR, Loureiro ME (2002) Photochemical responses and oxidative stress in two clones of *Coffea canephora* under water deficit conditions. Environ Exp Bot 47:239–247

Liu ZJ, Zhang XL, Bai JG, Suo BX, Xu PL, Wang L (2009) Exogenous paraquat changes antioxidant enzyme activities and lipid peroxidation in drought-stressed cucumber leaves. Sci Hort 121:138–143

Logan BA (2005) Reactive oxygen species and photosynthesis. In: Smirnoff N (ed) Antioxidants and reactive oxygen species in plants. Blackwell Publishing, Oxford, UK, pp 250–267

Lu P, Sang WG, Mal KP (2007) Activity of Stress-related antioxidative enzymes in the invasive plant crofton weed (*Eupatorium adenophorum*). J Integr Plant Biol 49:1555–1564

Manivannan P, Jaleel CA, Kishorekumar A, Sankar B, Somasundaram R, Sridharan R, Panneerselvam R (2007) Changes in antioxidant metabolism of *Vigna unguiculata* (L.) Walp. by propiconazole under water deficit stress. Colloid Surf B 57:69–74

Manivannan P, Jaleel CA, Somasundaram R, Panneerselvam R (2008) Osmoregulation and antioxidant metabolism in drought-stressed *Helianthus annuus* under triadimefon drenching. CR Biol 331:418–425

Marulanda A, Porcel R, Barea JM, Azcón R (2007) Drought tolerance and antioxidant activities in lavender plants colonized by native drought-tolerant or drought-sensitive *Glomus* species. Microb Ecol 54:543–552

Mehlhorn H, Lelandais M, Korth HG, Foyer CH (1996) Ascorbate is the natural substrate for plant peroxidases. FEBS Lett 378:203–206

Mittler R, Merquiol E, Hallak-Herr E, Rachmilevitch S, Kaplan A, Cohen M (2001) Living under a 'dormant' canopy: a molecular acclimation mechanism of the desert plant *Retama raetam*. Plant J 25:407–416

Mittler R, Poulos TL (2005) Ascorbate peroxidase. In: Smirnoff N (ed) Antioxidants and reactive oxygen species in plants. Blackwell Publishing, Oxford, UK, pp 87–100

Morales F, Abadía A, Abadía J (2006) Photoinhibition and photoprotection under nutrient deficiencies, drought and salinity. In: Demmig-Adams B, Adams WW III, Matoo AK (eds) Photoprotection, photoinhibition, gene regulation, and environment, advances in photosynthesis and respiration. Springer, The Netherlands, pp 65–85

Morell S, Follmann H, De Tullio M, Häberlein I (1997) Dehydroascorbate and dehydroascorbate rediuctase are phantom indicators of oxidative stress in plants. FEBS Lett 414:567–570

Munné-Bosch S, Mueller M, Schwarz K, Alegre L (2001) Diterpenes and antioxidative protection in drought-stressed *Salvia officinalis* plants. J Plant Physiol 158:1431–1437

Munné-Bosch S, Peñuelas J (2004) Drought-induced oxidative stress in strawberry tree (*Arbutus unedo* L.) growing in Mediterranean field conditions. Plant Sci 166:1105–1110

Nasibi F, Kalantari KM (2009) Influence of nitric oxide in protection of tomato seedling against oxidative stress induced by osmotic stress. Acta Physiol Plant 31:1037–1044

Nayyar H, Gupta D (2006) Differential sensitivity of C_3 and C_4 plants to water deficit stress: association with oxidative stress and antioxidants. Environ Exp Bot 58:106–113

Noctor G, Foyer CH (1998) Ascorbate and glutathione: keeping active oxygen under control. Annu Rev Plant Physiol Plant Mol Biol 49:249–729

Ozkur O, Ozdemir F, Bor M, Turkan I (2009) Physiochemical and antioxidant responses of the perennial xerophyte *Capparis ovata* Desf. to drought. Environ Exp Bot 66:487–492

Pallardy SG (2008) Physiology of woody plants. Academic, San Diego, CA, pp 1–7

Pinheiro HA, DaMatt FM, Chaves ARM, Fontes EPB, Loureiro ME (2004) Drought tolerance in relation to protection against oxidative stress in clones of *Coffea canephora* subjected to long-term drought. Plant Sci 167:1307–1314

Posch S, Bennett LT (2009) Photosynthesis, photochemistry and antioxidative defence in response to two drought severities and with re-watering in *Allocasuarina luehmannii*. Plant Biol 11:83–93

Pukacka S, Ratajczak E (2006) Antioxidative response of ascorbate-glutathione pathway enzymes and metabolites to desiccation of recalcitrant *Acer saccharinum* seeds. J Plant Physiol 163:1259–1266

Qiu ZB, Liu X, Tian XJ, Yue M (2008) Effects of CO_2 laser pretreatment on drought stress resistance in wheat. J Photochem Photobiol B 90:17–25

Quan LJ, Zhang B, Shi WW, Li HY (2008) Hydrogen peroxide in plants: a versatile molucule of the reactive oxygen species network. J Integr Plant Biol 50:2–18

Rapala-Kozik M, Kowalska E, Ostrowska K (2008) Modulation of thiamine metabolism in *Zea mays* seedlings under conditions of abiotic stress. J Exp Bot 59:4133–4143

Ratnayaka HH, Molin WT, Sterling TM (2003) Physiological and antioxidant responses of cotton and spurred anoda under interference and mild drought. J Exp Bot 54:2293–22305

Reddy AR, Chaitanya KV, Jutur PP, Sumithra K (2004a) Differential antioxidative responses to water stress among five mulberry (*Morus alba* L.) cultivars. Environ Exp Bot 52:33–42

Reddy AR, Chaitanya KV, Vivekanandan M (2004b) Drought-induced responses of photosynthesis and antioxidant metabolism in higher plants. J Plant Physiol 161:1189–1202

Ren J, Dai W, Xuan Z, Yao Y, Korpelainen H, Li C (2007) The effect of drought and enhanced UV-B radiation on the growth and physiological traits of two contrasting poplar species. Forest Ecol Manag 239:112–119

Riekert van Heerden PD, Krüger GHJ (2002) Separately and simultaneously induced dark chilling and drought stress effects on photosynthesis, proline accumulation and antioxidant metabolism in soybean. J Plant Physiol 159:1077–1086

Rossel JB, Walter PB, Hendrickson L, Chow WS, Poole A, Mullineaux PM, Pogson BJ (2006) A mutation affecting ascorbate peroxidase 2 gene expression reveals a link between responses to high light and drought tolerance. Plant Cell Environ 29:269–821

Sairam RK, Saxena DC (2000) Oxidative stress and antioxidants in wheat genotypes: possible mechanism of water stress tolerance. J Agron Crop Sci 184:55–61

Sairam RK, Srivastava GC (2001) Water stress tolerance of wheat (*Triticum aestivum* L.): variations in hydrogen peroxide accumulation and antioxidant activity in tolerant and susceptible genotypes. J Agron Crop Sci 186:63–70

Salekdeh GH, Siopongco J, Wade LJ, Ghareyazie B, Bennett J (2002) A proteomics approach to analyzing drought- and salt-responsiveness in rice. Field Crop Res 76:199–219

Sánchez-Díaz M, Tapia C, Antolín MC (2007) Drought-induced oxidative stress in Canarian laurel forest tree species growing under controlled conditions. Tree Physiol 27:1415–1422

Sankar B, Jaleel CA, Manivannan P, Kishorekumar A, Somasundaram R, Panneerselvam R (2007) Effect of paclobutrazol on water stress amelioration through antioxidants and free radical scavenging enzymes in *Arachis hypogaea* L. Colloid Surf B 60:229–235

Scebba F, Sebastiani L, Vitagliano C (2001) Activities of antioxidant enzymes during senescence of *Prunus armeniaca* leaves. Biol Plant 44:41–46

Shikanai T, Takeda T, Yamauchi H, Sano S, Tomizawa KI, Yokota A, Shigeoka S (1998) Inhibition of ascorbate peroxidase under oxidative stress in tobacco having bacterial catalase in chloroplasts. FEBS Lett 428:47–51

Shvaleva AL, Silva FCE, Breia E, Jouve L, Hausman JF, Almeida MH, Maroco JP, Rodrigues ML, Pereira JS, Chaves MM (2005) Metabolic responses to water deficit in two *Eucalyptus globulus* clones with contrasting drought sensitivity. Tree Physiol 26:239–248

Šircelj H, Tausz M, Grill D, BatiÇF (2005) Biochemical responses in leaves of two apple tree cultivars subjected to progressing drought. J Plant Physiol 162:1308–1318

Šircelj H, Tausz M, Grill D, BatiÇF (2007) Detecting different levels of drought stress in apple trees (*Malus domestica* Borkh.) with selected biochemical and physiological parameters. Sci Hortic – Amsterdam 113:362–369

Smirnoff N (1993) The role of active oxygen in the response to water deficit and desiccation. New Phytol 125:27–58

Smirnoff N (1996) The function and metabolism of ascorbic cid in plants. Ann Bot 78:661–669

Smirnoff N (1998) Plant resistance to environmental stress. Curr Opin Biotech 9:214–219

Smirnoff N (2000) Ascorbic acid: metabolism and functions of a multi-facetted molecule. Curr Opin Plant Biol 3:229–235

Smirnoff N (2005) Ascorbate, tocopherol and carotenoids: metabolism, pathway engineering and functions. In: Smirnoff N (ed) Antioxidants and reactive oxygen species in plants. Blackwell Publishing, Oxford, UK, pp 53–66

Sofo A, Dichio B, Xiloyannis C, Masia A (2004) Effects of different irradiance levels on some antioxidant enzymes and on malondialdehyde content during rewatering in olive tree. Plant Sci 166:293–302

Sofo A, Dichio B, Xiloyannis C, Masia A (2005a) Antioxidant defenses in olive trees during drought stress: changes in activity of some antioxidant enzymes. Funct Plant Biol 32:45–53

Sofo A, Tuzio AC, Dichio B, Xiloyannis C (2005b) Influence of water deficit and rewatering on the components of the ascorbate-glutathione cycle in four interspecific *Prunus* hybrids. Plant Sci 169:403–412

Sofo A, Dichio B, Montanaro G, Xiloyannis C (2009) Shade effect on photosynthesis and photoinhibition during drought and rewatering. Agric Water Manage 96:1201–1206

Srivalli B, Sharma G, Khanna-Chopra R (2003) Antioxidative defense system in an upland rice cultivar subjected to increasing intensity of water stress followed by recovery. Physiol Plant 119:503–512

Stevens RG, Creissen GP, Mullineaux PM (1997) Cloning and characterisation of a cytosolic glutathione reductase cDNA from pea (*Pisum sativum* L.) and its expression in response to stress. Plant Mol Biol 35:641–654

Synková H, Valcke R (2001) Response to mild water stress in transgenic P*ssu-ipt* tobacco. Physiol Plant 112:513–23

Tahkokorpi M, Taulavuori K, Laine K, Taulavuori E (2007) After-effects of drought-related winter stress in previous and current year stems of *Vaccinium myrtillus* L. Environ Exp Bot 61:85–93

Tamaoki M, Matsuyama T, Nakajima N, Aono M, Kubo A, Saji H (2004) A method for diagnosis of plant environmental stresses by gene expression profiling using a cDNA macroarray. Environ Pollut 131:137–145

Tausz M, Wonisch A, Peters J, Jiménez MS, Morales D, Grill D (2001) Short-term changes in free radical scavengers and chloroplast pigments in *Pinus canariensis* needles as affected by mild drought stress. J Plant Physiol 158:213–219

Tian XR, Lei YB (2007) Physiological responses of wheat seedlings to drought and UV-B radiation.Effect of exogenous sodium nitroprusside application. Russ J Plant Physl 54:676–682

Torres-Franklin ML, Contour-Ansel D, Zuily-Fodil Y, Pham-Thi AT (2008) Molecular cloning of glutathione reductase cDNAs and analysis of GR gene expression in cowpea and common bean leaves during recovery from moderate drought stress. J Plant Physiol 165:514–521

Vadassery J, Tripathi S, Prasad R, Varma A, Oelmüller R (2009) Monodehydroascorbate reductase and dehydroascorbate reductase are crucial for a mutualistic interaction between *Piriformospora indica* and *Arabidopsis*. J Plant Physiol 166:1263–1274

Van Breusegem F, Vranovà E, Dat JF, Inzé D (2001) The role of active oxygen species in plant signal transduction. Plant Sci 161:405–414

Vranová E, Inzé D, Van Breusegem F (2002) Signal transduction during oxidative stress. J Exp Bot 53:1227–1236

Wang WB, Kim YH, Lee HS, Kim KY, Deng XP, Kwak SS (2009) Analysis of antioxidant enzyme activity during germination of alfalfa under salt and drought stresses. Plant Physiol Bioch 47:570–577

Wu QS, Xia RX, Zou YN (2006) Reactive oxygen metabolism in mycorrhizal and non-mycorrhizal citrus (*Poncirus trifoliata*) seedlings subjected to water stress. J Plant Physiol 16:1101–1110

Yang XD, Dong CJ, Liu JY (2006) A plant mitochondrial phospholipid hydroperoxide glutathione peroxidase: its precise localization and higher enzymatic activity. Plant Mol Biol 62:951–962

Yang Y, Han C, Liu Q, Lin B, Wang J (2008) Effect of drought and low light on growth and enzymatic antioxidant system of *Picea asperata* seedlings. Acta Physiol Plant 30:433–440

Zabalza A, Gálvez L, Marino D, Royuela M, Arrese-Igor C, González EM (2008) The application of ascorbate or its immediate precursor, galactono-1, 4-lactone, does not affect the response of nitrogen-fixing pea nodules to water stress. J Plant Physiol 165:805–812

Zhang J, Kirkkham MB (1996a) Antioxidant responses to drought in sunflower and sorghum seedlings. New Phytol 132:361–373

Zhang J, Kirkkham MB (1996b) Enzymatic responses of the ascorbate-glutathione cycle to drought in sorghum and sunflower plants. Plant Sci 113:139–147

Zhang Y, Tan J, Guo Z, Lu S, He S, Shu W, Zhou B (2009) Increased abscisic acid levels in transgenic tobacco over-expressing 9 *cis*-epoxycarotenoid dioxygenase influence H_2O_2 and NO production and antioxidant defences. Plant Cell Environ 32:509–519

Zhu Z, Liang Z, Han R (2009) Saikosaponin accumulation and antioxidative protection in drought-stressed *Bupleurum chinense* DC. plants. Environ Exp Bot 66:326–333

Chapter 6
Glutathione and Herbicide Resistance in Plants

Zornitsa Ivanova Katerova and Lyuba Petar-Emil Miteva

Abstract Pesticide use is inseparable part of food production. The efficacy of modern agriculture is quite dependent on the chemicals used to fight with pests, including weeds, fungi and insects. Herbicides are chemicals which destroy weeds. According to their mode of action herbicides are divided on 24 groups (Herbicide Resistance Action Committee). The balance between toxicity on weeds and resistance of crops defines herbicide selectivity. Herbicide tolerance depends on the plants variety, development phase, climate, mode of action, dose and the way plants were treated with herbicides. Glutathione is one of the major defense substances of plants. It takes part in many detoxifying mechanisms, like active oxygen species reducing, and also regulates cell defense systems. Glutathione has key role in detoxifying of toxic xenobiotics, including herbicides. In some crops the resistance against herbicides is due to its direct detoxification by forming conjugates with glutathione. The process can be catalyzed by the enzyme glutathione S-transferase. After their forming, conjugates can be metabolized and excreted or can be stored in vacuoles and dead tissues. Many herbicides, such as atrazine, paraquat, etc., induce oxidative events in plant cell. Glutathione takes part in detoxifying active oxygen species and this is a way for indirect enhancement of plant resistance against herbicides. In the current review the mode of action of herbicides inducing oxidative stress will be discussed. Examples of plant antioxidant system response against herbicide action will be presented. The role of glutathione in direct and indirect detoxification of herbicides and increase of plants sustainability will be deeply reviewed.

Keywords Glutathione • Herbicides • Herbicide resistance

Z.I. Katerova (✉)
Acad. M. Popov Institute of Plant Physiology, Bulgarian Academy of Sciences, Acad. G. Bonchev Str., Bl. 21, 1113 Sofia, Bulgaria
e-mail: katerovazor@yahoo.com

L.P.-E. Miteva
Resbiomed EOOD, 81 B, Bulgaria Blvd, Sofia, Bulgaria

N.A. Anjum et al. (eds.), *Ascorbate-Glutathione Pathway and Stress Tolerance in Plants*, DOI 10.1007/978-90-481-9404-9_6, © Springer Science+Business Media B.V. 2010

1 Herbicides as Stress Factors

The natural metabolism of plants might be violated by natural and anthropogenic phenomena (pathogen attack, wounding or other mechanical effect, UV irradiation, drought, water shortage, salinity, xenobiotics, etc.). Animals can escape the harmful agent but because of their sedentary life plants do not have such possibility. Plants are constantly exposed to various chemical compounds. For example, xenobiotics (such as heavy metals, pesticides, herbicides etc.) which can be absorbed and accumulated in plants are neither source of energy or nutrients and could have strong toxic influence. Herbicides are xenobiotics and belong to the group of anthropogenic abiotic factors. They are applied for growing of important cultures like wheat, maize, soybean, etc. in order to protect them from weeds. Herbicides have been used less than a century: the first widely used herbicide was 2,4-dichlorophenoxyacetic acid (2,4-D), which was commercialized and has been in use since the late 1940s (Grabińska-Sota et al. 2003). Growing flux of chemical compounds as herbicides, implicated by humans impact all organisms, including plants. The noxious effect of these factors provoked stress condition in plants. Many herbicides induce oxidative events in plant cells additional to their main mode of action, such as glyphosate, 2,4-D, etc. (Miteva et al. 2003, 2005; Sergiev et al. 2006). Plants have built up mechanisms to detoxify detrimental substances and to cope with oxidative events. These mechanisms in plants are associated mainly with transformation of the toxic component (either the xenobiotic itself or oxygen radical) to harmless molecules and glutathione plays main role in these processes. The level of chemical transformation depends on plants, which determines different tolerance to herbicides or other pollutants. Important consequence of the chemical transformation of toxic molecules is decreasing their biological half-life, therefore the time for plants exposure to the noxious impact of the given molecule.

1.1 Herbicide Classifications

The compounds operating like herbicides (over 500) possess different modes of action and are classified accordingly in 24 groups (Acetyl coenzyme A carboxylase – ACCase inhibitors; acetolactate synthase – ALS inhibitors; enolpyruvylshikimate 3-phosphate synthase – EPSPS inhibitors; synthetic auxins; photosystem II inhibitors, etc.). However, these groups could be divided in four general groups: herbicides that block photosynthesis (atrazine), auxine-like type (2,4-D), inhibitors of amino-acid biosynthesis (glyphosate) and inhibitors of lipid and fatty acid biosynthesis.

The herbicide classification could be done on the base of their selectivity to certain or broad types of plants (selective as chlorsulfuron, 2,4-D and non-selective or total as glyphosate), by herbicide translocation in plant (contact – acts only at the place of the application, and systemic – translocates from the sites of contact to other parts of the plant). Some plants produce natural herbicides, such as the genus *Juglans* (walnuts), or the tree of heaven producing hydroquinone, which is oxidised in the environment to juglone, the compound responsible for the toxic effects on other plants (Vyvyan 2002). Organic herbicides such as spices and D-limonene (citrus oil) are less effective and more expensive than synthetic ones (Dayan et al. 2009).

1.2 Herbicide Resistance

Herbicide resistance is the inherited ability of a plant to survive and reproduce following exposure to a dose of herbicide that would normally be lethal to the wild type. In a plant, resistance may occur naturally due to selection or it may be induced through genetic engineering techniques.

Herbicides destroy weed but they could have toxic effect towards culture plants as well, for example wheat and pea cultures are sensitive to glyphosate, pea plants are sensitive to 2,4-D, etc. Moreover, there are weeds that acquire herbicide-resistance and more than 300 weed biotypes (both dicotyledons and monocotyledons) have evolved resistance to one or more of all the main groups of herbicides (Tharayil-Santhakumar 2003; Yuan et al. 2007, http://www.weedscience.org). Yuan et al. (2007) reported that herbicide-resistant weed biotypes have been found in more than 270,000 fields between 56 different countries all around the world (http://www.weedscience.org). Such herbicide-resistant weeds have become an increasing global menace to agriculture (Preston 2004; Owen and Zelaya 2005). Understanding the basis of resistance, mainly that associated with non-target metabolic and transport mechanisms, is essential for the development of modern agriculture. The mechanisms for deactivation of toxic compounds in plants and knowledge of the defense plants systems are also very important. Good knowledge of cell defense systems suggests effective approach to improve present-day horticulture.

In general, mechanisms of herbicide resistance in weeds could be divided as: (a) target-site resistance or resistance caused by mutations in target sites of the herbicide (photosystem II, acetolactate synthase, 5-enolpyruvyl-shikimate-3-phosphate synthase and acetyl coenzyme-A carboxylase) and (b) non-target site resistance or resistance caused by mutations in non-target sites (Yuan et al. 2007). For example glutathione S-transferases (GSTs, EC. 2.5.1.18) are among the gene families that participate in non-target herbicide resistance. There are two mechanisms that account for the non-target site herbicide resistance: (1) enhanced metabolism and (2) compartmentalization of the herbicide (Nemat Alla and Hassan 2008). The enzymatic conjugation of herbicide with GSH, catalyzed by GSTs, is the main detoxification pathway and belongs to the first mechanism for herbicide resistance.

2 Glutathione

2.1 Glutathione – Short Historical Survey and Evolution

In plants glutathione pool represents most of the non-protein reduced sulphur and it is the basic plant cell reservoir of such sulphur (Crawford et al 2000; Foyer and Noctor 2001). GSH was discovered in 1888. Working with yeasts and plant cells De Rey-Pailhade discovered component that spontaneously react with elementary sulphur and produce hydrogen sulphide (Rennenberg 2001). De Rey-Pailhade found one of the most spread sulphur-containing compounds in the nature and called it philothion (from the Greek words φίλος – friend and θείο – sulphur). In 1924, the same compound was rediscovered from Hopkins who chose the name "glutathione" (Meister 1988). The earliest experiments clarifying the significance of GSH were made by Rapkin (1930), who suppose its role in cellular division (Barron 1951). These suggestions were confirmed by May et al. (1998). Lalova (1967) reported one of the first scientific studies made in order to clarify the role of glutathione in plants subjected to triazine herbicides. The author found that all of the triazine herbicides used in the study decreased the number of the sulfhydril groups in *Fabaceae* plants and this corresponded in a direct ratio to the degree of the plants damage.

Numbers of biochemical processes were described in due to *Arabidopsis* mutants (Todorova 2003; Ohkama-Ohtsu et al. 2008). The essential role of GSH for plants, living in oxygen atmosphere was also proved by the use of *Arabidopsis* mutants. The rml1 mutant contains minimal (below the detection limit of the methods) quantities of GSH, because of the lack of enzyme γ-glutamilcysteinsynthetase. This mutant strain survives only in vitro as tissue culture, supplied with exogenic GSH and has typical phenotype: undeveloped root and small shoots (May et al. 1998). *E. coli* mutants which do not express enzymes, necessary for the GSH synthesis are extremely vulnerable to oxygen and to vast number of chemical reagents. In the lack of oxygen and chemical substances the same mutants develop as the genuine strain (Rennenberg 2001).

As GSH is found in prokaryotes, it could be concluded that it has developed in anaerobic environment but it became extremely necessary for the life at the oxygen environment. In evolutionary aspect GSH was prerequisite not only for the survival in oxygen atmosphere but also as defense factor against multiple stress factors (Rennenberg 2001).

High thiol concentrations are detected into the most primitive organisms. Some of them, in particular anaerobic bacteria, contain very low GSH concentrations and instead the basic low-molecular thiol is the GSH precursor (γ-glutamilcystein or co-enzyme A) (Rennenberg 2001). In some plants homologue forms of GSH are discovered, where the carboxyl residue is replaced with other amino acids. For example, in some legume species including *Glycine max* L. homoglutathione (hGSH or γ-glutamyl-L-cysteinyl-β-alanine) is the dominant thiol (Klapheck 1988), in cereals of the Poaceae family as *Triticum aestivum* – hydrohymethylglutathione (hmGSH or γ-glutamyl-L-cysteinyl-serine) co-accumulates with GSH

(Klapheck et al. 1992) and in *Zea mais* γ-ECE (γ-glutamyl-L-cysteinyl-glutamic acid) might be accumulated (Meuwly et al. 1993). Plants that synthesize the analogues contain GSH as well (Crawford et al. 2000). A number of experiments prove that homologues possess similar physiological functions as those of GSH (Wonisch and Schaur 2001). The homologues are evolutionary newer metabolites but up to now their advantage over GSH is unknown (Rennenberg 2001).

2.2 Precursors, Biosynthesis and Degradation of Glutathione

Glutathione is not DNA-coded molecule – it is synthesized by two enzyme-catalyzed steps in ATP-dependent process (Fig. 1), however, both enzymes are encoded in the nuclear genome (Meyer 2008). Studies on subcellular compartmentalization of GSH showed its biosynthesis is possible in the cytosol and in the chloroplasts (Leustek et al. 2000, reviewed by Meyer 2008, Szalai et al. 2009). At the first

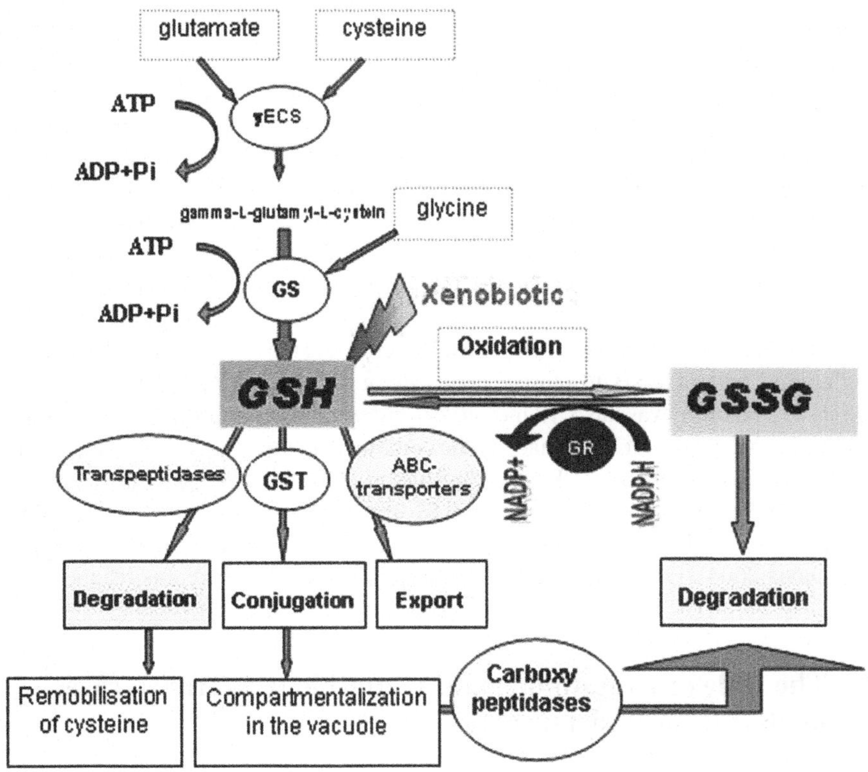

Fig. 1 General scheme of glutathione metabolism (According to Foyer and Noctor 2001, with modifications). γECS – γ-glutamylcysteine synthetase, GS – glutathione synthetase, GST – glutathione S-transferase, GR – glutathione reductase

stage the enzyme glutamate cysteine ligase or γ-glutamylcysteine synthetase (γECS, EC 6.3.2.2), encoded by *GSH1* in *Arabidopsis* (May and Leaver 1994), catalyze formation of peptide bond between L-glutamate and L-cysteine and produce γ-L-glutamyl-L-cysteine (γEC). At the next stage glycine is added to the dipeptide (L-glutamyl-L-cysteine) with the participation of glutathione synthetase enzyme (GS, EC 6.3.2.3), which is encoded by *GSH2* in *Arabidopsis thaliana* (Wang and Oliver 1996). The first-stage enzyme, γECS, is the regulatory enzyme of GSH synthesis (Rüegsegger and Brunold 1992; Rennenberg et al. 2007). It is localized entirely in the cytosol (Pasternak et al. 2008), however, for *Arabidopsis* γECS is localized exclusively in the plastids (Wachter et al. 2005). The activity of γECS is regulated by GSH in feedback mechanism (Jez et al. 2004).

While glutathione synthesis is executed in the chloroplast and cytosol, the degradation of GSH and its conjugates is limited to the vacuole and probably to the apoplast (Foyer et al. 2001; Ohkama-Ohtsu et al. 2007). Two possibilities exist to start GSH degradation. One of them is from the N-terminus through breaking the γ-glutamyl bond and generating glutamate and cysteine-glycine. γ-Glutamyl transpeptidases (GGT, EC 2.3.2.2) are responsible for this reaction. GGTs are essential for GSH hydrolysis (Martin et al. 2007). The second possibility is to start a reaction from the C-terminus, catalyzed by carboxypeptidase or phytochelatin synthase to produce γ-EC and glycine (Beck et al. 2003).

Four GGT genes were identified in *Arabidopsis* plants and their temporal and spatial pattern of GSH and its metabolite degradation were determined (Martin et al. 2007). Ohkama-Ohtsu et al. (2007) studied the *Arabidopsis thaliana* knockout mutant *ggt3*. The authors found that GGT3 catalyses the first step in catabolism of GSH conjugates in the vacuole and GGT3 activity complements GGT1 and GGT2 activities, implicated in GSH catabolism in the apoplast. It was reported that in either roots or leaves, the apoplastic GST enzymes have limited capacity in GSH conjugate catabolism.

Glutathione degradation is performed by series of steps, where GGT transfer the γ-glutamyl part of GSH to aminoacid acceptor. A GGT-independent pathway of GSH degradation was also found in *Arabidopsis* where GSH is converted to 5-oxoproline and then to glutamate by the combined action of γ-glutamyl cyclotransferase and 5-oxoprolinase (Ohkama-Ohtsu et al. 2008).

In plants, GSH conjugation with lipid hydroperoxides, toxic metabolite products or xenobiotics influence GSH concentration as well (Dixon et al. 2002; Anderson and Davis 2004). GST, induced by abiotic stresses may catalyze this reaction.

3 The Role of Glutathione and Glutathione S-Transferases in Herbicide Resistance

Glutathione functions in plant metabolism can be generalized as two main functions: it participates in the exchange of sulphur and it functions as endogenous defense constituent against biotic and abiotic stress (Noctor and Foyer 1998). As defence factor glutathione has multiple roles: against oxidative stress caused by different

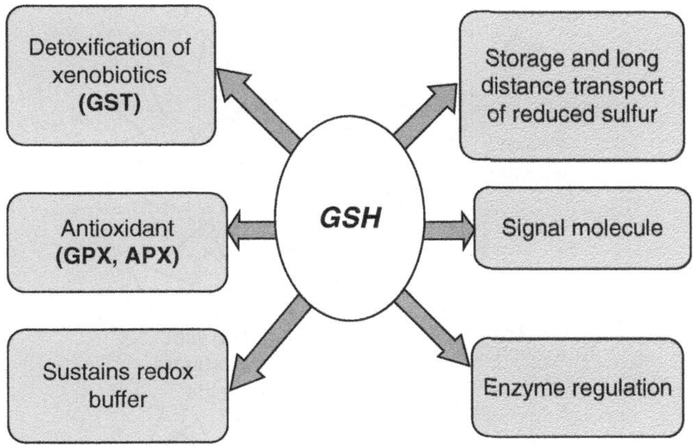

Fig. 2 Functions of glutathione in plants. GPX – guaiacol peroxidase, APX – ascorbate peroxidase, GSH – glutathione

stress factors, including herbicides, as regulator of redox state of the cells and in direct detoxification of xenobiotics, including herbicides. In general the functions of GSH in plants are summarized in Fig. 2 (Miteva 2005).

Glutathione is main ingredient of the system that reduces toxic compounds. The GSH role in defense system is the direct detoxification of xenobiotics also antioxidant against oxidative stress, provoked from herbicides like atrazin, paraquat etc. The detoxification of xenobiotics is made through conjugation process of GSH with the hazardous chemical molecule. The process can involve the enzyme GST or can be spontaneous and it has three stages: activation (preparation), conjugation and metabolism of conjugates (Fig. 3). This process will be reviewed more deeply in next section.

3.1 Oxidative Stress and the Role of Glutathione in Herbicide Defense System

GSH role as antioxidant is associated with direct detoxification of the active oxygen forms (AOS: hydrogen peroxide, singlet oxygen, superoxide and hydroxil radical) as well as with participation in glutathione–ascorbate shuttle (Haliwell–Asada cycle). At the process of AOS deactivation the GSH is oxidized to its dimer form or GSSG. An important component of antioxidant system is the enzyme glutathione reductase (GR, EC 1.6.4.2). It might restore the reduced form of glutathione. The process is important when homeostasis is supported as the accumulation of oxidized glutathione inhibits protein synthesis on the level of translation. For example after paraquat treatment, leading to strong oxidative stress a GR inhibition was observed (Ananieva et al. 2004). Sergiev et al. (2006) observed that glyphosate application in maize plants also caused oxidative stress, exhibited by amplified biomembrane lipid peroxidation (malondialdehyde), leaf tissue ion fluxes and the activities of catalase, peroxidase and

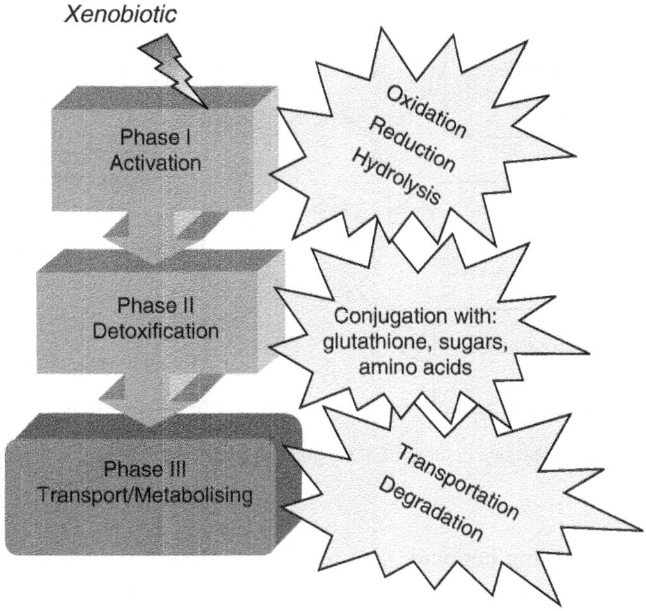

Fig. 3 Main phases of the process of detoxification of xenobiotics in plants (According to Schröder 2001, with modifications)

GST. The results of Miteva et al. (2005) also showed development of oxidative events in pea and wheat plants treated with glyphosate. Application of atrazine at low concentrations imitating residual concentrations increased the susceptibility of *Arabidopsis thaliana* to further treatments with the same herbicide applied at the recommended field amount, mainly through oxidative damages (Ivanov et al. 2005).

3.2 The Role of Reduced Glutathione and GSH/GSSG Ratio in Herbicide Resistance

GSH participates not only in the direct detoxification of AOS but it could protect cells against unfavorable effects of stress trough the activation of different defense mechanisms because of its involvement in redox signalling (Foyer et al. 1997; Apel and Hirt 2004; Foyer and Noctor 2005; Szalai et al. 2009). Overgeneration of AOS in the cells is known as "oxidative stress". The main mechanism of action of herbicides like atrazine and paraquat is generation of AOS and that led to the death of treated plant. GSH has important role in antioxidant system as part of the glutathione–ascorbate shuttle (Halliwell–Asada cycle). Its function there is to provide electrons for reduction of ascorbate, which also takes part in AOS detoxification. Reduced glutathione form could be restored by the enzyme GR. Culture plants with enhanced levels of GSH were found more resistant to similar herbicides.

The cysteine thiol-group is a functional group that is very sensitive to oxidation, making cysteine remarkably useful for structural functions (Meyer 2008). The redox characteristics of cysteine permit formation of intra- and inter-molecular disulfide bridges that allow appropriate protein folding and formation of stable multiprotein complexes. The oxidation of thiol-group frequently changes structural and catalytic protein features and finally activates downstream processes. Protein thiols are also vulnerable to oxidative damage due to their high sensitivity to oxidation. To evade permanent loss of proteins caused by oxidation, complicated mechanisms allowing recovery of such proteins via reduction of thiol residues have been developed in cells. The millimolar concentrations of GSH in metabolically active tissues, act as a key redox buffer, forming a blockage between protein cysteine groups and AOS (Foyer and Noctor 2005). Glutathione may endure reversible oxidation and reduction thanks to the central cysteine, and could exist in either the reduced form (GSH) or oxidized glutathione disulfide (GSSG).

Two GSH molecules form one molecule of GSSG through oxidation, and the concentration of GSH consequently enters the Nernst equation as a squared term. Therefore, the glutathione redox potential depends on the total glutathione concentration and the degree of oxidation (Meyer and Hell 2005). Changes in GSH redox state are responsible for the protective and regulatory roles of GSH. The redox state of GSH is defined by its reducing capacity (GSH concentration) and the half-cell reduction potential of the GSH/GSSG couple (or $E_{GSSG/2GSH}$) and differs in different organs, tissues, cells, and compartments, changing during plant growth and development. The value of $E_{GSSG/2GSH}$ can be calculated from the concentrations of GSH and GSSG using the Nernst equation (Schafer and Buettner 2001).

The synthesis, degradation, oxidation and transport of GSH as well as its conjugation with other sulfhydryl groups or other substances should be regarded when characterizing GSH/GSSG couple. A high GSH/GSSG ratio, maintained by increased GSH synthesis or/and GSSG reduction might be necessary for efficient protection against AOS accumulation, induced by abiotic stress and was shown by studies on several plants, subjected to different abiotic stresses (Szalai et al. 2009). The half-cell reduction potential of GSH/GSSG couple or $E_{GSSG/2GSH}$ value, calculated from the concentration of GSH and GSSG is more closely related with the biological cell status, than the GSH/GSSG ratio (Schafer and Buettner 2001). The value of $E_{GSSG/2GSH}$ can be calculated using the Nernst equation. The same parameter was found to be a universal marker of cell viability and could be used to monitor stress-induced damage (Kranner et al. 2006).

3.3 The Role of GST and the Process of Herbicide Conjugation for Plants Resistance to Herbicide Stress

A number of authors emphasize the significance of conjugation with GSH for herbicide selectivity as the cereal cultures. Using two herbicides, pretilachlor and metribuzin, the less and the most phytotoxic to maize, it was shown that the herbicide tolerance in *Zea mays* is related to the increased levels of GSH and GSH-associated enzymes (GSTs, γ-glutamyl-cysteine synthetase, GSH-synthetase and GR) (Nemat Alla et al. 2008). Using the transgenic rice plant over-expressing *OsGSTL1*, a GST gene from rice, Hu et al. (2009) demonstrated the detoxification role of OsGSTL1 for the growth and development. Something more, the accumulation of the OsGSTL1 protein in the vegetative tissues of transgenic rice plants enhanced their tolerance to chlorsulfuron and glyphosate. In addition, *OsGSTL1* over-expression may enhance the resistance to chlorsulfuron and glyphosate in rice plants.The results of Miteva et al. (2003, 2004) also showed that atrazine and 2,4-D increase GST enzyme activity in wheat, however, opposite tendency was observed in more sensitive pea plants. It was reported that in the more sensitive to 2,4-D plant, pea, the activity of protective enzyme systems GST and GR decreased but the content of the oxidized glutathione form was elevated (Miteva et al. 2003). In opposite, in wheat, which is more tolerant to 2,4-D treatment, normally functioning GR and GST enzymes were observed with increasing or constant enzyme activity levels. The total glutathione content was constant during the stress programme and GSSG/TG ratio was under the control values. The lack of accumulation of oxidized form of glutathione led to the assumption that the wheat plant subjected to 2,4-D treatment manage to keep its red-ox status and therefore its homeostasis.

3.4 *Phases of the Process of Xenobitic Detoxification*

The cellular detoxification system for xenobiotics, including herbicides, consists of three consecutive sequential processes (Coleman et al. 1997). Phase I and II – chemical modification (transformation) of the xenobiotic; Phase III – compartmentalization and metabolism of conjugates (Fig. 3).

Activation (Phase I). The chemical transformation at phase I includes one or more enzyme-catalyzed reactions, which aim is to prepare the given molecule. Reactive functional groups develop at the molecule of xenobiotic and facilitate the following process of conjugation in result of oxidation, reduction or hydrolysis reactions, catalyzed by esterases, amidases, peroxidases, reductases, dehydrogenases, by the cytohrom P-450 system or by mixed function oxidases (Coleman et al. 1997; Doran 2005; Yuan et al. 2007). Phase I is not obligatory and can be passed if the necessary functional groups are available in the initial substance. This phase is very important for lipophilic xenobiotics. As a result molecule different than the starting one is yielded. This molecule might be with higher toxicity because the designation of phase I is to prepare the compound for phase II reactions through development of reactive sites in the pollutant by addition or exposure of the functional groups (Coleman et al. 1997).

Conjugation (Phase II). This phase includes reactions, where the xenobiotic or the activated Phase I-product is deactivated through covalent bonding with endogenous hydrophilic molecule, as glucose, malonate, aminoacids or glutathione. However, GSH is the main substrate at the conjugation process. The resulted water-soluble conjugates are hydrophilic and less-toxic molecule than the initial compound. Hydrophility is an important molecular property as it facilitates metabolite transport through cellular membranes as well as their compartmentalization. Conjugation reactions could be spontaneous or catalyzed by enzymes as glucosyl-, malonyl- and glutathione S-transferases. Enzymes that detoxify xenobiotics exist in prokariotic and eukariotic organisms more than 2.5 milliard years and GSTs has the main numerical superiority (Sheehan et al. 2001).

Phase III. In Phase III, conjugated metabolites are removed from the cytoplasm and deposited in vacuoles or cell walls (Coleman et al. 1997; Hatzioz 1997). It has been classified in two independent phases (Abhilash 2009). The first one is restricted to transport and storage of the conjugates in the vacuole and/or the extracellular space by active transport, where ABC transporters are the most commonly involved (Rea 1999; Yuan et al. 2007; Schröder 2007; Abhilash 2009). The second one involves further degradation of the conjugation molecule in the vacuole or extracellular spaces taking final reactions like cell wall bindings or excretion.

3.5 The Significance of GST for Herbicide Resistance

In the 1970s, a relationship was clarified between a GSH conjugate and atrazine resistance in some grasses and GSTs were the first implicated in herbicide resistance in weeds (Jensen et al. 1977). Supplementary evidence that GSTs are associated in non-target-site herbicide resistance came from the assays of GST activity in herbicide-resistant weeds. In the resistant biotype of velvetleaf (*Abutilon theophrasti*) increased glutathione conjugation with atrazine was observed through amplified GST activity, which correlates with herbicide resistance in the weed (Anderson and Gronwald 1991). Moreover, Marcacci et al. (2006) showed that conjugation to GSH is a major metabolic pathway that detoxifies atrazine in *Chrysopogon zizanioides* Nash, which is species resistant to this herbicide. Atrazine tolerance can be due to the high intensity of one metabolic pathway or by involvement of several metabolic pathways, for example in *Zea mays* (Cherifi et al. 2001). On the contrary, lack of atrazine metabolism can be observed in the sensitive species (wheat and soybean), while dealkylation only confers moderate tolerance of atrazine in pea.

GST-mediated herbicide resistance sometimes can target multiple herbicides (Hall et al. 1997; Hatton et al. 1999; Cummins et al. 1999; Cocker et al. 2001; Letouze and Gasquez 2003). An increase in GST activity could be accompanied by enhanced GST gene expression (Cummins et al. 1997b, 1999). Safener application data gives complementary evidence for GST-mediated non-target herbicide resistance through induced GST gene expression (Hatzios and Burgos 2004; Smith et al. 2004; Zhang and Riechers 2004; Nutricati et al. 2006; DeRidder and Goldsbrough 2006). The supposed sequences of steps necessary for herbicide safeners to induce plant tolerance towards xenobiotics also include increased transcription of GSH synthesis and GST genes (Verkleij et al. 2009). In crops, functional characterizations of GST genes also showed that GSTs have a substantial role in herbicide metabolism (Cho and Kong 2005; Yang et al. 2009). Many GST enzymes were purified and characterized for their activity on diverse herbicides in different crop and non-crop (eg. perennial grasses) cultures as soybean, wheat, maize, rice, tall fescue etc. (Cummins et al. 1997a; Dixon et al. 1997; Andrews et al. 2005; Del Buono et al. 2007; Yang et al. 2009). A number of GSTs seem to be involved in herbicide metabolism in plants and their role is very important in protecting plants against herbicide stress (Kawahigashi 2009). In order to study the overexpression of GST genes a transgenic approach has been useful and herbicide-resistant transgenic plants were obtained (Milligan et al. 2001; Karavangeli et al. 2005; Skipsey et al. 2005). It is known that maize GSTs detoxify triazine and chloroacetanilide herbicides and transgenic tobacco plants expressing maize GST I have been shown to remediate alachlor (Karavangeli et al. 2005). The potential for phytoremediation of herbicides seems to be enhanced through overexpression of GST genes. Not only GST genes but these connected with gluthatione are also involved. Noctor and Foyer (1998) showed that a rate-limiting step in the biosynthesis of glutathione is catalyzed by γ-glutathione synthetase. In poplar plants introduction of γ-glutathione synthetase gene leads to higher concentrations of glutathione, and the plants became tolerant toward two chloroacetanilide herbicides, acetochlor, and metolachlor (Gullner et al. 2001). *Brassica juncea* expressing the gene of γ-glutathione synthetase also increased its tolerance to atrazine, 1-chloro-2,4-dinitrobenzene (CDNB), metolachlor, and phenanthrene (Flocco et al. 2004). It should be noted that the combined importance of specific GSTs and their preferred thiol co-substrates in conferring herbicide selectivity traits in plants was shown to be of great necessity (Skipsey et al. 2005). The same authors found that increased resistance resulted from coordinated overexpression of both GST and homoglutathione synthase genes because the particular GST activity requires available homoglutathione thiol.

However, several environmental problems are associated with application of genetically engineered plants for phytoremediation purposes (Nemat Alla and Hassan 2008; Abhilash et al. 2009). For example, the escape of the foreign gene/s from transgenic plants, via pollen dispersal to their weed relatives might create super-weeds. To resolve the invasiveness and interbreeding problem the uses of sterile clones have been suggested. Another problem is the small biomass and growth rates of some of the engineered plants, which makes them ineffective for application in polluted field. Up to now, greater part of the transgenic plants never has been used in real contaminated conditions (Abhilash et al. 2009).

References

Abhilash PC, Jamil S, Singh N (2009) Transgenic plants for enhanced biodegradation and phytoremediation of organic xenobiotics. Biotech Adv 27:474–488

Ananieva E, Christov K, Popova L (2004) Exogenous treatment with salicylic acid leads to increased antioxidant capacity in leaves of barley plants exposed to paraquat. J Plant Physiol 161:319–328

Anderson JV, Davis DG (2004) Abiotic stress alters transcript profiles and activity of glutathione S-transferase, glutathione peroxidase, and glutathione reductase in *Euphorbia esula*. Physiol Plant 120:421–433

Anderson MP, Gronwald JW (1991) Atrazine resistance in a velvetleaf (*Abutilon theophrasti*) biotype due to enhanced glutathione S-transferase activity. Plant Physiol 96:104–109

Andrews CJ, Cummins I, Skipsey M, Grundy NM, Jepson I, Townson J, Edward R (2005) Purification and characterisation of a family of glutathione transferases with roles in herbicide detoxification in soybean (*Glycine max* L.); selective enhancement by herbicides and herbicide safeners. Pest Biochem Physiol 82:205–219

Apel K, Hirt H (2004) Reactive oxigen species: metabolism, oxidative stress, and signal transduction. Annu Rev Plant Biol 55:373–99

Barron E (1951) Thiol groups of biological importance. Adv Enzymol 11:201–266

Beck A, Lendzian K, Oven M, Christmann A, Grill E (2003) Phytochelatin synthase catalyzes key step in turnover of glutathione conjugates. Phytochemistry 62:423–431

Brazier-Hicks M, Evans KM, Cunningham OD, Hodgson DRW, Steel PG, Edwards R (2008) Catabolism of Glutathione Conjugates in *Arabidopsis thaliana*: role in metabolic reactivation of the herbicide safener fenclorim. J Biol Chem 283:21102–21112

Cherifi M, Raveton M, Picciocchi A, Ravanel P, Tissut M (2001) Atrazine metabolism in corn seedlings. Plant Physiol Biochem 39:665–672

Cho HY, Kong KH (2005) Molecular cloning, expression, and characterization of a phi-type glutathione S-transferase from *Oryza sativa*. Pest Biochem Physiol 83:29–36

Cocker KM, Northcroft DS, Coleman JOD, Moss SR (2001) Resistance to ACCase-inhibiting herbicides and isoproturon in UK populations of *Lolium multiflorum*: mechanisms of resistance and implications for control. Pest Manag Sci 57:587–597

Coleman J, Blake-Kalff M, Davies E (1997) Detoxification of xenobiotics by plants: chemical modulation and vacuolar compartmentation. Trends Plant Sci 2:144–151

Crawford N, Kahn M, Leustek T, Long S (2000) Chapter 16 Nitrogen and sulfur. In: Buchanan B, Gruissem W, Jones R (eds) Biochemistry and molecular biology of plants. American Society of Plant Physiologists, Rockville, MD, pp 824–844

Cummins I, Cole DJ, Edwards R (1997a) Purification of multiple glutathione transferases involved in herbicide detoxification from wheat (*Triticum aestivum* L.) treated with the safener fenchlorazole-ethyl. Pest Biochem Physiol 59:35–49

Cummins I, Cole DJ, Edwards R (1999) A role for glutathione transferases functioning as glutathione peroxidases in resistance to multiple herbicides in black-grass. Plant J 18:285–292

Cummins I, Moss S, Cole DJ, Edwards R (1997b) Glutathione transferases in herbicide-resistant and herbicide-susceptible black-grass (*Alopecurus myosuroides*). Pest Sci 51:244–250

Dayan FE, Cantrell CL, Duke SO (2009) Natural products in crop protection. Bioorg Med Chem 17:4022–4034

De Ridder BP, Goldsbrough PB (2006) Organ-specific expression of glutathione *S*-transferases and the efficacy of herbicide safeners in Arabidopsis. Plant Physiol 140:167–175

Del Buono D, Scarponi L, Espen L (2007) Glutathione S-transferases in Festuca arundinacea: Identification, characterization and inducibility by safener benoxacor. Phytochemistry 68:2614–2624

Dixon D, Cole DJ, Edwards R (1997) Characterisation of multiple glutathione transferases containing the GST I subunit with activities toward herbicide substrates in maize (*Zea mays*). Pest Sci 50:72–82

Dixon DP, Lapthorn A, Edwards R (2002) Plant glutathione transferases. Genome Biol 3:30004.1–30004.10

Doran PM (2005) Application of plant tissue cultures in phytoremediation research: incentives and limitations. Biotechnol Bioeng 103:60–76

Flocco CG, Lindblom SD, Smits EA (2004) Overexpression of enzymes involved in glutathione synthesis enhances tolerance to organic pollutants in *Brassica juncea*. Int J Phytoremediation 6:289–304

Foyer C, Noctor G (2001) The molecular biology and metabolism of glutathione. In: Grill D, Tausz M, De Kok L (eds) Significance of glutathione in plant adaptation to the environment, vol 2, Handbook series of plant ecophysiology. Kluwer, Dordrecht, The Netherlands, pp. 27–56

Foyer CH, Lopez-Delgado H, Dat JF, Scott IM (1997) Hydrogen peroxide and glutathione-associated mechanisms of acclamatory stress tolerance and signaling. Physiol Plant 100:241–254

Foyer CH, Noctor G (2005) Redox homeostasis and antioxidant signalling: a metabolic link between stress perception and physiological responses. Plant Cell 17:1866–1875

Foyer CH, Theodoulou FL, Delrot S (2001) The functions of inter- and intracellular glutathione transport systems in plants. Trends Plant Sci 6:486–492

Grabińska-Sota E, Wiśniowska E, Kalka J (2003) Toxicity of selected synthetic auxines – 2, 4-D and MCPA derivatives to broad-leaved and cereal plants. Crop Prot 22:55–360

Gullner G, Komives T, Rennenberg H (2001) Enhanced tolerance of transgenic poplar plants overexpressing gammaglutamylcysteine synthetase towards chloroacetanilide herbicides. J Exp Bot 52:971–979

Hall LM, Moss SR, Powles SB (1997) Mechanisms of resistance to aryloxyphenoxypropionate herbicides in two resistant biotypes of *Alopecurus myosuroides* (blackgrass): herbicide metabolism as a cross-resistance mechanism. Pest Biochem Physiol 57:87–98

Hatton PJ, Cummins I, Cole DJ, Edwards R (1999) Glutathione transferases involved in herbicide detoxification in the leaves of *Setaria faberi* (giant foxtail). Physiol Plant 105:9–16

Hatzios KK, Burgos N (2004) Metabolism-based herbicide resistance: regulation by safeners. Weed Sci 52:454–467

Hatzioz KK (1997) Regulation of enzymatic systems detoxifying xenobiotics in plants: a brief overview and directions for future research. Kluwer, Dordrecht, The Netherlands

Hu T, Qv X, Xiao G, Huang X (2009) Enhanced tolerance to herbicide of rice plants by overexpression of a glutathione *S*-transferase. Mol Breed 24:409–418

Ivanov SV, Alexieva VS, Karanov EN (2005) Cumulative effect of low and high atrazine concentrations on *Arabidopsis thaliana* plants. Russ J Plant Physiol 52(2):213–219

Jaleel CA, Riadh K, Gopi R, Manivannan P, Inès J, Al-Juburi HJ, Hong-Bo ZS, Panneerselvam R (2009) Antioxidant defense responses: physiological plasticity in higher plants under abiotic constraints. Acta Physiol Plant 31:427–436

Jensen KIN, Stephenson GR, Hunt LA (1977) Detoxification of atrazine in three Gramineae subfamilies. Weed Sci 25:212–220

Jez JM, Cahoon RE, Chen S (2004) Arabidopsis thaliana glutamate-cysteine ligase: functional properties, kinetic mechanism, and regulation of activity. J Biol Chem 279:33463–33470

Jimenez A, Hernandez JA, del Rio LA, Sevilla F (1997) Evidence for the presence of the ascorbate–glutathione cycle in mitochondria and peroxisomes of pea leaves. Plant Physiol 114:275–284

Karavangeli M, Labrou NE, Clonis YD, Tsaftaris A (2005) Development of transgenic tobacco plants overexpressing maize glutathione S-transferase I for chloroacetanilide herbicides phytoremediation. Biomol Eng 22:121–128

Kawahigashi H (2009) Transgenic plants for phytoremediation of herbicides. Curr Opin Biotechnol 20:225–230

Klapheck S (1988) Homoglutathione: isolation, quantification and occurrence in legumes. Physiol Plant 74:727–732

Klapheck S, Chrost B, Starke J, Zimmermann H (1992) Gammaglutamylcysteinylserine – a new homologue of glutathione in plants of the family Poaceae. Bot Acta 105:174–179

Kömives T, Gulner G (2000) Phytoremediation. In: Wilkinson R (ed) Plant–environment interactions. Marcel Dekker Inc., New York, pp 437–452

Kranner I, Birtic S, Anderson KM, Pritchard HW (2006) Glutathione half-cell reduction potential: a universal stress marker and modulator of programmed cell death? Free Radic Biol Med 40:2155–2165

Lalova M (1967) The effect of triazine herbicides on the SH groups in plants. Compt rend Acad Bulg Sci 20:229–231

Letouze A, Gasquez J (2003) Enhanced activity of several herbicide-degrading enzymes: a suggested mechanism responsible for multiple resistance in blackgrass (*Alopecurus myosuroides* Huds.). Agronomie 23:601–608

Leustek T, Martin M, Bick J, Davies J (2000) Pathways and regulation of sulfur metabolism revealed through molecular and genetic studies. Ann Rev Plant Physiol Plant Mol Biol 51:141–165

Liu Z-J, Zhang X-L, Bai J-G, Suo B-X, Xu P-L, Wang L (2009) Exogenous paraquat changes antioxidant enzyme activities and lipid peroxidation in drought-stressed cucumber leaves. Sci Hort 121:138–143

Loyall L, Uchida K, Braun S, Furuya M, Frohnmeyer H (2000) Glutathione and a UV light-induced glutathione S-transferase are involved in signaling to chalcone synthase in cell cultures. Plant Cell 12:1939–1950

Marcacci S, Raveton M, Ravanel P, Schwitzgúebel J-P (2006) Conjugation of atrazine in vetiver (*Chrysopogon zizanioides* Nash) grown in hydroponics. Environ Exp Bot 56:205–215

Martin NM, Saladores PH, Lambert E, Hudson AO, Leustek T (2007) Localization of members of the γ-glutamyl transpeptidase family identifies sites of glutathione and glutathione *S*-conjugate hydrolysis. Plant Physiol 144:1715–1732

May M, Vernoux T, Leaver C, Van Montagu M, Inze D (1998) Glutathione homeostasis in plants: implications for environmental sensing and plant development. J Exp Bot 49:649–667

May MJ, Leaver CJ (1994) *Arabidopsis thaliana* cγ-glutamylcysteine synthetase is structurally unrelated to mammalian, yeast, and *Escherichia coli* homologs. Proc Natl Acad Sci USA 91:10059–10063

Meister A (1988) On the discovery of glutathione. Trends Biochem Sci 13(5):185–188

Meuwly P, Thibault P, Rauser WE (1993) Gamma-glutamylcysteinylglutamic acid – a new homolog of glutathione in maize seedlings exposed to cadmium. FEBS Lett 336:472–476

Meyer A, Hell R (2005) Glutathione homeostasis and redox-regulation by sulfhydryl groups. Photosynth Res 86:435–57

Meyer AJ (2008) The integration of glutathione homeostasis and redox signaling. J Plant Physiol 165:1390–1403

Milligan AS, Daly A, Parry MAJ, Lazzeri PA, Jepson I (2001) The expression of a maize glutathione S-transferase gene in transgenic wheat confers herbicide tolerance, both in planta and in vitro. Mol Breed 7:301–315

Miteva L (2005) Glutathione and glutathione S-transferase – physiological function in plants, grown on normal and stress conditions. Plant Sci 42:483–494 (In Bulgarian)

Miteva L, Ivanov S, Alexieva V (2003) Comparative effect of 2, 4-D on the glutathione levels, glutathione-S-transferase and glutathione reductase activities in pea (*Pisum sativum* L.) and wheat (*Triticum aestivum* L.). Compt rend Acad Bulg Sci 56:53–58

Miteva L, Ivanov S, Alexieva V, Karanov E (2004) Effect of Atrazine on Glutathione Levels, Glutathione-S-Transferase and Glutathione Reductase Activities in Pea and Wheat Plants. Plant Prot Sci 40:16–20

Miteva L, Tsoneva J, Ivanov S, Alexieva V (2005) Alterations of the content of hydrogen peroxide and malondialdehyde and the activity of some antioxidant enzymes in the roots and leaves of pea and wheat plants exposed to glyphosate. Compt rend Acad Bulg Sci 58:733–738

Miteva LP-E, Ivanov SV, Alexieva VS (2010) Alterations in glutathione pool and some related enzymes in leaves and roots of pea plants treated with the herbicide glyphosate. Russ J Plant Physiol 57:131–136

Moskova I, Todorova D, Alexieva V, Ivanov S, Sergiev I (2009) Effect of exogenous hydrogen peroxide on enzymatic and nonenzymatic antioxidants in leaves of young pea plants treated with paraquat. Plant Growth Regul 57:193–202

Nemat Alla M, Badawi A-H M, Hassan NM, El-Bastawisy ZM, Badran EG (2008) Herbicide tolerance in maize is related to increased levels of glutathione and glutathione-associated enzymes. Acta Physiol Plant 30:371–379

Nemat Alla M, Hassan NM (2008) Recognition, implication and management of plant resistance to herbicides. Am J Plant Physiol 3:50–66

Nemat Alla MM, Hassan NM (2006) Changes of antioxidants levels in two maize lines following atrazine treatments. Plant Physiol Biochem 44:202–210

Noctor G, Foyer CH (1998) Simultaneous measurement of foliar glutathione, gamma-glutamylcysteine, and amino acids by high-performance liquid chromatography: comparison with two other assay methods for glutathione. Anal Biochem 264:98–110

Nutricati E, Miceli A, Blando F, De Bellis L (2006) Characterization of two Arabidopsis thaliana glutathione-S-transferases. Plant Cell Rep 25:997–1005

Ohkama-Ohtsu N, Oikawa A, Zhao P, Xiang C, Saito K, Oliver DJ (2008) A γ-glutamyl transpeptidase-independent pathway of glutathione catabolism to glutamate via 5-oxoproline in Arabidopsis. Plant Physiol 148:1603–1613

Ohkama-Ohtsu N, Zhao P, Xiang C, Oliver DJ (2007) Glutathione conjugates in the vacuole are degraded by γ-glutamyl transpeptidase GGT3 in Arabidopsis. Plant J 49:878–888

Owen MD, Zelaya IA (2005) Herbicide-resistant crops and weed resistance to herbicides. Pest Manag Sci 61:301–311

Pasternak M, Lim B, Wirtz M, Hell R, Cobbett CS, Meyer AJ (2008) Restricting glutathione biosynthesis to the cytosol is sufficient for normal plant development. Plant J 53:999–1012

Preston C (2004) Herbicide resistance in weeds endowed by enhanced detoxification: complications for management. Weed Sci 52:448–453

Rea PA (1999) MRP subfamily ABC transporters from plants and yeast. J Exp Bot 50:895–913

Rennenberg H (2001) Glutathione – an ancient metabolite with modern tasks. In: Grill D, Tausz M, De Кок L (eds) Significance of glutathione in plant adaptation to the environment, vol 2, Handbook series of plant ecophysiology. Kluwer, Dordrecht, The Netherlands, pp 1–12

Rennenberg H, Herschbach C, Haberer K, Kopriva S (2007) Sulfur metabolism in plants: are trees different? Plant Biol 9:620–637

Rüegsegger A, Brunold C (1992) Effect of cadmium on γ-glutamylcysteine synthesis in maize seedlings. Plant Physiol 99:428–433

Schafer FO, Buettner GR (2001) Redox environment of the cell as viewed through the redox state of the glutathione disulphide/glutathione couple. Free Radic Biol Med 30:1191–1212

Schröder P (2001) The role of glutathione and glutathione S-transferases in plant reaction and adaptation to xenobiotics. In: Grill D, Tausz M, De Kok L (eds) Significance of glutathione in plant adaptation to the environment, vol 2, Hand book series of plant ecophysiology. Kluwer, Dordrecht, The Netherlands, pp 155–183

Schröder P (2007) Exploiting plant metabolism for the phytoremediation of organic xenobiotics. In: Willey N (ed) Phytoremediation: methods and reviews. Humana, Totowa, NJ, pp 254

Sergiev IG, Alexieva VS, Ivanov SV, Moskova II, Karanov EN (2006) The phenylurea cytokinin 4PU-30 protects maize plants against glyphosate action. Pest Biochem Physiol 85:139–146

Sheehan D, Meade G, Foley V, Dowd C (2001) Structure, function and evolution of glutathione transferases: implications for classification of non-mammalian members of an ancient enzyme superfamily. Biochem J 360:1–16

Skipsey M, Cummins I, Andrews CJ, Jepson I, Edwards R (2005) Manipulation of plant tolerance to herbicides through co-ordinated metabolic engineering of a detoxifying glutathione transferase and thiol cosubstrate. Plant Biotech J 3:409–420

Smith AP, DeRidder BP, Guo W-J, Seeley EH, Regnier FE, Goldsbrough PB (2004) Proteomic analysis of Arabidopsis glutathione S-transferases from benoxacor- and copper-treated seedlings. J Biol Chem 279:26098–26104

Szalai G, Kell s T, Kocsy G (2009) Glutathione as an antioxidant and regulatory molecule in plants under abiotic stress conditiond. Plant Growth Regul 28:66–80

Tharayil-Santhakumar N (2003) Mechanism of herbicide resistance in weeds. WeedScience.org, pp. 2. http://www.weedscience.org/in.asp)

Todorova D (2003) Ethilene mutants of *Arabidopsis thaliana* L *Heynh*. Plant Sci 40:483–490 (In Bulgarian)

Verkleij JAC, Golan-Goldhirsh A, Antosiewisz DM, Schwitzguébel J-P, Schröder P (2009) Dualities in plant tolerance to pollutants and their uptake and translocation to the upper plant parts. Environ Exp Bot 67:10–22

Vyvyan JR (2002) Allelochemicals as leads for new herbicides and agrochemicals. Tetrahedron 58:1631–1646

Wachter A, Wolf S, Steininger H, Bogs J, Rausch T (2005) Differential targeting of GSH1 and GSH2 is achieved by multiple transcription initiation: implications for the compartmentation of glutathione biosynthesis in the Brassicaceae. Plant J 41:15–30

Wang CL, Oliver DJ (1996) Cloning of the cDNA and genomic clones for glutathione synthetase from *Arabidopsis thaliana* and complementation of a *gsh2* mutant in fission yeast. Plant Mol Biol 31:1093–1104

Wonisch W, Schaur R (2001) Chemistry of glutathione. In: Grill D, Tausz M, De Kok L (eds) Significance of glutathione in plant adaptation to the environment, vol 2, Handbook series of plant ecophysiology. Kluwer, Dordrecht, The Netherlands, pp 13–26

Yang X, Sun W, Liu J-P, Liu Y-J, Zeng Q-Y (2009) Biochemical and physiological characterization of a tau class glutathione transferase from rice (*Oryza sativa*). Plant Physiol Biochem 47:1061–1068

Yuan JS, Tranel PJ, Stewart CN Jr (2007) Non-target-site herbicide resistance: a family buisiness. Trends Plant Sci 12:6–13

Zhang O, Riechers DE (2004) Proteomic characterization of herbicide safener-induced proteins in the coleoptile of *Triticum tauschii* seedlings. Proteomics 4:2058–2071

Chapter 7
Ascorbate and Glutathione: Protectors of Plants in Oxidative Stress

Qaisar Mahmood, Raza Ahmad, Sang-Soo Kwak, Audil Rashid, and Naser A. Anjum

Abstract Reactive oxygen species (ROS) are produced naturally in plants during normal growth conditions. However, their production is accelerated manifold during various abiotic and biotic stresses. Rapid and efficient detoxification of ROS is vital to avoid any damage at cellular level. This is done by a well defined antioxidative system which comprises of various enzymes (superoxide dismutase, catalases and peroxidases) and low molecular weight compounds such as; praline, betaine, ascorbate and glutathione. Among these, ascorbate and glutathione are directly involved in scavenging of ROS. The present article will emphasize on the biosynthesis and role of ascorbate and glutathione during oxidative stress.

Keywords Abiotic stress • Detoxification • Reactive Oxygen species • Oxidative stress

Q. Mahmood (✉) and R. Ahmad
Department of Environmental Sciences, COMSATS Institute of Information Technology, Abbottabad 22060, Pakistan
e-mail: mahmoodzju@gmail.com

S.-S. Kwak
Environmental Biotechnology Research Center, Korea Research Institute of Bioscience and Biotechnology (KRIBB), Eoeun-dong 52, Yuseong, Daejon, 305-806, South Korea

A. Rashid
Department of Environmental Sciences, PMAS Arid Agriculture University, Rawalpindi, Pakistan

N.A. Anjum
Centre for Environmental and Marine Studies (CESAM) & Department of Chemistry University of Aveiro, 3810–193 Aveiro, Portugal
Academia Sinica
Agricultural Biotechnology Research Center (ABRC), Taiwan
and
Department of Botany, Division of Plant Physiology and Biochemistry, Faculty of Life Sciences, Aligarh Muslim University, Aligarh - 202 002, UP, India

N.A. Anjum et al. (eds.), *Ascorbate-Glutathione Pathway and Stress Tolerance in Plants*,
DOI 10.1007/978-90-481-9404-9_7, © Springer Science+Business Media B.V. 2010

Abbreviations

AsA	ascorbic acid
APX	ascorbate peroxidase
CAT	catalase
DHAR	dehydroascorbate, reductase
EBR	2,4-epibrassinosteroid
FTS	ferredoxin thioredoxin system
GR	glutathione reductase
Grx	glutaredoxin
GSH	reduced glutathione
GSSG	oxidized glutathione
GPX	guaiacol peroxidase
LOX	lipoxygenase
MDA	malondialdehyde
NGS	NADPH glutaredoxin system
NTS	NADPH thioredoxin system
NTR	Trx reductase
PAPS	phosphoadenylyl sulfate
RNR	ribonucleotide reductase
SOD	superoxide dismutase
Trx	thioredoxin

1 Antioxidant Defense System of Plants

This antioxidant defense system mainly includes the antioxidative enzymes such as superoxide dismutase (SOD), catalase (CAT), guaiacol peroxidase (GPX), ascorbate peroxidase (APX) and glutathione reductase (GR) and non-enzymatic antioxidant compounds such as glutathione (GSH), ascorbic acid (AsA), carotenoids, cysteine, etc. SOD is the key enzyme responsible for catalyzing the dismutation of highly reactive $O_2\cdot^-$ to O_2 and H_2O_2. The resulting H_2O_2 is further decomposed to water and oxygen either by APX of the ascorbate–glutathione cycle or by GPX and CAT localized in the cytoplasm and other cellular compartments. GR is the complementary enzyme of the ascorbate–glutathione cycle, maintaining a high GSH/oxidized glutathione (GSSG) ratio for protection against oxidative damage (May et al. 1998a; Noctor and Foyer 1998; Aravind and Prasad 2005). GSH (γ-glutamylcysteinyl glycine) is a major non-enzymatic scavenger of reactive oxygen species (ROS) due to its unique structural properties, broader redox potential, abundance and wide distribution in plants.

Reversible dithiol disulfide exchange is now well acknowledged as an important biological mechanism regulating various aspects of the metabolism, development, and adaptation to changing environmental situations. Several reductases, including ribonucleotide reductases (RNR), peroxiredoxins, and methionine sulfoxide reductases, gain their reduction power via a pair of cysteines switching from the disulfide to the dithiol state. In numerous proteins that do not need a redox flux for activity, including transcription factors, the disulfide to dithiol switch results in a conformational modification allowing or restricting activity. The reduction of disulfide bridges is performed by Trxs or Grxs, two phylogenetically unrelated proteins but which present a similar three-dimensional conformation, including at their surface a redox-active site CxxC with two cysteines separated by two amino acids. Almost all organisms encode several Trxs and Grxs. Trx and Grx have they own reduction system: in most organisms and in the cytosol and mitochondria of plants the Trx pathway comprises a redox cascade including NADPH, Trx reductase (NTR), and Trx, while the Grx pathway is composed of NADPH, glutathione reductase (GR), glutathione (GSH), and Grx. These two systems act in parallel and have several common target proteins as shown by biochemical and genetic studies. Two important redoxin targets are the subject of an abundant literature: (i) there are several types of RNR, which provide cells with the deoxyribonucleotides necessary for DNA. All organisms growing aerobically use Trx- or Grx-dependent RNR. (ii) Sulfate assimilation features a reduction step performed by a phosphoadenylyl sulfate (PAPS) reductase in bacteria and yeast and by an adenylyl sulfate (APS) reductase in plants, and these enzymes are also Trx dependent.

Recent genetic experiments in Arabidopsis show that the cytosolic Trx and Grx have certain common target proteins as is the case in most organisms, but, in addition, their reduction systems are in fact more complex. The reduction of cytosolic Trx and Grx can occur via alternative pathways. In the absence of NTR, cytosolic Trxs are reduced by a GSH-dependent pathway, while in the absence of cytosolic GR oxidized glutathione (GSSG) is reduced by the NTR Trx pathway. By contrast, sulfate reduction, which in plants is carried out in chloroplasts, is performed by a two-domain APS reductase with the APS domain associated to a Trx homolog reduced by glutathione. Sulfate reduction in plants is thus independent from external Trx and Grx and their reductases.

Plants are distinguishable from other eukaryotes by the presence of the chloroplast. In addition, they have a complex redoxin system as revealed by sequencing the *Arabidopsis thaliana* genome (Alonso et al. 2000). About 40 genes coding for Trx and Trx-related proteins have been identified (Meyer et al. 2006). Among them, at least eleven belong to the cytosolic Trxh group, but additional Trx-like proteins are predicted to have a cytosolic localization. A functional mitochondrial NTS system has been identified in Arabidopsis (Laloi et al. 2001). Moreover, the chloroplast Trx system has been extensively studied in plants (Schürmann and Jacquot 2000; Buchanan et al. 2002; Schürmann and Buchanan 2008). The genome of *A. thaliana* encodes 40 genes of Grx and Grx-like proteins.

2 Glutathione Synthesis and Reduction in Plants

Glutathione is a tripeptide (α-glutamyl cysteinylglycine), which has been detected virtually in all cell compartments such as cytosol, chloroplasts, endoplasmic reticulum, vacuoles, and mitochondria (Foyer and Noctor 2003). Glutathione is the major source of non-protein thiols in most plant cells. The chemical reactivity of the thiol group of glutathione makes it particularly suitable to serve a broad range of biochemical functions in all organisms. The nucleophilic nature of the thiol group is also important in the formation of mercaptide bonds with metals and for reacting with selected electrophiles. This reactivity along with the relative stability and high water solubility of GSH makes it an ideal biochemical to protect plants against stress including oxidative stress, heavy metals and certain exogenous and endogenous organic chemicals (Foyer and Noctor 2005a, b). Glutathione takes part in the control of H_2O_2 levels (Shao et al. 2005c, 2007a). The change in the ratio of its reduced (GSH) to oxidized (GSSG) form during the degradation of H_2O_2 is important in certain redox signaling pathways (Li and Jin 2007). It has been suggested that the GSH/GSSG ratio, indicative of the cellular redox balance, may be involved in ROS perception (Shao et al. 2005c; Li and Jin 2007). Reduced glutathione (GSH) acts as an antioxidant and is involved directly in the reduction of most active oxygen radicals generated due to stress. There was a study reporting that glutathione, an antioxidant, helped to withstand oxidative stress in transgenic lines of tobacco (Shao et al. 2005c).

The metabolism of glutathione has been extensively characterized in plants as in other organisms. Glutathione is synthesized in two steps catalyzed by the gamma-glutamyl cysteine synthase (GSH1) and the glutathione synthase (GSH2) (Fig. 1). In

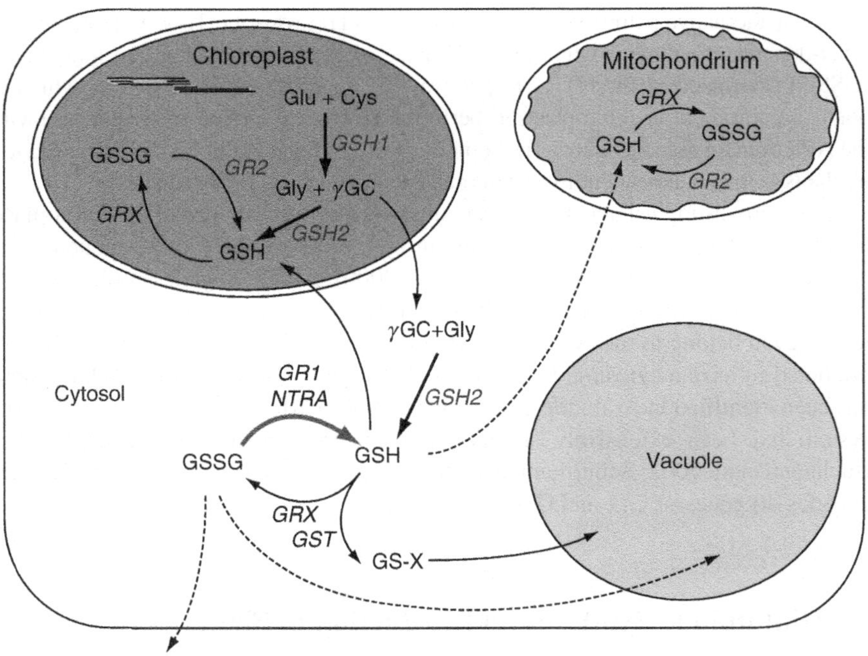

Fig. 1 Glutathione synthesis and reduction in plants. The GSH synthesis enzymes GSH1 and GSH2 are indicated in *grey* (*brown* in the web version). The GSSG reduction enzymes GR1 and GR2 are indicated in *dark grey* (*blue* in the web version). The alternative reduction pathway is represented by the *grey arrow* (*red* in the web version). The *dotted arrows* indicate putative GSH fluxes (Adapted from Reichheld et al. 2009)

higher plants, GSH1 is exclusively located in plastids, while GSH2 is dually targeted to plastids and cytosol (Wachter et al. 2005; Pasternak et al. 2008). As in other eukaryotes, GSH is required for plant development as shown by the embryo lethality of gsh1 knockout lines (Tzafrir et al. 2004; Cairns et al. 2006). In Arabidopsis, forward genetic screens allowed the isolation of several mutants which after cloning and sequencing of the mutated genes appears to be weak alleles of GSH1. The first discovered mutant allele was rml1 (rootmeristemless1) (Cheng et al. 1995) which encodes an inefficient GSH1, allowing approximately only 5% GSH in comparison to the wild type (Vernoux et al. 2000). Homozygote seeds are formed with a normal embryo, but the root meristem fails to grow during germination. In contrast the shoot meristem develops at least in the initial stage. Two other GSH1 alleles are available: pad2-1 with about 16% GSH and cad2 with about 20% GSH present a normal development under standard conditions and are fertile (Cobbett et al. 1998). The pad2-1 (Phytoalexin-deficient2-1) mutant was isolated on the basis of its sensitivity to several pathogens, possibly due to inefficient production of phytoalexins and glucosinolates (Parisy et al. 2007). Cad2 (cadmiumsensitive2) is hypersensitive to Cd possibly due to a limited synthesis of phytochelatins (Howden et al. 1995).

In contrast to the gsh1 KO mutant, a null mutant in GSH2 shows normal embryogenesis, but the seeds do not germinate. The gsh2 mutant accumulates high levels of the substrate of GSH2, gamma-GC, suggesting that this low molecular weight thiol only very partially compensates GSH in the early stage of plant development (Pasternak et al. 2008). This contrasts with *E. coli* and yeast in which elevated gamma-GC in the gsh2 mutant was shown to compensate the absence of GSH (Faulkner et al. 2008; Grant et al. 1996). Interestingly, Pasternak et al. (2008) have shown by complementation of a gsh2 mutant that the cytosolic GSH2 is sufficient for GSH synthesis, showing that gamma-GC is exported from the plastids to supply the cytosol with the precursor of GSH biosynthesis and that GSH is efficiently reimported into the plastids and mitochondria (Fig. 1). As in most organisms, oxidized glutathione is reduced by GRs which are present in distinct cellular compartments.

In Arabidopsis and other higher plants, two genes encode GRs. GR2 was found to encode a dual-targeted chloroplast and mitochondrial GR, while GR1 encodes a cytosolic protein. Although both GRs have similar reduction activities with GSSG, only GR2 is essential for plant development (Tzafrir et al. 2004). It is still unknown whether the embryo lethality of the gr2 mutant is due to inactivation of the chloroplast and/or mitochondrial isoform of GR2. Nevertheless, it clearly indicates that cytosolicGR1 is not able to compensate for the lack of organellar glutathione reduction. As previously stated, GSH synthesis takes place both in the chloroplast and the cytosol, but engineered Arabidopsiswith only a cytosolic GSH2 are fully viable. Thus GSH import from the cytosol to organelles is sufficient, but GSSG reduction should take place in the organelles. Biochemical characterization of gr1 mutant plants has established that GR1 shows 65% of total GR activity of leaf extracts. Mutant gr1 accumulates high levels of GSSG. However, in contrast to gr2 mutants, insertion mutants in gr1 are aphenotypic even under stress conditions, indicating that cytosolic GR is dispensable for plant development (Marty et al. 2009). This finding suggests that accumulation of GSSG in the cytosol is either not toxic for plants or that it is exported out of the cytosol. Export to the organelles seems unlikely because chloroplast and mitochondrial redox state of glutathione was shown to be unchanged in the gr1 mutant (Marty et al. 2009). However, vacuolar or apoplastic export cannot be excluded.

3 Glutathione and Oxidative Stress

Oxidative stress from environmental sources and plant developmental processes generates various oxidants, such as reactive oxygen, sulfur, and nitrogen, which can cause serious damage to biological macromolecules. In higher plants, chloroplasts and mitochondria are two major organelles that contribute to the production of ROS during photosynthesis and carbon metabolism. To prevent oxidative damage, plants have evolved effective mechanisms comprising nonenzymatic antioxidant molecules and enzymes (Michelet et al. 2006). Oxidants have been traditionally considered

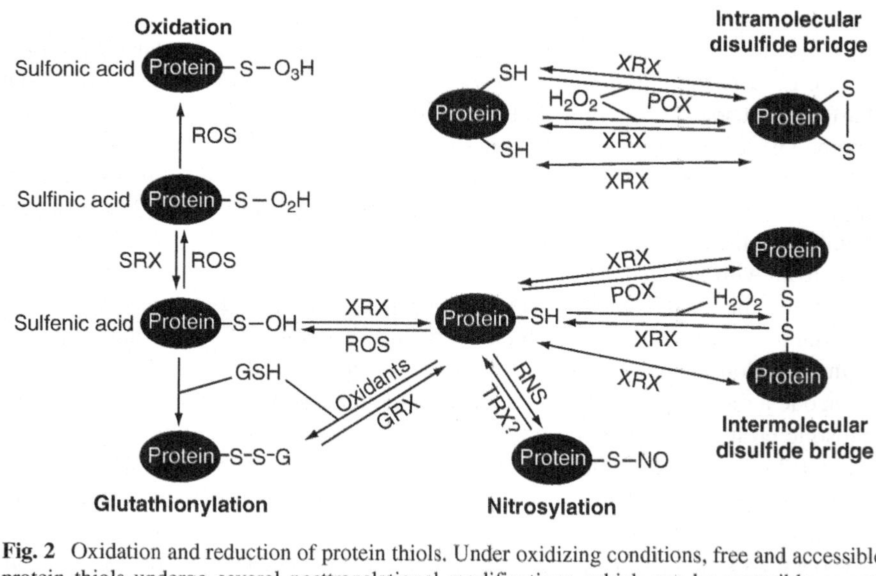

Fig. 2 Oxidation and reduction of protein thiols. Under oxidizing conditions, free and accessible protein thiols undergo several posttranslational modifications, which can be reversible or not. Protein cysteines can be oxidized by ROS into sulfenic acid (SOH), which can be reduced by XRX (GRX or TRX). Further oxidation of sulfenic acid by ROS can result in the formation of sulfinic acid (SO_2H), which can be reversed by sulfiredoxins (SRX), or irreversibly oxidized to sulfonic acid (SO_3H). The presence of oxidants and/or GSH allows glutathionylation of protein cysteines to occur via different mechanisms. Deglutathionylation can be catalyzed by GRX. Reversible formation of intra- and intermolecular disulfide bridges is mediated by XRX. Direct H_2O_2-dependent oxidation of cysteines to intra- and intermolecular disulfides and peroxidase (POX)-catalyzed H_2O_2 sensing can be reversed by XRX. In addition, protein cysteines also undergo nitrosylation in the presence of reactive nitrogen species (RNS), a reversible process that could be catalyzed by TRX (Adapted from Li and Zachgo 2009)

as poisonous molecules with deleterious effects on plant cells. However, recent studies suggest that these reactive species, particularly ROS and reactive nitrogen species (RNS), are involved in redox signaling through reversible posttranslational modifications of protein thiols (Foyer and Noctor 2005a; Michelet et al. 2006). Free protein thiols can be oxidized into different reversible states, such as S-glutathionylation, S-nitrosylation, sulfenic (SOH) or sulfinic acids (SO2H), and intra- or interprotein disulfide bonds (Fig. 2). S-Glutathionylation of protein sulfhydryl groups protects cysteinyl residues from irreversible oxidation to sulfonic acids. Apart from sulfinic acids that are reduced by sulfiredoxins (SRXs), these oxidized cysteines can be efficiently reduced by GRXs, providing evidence that GRXs play an important role in oxidative stress signaling (Foyer and Noctor 2005a; Michelet et al. 2006). Antioxidant defense involves the activation of redox-responsive transcription factors in *E. coli*, yeast, and mammals (Delaunay et al. 2000; Rahman and MacNee 2000). In apoptosis signaling of mammalian cells, TRXs are known to catalyze denitrosylation of cysteine residues of caspase-3, a process required for caspase-3 activation (Benhar et al. 2008). Higher plants are also thought to acclimate to oxidative stress through coordinate modulation of a battery of antioxidant genes. Tsukamoto et al. (2005) identified a novel conserved 28-bp long cis-element on the promoter regions of three antioxidant rice genes, one of which codes for a rice GRX. This short regulatory sequence is responsive to oxidative stress induced by methyl viologen treatment and regulates expression of these antioxidant genes. In Arabidopsis, a network of at least 152 genes is involved in producing and scavenging ROS, among which 27 encode GRXs functioning as ROS-scavenging enzymes (Mittler et al. 2004a). As mentioned earlier, overexpression of ROXY1 and rice OsROXY1/2 in Arabidopsis leads to accumulation of H_2O_2 and increased susceptibility of transgenic plants to pathogen infection, indicating a link between oxidative stress, plant development, and pathogen responses (Wang et al. 2009). The use of ROS as signaling molecules by plant cells suggests that plants are able to achieve a high degree of control over ROS toxicity during the course of evolution. The first piece of biochemical evidence indicating the possible involvement of plant GRXs in oxidative stress responses was obtained from a poplar CPYC GRX that is able to reduce a type II peroxiredoxin (PRX; Rouhier et al. 2001, 2002). Further independent observations have shown that GRXs and PRXs are abundantly colocalized in the phloem sap, allowing in planta reduction reactions to occur and preventing oxidative damage of sieve-tube proteins (Balachandran et al. 1997; Ishiwatari et al. 1995; Szederkenyi et al. 1997). A redox-sensitive green fluorescent protein (roGFP) has been constructed and is now available for nondestructive, real-time measurement of the redox potential in both animal and plant cells (Hanson et al. 2004; Jiang et al. 2006). The CPYC GRXC1 from Arabidopsis is able to reduce disulfide bonds of roGFPs (Meyer et al. 2007), demonstrating the potential of this novel tool to monitor plant redox changes and indicating a role for GRXC1 in maintaining the cellular redox homeostasis.

GSH is a key component of plant antioxidant networks and is required for both the dithiol and monothiol reduction mechanism mediated by GRXs. In the dithiol mechanism, GSH is necessary for the regeneration of GRX. In contrast, in the

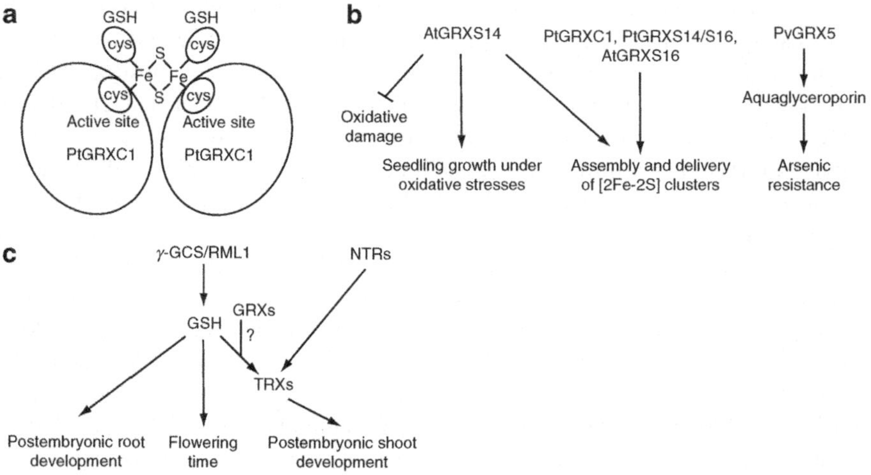

Fig. 3 Other redox-sensitive cellular processes in plants. (**a**) Coordination patterns of a [2Fe–2S] cluster for poplar dimeric holo PtGRXC1. Cytosolic poplar PtGRXC1 with a CGYC active site is present as either a monomeric apoprotein or a dimeric holoprotein with a subunit-bridging [2Fe–2S] cluster that is ligated by catalytic cysteines of two PtGRXC1 molecules and cysteines of two GSH molecules. (**b**) AtGRXS14 with a CGFS active site protects cells against oxidative damage and is necessary for early seedling growth under oxidative stresses. As PtGRXC1, AtGRXS14/S16 and PtGRXS14/S16 also play a role in the assembly and delivery of [2Fe–2S] clusters. PvGRX5 with a CGFS active site likely regulates intracellular arsenite levels, either by directly or indirectly modulating aquaglyceroporins. (**c**) RML1 codes for γ-GCS, the first enzyme of GSH biosynthesis. Analyses of Arabidopsis rml1 mutants have revealed that GSH promotes flowering and positively regulates postembryonic root development. Inactivation of two NTRs and further analyses of ntra ntrb rml1 triple mutants allowed characterization of a complex interplay between TRXs and a GSH-dependent pathway in Arabidopsis postembryonic shoot development. Biochemical data indicate that GRXs might participate in alternative reduction TRXs by the GSH pathway. Arrows and hatchets denote positive and negative regulation, respectively. Abbreviations: At, *Arabidopsis thaliana*; Pt, *Populus trichocarpa*; Pv, *Pteris vittata*; SA, salicylic acid; GRX, glutaredoxin; Đ-GCS, Đ-glutamylcysteine synthase; GSH, glutathione; NTR, NADPH thioredoxin reductase; TRX, thioredoxin (Adapted from Li and Zachgo 2009)

monothiol mechanism, GSH is directly attached to a cysteine and thereby reversibly modifies target proteins posttranslationally. GSH is synthesized in two ATP-dependent steps that are sequentially catalyzed by γ-glutamylcysteine synthase (γ-GCS) and glutathione synthase. GSH is present at millimolar concentrations in plant cells and participates in stress resistance and adaptation, such as detoxification of heavy metals and scavenging of reactive oxygen species (ROS) generated during normal cell metabolism and induced by abiotic factors such as UV light, drought, and chilling as well as by biotic factors such as pathogens (Noctor and Foyer 1998; Cobbett 2000). However, besides playing critical roles in stress responses, evidence is currently emerging that GSH is also crucial for many plant developmental processes (Fig. 3c).

Recent data indicate that GSH plays an important role in determining flowering time. Ogawa et al. (2001, 2004) examined the relationship between GSH levels, photosynthesis, and flowering in Arabidopsis and found that flowering is regulated by the key reaction of GSH biosynthesis catalyzed by γ-GCS. Arabidopsis mutants, defective in glutathione biosynthesis or possessing nonfunctional light-harvesting antenna of photosystem II, produce reduced GSH levels and hence develop a late flowering phenotype (Cobbett et al. 1998; Ogawa et al. 2001, 2004).

Embedded within root meristems is a population of slowly dividing cells collectively designated as the quiescent center (QC). Postembryonic root development entails activation of cell division in this center. Analysis of Arabidopsis plants homozygous for a mutation in ROOT MERISTEMLESS1 (RML1)/CADMIUM SENSITIVE2 (CAD2), which encodes the γ-GCS, identified a GSH-dependent pathway involved in establishing the postembryonic root meristem (Cobbett et al. 1998; Vernoux et al. 2000). The rml1/cad2 mutant is almost devoid of GSH due to strongly reduced GSH synthesis, thereby causing an arrest in root development due to abolished cell division after germination in the root but not in the shoot (Vernoux et al. 2000). In maize roots, quiescent center formation is associated with an auxin-regulated oxidizing environment. High levels of auxin and oxidized GSH occur in the QC, contrasting with reduced GSH and low levels of auxin favored in adjacent, rapidly dividing cells in the root meristem. Decreasing auxin levels in the QC by perturbing polar auxin transport leads to a reduced environment and thereby activates the QC (Jiang et al. 2003). These experiments corroborate earlier observations that GSH participates in the regulation of cell division in the apical meristem of Arabidopsis roots (Sánchez-Fernández et al. 1997).

NADPH thioredoxin reductases (NTRs) are key regulatory enzymes determining the redox state of the TRX system. The Arabidopsis genome contains two genes encoding NTRs (NTRA and NTRB) and one gene coding NTRC, a chloroplastic hybrid enzyme containing both an NTR and a TRX modules. The ntra ntrb double mutant is viable and fertile, and the cytosolic TRXh3 in this mutant is only partially oxidized. Crossing ntra ntrb with rml1/cad2 leads to complete inhibition of both shoot and root growth, indicating that a GSH-dependent pathway is implicated in the alternative reduction of TRXh3 and is indispensable for postembryonic activity in the shoot and root apical meristems (Reichheld et al. 2007). However, a direct reduction of type-h TRXs by GSH is excluded by biochemical data, suggesting that GRXs may be responsible for the alternative reduction of TRXs in the ntra ntrb double mutant (Reichheld et al. 2007), as observed in vitro for poplar TRX h4 (Gelhaye et al. 2003; Koh et al. 2008).

Complexation of As(III) with PCs or GSH is an efficient way to detoxify arsenic, probably because the complexes are pumped and sequestered in the vacuole catalyzed by the homologs of multidrug resistance proteins (MRPs), members of the ABC superfamily (Lu et al. 1997; Tommasini et al. 1998). Targeting increased accumulation or synthesis of PCs and/or GSH may be one way to develop arsenic phytoremediation. Increased expression of phytochelatin synthase (PCS), the rate-limiting step in PC biosynthesis, has been attempted to increase plant tolerance to and accumulation of arsenic. Gasic and Korban (2007) found that the overexpression of PCS in Indian mustard increased its tolerance to arsenic but did not enhance arsenic accumulation significantly. The lack of response in accumulation could be due to the fact that PC synthesis is also limited by the production of GSH. A more recent study by Guo et al. (2008) showed that overexpressing AtPCS1 and GSH1, which encodes γ-glutamylcysteine synthetase (γ-ECS), the rate-limiting step in GSH biosynthesis, individually in *Arabidopsis thaliana* increased arsenic tolerance and accumulation. Although these studies indicated the feasibility of overexpressing PCS and/or γ-ECS for increasing arsenic accumulation and concomitantly tolerance, there are no direct data on the site of arsenic storage in these transgenic lines; thus it remains unclear whether the complexed As(III) is primarily vacuolar or remains in the cytoplasm. It is possible that transport of complexed As(III) or even free As(III) across the tonoplast membrane is potentially the rate-limiting step in overall arsenic tolerance and accumulation. Yet, to date, there are no reports of genetic engineering of tonoplast transport.

4 Coordinated Role of Ascorbate and Glutathione During Oxidative Stress

Ascorbate is one of the most extensively studied antioxidants and has been detected in the majority of plant cell types, organelles and apoplast (Aro and Ohad 2003; Borland et al. 2006). Ascorbate is synthesized in the mitochondria and is transported to the other cell components through a proton-electrochemical gradient or through facilitated diffusion. Further, ascorbic acid also has been implicated in regulation of cell elongation (Noctor 2006; Yabuta et al. 2004). In the ascorbate–glutathione cycle, two molecules of ascorbic acid are utilized by APX to reduce H_2O_2 to water with concomitant generation of monodehydroascorbate. Monodehydroascorbate is a radical with a short life time and can disproportionate into dehydroascorbate and ascorbic acid. The electron donor is usually NADPH and catalyzed by monodehydroascorbate reductase or ferredoxin in water–water cycle in the chloroplasts (Gapper and Dolan 2006). In plant cells, the most important reducing substrate for H_2O_2 removal is ascorbic acid (Foyer et al. 1994, 1997; Biehler and Fock 1996; Shao and Chu 2005; Shao et al. 2005a, b, 2006, 2007a, b, c; Sagi and Fluhr 2006; Wu et al. 2007). A direct protective role for ascorbic acid has also been demonstrated in rice.

Ascorbate (vitamin C) occurs in all plant tissues, usually being higher in photosynthetic cells and meristems (and some fruits). About 30–40% of the total ascorbate is in the chloroplast and stromal concentrations as high as 50 mM have been reported (Shao et al. 2005b). It is highest in the mature leaf, where the chloroplasts are fully developed and the chlorophyll levels are highest. Although it has been determined that D-glucose is the precursor of L-ascorbic acid, the synthetic pathway has not been totally understood (Shao et al. 2005a, c). Ascorbic acid has effects on many physiological processes, including the regulation of growth, differentiation and metabolism of plants. A fundamental role of AsA in the plant defense system is to protect metabolic processes against H_2O_2 and other toxic derivatives of oxygen.

Acting essentially as a reductant and reacting with and scavenging many types of free radicals, AsA reacts nonenzymatically with superoxide, hydrogen peroxide, and singlet oxygen. It can react indirectly by regenerating α-tocopherol or in the synthesis of zeaxanthin in the xanthophyll cycle. Therefore, AsA influences many enzyme activities, and minimizes the damage caused by oxidative process through synergic function with other antioxidants (Li and Jin 2007; Shao et al. 2005a, b, c, 2006, 2008).

A continuous oxidative assault on plants during drought stress has led to the presence of an arsenal of enzymatic and non-enzymatic plant antioxidant defenses to counter the phenomenon of oxidative stress in plants (Shao and Chu 2005). Ascorbic acid (AsA) is an important antioxidant, which reacts not only with H_2O_2, but also with O^{2-}, OH and lipid hydroperoxidases (Igamberdiev and Hill 2004; Shao et al. 2005). AsA is water soluble and also has an additional role in protecting or regenerating oxidized carotenoids or tocopherols (Shao et al. 2007a; Wu et al. 2007). Water stress resulted in a significant increase in antioxidant AsA concentration in turf grass (Shao et al. 2006). Ascorbic acid showed a reduction under drought stress in maize and wheat, suggesting its vital involvement in deciding the oxidative response (Shao et al. 2007b). Some studies reported a decrease in the level of antioxidants, including ascorbic acid, with an increase in stress intensity in wheat. Ascorbic acid can also directly scavenge $1O_2$, O_2^- and •OH and regenerate tocopherol from tocopheroxyl radicals, thus providing membrane protection (Shao et al. 2007b, c). Ascorbic acid also acts as a co-factor of violaxanthin de-epoxidase, thus sustaining dissipation of excess excitation energy (Shao et al. 2006, 2007c). Antioxidants such as ascorbic acid and glutathione are involved in the neutralization of secondary products of ROS reactions (Hare et al. 1998; Shao et al. 2005b, c; Li and Jin 2007).

Two key metabolites of this pathway, ascorbate and glutathione, are major water-soluble antioxidants and redox buffers in plant cells but also have crucial functions in stress responses and organ development (Arrigoni and De Tullio 2002; Noctor et al. 2002). Ascorbate can be oxidized to monodehydroascorbate by ascorbate oxidase (AO) in the apoplast. This enzyme can therefore modulate the redox state of the apoplastic ascorbate pool, which is important for controlling cell elongation and triggering signal transduction cascades in response to external stimuli (Kato and Esaka 2000; Pignocchi et al. 2003). Glutathione participates in the induction of defense genes, sulphur transport and storage, and heavy metal detoxification, and its concentration needs to be tightly controlled at several levels. Major mechanisms are the transcriptional, translational, and post-translational regulation of the enzyme δ-glutamylcysteine synthetase (γ-ECS), which catalyses the first step of glutathione biosynthesis in all organisms (May et al. 1998b; Xiang and Oliver 1998; Noctor et al. 2002; Hicks et al. 2007).

The antioxidants of legume leaves and nodules have been examined in considerable detail (Matamoros et al. 2003; Palma et al. 2006). The most extensive studies on the role of ROS and antioxidants in fruit development and maturation (ripening) have been conducted on climacteric fruits such as pear (*Pyrus communis*; Brennan and Frenkel 1977), saskatoon fruit (*Amelanchier alnifolia*; Rogiers et al. 1998), and tomato (*Lycopersicon esculentum*; Jiménez et al. 2002a). The involvement of ROS and antioxidants during fruit ripening has been also investigated in the chloroplasts (Bouvier et al. 1998) and mitochondria (Jiménez et al. 2002b) of pepper (*Capsicum annuum*) fruit, which exhibits non-climacteric physiology. The study of antioxidants in fruits is important for several reasons. First, antioxidants may protect fruit tissues from potentially toxic ROS and thereby contribute to the stress tolerance of crops (Mittler et al. 2004b; Van Breusegem et al. 2008). Second, fruits may have nutritional value for animal and/or human consumption. Third, in many cases fruits have a relatively short shelf-life following harvest, during which they undergo changes in texture, colour, and flavour, which may be accompanied by a decline in antioxidants (Davey and Keulemans 2004; Malacrida et al. 2006; Stevens et al. 2008). The function of antioxidant enzymes and metabolites during maturation of pea fruits was investigated by Matamoros et al. (2010). It is concluded that symbiotic nitrogen fixation is as effective as nitrogen fertilization in maintaining the antioxidant capacity of pea fruits and that, contrary to climacteric fruits; a general decrease in antioxidants during maturation does not involve oxidative stress. Results underscore the importance of the antioxidant system in reproductive organs and point to ascorbate–glutathione metabolism and cytosolic peroxiredoxin as key players in pea fruit development (Matamoros et al. 2010).

The coordinated role of glutathione–ascorbate to alleviate metal stress has been demonstrated in many plant species. Being a major contributor to the cellular redox state, AsA is involved in oxidative defense in plants (Noctor and Foyer 1998). In common bean, genotypic Mn tolerance may be associated with the maintenance of higher ASC concentrations in leaf tissue under Mn stress (González et al. 1998). Moreover, it is an effective scavenger for phenoxy radicals, which may lead to oxidation of phenols, thus forming necrosis in plants (Hai-hua et al. 2009). The oxidative stress caused by excess Mn, as reflected by the increase in malondialdehyde (MDA) content and lipoxygenase (LOX, EC 1.13.11.12) activity, accumulation of superoxide radical and H_2O_2, was greatly decreased by EBR treatment. Further investigations revealed that EBR application enhanced the activities of superoxide dismutase (SOD, EC 1.15.1.1), peroxidase (POD, EC 1.11.1.7), catalase (EC 1.11.1.6), ascorbate peroxidase (APX, EC 1.11.1.11), dehydroascorbate reductase (DHAR, EC 1.8.5.1), and glutathione reductase (GR, EC 1.6.4.2), and the contents of reduced ascorbate and glutathione, compared with the plants without EBR treatment. It was concluded that the ameliorative effects of EBR on Mn toxicity were due to the upregulation of antioxidative capacity in maize under Mn stress (Hai-hua et al. 2009).

The predominantly accepted model for plant detoxification of arsenic in plants is complexation of As(III) with glutathione(GSH) and/or phytochelatins(PCs) through –SH coordination. PCs have a general structure [(γ-Glu–Cys)n–Gly) (n = 2–11)] and are synthesized non-translationally by PC synthase using glutathione as a substrate (Schmöger et al. 2000). Reported the effect of arsenic exposure on maize (*Zea mays*) root proteome and concluded that the induction of oxidative stress was the main process underlying arsenic toxicity in plants. The accumulation of H_2O_2 was prevented in cells by the ascorbate–glutathione cycle where APX reduced it to H_2O (Shri et al. 2009). Significant increase in APX activity was observed in As(III) treatment in roots. One APX isoenzyme was detected markedly, especially in roots of 100 mM As(III) treated seedlings. However, in shoots maximum APX activity was observed at 50 mM As(III), which decreased at higher concentration. At higher As(III) concentration, roots experienced high ROS buildup and detoxification by increased antioxidant systems before As(III) was transported to shoots. According to POD isozyme chart, with the increase of the concentration of As, POD-2 isozyme band increased gradually, which coincided with the measured results of POD activity in roots of As(III) treatment. It was also observed that new isoforms of GR induced during As(III) stress. It was possible that As(III) significantly decreased the GSH content in rice roots, due to its conversion to phytochelatins. The enhanced demand for GSH in response to arsenic-induced oxidative stress might be met by the increased GR activity and elevated GSH turn-over. The results suggested that the anti oxidative enzymes were highly up regulated in As(III) stress than As(V) (Shri et al. 2009).

Hydrilla verticillata plants were exposed to arsenate (AsV;50 mM) and arsenite (AsIII;5 mM) under variable S supply: deficient (2 mM S,–S), normal (1 mMS, +S) and excess (2 mMS, +HS) (Srivastava and D'Souza 2010). Arsenic accumulation (mg g^{-1} DW) in +HS plants was about twofold higher upon exposure to both AsV (30) and AsIII (50) than that observed in +S (12 and 24) and –S (14 and 26) plants. Despite lower As accumulation, –S plants experienced the maximum oxidative stress owing to an inadequate response of enzymatic and molecular antioxidants and significant decline in total thiols and the ratio of reduced glutathione (GSH) to oxidized glutathione (GSSG). By contrast +HS plants had significant increase in total thiols and an improved redox status, did not demonstrate any negative impact to antioxidants except catalase and hence experienced the least increase in oxidative stress parameters. In conclusion, an increase in S supply to plants may improve their accumulation capacity for As through enhanced tolerance caused by a positive effect on thiol metabolism and antioxidant status of the plants (Srivastava and D'Souza 2010).

Plant growth, ultrastructural and antioxidant adaptations and glutathione biosynthesis in Cd-hyperaccumulating ecotype *Sedum alfredii* Hance (HE) countering high Cd environment were investigated and compared with its non Cd-hyperaccumulating ecotype (NHE) by Jin et al. (2008a). Cadmium stress damaged chloroplasts causing imbalanced lamellae formation coupled with early leaf senescence. Histochemical results revealed that glutathione (GSH) biosynthesis inhibition led to overproduction of hydrogen peroxide (H_2O_2) and superoxide radical ($O_2{\cdot}^-$) in HE but not in NHE. Differences were noted in both HE and NHE for catalase (CAT), guaiacol peroxidase (GPX), ascorbate peroxidase (APX) and glutathione reductase (GR) activities under various Cd stress levels. No relationship was found between antioxidative defense capacity including activities of superoxide dismutase (SOD), CAT, GPX, APX and GR as well as ascorbic acid (AsA) contents and Cd tolerance in the two ecotypes of *S. alfredii*. The GSH biosynthesis induction in root and shoot exposed to elevated Cd conditions may be involved in Cd tolerance and hyperaccumulation in HE of *S. alfredii* H (Jin et al. 2008a).

Ultrastructural changes, zinc hyperaccumulation and its relation with antioxidants in two ecotypes of *Sedum alfredii* Hance were investigated by Jin et al. (2008b). Zn stress may have involved NADPH oxidase, protein phosphatases and intracellular Ca2Þ to activate the reactive oxygen species production. Inhibition of glutathione synthesis may have led to increased H_2O_2 and $O_2 \cdot^-$ accumulations in leaves of HE. In response to higher Zn concentrations, ascorbic acid significantly increased in both ecotypes and levels of glutathione increased in both leaves and roots of HE and in roots of NHE without any change in the leaves of NHE. The enzymatic activities like those of superoxide dismutase (SOD, EC 1.15.1.1), catalase (CAT, EC 1.11.1.6), guaiacol peroxidase (GPX, EC 1.11.1.7), ascorbate peroxidase (APX, EC 1.11.1.11), dehydroascorbate reductase (DHAR, EC 1.8.5.1), and glutathione reductase (GR, EC 1.6.4.2) in leaves of HE were all enhanced at supplied Zn concentration of 500 mM, which may account for its better growth (Jin et al. 2008b).

Effects of zinc on root morphology and antioxidant adaptations of cadmium-treated *Sedum alfredii* H were investigated by Jin et al. (2009). The study demonstrated *S. alfredii* is an excellent cadmium (Cd)/zinc (Zn) hyperaccumulator as Cd and Zn concentrations in leaves reached 2,183 and 13,799 mg kg^{-1} DW, respectively. There was a significant increase in root morphological parameters induced by 50 and 500 μM Zn supplement; however, a sharp decrease in these parameters occurred when treated with 100 μM Cd + 1,000 μM Zn. The inhibited root dehydrogenase activity in 100 μMCd treated plants was restored to control levels when supplemented with 500 μM Zn. Moderate Zn supplement did not produce significant changes in (malondialdehyde)MDA concentrations as compared with those treated with Cd alone. Variations of the antioxidative enzymes proved an ineffective role in coping with metal-stress in *S. alfredii*. Combined Cd and Zn treatment significantly enhanced ascorbic acid (AsA) and glutathione (GSH) contents in leaves of *S. alfredii*, as compared with those treated with Cd alone. Thus, Zn may rely on the involvement of GSH in detoxification and tolerance (Jin et al. 2009).

Salinity stress-induced decrease in photosynthetic potential and dry mass of plants also results from imbalance between ROS and antioxidant defense (Vital et al. 2008). The protection against oxidative stress by scavenging ROS is mediated by components of enzymatic and non-enzymatic antioxidant system of ascorbate–glutathione pathway. The first step of sulfur assimilation in plants is the activation of sulfur by the enzyme ATP-sulfurylase, and in a cascade of enzymatic reactions inorganic sulfur is converted to non-protein thiol glutathione (Tausz et al. 2004). The increased content of glutathione with sulfur application has been found to protect dry mass and photosynthesis of cadmium-treated mustard (*Brassica campestris*) (Anjum et al. 2008). Khan et al. (2009) studied the effect of increasing salinity levels (0, 50, 100 mM NaCl) on growth, photosynthetic traits, leaf water potential, oxidative stress, enzymatic and non-enzymatic antioxidants was studied in Pusa Jai Kisan and SS2 cultivars of mustard (*Brassica juncea*) differing in ATP-sulfurylase activity at 30 days after sowing (DAS). The cultivar SS2 (low ATP-sulfurylase activity) accumulated higher content of Na^+ and Cl^- in leaf than root. SS2 also showed greater content of thiobarbituric acid reactive substances (TBARS) and H_2O_2 and higher decrease in growth, photosynthetic traits and leaf water potential than Pusa Jai Kisan with increasing salinity levels. Contrarily, Pusa Jai Kisan (high ATP-sulfurylase activity) exhibited higher Na^+ and Cl^- content in root than leaf, lower TBARS and H_2O_2 content and higher activity of catalase, ascorbate peroxidase and glutathione reductase. However, the activity of superoxide dismutase was greater in SS2 than Pusa Jai Kisan. Higher activity of ATP-sulfurylase in Pusa Jai Kisan resulted in an increased content of glutathione, a reduced form of inorganic sulfur and an essential component of cellular antioxidant defense system. The lesser decrease in growth and photosynthesis in Pusa Jai Kisan was the result of lesser Na^+ and Cl^- in leaf, higher turgidity and increased activity of antioxidant enzymes and glutathione content (Khan et al. 2009).

Piotrowska et al. (2009) tested the influence of exogenously applied jasmonic acid (JA) upon the growth and metabolism of *Wolffia arrhiza* (Lemnaceae), the smallest vessel aquatic plant exposed to lead (Pb) stress. The presence of exogenous Pb resulted in non-enzymatic antioxidant response in *W. arrhiza* fronds expressed as the increase in ascorbate and glutathione level in relation to the control culture. Application of both 100 μM Pb and JA at the most stimulating concentration 0.1 μM showed an additive effect on the content of antioxidants in fronds. The threefold higher increase in Ascorbate and glutathione accumulation was observed after 100 μM Pb and 0.1μM JA supply during 2 weeks of cultivation. The results revealed that JA at 0.1 μM may be involved in inhibition of lipid peroxides formation which was obvious from lower MDA content in *W. arrhiza* exposed to Pb. JA diminishes lipid peroxidation through the stimulation of non-enzymatic (ascorbate, glutathione) antioxidant machinery responsible for a tight regulation of ROS homeostasis during heavy metal stress. Ascorbate derives its role from its sensitivity to ROS and from the fact that its oxidation affects the redox balances of other metabolites, such as glutathione which being themselves involved in the perception of the cellular redox unbalance contributed to signal amplification (Apel and Hirt 2004). The glutathione level has been shown to correlate with the plant adaptation to extreme heavy metal stress and reduced glutathione pool shows marked alterations in response to heavy metal stress (Jin et al. 2008a; Singh et al. 2006). Moreover, glutathione is also a precursor of phytochelatins, low molecular mass peptides produced by plants to immobilize heavy metals. Xiang and Oliver (1998) indicated that JA stimulated the expression of enzymes of the glutathione biosynthetic pathways leading to an increase in phytochelatin biosynthesis and heavy metal tolerance in *A. thaliana*. Therefore, the results ruled out the possibility that JA-induced increase in ascorbate and glutathione content plays a key role in *W. arrhiza* tolerance to Pb providing the protection against oxidative stress and that glutathione accumulation may precede the phytochelatin synthesis (Piotrowska et al. 2009).

5 Conclusions

The coordinated role of glutathione–ascorbate to alleviate metal stress has been demonstrated in many plant species. Ascorbate and glutathione play an essential role in chloroplastic protection against the potentially deleterious effects of ROS, acting as direct antioxidants and fulfilling other functions related to redox sensing and signaling. Glutathione is also a precursor of phytochelatins, low molecular mass peptides produced by plants to immobilize heavy metals. Salinity stress-induced decrease in photosynthetic potential and dry mass of plants also results from imbalance between ROS and antioxidant defense. The protection against oxidative stress by scavenging ROS is mediated by components of enzymatic and non-enzymatic antioxidant system of ascorbate–glutathione pathway.

References

Alonso JM et al (2000) Genome-wide insertional mutagenesis of *Arabidopsis thaliana*. Science 301:653–657

Anjum NA, Umar S, Ahmad A, Iqbal M, Khan NA (2008) Sulphur protects mustard (*Brassica campestris* L.) from cadmium toxicity by improving leaf ascorbate and glutathione. Plant Growth Regul 54:271–279

Apel K, Hirt H (2004) Reactive oxygen species: metabolism, oxidative stress and signal transduction. Annu Rev Plant Biol 55:373–399

Aravind P, Prasad MNV (2005) Modulation of cadmium-induced oxidative stress in *Ceratophyllum demersum* by zinc involves ascorbate-glutathione cycle and glutathione metabolism. Plant Physiol Biochem 43:107–116

Aro EM, Ohad I (2003) Redox regulation of thylakoid protein phosphorylation. Antioxid Redox Signal 5:56–67

Arrigoni O, De Tullio MC (2002) Ascorbic acid, much more than just an antioxidant. Biochim Biophys Acta 1569:1–9

Balachandran S, Xiang Y, Schobert C, Thompson GA, Lucas WJ (1997) Phloem sap proteins from *Cucurbita maxima* and *Ricinus communis* have the capacity to traffic cell to cell through plasmodesmata. Proc Natl Acad Sci USA 94:14150–14155

Benhar M, Forrester MT, Hess DT, Stamler JS (2008) Regulated protein denitrosylation by cytosolic and mitochondrial thiredoxins. Science 320:1050–1054

Biehler K, Fock H (1996) Evidence for the contribution of the Mehlerperoxidase reaction in dissipating excess electrons in droughtstressed wheat. Plant Physiol 112:265–272

Borland A, Elliott S, Patterson S (2006) Are the metabolic components of crassulacean acid metabolism up-regulated in responses to an increase in oxidative burden? J Exp Bot 57:319–328

Bouvier F, Backhaus RA, Camara B (1998) Induction and control of chromoplast-specific carotenoid genes by oxidative stress. J Biol Chem 273:30651–30659

Brennan T, Frenkel C (1977) Involvement of hydrogen peroxide in the regulation of senescence in pear. Plant Physiol 59:411–416

Buchanan BB, Schürmann P, Wolosiuk RA, Jacquot JP (2002) The ferredoxin/thioredoxin system: from discovery to molecular structures and beyond. Photosynth Res 73:215–222

Cairns NG, Pasternak M, Wachter A, Cobbett CS, Meyer AJ (2006) Maturation of Arabidopsis seeds is dependent on glutathione biosynthesis within the embryo. Plant Physiol 141:446–455

Cheng JC, Seeley KA, Sung ZR (1995) RML1 and RML2, Arabidopsis genes required for cell proliferation at the root tip. Plant Physiol 107:365–376

Cobbett CS, May MJ, Howden R, Rolls B (1998) The glutathionedeficient, cadmium-sensitive mutant, cad2–1, of Arabidopsis thaliana is deficient in gamma-glutamylcysteine synthetase. Plant J 16:73–78

Cobbett CS (2000) Phytochelatin biosynthesis and function in heavy-metal detoxification. Curr Opin Plant Biol 3:211–216

Davey MW, Keulemans J (2004) Determining the potential to breed for enhanced antioxidant status in Malus: mean inter- and intravarietal fruit vitamin C and glutathione contents at harvest and their evolution during storage. J Agric Food Chem 52:8031–8038

Delaunay A, Isnard AD, Toledano MB (2000) H_2O_2 sensing through oxidation of the Yap1 transcription factor. EMBO J 19:5157–5166

Faulkner MJ, Veeravalli K, Gon S, Georgiou G, Beckwith J (2008) Functional plasticity of a peroxidase allows evolution of diverse disulfidereducing pathways. Proc Natl Acad Sci USA 105:6735–6740

Foyer CH, Lelandais M, Kunert KJ (1994) Photooxidative stress in plants. Physiol Plant 92:696–717

Foyer CH, Lopez-Delgado H, Dat JF et al (1997) Hydrogen peroxide-and glutathione-associated mechanisms of acclamatory stress tolerance and signaling. Physiol Plant 100:241–254

Foyer CH, Noctor G (2003) Redox sensing and signaling associated with reactive oxygen in chloroplasts, peroxisomes and mitochondria. Physiol Plant 119:355–364

Foyer CH, Noctor G (2005a) Redox homeostasis and antioxidant signaling: a metabolic interface between stress perception and physiological responses. Plant Cell 17:1866–1875

Foyer CH, Noctor G (2005b) Oxidant and antioxidant signaling in plants: a re-evaluation of the concept of oxidative stress in a physiological context. Plant Cell Environ 28:1056–1071

Gasic K, Korban SS (2007) Transgenic Indian mustard (*Brassica juncea*) plants expressing an Arabidopsis phytochelatin synthase (AtPCS1) exhibit enhanced As and Cd tolerance. Plant Mol Biol 64:361–369

Gapper C, Dolan L (2006) Control of plant development by reactive oxygen species. Plant Physiol 141:341–345

Gelhaye E, Rounier N, Jacquot JP (2003) Evidence for a subgroup of thioredoxin h that requires GSH/GRX for its reduction. FEBS Lett 555:443–448

González A, Steffen KL, Lynch JP (1998) Light and excess manganese, implications for oxidative stress in common bean. Plant Physiol 118:493–504

Grant CM, MacIver FH, Dawes IW (1996) Glutathione is an essential metabolite required for resistance to oxidative stress in the yeast *Saccharomyces cerevisiae*. Curr Genet 29:511–515

Guo JB, Dai XJ, Xu WZ, Ma M (2008) Overexpression of GSH1 and AsPCS1 simultaneously increase the tolerance and accumulation of cadmium ad arsenic in *Arabidopsis thaliana*. Chemosphere 72:1020–1026

Hai-hua W, Tao F, Xi-xu P, Ming-li Y, Ping-lan Z, Xin-ke T (2009) Ameliorative Effects of Brassinosteroid on excess manganese-induced oxidative stress in *Zea mays* L. leaves. Agric Sci China 8:1063–1074

Hanson GT, Aggeler R, Oglesbee D, Cannon M, Capaldi RA, Tsien RY, Remington SJ (2004) Investigating mitochondrial redox potential with redox-sensitive green fluorescent protein. J Biol Chem 279:13044–13053

Hare PD, Cress WA, Van Staden J (1998) Dissecting the roles of osmolyte accumulation during stress. Plant Cell Environ 21:535–553

Hicks LM, Cahoon RE, Bonner ER, Rivard RS, Sheffield J, Jez JM (2007) Thiol-based regulation of redox-active glutamatecysteine ligase from *Arabidopsis thaliana*. Plant Cell 19:2653–2661

Howden R, Goldsbrough PB, Andersen CR, Cobbett CS (1995) Cadmium-sensitive, cad1 mutants of *Arabidopsis thaliana* are phytochelatin deficient. Plant Physiol 107:1059–1066

Igamberdiev AU, Hill RD (2004) Nitrate, NO and haemoglobin in plant adaptation to hypoxia: an alternative to classic fermentation pathways. J Exp Bot 55:2473–2482

Ishiwatari Y, Honda C, Kawashima I, Nakamura S, Hirano H, Mori S, Fujiwara T, Hayashi H, Chino M (1995) Thioredoxin h is one of the major proteins in rice phloem sap. Planta 195:456–463

Jiang K, Meng YL, Feldman LJ (2003) Quiescent center formation in maize roots is associated with an auxin-regulated oxidizing environment. Development 130:1429–1438

Jiang K, Schwarzer C, Lally E, Zhang S, Ruzin S, Machen T, Remington SJ, Feldman L (2006) Expression and characterization of a redox-sensing green fluorescent protein (reduction-oxidation-sensitive green fluorescent protein) in Arabidopsis. Plant Physiol 141:397–403

Jiménez A, Creissen G, Kular B, Firmin J, Robinson S, Verhoeyen M, Mullineaux P (2002a) Changes in oxidative processes and components of the antioxidant system during tomato fruit ripening. Planta 214:751–758

Jiménez A, Gómez JM, Navarro E, Sevilla F (2002b) Changes in the antioxidative systems in mitochondria during ripening of pepper fruits. Plant Physiol Biochem 40:515–520

Jin X, Yang X, Islam E, Liu D, Mahmood Q (2008a) Effects of cadmium on ultrastructure and antioxidative defense system in hyperaccumulator and non-hyperaccumulator ecotypes of *Sedum alfredii* Hance. J Hazard Mat 156:387–397

Jin X, Yang X, Islam E, Liu D, Mahmood Q, Li H, Li J (2008b) Ultrastructural changes, zinc hyperaccumulation and its relation with antioxidants in two ecotypes of *Sedum alfredii* Hance. Plant Physiol Biochem 46:997–1006

Jin X, Liu D, Islam E, Mahmood Q, Yang X, He ZL, Stoffella PJ (2009) Effects of zinc on root morphology and antioxidant adaptations of cadmium treated *Sedum alfredii* H. J Plant Nutr 32:1642–1656

Kato N, Esaka M (2000) Expansion of transgenic tobacco protoplasts expressing pumpkin ascorbate oxidase is more rapid than that of wild-type protoplasts. Planta 210:1018–1022

Khan NA, Nazar R, Anjum NA (2009) Growth, photosynthesis and antioxidant metabolism in mustard (*Brassica juncea* L.) cultivars differing in ATP-sulfurylase activity under salinity stress. Sci Hortic 122:455–460

Koh CS, Navrot N, Didierjean C, Rouhier N, Hirasawa M, Knaff DB, Wingsle G, Samian R, Jacquot JP, Corbier C, Gelhaye E (2008) An atypical catalytic mechanism involving three cysteines of thioredoxin. J Biol Chem 283:23062–23072

Laloi C, Rayapuram N, Chartier Y, Grienenberger JM, Bonnard G, Meyer Y (2001) Identification and characterization of a mitochondrial thioredoxin system in plants. Proc Natl Acad Sci USA 98:14144–14149

Li JM, Jin H (2007) Regulation of brassinosteroid signaling. Trends Plant Sci 12:37–41

Li S, Zachgo S (2009) Glutaredoxins in development and stress responses of plants. Adv Bot Res 52:333–361

Lu YP, Li ZS, Rea PA (1997) AtMRP1 gene of Arabidopsis encodes a glutathione S-conjugate pump: isolation and functional definition of a plant ATP-binding cassette transporter gene. Proc Natl Acad Sci USA 94:8243–8248

Malacrida C, Valle E, Boggio S (2006) Postharvest chilling induces oxidative stress response in the dwarf tomato cultivar Micro-Tom. Physiol Plant 127:10–18

Marty L, Siala W, Schwarzländer M, Fricker MD, Wirtz M, Sweetlove LJ, Meyer Y, Meyer AJ, Reichheld J, Hell R (2009) The NADPH dependent thioredoxin system constitutes a functional backup for cytosolic glutathione reductase in Arabidopsis. Proc Natl Acad Sci USA 106:9109–9114

Matamoros MA, Loscos J, Dietz KJ, Aparicio-Tejo PM, Becana M (2010) Function of antioxidant enzymes and metabolites during maturation of pea fruits. J Exp Bot 61:87–97

Matamoros MA, Dalton DA, Ramos J, Clemente MR, Rubio MC, Becana M (2003) Biochemistry and molecular biology of antioxidants in the rhizobia–legume symbiosis. Plant Physiol 133:499–509

May M, Vernoux T, Leaver C, Van Montagu M, Inze D (1998a) Glutathione homeostasis in plants: implications for environmental sensing and plant development. J Exp Bot 49:649–667

May MJ, Vernoux T, Sánchez-Fernández R, Van Montagu M, Inzé D (1998b) Evidence for post-transcriptional activation of γ-glutamylcysteine synthetase during plant stress responses. Proc Natl Acad Sci USA 95:12049–12054

Meyer Y, Riondet C, Constans L, Abdelgawwad MR, Reichheld JP, Vignols F (2006) Evolution of redoxin genes in the green lineage. Photosynth Res 89:179–192

Meyer AJ, Brach T, Marty L, Kreye S, Rouhier N, Jacquot JP, Hell R (2007) Redox-sensitive GFP in *Arabidopsis thaliana* is a quantitative biosensor for the redox potential of the cellular glutathione redox buffer. Plant J 52:973–986

Michelet L, Zaffagnini M, Massot V, Keryer E, Vanacker H, Miginiac-Maslow M, Issakidis-Bourguet E, Lemaire SD (2006) Thioredoxins, glutaredoxins, and glutathionylation: new crosstalk's to explore. Photosynth Res 89:225–245

Mittler R, Vanderauwera S, Gollery M, Breusegem FV (2004a) Reactive oxygen gene network of plants. Trends Plant Sci 10:1360–1385

Mittler R, Vanderauwera S, Gollery M, Van Breusegem F (2004b) Reactive oxygen gene network of plants. Trends Plant Sci 9:490–498

Noctor G, Foyer CH (1998) Ascorbate and glutathione: keeping active oxygen under control. Annu Rev Plant Physiol Plant Mol Biol 49:249–279

Noctor G, Gomez L, Vanacker H, Foyer CH (2002) Interactions between biosynthesis, compartmentation and transport in the control of glutathione homeostasis and signalling. J Exp Bot 53:1283–1304

Noctor G (2006) Metabolic signaling in defence and stress: the central roles of soluble redox couples. Plant Cell Environ 29:409–425

Ogawa K, Tasaka Y, Mino M, Tanaka Y, Iwabuchi M (2001) Association of glutathione with flowering in *Arabidopsis thaliana*. Plant Cell Physiol 42:524–530

Ogawa K, Hatano-Iwasaki A, Yanagida M, Iwabuchi M (2004) Level of glutathione is regulated by ATP-dependent ligation of glutamateand cysteine through photosynthesis in *Arabidopsis thaliana*: mechanism of strong interaction of light intensity with flowering. Plant Cell Physiol 45:1–8

Palma JM, Jiménez A, Sandalio LM, Corpas FJ, Lundqvist M, Gómez M, Sevilla F, del Río A (2006) Antioxidative enzymes from chloroplasts, mitochondria, and peroxisomes during leaf senescence of nodulated pea plants. J Exp Bot 57:1747–1758

Parisy V, Poinssot B, Owsianowski L, Buchala A, Glazebrook J, Mauch F (2007) Identification of PAD2 as a gamma-glutamylcysteine synthetase highlights the importance of glutathione in disease resistance of Arabidopsis. Plant J Cell Mol Biol 49:159–172

Pasternak M, Lim B, Wirtz M, Hell R, Cobbett CS, Meyer AJ (2008) Restricting glutathione biosynthesis to the cytosol is sufficient for normal plant development. Plant J Cell Mol Biol 53:999–1012

Pignocchi C, Fletcher JM, Wilkinson JE, Barnes JD, Foyer CH (2003) The function of ascorbate oxidase in tobacco. Plant Physiol 132:1631–1641

Piotrowska A, Bajguz A, Godlewska-Z B, Czerpaka R, Kamińska M (2009) Jasmonic acid as modulator of lead toxicity in aquatic plant *Wolffia arrhiza*(Lemnaceae). Environ Exp Bot 66:507–513

Rahman I, MacNee W (2000) Regulation of redox glutathione levels and gene transcription in lung inflammation: therapeutic approaches. Free Radic Biol Med 28:1405–1420

Reichheld JP, Khafif M, Riondet C, Droux M, Bonnard G, Meyer Y (2007) Inactivation of thioredoxin reductases reveals a complex interplay between thioredoxin and glutathione pathways in Arabidopsis development. Plant Cell 19:1851–1865

Reichheld JP, Bashandy T, Siala W, Riondet C, Delorme V, Meyer A, Meyer Y (2009) Redundancy and crosstalk within the thioredoxin and glutathione pathways: a new development in plants. Adv Bot Res 52:253–276

Rogiers SY, Kumar GNM, Knowles NR (1998) Maturation and ripening of fruit of *Amelanchier alnifolia* Nutt. are accompanied by increasing oxidative stress. Ann Bot 81:203–211

Rouhier N, Gelhaye E, Sautiere PE, Brun A, Laurent P, Tagu D, Gérard J, de Faÿ E, Meyer Y, Jacquot JP (2001) Isolation and characterization of a new peroxiredoxin from poplar sieve tubes that uses either glutaredoxin or thioredoxin as a proton donor. Plant Physiol 127:1299–1309

Rouhier N, Gelhaye E, Jacquot JP (2002) Glutaredoxin-dependent peroxiredoxin from poplar: protein–protein interaction and catalytic mechanism. J Biol Chem 277:13609–13614

Sagi M, Fluhr R (2006) Production of reactive oxygen species by plant NAPDH oxidases. Plant Physiol 141:336–340

Sánchez-Fernández R, Fricker M, Corben LB, White NS, Sheard N, Leaver CJ, Montagu MV, Inzé D, May MJ (1997) Cell proliferation and hair tip growth in the Arabidopsis root are under mechanistically different forms of redox control. Proc Natl Acad Sci USA 94:2745–2750

Schürmann P, Buchanan BB (2008) The ferredoxin/thioredoxin system of oxygenic photosynthesis. Antioxid Redox Signal 10:1235–1274
Schürmann P, Jacquot JP (2000) Plant thioredoxin systems revisited. Annu Rev Plant Physiol Plant Mol Biol 51:371–400
Schmöger ME, Oven M, Grill E (2000) Detoxification of arsenic by phytochelatins in plants. Plant Physiol 122:793–801
Shao HB, Chu LY (2005) Plantmolecular biology in China: opportunities and challenges. Plant Mol Biol Rep 23:345–358
Shao HB, Liang ZS, Shao MA et al (2005a) Investigation on dynamic changes of photosynthetic characteristics of 10 wheat (*Triticum aestivum* L.) genotypes during two vegetative-growth stages at water deficits. Biointerfaces 43:221–227
Shao HB, Liang ZS, Shao MA (2005b) Adaptation of higher plants to stresses and stress signal transduction. Acta Ecol Sin 25:1871–1882
Shao HB, Liang ZS, Shao MA (2005c) Dynamic changes of antioxidative enzymes of 10 wheat genotypes at soil water deficits. Biointerfaces 42:187–195
Shao HB, Chu LY, Zhao CX et al (2006) Plant gene regulatory net work system under abiotic stress. Acta Biol Sezeged 50:1–9
Shao HB, Chu LY, Wu G et al (2007a) Changes of some antioxidative physiological indices under soil water deficits among 10 wheat (*Triticum aestivum L.*) genotypes at tillering stage. Biointerfaces 54:143–149
Shao HB, Guo QJ, Chu LY et al (2007b) Understanding molecular mechanism of higher plant plasticity under abiotic stress. Biointerfaces 54:37–45
Shao HB, Jiang SY, Li FM et al (2007c) Some advances in plant stress physiology and their implications in the systems biology era. Biointerfaces 54:33–36
Shao HB, Chu LY, Lu ZH, Kang CM (2008) Primary antioxidant free radical scavenging and redox signaling pathways in higherplant cells. Int J Biol Sci 4:8–14
Shri M, Kumar S, Chakrabarty D, Trivedi PK, Mallick S, Misra P, Shukla D, Mishra S, Srivastava S, Tripathi RD, Tuli R (2009) Effect of arsenic on growth, oxidative stress, and antioxidant system in rice seedlings. Ecotoxicol Environ Saf 72:1102–1110
Singh S, Eapen S, D'Souza SF (2006) Cadmium accumulation and its influence on lipid peroxidation and antioxidative system in an aquatic plant *Bacopa monnieri* L. Chemosphere 62:233–246
Srivastava S, D'Souza SF (2010) Effect of variable sulfur supply on arsenic tolerance and antioxidant responses in *Hydrilla verticillata* (L.f.) Royle. Ecotoxicol Environ Saf. doi:10.1016/j.ecoenv.2009.12.023
Stevens R, Page D, Gouble B, Garchery C, Zamir D, Causse M (2008) Tomato fruit ascorbic acid content is linked with monodehydroascorbate reductase activity and tolerance to chilling stress. Plant Cell Environ 31:1086–1096
Szederkenyi J, Komor E, Schobert C (1997) Cloning of the cDNA for glutaredoxin, an abundant sieve-tube exudate protein from *Ricinus communis* L. and characterisation of the glutathione-dependent thiol-reduction system in sieve tubes. Planta 202:349–356
Tausz M, Sircelj H, Grill D (2004) The glutathione system as a stress marker in plant ecophysiology: is a stress-response concept valid? J Exp Bot 55:1955–1962
Tommasini R, Vogt E, Fromenteau M, Hortensteiner S, Matile P, Amrhein N, Martinoia E (1998) An ABC-transporter of *Arabidopsis thaliana* has both glutathione-conjugate and chlorophyll catabolite transport activity. Plant J 13:773–780
Tsukamoto S, Morita S, Hirano E, Yokoi H, Masumura T, Tanaka K (2005) A novel cis-element that is responsive to oxidative stress regulates three antioxidant defense genes in rice. Plant Physiol 137:317–327
Tzafrir I, Pena-Muralla R, Dickerman A, Berg M, Rogers R, Hutchens S, Sweeney TC, McElver J, Aux G, Patton D, Meinke D (2004) Identification of genes required for embryo development in Arabidopsis. Plant Physiol 135:1206–1220
Van Breusegem F, Bailey-Serres J, Mittler R (2008) Unraveling the tapestry of networks involving reactive oxygen species in plants. Plant Physiol 147:978–984

Vernoux T, Wilson RC, Seeley KA, Reichheld JP, Muroy S, Brown S, Maughan SC, Cobbett CS, Van Montagu M, Inzé D, May MJ, Sung ZR (2000) The ROOT MERISTEMLESS1/CADMIUM SENSITIVE2 gene defines a glutathione-dependent pathway involved in initiation and maintenance of cell division during postembryonic root development. Plant Cell 12:97–110

Vital SA, Fowler RW, Virgen A, Gossett DR, Banks SW, Rodriguez J (2008) Opposing roles for superoxide and nitric oxide in the NaCl stress-induced upregulation of antioxidant enzyme activity in cotton callus tissue. Environ Exp Bot 62:60–68

Wang Z, Xing S, Birkenbihl RP, Zachgo S (2009) Conserved functions of Arabidopsis and rice CC-type glutaredoxins in flower development and pathogen response. Mol Plant 2:323–335

Wu G, Wei ZK, Shao HB (2007) The mutual responses of higher plants to environment: physiological and microbiological aspects. Biointerfaces 59:113–119

Xiang C, Oliver DJ (1998) Glutathione metabolic genes coordinately respond to heavy metals and jasmonic acid in Arabidopsis. Plant Cell 10:1539–1550

Wachter A, Wolf S, Steininger H, Bogs J, Rausch T (2005) Differential targeting of GSH1 and GSH2 is achieved by multiple transcription initiation: implications for the compartmentation of glutathione biosynthesis in the Brassicaceae. Plant J Cell Mol Biol 41:15–30

Yabuta Y, Maruta T, Yoshimura K (2004) Two distinct redox signaling pathways for cytosolic APX induction under photooxidative stress. Plant Cell Physiol 45:1586–1594

Chapter 8
Changes in the Glutathione and Ascorbate Redox State Trigger Growth During Embryo Development and Meristem Reactivation at Germination

Claudio Stasolla

Abstract In vitro embryogenesis via androgenesis and somatic embryogenesis represents an efficient propagation tool as well as a suitable model system for investigating structural, physiological, and molecular events governing embryo development. One of the major problems encountered by tissue culturists is the poor efficiency of embryos produced in vitro and their inability to regenerate viable plants, which denote sub-optimal culture conditions. Judicious alterations of the glutathione (GSH) and ascorbate (AsA) redox state have a profound effect on morphogenesis and improve the quality of the embryos by promoting a zygotic-like histodifferentiation pattern and producing well organized meristems. This approach has been investigated successfully during the development and germination of both *Brassica napus* (canola, an angiosperm) microspore-derived embryos (MDEs) and *Picea glauca* (white spruce, a conifer) somatic embryos. The imposition of a reduced glutathione environment during the early embryonic phases induces cellular proliferation and increases the number of immature embryos, possibly by promoting the synthesis of nucleotides required for energetic processes and mitotic activity. Continuation of embryo development is best conducted if the glutathione pool is experimentally switched towards an oxidized state; a condition favoring histodifferentiation and post-embryonic growth in both canola and spruce. Structural analyses showed that the oxidized glutathione environment favors the proper formation of the shoot apical meristem (SAM), which acquires a "zygotic-like" appearance. The apical poles of glutathione-treated embryos are well organized and display a proper expression and localization of meristem marker genes. These conditions are not found in control embryos which develop abnormal SAMs characterized by the presence of intercellular spaces and differentiation of meristematic cells. Such meristems fail to reactivate at germination resulting in embryo abortion. Physiological and molecular studies have further demonstrated that the oxidized glutathione environment induces several responses, including changes in ascorbate metabolism, abscisic acid and ethylene synthesis, as well as alterations in storage product deposition patterns. If an oxidized

C. Stasolla (✉)
Department of Plant Science, University of Manitoba, Winnipeg R3T 2N2, Manitoba, Canada
e-mail: stasolla@ms.umanitoba.ca

N.A. Anjum et al. (eds.), *Ascorbate-Glutathione Pathway and Stress Tolerance in Plants*, DOI 10.1007/978-90-481-9404-9_8, © Springer Science+Business Media B.V. 2010

glutathione environment favors embryo formation, a reduced ascorbate redox state is required to promote germination and conversion, i.e. the emergence of functional shoots and roots. Specifically, a highly reduced ascorbate environment induces cell proliferation and de-novo meristem formation within those SAMs with an abnormal architecture. Overall it appears that precise changes in glutathione and ascorbate metabolism are required to ensure proper embryo formation and regeneration.

Keywords Ascorbate • Glutathione • Meristem • Microspore-derived and somatic embryos • Germination

1 Introduction

Over the past few years the cellular redox state has emerged as an important determinant for plant growth and development. Despite this realization, however, not many studies have emerged on the topic due to the inability to define changes in redox environment and most importantly to assess their effect on specific cellular and metabolic processes. One of the first acceptable definitions of redox environment is attributed to Shafer and Buettner (2001) who defined the concept as the resulting action of several redox couples which are able to interconvert between oxidized and reduced forms. Examples of these couples include glutharedoxins, thioredoxins, and a variety of thiol/disulfide-containing proteins. The importance of these compounds is further confirmed by the observation that their interconversion is carefully modulated within the cellular environment and unexpected offsets of this equilibrium, which are usually caused by external stimuli, may result in severe cell damage or death. Alterations of the cellular redox environment usually occur during growth and development, and are perceived by unique sensors which trigger specific metabolic pathways leading to short and long term responses. Examples of responses modulated by the redox environment include changes in photosynthetic rate, defense mechanisms following biotic and abiotic stresses, and cell death (Buchanan 1980; Grant and Loake 2001; Schutzendubel and Polle 2002; Mahalingam and Fedoroff 2003). A key factor affecting the cellular redox environment is the synergistic interaction of two sets of redox couples: the glutathione disulfide–glutathione (GSH/GSSG) couple and the reduced-oxidized ascorbate couple. The GSH/GSSG couple, defined as the best indicator of the overall cellular environment (Shafer and Buettner 2001) is closely linked to the ascorbate system via the well characterized Halliwell–Asada cycle (see Potters et al. 2002). It is therefore not surprising that both glutathione and ascorbate have been the subject of many studies leading to exciting insights into the general function of these two molecules in relation to cell behavior.

Studies on the cellular redox state during in vivo growth and development are not easy to perform due to the fact that environmental factors, difficult to predict and/or control, may cause metabolic disturbances which directly or indirectly alter the redox environment. An alternative model used by many labs is represented by in vitro systems in which cells, tissues, and organs can grow and develop in a highly controlled environment in which physical and chemical conditions can be carefully manipulated. Specifically, addenda in the culture medium that alter the redox state can be used to investigate the role of glutathione and ascorbate metabolism during tissue patterning and morphogenesis. This approach has been employed successfully in my lab to elucidate the role played by the cellular redox state during in-vitro embryogenesis and germination. The following sections will address this topic and will demonstrate how specific and controlled alterations of the glutathione and ascorbate pools can be used to improve embryo yield and quality in both angiosperm and gymnosperm species.

2 A Description of the In vitro Systems: *Brassica napus* and *Picea glauca*

Over the past years several protocols have been developed for inducing embryo formation in culture and two fundamental processes are used routinely. The first is gametophytic androgenesis in which embryos can be generated from cells of either the male or female gametophyte. Production of embryos from the male gametophyte, i.e. androgenesis, has been applied to many species and uses immature pollen grains, or microspores, as a starting material for generating microspore-derived embryos (MDEs). Through this system the endogenous genetic variation can indeed be captured through the recovery of diploid homozygous plants after doubling the chromosome number with colchicine treatments (Yao et al. 1997). The second process for generating embryos is somatic embryogenesis in which embryos arise from somatic cells, that is cells other than gametes. Somatic embryos are very similar in morphology and physiology to their zygotic counterparts and therefore represent a very suitable system for embryogenic studies. This technique was developed more than 60 years ago (Levine 1947) and since then there have been tremendous advancements in the design of optimized media and culture conditions (reviewed by Thorpe and Stasolla 2001).

Studies on the role played by glutathione and ascorbate during in vitro embryogenesis have mainly been conducted in two systems: *Brassica napus* (canola), an angiosperm species propagated via androgenesis, and *Picea glauca* (white spruce) a conifer able to produce somatic embryos. Protocols for generating MDEs in canola require the isolation of microspores from young buds and a heat shock treatment needed for re-routing their gametophytic pathway into an embryogenic fate. Embryo growth is achieved in a medium devoid of plant growth regulators and many fully developed cotyledonary embryos can be generated after 4 weeks in culture (Fig. 1a). Such embryos can then be transferred onto germination medium for conversion, i.e. formation of a fully functional shoot and root system. A detailed description of the process with the structural events associated to the different phases of MDE development can be found in Yeung (2002).

Somatic embryogenesis in spruce is executed using specific media and unique culture conditions. The process, summarized in Fig. 1b can be divided into five steps; namely induction, maintenance, development, desiccation, and conversion. During the induction phase immature zygotic embryos are cultured with high levels

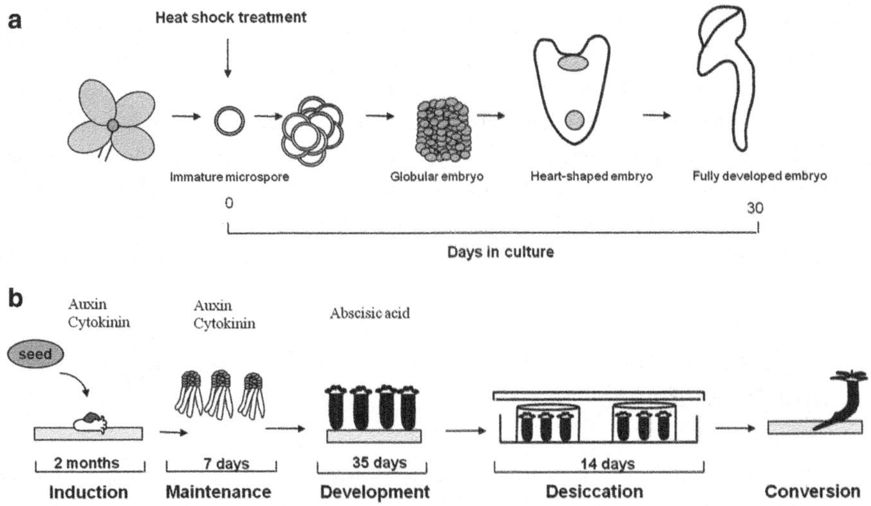

Fig. 1 Schematic representation of the *Brassica napus* microspore-derived embryogenesis system (**a**) and the spruce (*Picea glauca*) somatic embryogenic system (**b**). Production of *Brassica napus* microspore-derived embryos (MDEs) is initiated from immature microspores collected from developing buds. Under appropriate culture conditions (see text for details) the microspores divide and form embryos similar in morphology to their zygotic counterparts. At the end of the process, which lasts 30 days, fully developed MDEs, characterized by two extended cotyledons are produced. These embryos can germinate and convert into viable plants without a desiccation period (**a**) In spruce, embryogenic tissue is induced from zygotic embryos and proliferated in a maintenance medium. Proliferation of the tissue is perpetuated by replacing fresh maintenance medium every 7 days. Embryonic development is carried out on solid medium and culminates with the production of cotyledonary embryos after 35 days in culture. Fully developed somatic embryos are partially dried for 14 days and then placed on germination medium where conversion occurs (see text for further details). Hormonal requirements for each step are also illustrated (**b**)

of auxins and cytokinins to produce embryogenic tissue which is a translucent mass of cells including immature filamentous-shaped somatic embryos. Embryogenic tissue is then proliferated on a maintenance medium which also contains auxins and cytokinins (albeit at lower concentrations than those used in the induction step) and sub-cultured every 7–10 days. Development of somatic embryos is achieved in the presence of abscisic acid (ABA) and lasts 3–4 weeks. At the end of the development period fully grown embryos characterized by a ring of cotyledons surrounding the apical pole are produced. Such embryos must experience desiccation prior to germination. In spruce the desiccation process is performed by incubating the embryos for 14 days in a relatively high humidity environment where they undergo a moderate water loss (about 20%). In this step embryos are placed in the central wells of a 12-multi well plate in which the outer wells are filled with sterile water. Following the drying period the somatic embryos are transferred onto a hormone-free germination medium where post-embryonic growth is encouraged. The percentage of embryos able to generate viable plants is strictly genotype dependent. A complete description of the process is available in Stasolla et al. (2003).

3 The Glutathione Redox State: Metabolism and Modulation of Growth and Development

Glutathione is a ubiquitous thiol which is commonly found in a large variety of organisms, including plants, animals and fungi. Its de novo synthesis occurs through two well characterized steps. The first, involving the formation of γ-glutamylcysteine from glutamate and cysteine, is catalyzed by the enzyme γ-glutamylcysteine synthatase. In the second step γ-glutamylcysteine is further converted to glutathione with glutathione synthase catalyzing the reaction (Noctor and Foyer 1998). Given the importance of glutathione for many cellular processes it is not surprising that the de novo synthesis is regulated by feedback mechanisms preventing its overproduction (May et al. 1998). As reviewed by Potters et al. (2002), glutathione concentration in chloroplasts is estimated around 1–4.5 mM with no information available for the other cellular compartments. Many studies on the enzymes involved in glutathione biosynthesis can be found in bacteria, whereas little information is currently available in plants (May and Leaver 1993, 1994; Rawlins et al. 1995). The glutathione pool consists of a reduced (GSH) and an oxidized (GSSG) form, and their respective proportions can be modulated by the enzyme glutathione reductase which reduces GSSG to GSH. Besides having a direct effect on several metabolic processes, this inter-conversion is also responsible for affecting the ascorbate pool since GSH is the substrate of a reaction reducing dehydroascorbate to ascorbate. In view of this intimate relationship, glutathione metabolism is tightly linked to the recycling of ascorbate in several fundamental processes (reviewed by Potters et al. 2002). Other enzymes, including thioltransferase and protein disulphide isomerases are also affected by a change in the relative amount of GSH and GSSG within the glutathione pool (Noctor et al. 1998). Fluctuations in glutathione metabolism are common during plant development and in response to stress conditions. Several studies have documented an increase in the overall glutathione pool in response to heavy metals and exposure to high levels of ozone (May and Leaver 1993; Chen and Goldsborough 1994; Luwe 1996). Besides an increase in the total glutathione pool, changes in the GSH/GSSG ratio also occur under sub-optimal growth conditions. Based on this observation, Noctor et al. (1998) proposed that alterations in the GSH/GSSG ratio are more important than the changes in the overall glutathione pool during stress responses.

In pea tissue, there is a very close relationship between cellular proliferation and glutathione levels with high GSH concentrations found in meristematic tissue characterized by rapid growth (Bielawski and Joy 1986). The involvement of GSH in cell division processes was further explored by Sanchez-Fernandez et al. (1997) who showed that while applications of exogenous GSH stimulated *Arabidopsis* root growth, a depletion of this metabolite, affected by applications of L-buthione sulfoximine (BSO) a specific inhibitor of GSH synthesis, prevented cell proliferation. The effects of glutathione on root development, especially in relation to the formation and maintenance of the quiescent center were further investigated by Jiang et al. (2004) and Jiang and Feldman (2005)). These studies showed that quiescent cells are kept in a non-proliferative state (they are arrested in the G1 phase of the cell cycle) by the presence of an oxidized glutathione and ascorbate environment created by the basipetal flow of auxin. Experimental perturbations of this oxidized environment affect cell division and ultimately tissue patterning within the root.

In a simplistic view it appears that while GSH favors cell division and proliferation through a direct involvement with the cell cycle machinery, GSSG levels might be related to differentiation processes. Using one of the best characterized experimental systems to study cell differentiation Henmi et al. (2001) showed that when applied early in culture GSSG promotes the trans-differentiation of *Zinnia elegans* mesophyll cells into tracheary elements. The authors further speculated that this effect was primarily due to a change in the GSH/GSSG ratio rather than to an absolute increase in GSSG levels. Studies on glutathione metabolism have been extended to many systems and several developmental phases of plant development, from seed maturation to flowering. The importance of high GSSG levels and an oxidized glutathione environment during seed development was outlined by De Gara et al. (2003). The authors showed that kernel maturation is characterized by a slow, but steady increment in GSSG concentration which slows down the overall seed metabolism and ensures normal cell differentiation and tissue patterning. Flowering is also regulated by glutathione. By conducting comparative studies between wild type plants and the late flowering mutant *fca-1*, Ogawa et al. (2001) reported a correlation between flowering and biosynthesis of GSH generated from the stress-induced activation of γ-glutamylcysteine synthetase. These studies suggest that manipulations of cellular glutathione can be exploited in vitro to affect morphogenesis and ultimately improve plant propagation. The next sections will provide examples on how changes in glutathione metabolism modulate growth and development in culture.

4 Glutathione Affects In vitro Embryogenesis and Plant Regeneration

The notion that morphogenetic events in culture can be related to alterations in glutathione metabolism was proposed more than half a century ago by Marre and Arrigoni (1957) who demonstrated that the effect of auxin on tissue growth is modulated by glutathione. The auxin-promotive effect on proliferation in different tissue types correlated to high levels of GSH relative to GSSG (high GSH/GSSG ratio). An opposite change in this ratio, due to increasing levels of GSSG, was observed when growth was experimentally inhibited. A similar relationship was also reported during carrot somatic embryogenesis by Earnshaw and Johnson (1985, 1987). In this system the 2,4-D induced proliferation of the embryogenic tissue was associated to a high GSH concentration. During the following phases of embryo development, achieved on a medium devoid of auxins, GSSG slowly increased resulting in a low GSH/GSSG ratio. The notion that a removal of GSH acts as a trigger to promote embryo development was further explored by the authors who demonstrated that applications of BSO, which inhibits GSH synthesis, improves embryonic yield. Using Norway spruce (*Picea abies*) as a model, Jain et al. (1988) showed that a depletion of GSH, affected by BSO, promoted the formation of "embryogenic calli". From this study, however, several questions remained unanswered. Since this experiment was only conducted during the induction phase of the embryogenic process there was no indication on whether BSO can improve the embryogenic potential of individual cell lines. In addition no information was available on the effect of BSO on embryo quality, which is the ability to generate embryos which are morphologically and physiologically ready to convert at germination. Further work on a different species of spruce, white spruce (*Picea glauca*), demonstrated that early embryogeny is favored by a reduced glutathione environment (high GSH/GSSG ratio) whereas embryo development requires an oxidized glutathione environment (low GSH/GSSG ratio) (Belmonte et al. 2005a). Based on these observations a new protocol was developed whereby embryogenic tissue was initially exposed to GSH, which increased the number of immature embryos protruding from the tissue, and then to GSSG to promote the proper development of the embryos. Using this approach, Belmonte et al. (2005a) were able to double the number of white spruce somatic embryos produced and increase their conversion frequency from 22% to almost 70%. This enhanced conversion frequency reflects an improvement in embryo quality. Physiological studies revealed that a threshold of endogenous GSH is required to sustain the proliferation of the embryogenic tissue in spruce. As reported by Belmonte et al. (2003), applications of GSH in the maintenance medium (Fig. 1b) promoted mitotic activity and increased the growth of spruce embryogenic tissue by 25%. Furthermore, in control cultures the progression from the exponential growth to the stationary phase was accompanied by a decrease in endogenous GSH. The requirement of GSH in support of cell proliferation is not a novel concept since it was documented in a variety of plant and animal systems (reviewed by Potters et al. 2002).

In spruce, the imposition of an oxidized environment, effected by either applications of GSSG or BSO is needed to promote proper embryo development. Belmonte et al. (2005a) showed that when GSSG is applied to the development medium (Fig. 1b) the GSH/GSSG ratio declined from 17 to 8 and this resulted in an increased number of "good quality" embryos produced in culture. Such embryos had four or more cotyledons and displayed an improved post-embryonic performance. The beneficial effect of an oxidized glutathione environment is not unique to spruce as it was also demonstrated during microspore-derived embryogenesis in *Brassica napus* (Fig. 1a). The conversion and post-embryonic performance of MDEs treated with BSO increased more than three times (Belmonte et al. 2006). These improvements were only due to the imposed oxidized environment as they were abolished if GSH was added together with BSO. Several reasons can be accountable for the beneficial role of the oxidized glutathione environment during spruce embryogenesis. Firstly, as reported in other systems (Henmi et al. 2001) GSSG is involved in programmed cell death, which is an important obligatory event occurring during spruce embryo development. It is through marked changes in structure involving the dismantling and the elimination of specific cell types that immature somatic embryos develop from the embryogenic tissue (Filonova et al. 2000). The requirement of programmed cell death for embryo development was further demonstrated by Bozhkov et al. (2002) who induced an arrest in the embryonic developmental program when cell death was experimentally eliminated. Another effect of GSSG might be that to inhibit the cell cycle progression (de Pinto et al. 1999) thereby reducing the proliferation of the embryogenic tissue in preparation for embryo growth and differentiation. Based on the above observations it is predicted that the imposed oxidized environment results in profound structural, physiological, and molecular changes, which all contribute to the improved embryonic performance.

5 Structural, Physiological and Molecular Changes Induced by Glutathione

Structural studies documenting the ontogeny and development of spruce embryos are currently available (reviewed by Stasolla et al. 2003). Unlike non-embryogenic tissue, which is dark and compact, spruce embryogenic tissue is recognizable by its translucent mass of immature, filament-shaped embryos. Such embryos are composed by two different types of cells: small cytoplasmic cells composing the embryo proper and long, vacuolated cells forming the suspensor region (Fig. 1b). Cleavage polyembryony, the proliferation and separation of the immature embryos, is the mechanism responsible for the growth and perpetuation of the embryogenic tissue (see Stasolla et al. 2003). This process is accelerated in the presence of exogenous GSH, possibly by the ability of this metabolite to promote the synthesis of purine and pyrimidine nucleotides (Belmonte et al. 2003, 2005b), which are important building blocks of nucleic acids and key intermediates of bio-energetic reactions (Stasolla and Thorpe 2004). A large nucleotide pool is required to sustain rapid cell proliferation and tissue growth. In situ tracer experiments revealed that GSH increased the rate of pyrimidine nucleotide synthesis and reduced their catabolism. When GSH was added to the maintenance medium (Fig. 1b), the activity of orotate phoshoribosyltransferase, the enzyme responsible for the conversion of orotic acid (a precursor of the de novo pathway) into UMP increased rapidly during the first 4 days. This resulted in increasing levels of UMP in the embryogenic tissue. These changes were also accompanied by a reduction in pyrimidine nucleotide degradation, as estimated by the decreased rate of uracil catabolism (Belmonte et al. 2005b). The metabolism of purine nucleotides was also affected by GSH applications. Inclusions of GSH in the maintenance medium resulted in an increased uptake of adenine and adenosine, intermediates of the salvage pathway, and increased conversion of inosine to ATP (Belmonte et al. 2003). Based on these observations it was suggested that the imposition of a reduced glutathione environment (high GSH/GSSG ratio) accelerates tissue proliferation in spruce by promoting the synthesis of pyrimidine and purine nucleotides via de novo and salvage pathways.

If GSH is required to improve the number of immature embryos formed, the imposition of an oxidized environment, effected by either GSSG or BSO applications in the development medium (Fig. 1b) is crucial for proper development. One of the major problems encountered by tissue culturists is the formation of embryos with abnormal shoot apical meristems (SAMs). Control spruce embryos often develop poorly organized SAMs which are disrupted by the presence of intercellular spaces. The meristematic cells within these meristems show signs of differentiation, as estimated by the formation of large vacuoles and the accumulation of storage products (Stasolla and Yeung 2003; Thair and Stasolla 2005). Abnormal meristems fail to reactivate at germination resulting in poor regeneration and in many instances embryo abortion. Similar problems were also encountered in *Brassica napus* MDEs suggesting that meristem formation in vitro occurs under sub-optimal conditions. The observation that SAMs can deteriorate in culture suggests that their fate is not "determined" but it can be altered by changing the culture conditions (reviewed by Thair and Stasolla 2005). This is in contrast to SAMs produced in vivo which are stable entities. Independent in vivo and in vitro experiments confirm this notion (Raghavan 1976; Ramesar-Fortner and Yeung 2001). A significant improvement in SAM architecture can be achieved in the presence of an oxidized glutathione environment. In Brassica, MDEs cultured in the presence of GSSG or BSO (both compounds lower the GSH/GSSG ratio) exhibit meristems with a "zygotic-like" architecture. These meristems are devoid of intercellular spaces, have a characteristic dome-shaped appearance, and are composed of small meristematic cells (Belmonte et al. 2005a). Culture experiments have further demonstrated that these improvements are only the result of the imposed oxidized environment since applications of GSH produce an opposite effect (Belmonte et al. 2005a).

In light of the structural and morphological improvements described above it is therefore suggested that the imposition of an oxidized glutathione environment elicits major responses at a molecular levels. This concept was tested by conducting cDNA microarray analyses during Brassica MDE development. These experiments revealed that out of a total of 15,000 elements (clones) analyzed, only 18 were differentially expressed during the progression from globular to heart-shaped embryos (Stasolla et al. 2008). In particular, the transcription of two genes, ARGONAUTE 1 and CLAVATA 3 ESR-related 27 (CLE 27) were induced in the presence of BSO. Several studies in *Arabidopsis* have characterized ARGONAUTE 1, which is a member of the ARGONAUTE family containing by two conserved domains: the PAZ and PIWI which are involved in protein–protein interaction (Cerrutti et al. 2000). Genetic analyses revealed the involvement of ARGONAUTE 1 in SAM formation and maintenance, since a mutation of this gene produces in pin-shaped SAMs (Lynn et al. 1999). An important role of ARGONAUTE 1 is that to ensure the proper accumulation and localization of SHOOTMERISTEMLESS and the specification of CUPSHAPED COTYLEDONS 1 and 2 boundaries. The key role played by these genes during SAM formation has been reviewed by Thair and Stasolla (2005)). The other gene, i.e. CLE 27, up-regulated by applications of BSO belongs to a large gene family which in Arabidopsis is involved in the regulation of meristem size (Fiers et al. 2005; Ito et al. 2006).

During the middle and late phases of Brassica MDE development, two additional meristem-specific genes are induced by BSO: ZWILLE and SHOOOTMERISTEMLESS. Studies conducted in Arabidopsis showed that ZWILLE is responsible for maintaining the central cluster of cells within the meristem in an undifferentiated state (Moussian et al. 1998). Embryos with abnormal SAMs in which the meristematic cells differentiate precociously were produced by *zwille* mutant plants. As revealed by Stasolla et al. (2008) a similar phenotype was observed in control *Brassica napus* MDEs which developed in a reduced glutathione environment. This observation reinforces the notion that the cellular redox state affects the expression of this gene. An additional role for ZWILLE is that to specify the expression pattern of SHOOTMERISTEMLESS, which is also induced by BSO (Stasolla et al. 2008). SHOOTMERISTEMLESS plays a key role during SAM maintenance as it regulates cellular homeostasis ensuring a balance between cell division and differentiation (reviewed by Thair and Stasolla 2005). Localization studies during *Brassica* MDE development have further confirmed that in the presence of BSO both ZWILLE and SHOOTMERISTEMLESS have zygotic-like expression patterns. This is in contrast to control embryos and embryos grown in a reduced environment (created by exogenous applications of GSH) (M. Elhiti and C. Stasolla, unpublished data).

Molecular work conducted during spruce somatic embryogenesis also reveals the existence of genes induced by the imposition of an oxidized glutathione environment. The expression pattern of HBK1, a homeobox of the KNOTTED class which might exercise functions similar to those ascribed to SHOOTMERISTEMLESS (Sundås-Larsson et al. 1998), increased in embryos treated with GSSG where it encompassed the whole apical region. This was in contrast to control embryos where HBK1 localization was restricted to a few cells (Belmonte et al. 2005a). Unfortunately molecular studies on conifers, include spruce, are very scarce due to the lack of genetic information available for these species. However from our limited evidence it appears that similar mechanisms induced by an oxidized glutathione environment and related to SAM function might exist across species. In particular a switch of the glutathione pool towards an oxidized state might be required for the proper localization of meristem-marker genes leading to normal SAM formation and reactivation at germination. These conditions are not met in control embryos with deleterious consequences on embryo quality.

From the microarray experiments conducted during *Brassica* MDE development it emerges that the overall antioxidant machinery is also affected by the imposed oxidized glutathione environment. This is not surprising given the involvement of GSH with AsA metabolism (Potters et al. 2002). The expression and activity of ascorbate peroxidase, a key enzyme participating in detoxification processes, increased in BSO-treated MDEs. Besides its role in scavenging hydrogen peroxide produced during period of intense growth, such as embryo development, ascorbate peroxidase is require to maintain cell wall plasticity through the inhibition of peroxidases which crosslink wall components (De Gara et al. 1996). A reduction in cell wall cross-linkage was related to a better ability of SAMs to reactivate at germination (Stasolla and Yeung 2007). Based on the above it appears that alterations in cell wall plasticity, affected by a switch of the glutathione pool towards an oxidized state, might create favorable conditions for optimal meristem function. Another key enzyme induced by BSO treatments during the late stages of MDE development is monodehydroascorbate, which is responsible for the reduction of monodehydroascorbate to ascorbate. The increased activity of this enzyme correlated to high levels of its substrate, monodehydroascorbate (Stasolla et al. 2008). As suggested by the authors these metabolic alterations would ensure a prompt reduction of monodehydroascorbate to ascorbate at the onset of germination, prior to the reactivation of the ascorbate de novo machinery. This strategy, documented in both angiosperm and gymnosperm species (De Gara et al. 1997; Stasolla and Yeung 2003) would ensure available ascorbate in support of cell division and meristem reactivation at germination (discussed above). Therefore it is predicted that compared to their control counterparts, embryos cultured in an oxidized glutathione environment have higher ascorbate levels at the onset of germination.

The activity of four different glutathione S-transferase also increased during the early cotyledonary stage of *Brassica* MDEs cultured with BSO (Stasolla et al. 2008). Several studies have shown the presence of these proteins in cells and tissues undergoing rapid growth where they participate in detoxification processes by conjugating glutathione to a variety of substrates, such as reactive oxygen species. During embryonic and fetal growth in animal systems, for examples, high glutathione S-transferase activity is required to lower the content of toxic organic hydroperoxides, thereby ensuring proper development. An increase of several glutathione S-transferase transcripts was also reported in several culture systems, such as barely androgenesis (Maraschin et al. 2006).

Synthesis and degradation of plant growth regulators in culture affect morphogenesis, including embryogenesis. As shown in Fig. 1b, applications of abscisic acid (ABA) are required in the production of spruce somatic embryos. The function of this phytohormone during embryogenesis is multifaceted, ranging from the acquisition of desiccation tolerance and inhibition of precocious germination to the accumulation of storage products and the execution of several histodifferentiation events (reviewed by Stasolla et al. 2003). Hays (1996) showed that inhibition of ABA biosynthesis compromises embryo production and results in several morphological abnormalities. When *Brassica napus* MDEs were cultured in the presence of an oxidized glutathione environment, effected by BSO applications, the biosynthesis of ABA, its degradation to phaseic and dehydrophaseic acid via the 8'hydroxylation pathway, as well as its conjugation to sugars increased (Belmonte et al. 2006). These changes were the result of a faster ABA turnover suggesting a better utilization of this phytohormone in the presence of BSO. In the same study it was also shown that applications of ABA induced the same structural and morphological changes observed in embryos cultured with BSO and improved the overall embryo quality and conversion frequency. This observation, together with the BSO-induced up-regulation of several ABA-responsive proteins confirms the intimate relationship between glutathione redox state and ABA metabolism.

Ethylene is another growth regulator involved in stress responses which, due to its gaseous nature, tends to accumulate in culture affecting morphogenesis (Gaspar et al. 1996). Experimental systems altering the synthesis of this hormone have shown that over-production of ethylene results in the separation of the meristematic cells within the S-adenosylmethionine (SAM), thereby compromising the reactivation of the meristems at germination. It is suggested that the improved SAM architecture observed in embryos cultured in an oxidized glutathione environment is the result of a reduced synthesis of ethylene. Using white spruce somatic embryos Belmonte et al. (2005a) showed that compared to control conditions, where ethylene biosynthesis increased more than six times between day 10 and 20 in culture, no increase in ethylene level was observed when the glutathione pool was switched towards an oxidized state by the addition of exogenous GSSG. Regulation of ethylene production by glutathione might occur through the transcriptional regulation of the two biosynthetic enzymes: S-adenosylmethionine synthase, which converts S-adenosylmethionine to 1-aminocyclopropane-1-carboxylase (ACC) , and ACC oxidase, which generates ethylene from ACC. Molecular studies showed that the ACC oxidase transcripts increased in spruce somatic embryos grown in a reduced (GSH) environment (Stasolla et al. 2004) whereas S-adenosylmethionine synthase was repressed in *Brassica napus* MDEs cultured in an oxidized environment (Stasolla et al. 2008). These studies confirm the notion that a switch of the glutathione pool towards an oxidized state prevents the deterioration of the meristems by down-regulating the biosynthesis of ethylene and preventing its accumulation in the culture vessels.

Another important aspect associated to the imposition of an oxidized glutathione environment is the increased deposition of storage products. Accumulation of proteins, including late embryogenesis abundant (LEA) proteins is an important event occurring during the late phases of embryo development. In general the accumulation of storage products in vitro differs significantly from that in vivo (Stasolla et al. 2003). Histological analyses conducted by Belmonte et al. (2005a) revealed that during spruce somatic embryogenesis there are remarkable differences in the accumulation pattern of storage products between control embryos and embryos cultured in an oxidized glutathione environment. Compared to control embryos, where starch was the main storage product, spruce embryos treated with GSSG accumulated starch first, followed by proteins and lipid bodies. The authors showed that this pattern is similar to that occurring in zygotic embryos and suggested that GSSG-treated embryos undergo a "normal" maturation process which ensures successful regeneration. Similar improvements in storage deposition patterns to those described for spruce where also observed in Brassica MDEs cultured in the presence of BSO, which also switches the glutathione pool towards an oxidized state (Stasolla et al. 2008). The promotive effect of the oxidized glutathione environment on the accumulation of storage products might be both direct and/or indirect. A direct effect was proposed by De Gara et al. (2003) who showed that the creation of an oxidized environment stimulates the execution of post-translational mechanisms involved in the assembling of newly synthesized proteins. An indirect effect might be exercised through the stimulation of the synthesis of ABA (discussed above), since ABA has been shown to promote the accumulation of storage products during embryogenesis (Kermode 1995). The limited accumulation of storage products observed under control conditions, indicates that control embryos are morphologically competent to germinate and convert, but are not physiologically ready to carry on the process. Both morphological and physiological maturation is achieved with the imposition of an oxidized environment.

6 The Ascorbate Redox State: Metabolism and Modulation of Embryo Conversion

As indicated in the Section 5, ascorbate metabolism is tightly linked to glutathione metabolism through the Halliwell–Asada cycle which involves the common utilization of basic enzymes (see Potters et al. 2002). All plant cells so far investigated have the ability to produce ascorbate using a pathway proceeding from fructose 6-phosphate to l-galactono-1,4-lactone, which is the last precursor of ascorbate. Several enzymes in this pathway have been cloned and the respective proteins purified (see De Tullio and Arrigoni 2006 for a comprehensive review on ascorbate biosynthesis and utilization). As for the glutathione system, the reduced form of ascorbate (AsA) can be oxidized to form ascorbic acid free radical (AFR), also known as monodehydroascorbate, and dehydroascorbate (DHA). Enzymes responsible for this oxidation are ascorbate oxidase, ascorbate peroxidase, and ascorbate-dependent dioxygenases. The oxidized forms, AFR and DHA can be converted back to AsA by the respective NAD(P)H-dependent AFR reductase and the glutathione-dependent DHA reductase. DHA can also be catabolized to diketogulonic acid which is unable to produce AsA (Arrigoni and De Tullio 2002). Besides providing an efficient mechanism to increase the endogenous AsA pool, reduction of AFR and DHA to AsA might have the alternative function of depleting DHA which has been shown to inhibit the activity of several important enzymes (Fiorani et al. 2000).

Since AsA is a key metabolite required for critical processes of plant growth, it is not surprising that cells are able to regulate its endogenous content according to their metabolic requirements. It has been demonstrated for example, that AsA concentration varies during cell division and differentiation, with high levels required to support mitotic activity. In animal systems cancerous tissues of lung and larynx accumulate large quantity of AsA (Piyathilake et al. 2000). Supplementations of AsA have been found to amplify growth in chemically induced bladder (Fukushima et al. 1988), and forestomach tumors (Shibata et al. 1992). In plants, high levels of endogenous AsA correlate to period of intense growth, including embryo development (Arrigoni et al. 1992) and germination (De Gara et al. 1997; Stasolla and Yeung 2001). De Gara and Tommasi (1999) further demonstrated that in pea stems the highest levels of endogenous AsA were measured in the shoot apex, characterized by rapid cell division. The importance of ASC during growth has also been emphasized by other studies. Experimental alterations of the cellular AsA content, in fact, have been found to affect the rate of cell division in many systems, including tobacco cultured cells (de Pinto et al. 1999), corn and onion root cells (Kerk and Feldman 1995). In onion roots, for example, a decrease in the AsA level caused by addition of lycorine, an inhibitor of the AsA biosynthetic pathway, resulted in a profound inhibition of cell division. This effect was however reversed after the concentration of AsA in the tissue was increased (Kerk and Feldman 1995). Possible mechanisms relating AsA to cell cycle mechanisms are reviewed by Potters et al. (2002). During the cell cycle cells must cross two important checkpoints: one at the G1-S phase and the other at the G2-S phase. These transitions might be facilitated by AsA through two different events. The enzyme prolyl hydroxylase is required for the posttranslational hydroxylation of proline residues. Hydroxyproline-reach proteins are needed in the progression of the cell cycle as under-hydroxylation of prolines causes a block during metaphase of onion cells (De Tullio et al. 1999). Of interest AsA is a cofactor of prolyl hydroxylase and therefore its availability is associated to the execution of these events. A second hypothesis argues for the involvement of AsA in releasing iron from ferritin to be used for the synthesis of ribonucleotide reductase, the enzyme responsible for the production of ribonucleotides. A high activity of this enzyme would be required for the entry into the S phase of the cell cycle (see Potters et al 2002 for a comprehensive review).

Given the involvement of AsA in cell division processes several in vitro studies have investigated its requirement for inducing meristem reactivation during the early phases of germination. Using the spruce somatic embryogenesis system (Fig. 1b), Stasolla and Yeung (2006a) demonstrated that a shift of the ascorbate pool towards a reduced state (high AsA/DHA + AFR ratio) can rescue poorly organized SAMs , characterized by the presence of intercellular spaces. Applications of exogenous AsA in the germination medium stimulated cell division and meristemoid formation within the disrupted SAMs. Continual cell division and differentiation of the meristemoids resulted in the generation of new shoots through a pattern similar to shoot organogenesis described in coniferous species (Yeung et al. 1981; Stasolla et al. 2007). The effects of AsA applications on SAM reactivation may be the result of different processes involving cellular peroxidases and nucleotide metabolism.

At the onset of germination, with the resumption of mitotic activity in the SAM, a switch of cellular ascorbate pool towards its reduced form (AsA) reduced the activity of major cellular peroxidases, i.e. guaiacol and ferulic acid, which are implicated in the cross-linkage of cell wall polymers (Stasolla and Yeung 2007). A reduction in the activity of these enzymes induced the relaxation of wall components in the meristematic cells thereby promoting their reactivation. This would eventually lead to the formation of a functional shoot. In the same study it was also demonstrated that treatments which reduce the endogenous AsA level have opposite effects resulting in high peroxidase activity and reduced SAM conversion. The requirement for high AsA levels during meristematic cell reactivation is not a novel concept and it was indeed documented by others. Corn cells in the quiescent center of root apical meristems contain low levels of AsA compared to their mitotically active derivatives (Kerk and Feldman, 1995).

Besides altering wall-bound peroxidase activity, cellular ascorbate is also required for increasing the production of pyrimidine nucleotides in the meristematic cells of SAM. Pyrimidine nucleotides, required during period of intense cell proliferation, can be produced via a salvage mechanism which utilizes bases, including thymidine and uridine as precursors (Stasolla and Thorpe 2004). Metabolic studies conducted on dissected apical segments of germinating spruce somatic embryos show that utilization of thymidine and uridine for nucleotides (TMP + TDP + TTP and UMP + UDP + UTP), as well as for nucleic acid synthesis, is under the control of cellular AsA. High levels of endogenous AsA increased the salvage activity of both thymidine and uridine which are readily converted to nucleotides and nucleic acids. An opposite tendency was observed when the endogenous AsA levels were experimentally reduced with lycorine, which inhibits the de-novo AsA synthesis. Apical cells of lycorine-treated embryos were not able to anabolize pryrimidine precursors which were either non-metabolized or degraded. The increased utilization of thymidine and uridine for nucleic acids and nucleotides in the apical poles of the embryos is possibly due to the activities of their respective salvage enzymes, thymidine kinase and uridine kinase, which also increased in the presence of high AsA levels. In-situ autoradiographic studies performed with ^{3}H-labeled thymidine further corroborated the notion that DNA synthesis in the cells of SAM is under the control of cellular ascorbate (Stasolla and Yeung 2006b).

Based on the above information it is therefore clear that the imposition of a reduced ascorbate environment (high AsA/DHA + AFR) ratio is a requirement at the onset of germination where it ensures the reactivation of meristematic cells and guarantees proper SAM conversion. This strategy can indeed be used to reactivate and convert those abnormal meristems developed under sub-optimal conditions.

7 Concluding Remarks

The cellular redox state plays an important role during plant growth and development by modulating several cellular and metabolic processes. A key factor affecting the cellular redox environment is the interaction of two sets of redox couples: the glutathione disulfide–glutathione (GSH/GSSG) couple and the reduced (AsA)-oxidized (DHA + AFR) ascorbate couple. In vitro studies have revealed that manipulations of glutathione and ascorbate affect morphogenesis including embryonic and post-embryonic development. During embryogenesis a slow switch of the glutathione pool towards an oxidized state (low GSH/GSSG ratio) improves the quality of the embryos by promoting a zygotic-like histodifferentiation pattern and producing well organized meristems. These structural features, which are induced by a variety of molecular and metabolic changes would improve the quality of the embryos and ensure post-embryonic growth. The imposition of a reduced ascorbate environment (high AsA/DHA + AFR ratio) can help the reactivation of meristems at germination. This approach can be used to promote meristematic activity and meristemoid organization in those abnormal meristems developed under sub-optimal conditions. Taken together this strategy, i.e. imposition of an oxidized glutathione redox state during development and a reduced ascorbate redox state at germination, can be used to improve propagation of recalcitrant species thereby enhancing embryo yield and quality in culture. The improvement in embryo production and conversion documented in *Brassica napus* (an angiosperm) and *Picea glauca* (a conifer) indicates the presence of a similar ascorbate/glutathione regulation on embryogenesis operating across species.

References

Arrigoni O, De Gara L, Tommasi F, Liso R (1992) Changes in the ascorbate system during seed development in *Vicia faba* L. Plant Physiol 99:235–238

Arrigoni O, and De Tullio MC (2002) Ascorbic acid: much more than just an antioxidant. Biochim Biophys Acta 1569:1–9

Belmonte MF, Stasolla C, Loukanina N, Yeung EC, Thorpe TA (2003) Glutathione modulation of purine metabolism in cultured white spruce embryogenic tissue. Plant Sci 165:1377–1385

Belmonte MF, Donald G, Reid DM, Yeung EC, Stasolla C (2005a) Alterations of the glutathione redox state improve apical meristem structure and somatic embryo quality in white spruce (*Picea glauca*). J Exp Bot 56:2355–2364

Belmonte MF, Stasolla C, Katahira R, Loukanina N, Yeung EC, Thorpe TA (2005b) Glutathione-induced growth of embryogenic tissue of white spruce correlates with changes in pyrimidine nucleotide metabolism. Plant Sci 168:803–812

Belmonte MF, Ambrose SJ, Ross ARS, Abrams S, Stasolla C (2006) Improved development of microspore-derived embryo cultures of *B. napus* cv Topas following changes in glutathione metabolism. Physiol Plant 127:690–700

Bielawski W, Joy KW (1986) Reduced and oxidized glutathione and glutathione reductase activity in tissues of *Pisum sativum*. Planta 169:267–272

Bozhkov PV, Filonova LH, von Arnold S (2002) A key developmental switch during Norway spruce somatic embryogenesis is induced by withdrawal of growth regulators and is associated with cell death and extracellular acidification. Biotech Bioeng 77:658–667

Buchanan BB (1980) Role of light in the regulation of chloroplast enzymes. Annu Rev Plant Physiol 31:341–374

Cerrutti L, Mian N, Bateman A (2000) Domains in gene silencing and cell differentiation proteins: the novel PAZ domain and redefinition of the PIWI domain. Trends Biochem Sci 25:481–482

Chen J, Goldsborough PB (1994) Increased activity of glutamylcysteine synthetase in tomato cells selected for cadmium tolerance. Plant Physiol 10:233–239

De Gara L, de Pinto M, Paciolla A, Cappetti V, Arrigoni O (1996) Is ascorbate peroxidase only a scavenger of hydrogen peroxide? In: Obinger C, Burner U, Pennel C, Greppin H (eds) Plant peroxidases: biochemistry and physiology. University of Geneva, Switzerland, pp 157–162

De Gara L, de Pinto MC, Arrigoni O (1997) Ascorbate synthesis and ascorbate peroxidase activity during the early stage of wheat germination. Physiol Plant 100:894–900

De Gara L, Tommasi F (1999) Ascorbate redox enzymes: a network of reactions involved in plant development. Recent Res Dev Phytochem 3:1–5

De Gara L, de Pinto M, Moliterni VM, D'Egidio MG (2003) Redox regulation and storage processes during maturation in kernels of *Triticum durum*. J Exp Bot 54:249–258

de Pinto M, Francis D, De Gara L (1999) The redox state of the ascorbate-dehydroascorbate pair as a specific sensor of cell division in tobacco BY2 cells. Protoplasma 209:90–97

De Tullio MC, Paciolla C, Dalla Vecchia F, Rascio N, D'Emerico S, De Gara L, Liso R, Arrigoni O (1999) Changes in onion root development induced by the inhibition of peptidyl-prolyl hydroxylase and influence of the ascorbate system on cell division and elongation. Planta 21:424–434

De Tullio MC, Arrigoni O (2006) The ascorbic acid system in seeds: to protect and serve. Seed Sci Res 13:249–260

Earnshaw BA, Johnson MA (1985) The effect of glutathione on development of white carrot suspension cultures. Biochem Biophys Res Comm 133:988–993

Earnshaw BA, Johnson MA (1987) Control of wild carrot somatic embryo development by antioxidants. Plant Physiol 85:273–276

Fiers M, Golemiec E, Xu J, van der Geest L, Heidstra R, Stiekema W, Liu CM (2005) The 14-amino acid CLV3, CLE19, and CLE40 peptides trigger consumption of the root meristem in Arabidopsis through a CLAVATA2-dependent pathway. Plant Cell 17:2542–2553

Filonova LH, Bozhkov PV, Brukhin VB, Daniel G, Zhivotovsky B, von Arnold S (2000) Two waves of programmed cell death occur during formation and development of somatic embryos in the gymnosperm Norway spruce. J Cell Sci 113:4399–4411

Fiorani M, De Sanctis R, Scarlatti F, Vallorani L, De Bellis R, Serafini G, Bianchi M, Stocchi V (2000) Dehydroascorbic acid irreversibly inhibits hexokinase activity. Mol Cell Biochem 209:145–153

Fukushima S, Imaida K, Shibat MA, Tamano T, Kurata Y, Shirai T (1988) L-ascorbic acid amplification of second-stage bladder carcinogenesis promotion by $NaHCO_3$. Cancer Res 48:6317–6320

Gaspar T, Kevers C, Penel C, Greppin H, Reid DM, Thorpe TA (1996) Plant hormones and plant growth regulators in plant tissue culture. In Vitro Cell Dev Biol Plant 32:272–289

Grant JJ, Loake GJ (2001) Role of reactive oxygen intermediates and cognate redox signaling in disease resistance. Plant Physiol 124:21–29

Hays DB (1996) The role of hormones in *Brassica napus* embryo development. PhD thesis. University of Calgary, Calgary, Alberta, Canada

Henmi K, Tsuboi S, Demura T, Fukuda H, Iwabuchi M, Ogawa KA (2001) A possible role of glutathione and glutathione disulfide in tracheary element differentiation in the cultured mesophyll cells of *Zinnia elegans*. Plant Cell Physiol 42:673–676

Ito Y, Nakanomyo I, Motose H, Iwamoto K, Sawa S, Dohmae N, Fukuda H, Dodeca H (2006) CLE peptides as suppressors of plant stem cell differentiation. Science 313:842–845

Kermode AR (1995) Regulatory mechanisms involved in the transition from seed development to germination. Crit Rev Plant Sci 9:155–195

Jain SM, Newton RJ, Soltes EJ (1988) Enhancement of somatic embryogenesis in norway spruce (*Picea abies*). Theor Appl Genet 76:501–506

Jiang K, Meng YL, Feldman L (2004) Quiescent center formation in maize is associated with an auxin-related oxidizing environment. Development 130:1429–1438

Jiang K, Feldman LJ (2005) Regulation of root apical meristem development. Annu Rev Cell Dev Biol 21:485–509

Levine M (1947) Differentiation of carrot root tissue grown in culture. Bull Torrey Bot Club 74:321–323

Kerk NM, Feldman LJ (1995) A biochemical model for the initiation and maintenance of quiescent center: implications for organization of root meristems. Development 121:2825–2833

Luwe M (1996) Antioxidants in the apoplast and symplast of beech (*Fagus sylvatica* L.) leaves: seasonal variations and responses to changing ozone concentration in air. Plant Cell Environ 19:321–328

Lynn K, Fernandez A, Aida M, Sedbrook J, Tasaka M, Masson P, Barton MK (1999) The PINHEAD/ZWILLE gene acts pleiotropically in *Arabidopsis* development and has overlapping functions with the ARGONAUTE1 gene. Development 126:469–481

Mahalingam R, Fedoroff N (2003) Stress response, cell death and signaling: the many faces of reactive oxygen species. Physiol Plant 119:56–68

Maraschin SF, Caspers M, Potokinad E, Wulfert F, Granerd A, Spaink HP, Wang M (2006) cDNA array analysis of stress-induced gene expression in barely androgenesis. Physiol Plant 127:535–550

Marre E, Arrigoni O (1957) Metabolic reactions to auxin. I. The effects of auxin on glutathione and the effects of glutathione on growth of isolated plant parts. Physiol Plant 10:289–301

May MJ, Leaver CJ (1994) *Arabidopsis thaliana* γ-glutamylcysteine synthetase is structurally unrelated to mammalian, yeast, and *E.coli* homologs. Proc Natl Acad Sci USA 91:10059–10063

May MJ, Leaver CJ (1993) Oxidative stimulation of glutathione synthesis in *Arabidopsis thaliana* suspension cultures. Plant Physiol 103:621–627

May MJ, Vernoux T, Leaver CJ, Montagu V, Inez D (1998) Glutathione homeostasis in plants: implications for environmental sensing and plant development. J Exp Bot 49:649–667

Moussian B, Schoof H, Haecker A, Jurgens G, Laux G (1998) Role of ZWILLE gene in the regulation of central shoot meristem cell fate during *Arabidopsis* embryogenesis. EMBO J 17:1799–1809

Noctor G, Foyer CH (1998) Ascorbate and glutathione: keeping active oxygen under control. Annu Rev Plant Physiol Plant Mol Biol 49:249–279

Noctor G, Arisi A-CM, Jouanin L, Kunert KJ, Rennenberg H, Foyer CH (1998) Glutathione: biosynthesis, metabolism and relationship to stress tolerance explored in transformed plants. J Exp Bot 49:623–647

Ogawa K, Tasaka Y, Mino M, Tanaka T, Iwabuchi M (2001) Association of glutathione with flowering in *Arabidopsis thaliana*. Plant Cell Physiol 42:524–530

Piyathilake CJ, Bell WC, Johanning GL, Cornwell PE, Heimburger DC, Grizzle WE (2000) The accumulation of ascorbic aicd by squamous cell carcinomas of the lung and larynx is associated with global methylation of DNA. Cancer 89:171–176

Potters G, De Gara L, Asard H, Horemans N (2002) Ascorbate and glutathione: guardians of the cell cycle, pattern in crime? Plant Physiol Biochem 40:537–548

Raghavan V (1976) Experimental embryogenesis in vascular plants. Academic, London

Ramesar-Fortner NS, Yeung EC (2001) Tri-iodobenzoic acid affects shoot apical meristem formation in zygotic embryos of *Brassica napus*. Can J Bot 79:265–273

Rawlins MR, Leaver CJ, May MJ (1995) Characterization of *Arabidopsis thaliana* cDNA encoding glutathione synthetase. FEBS Lett 376:81–86

Sanchez-Fernandez R, Fricker M, Corben L, White NS, Sheard N, Leaver CJ, Van Montagu M, Inze D, May MJ (1997) Cell proliferation and hair tip growth in the Arabidopsis root under mechanistically different forms of redox control. Proc Natl Acad Sci USA 94:2745–2750

Shafer FQ, Buettner GR (2001) Redox environment of the cell as viewed through the redox state of the glutathione disulfide/glutathione couple. Free Radic Biol Med 30:1191–1198

Schutzendubel A, Polle A (2002) Plant responses to abiotic stresses: heavy metal-induced oxidative stress ad protection by micorrhization. J Exp Bot 53:1351–1365

Shibata MA, Fukushima S, Asakawa E, Hirose M, Ito N (1992) The modifying effects of indomethacin or ascorbic acid on cell proliferation induced by different types of bladder tumor promoters in rat urinary bladder and forestomach mucosal epithelium. Jpn J Cancer Res 83:31–39

Stasolla C, Yeung EC (2001) Ascorbic acid metabolism during white spruce somatic embryogenesis. Physiol Plant 111:196–205

Stasolla C, Yeung EC (2003) Recent advances in conifer somatic embryogenesis: improving somatic embryo quality. Plant Cell Tissue Org Cult 74:15–35

Stasolla C, Yeung EC, Thorpe TA (2003) Somatic embryogenesis in white spruce: physiology and biochemistry. In: Bhojawani SS (ed) Agrobiotechnology and plant tissue culture. Science Publishers Inc., Enfield, NH, pp 10–27

Stasolla C, Thorpe TA (2004) Purine and pyrimidine nucleotide synthesis and degradation during in vitro morphogenesis of white spruce (*Picea glauca*). Front Biosci 9:1506–1519

Stasolla C, Belmonte MF, van Zyl L, Craig DL, Liu W, Yeung EC, Sederoff R (2004) The effect of reduced glutathione on morphology and gene expression of white spruce (*Picea glauca*) somatic embryos. J Exp Bot 55:695–709

Stasolla C, Yeung EC (2006a) Exogenous applications of ascorbic acid induce shoot apical meristem growth in germinating white spruce (*Picea glauca*) somatic embryos. Int J Plant Sci 167:429–436

Stasolla C, Yeung EC (2006b) Endogenous ascorbic acid affects thymidine and uridine metabolism and modulates meristem reactivation in white spruce somatic embryos. Tree Physiol 26:1197–1206

Stasolla C, Yeung EC (2007) Cellular ascorbic acid regulates the activity of major peroxidases in the apical poles of germinating white spruce *(Picea glauca)* somatic embryos. Plant Physiol Biochem 45:188–198

Stasolla C, Loukanina N, Yeung EC, Thorpe TA (2007) Comparative studies on pyrimidine metabolism of Pinus radiata cotyledons cultured under shoot-forming and non shoot-forming conditions. J Plant Physiol 164:429–441

Stasolla C, Belmonte M, Tahir M, Elhiti M, Joosen R, Maliepaard C, Sharpe A, Boutilier K (2008) Buthionine sulfoximine (BSO)–mediated improvement in *in vitro* cultured embryo quality is associated with major changes in transcripts involved with ascorbate metabolism, meristem development and embryo maturation. Planta 228:255–272

Sundås-Larsson A, Svenson M, Liao H, Engström P (1998) A homeobox gene with potential developmental control function in the meristem of the conifer *Picea abies*. Proc Natl Acad Sci USA 95:15118–15122

Thair M, Stasolla C (2005) Shoot apical development during in vitro embryogenesis. Can J Bot 84:1650–1659

Thorpe TA, Stasolla C (2001) Somatic embryogenesis. In: Bhojwani SS, Soh WY (eds) Current trends in the embryology of angiosperms. Kluwer, Boston, MA, pp 279–336

Yao QA, Simion E, William M, Krochko J, Kasha KJ (1997) Biolistic transformation of haploid isolated microspores of barley (*Hordeum vulgare* L.). Genome 40:570–581

Yeung EC, Aitken C, Biondi S, Thorpe TA (1981) Shoot histogenesis in cotyledon explants of radiata pine. Bot Gaz 142:494–501

Yeung EC (2002) The canola microspore-derived embryo as a model system to study developmental processes in plants. J Plant Biol 45:119–133

Chapter 9
A Winning Two Pair: Role of the Redox Pairs AsA/DHA and GSH/GSSG in Signal Transduction

Günce Şahin and Mario C. De Tullio

Abstract For a few decades after its identification, the AsA/GSH (Halliwell–Asada) cycle has been almost exclusively considered as a scavenging system, more or less efficiently removing reactive oxygen species (ROS). The whole issue was simplistically viewed as a fight between "bad" ROS and "good" antioxidants until recently, when ROS (and reactive nitrogen species, RNS) were identified as important signalling molecules, inducing an array of defensive responses in both animal and plant systems. In this framework, antioxidants take a completely different role, becoming part of complex signalling modules. AsA and GSH are connected in many ways, forming a functional entity. Still they have their own specificity and control different aspects of plant metabolism and growth. It is increasingly clear that AsA and GSH, together with their oxidized forms, have a pivotal, multi-level regulatory role in protein function and gene expression. Remarkable progress has been recently achieved in the identification of transcription factors and other proteins, whose activity is regulated by changes in the relative amount of redox components. Additionally, participation of AsA as a co-substrate of reactions catalysed by dioxygenases (a large class of enzymes involved in hormone synthesis and post-translational protein modification), also has a relevant regulatory role.

Keywords Ascorbic acid • Glutathione • Redox regulation • Evolution of metabolism

1 Ascorbic Acid and Glutathione: Just Antioxidants?

In the last few decades we have been accustomed to think of ascorbic acid (AsA) and glutathione (GSH) mainly as antioxidants. However, this definition only highlights one of the many facets of these interesting molecules. GSH was first observed

G. Şahin
Abant Izzet Baysal University, Bolu, Turkey

M.C. De Tullio (✉)
Plant Biology and Pathology, Aldo Moro University, Bari, Italy
e-mail: detullio@botanica.uniba.it

N.A. Anjum et al. (eds.), *Ascorbate-Glutathione Pathway and Stress Tolerance in Plants*,
DOI 10.1007/978-90-481-9404-9_9, © Springer Science+Business Media B.V. 2010

as an organic sulphur-hydrogenating substance by de Rey Pailhade (1888)), and its structure was eventually identified, not without controversies, by Hopkins in the late 1920s (Hopkins 1929). In retrospective, the ironical title of a review published by the Kosowers in 1969 ("Lest I forget thee, glutathione...", Kosower and Kosower 1969) induce to think that in those days, before the widespread of the free radical theory of ageing and disease (Marx 1985), GSH was much less popular than it is now. The same applies to AsA. The name "ascorbic", given by Szent Györgyi and Haworth to the molecule previously known as "hexuronic acid" (Szent-Giörgyi 1963), literally means "against scurvy", the pathological condition of humans and other organisms unable to synthesize AsA due to the inactivation of some important AsA-dependent enzymes belonging to a subclass of dioxygenases, which catalyse the synthesis of several key molecules including different collagen forms, some neurotransmitters, and carnitine (De Tullio 2004). This has very little to do with the antioxidant role of AsA.

Identification of the actual biochemical role of AsA, GSH, tocopherols, xanthophylls, flavonoids, and many other molecules should of course take into consideration not only the fact that they are all capable of reducing reactive oxygen species (ROS), but also their specificity. Antioxidants are definitely not interchangeable. For example, AsA and GSH can affect different aspects of root development in different and sometimes even contrasting ways (Sanchez-Fernandez et al. 1997).

A special connection between AsA and GSH was discovered long time ago (Borsook and Jeffreys 1936; Hopkins and Morgan 1936), and has been further explored ever since (Mårtensson and Meister 1991). When the AsA–GSH cycle was discovered in plant chloroplasts, it seemed the perfect system to get rid of ROS unavoidably produced as by products of excess photosynthetic excitation energy (Foyer and Halliwell 1976; Groden and Beck 1979; Nakano and Asada 1980). Of course this view still holds true, but this is probably not the only function of the AsA–GSH cycle (De Tullio 2010).

The increasing complexity of living organisms induced the necessity of coordinating a huge number of different activities. This resulted in the establishment of signalling networks connecting all aspects of development and metabolism. In older days it was assumed that signalling in biological systems was restricted to specialized molecules, such as hormones or neurotransmitters, but we now know that virtually all cellular components contribute to efficient molecular communication at different levels (systemic, intercellular, intracellular). We tentatively introduce the concept of "informational content" as an intrinsic property of a given molecule, defined as the capability of directing specific responses by interacting with one or more molecules able to "sense" it. In this context, "specific responses" also means that no other molecule can take over, and produce the same effect. Informational content would be therefore directly proportional to specificity, "uniqueness" being the highest possible score (Fig. 1). In terms of molecular communication, little information is associated to the generic antioxidant function, since this activity is shared among many diverse and unrelated compounds. On the other hand, AsA and GSH individually have some informational content, in that they can direct specific responses.

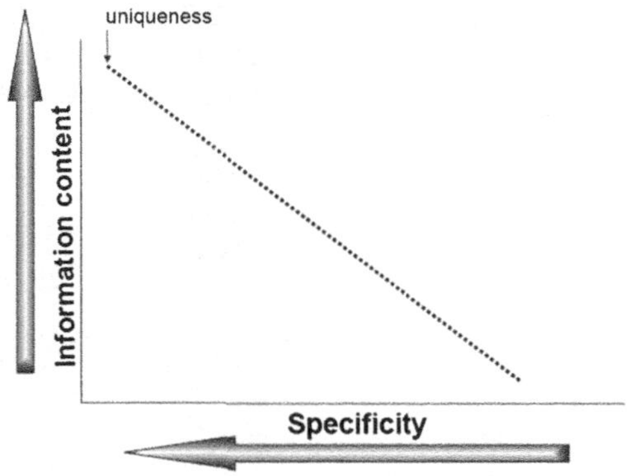

Fig. 1 The relationship between specificity and informational content of a molecule

In the next sections we will discuss separately the signalling-related implications of the single AsA and GSH "modules", defined as the sequence of events spanning from biosynthesis to utilization in biochemical reactions. In our view, the AsA and GSH "vertical" modules have distinct modes of operation. However, the two modules are not isolated. They are "horizontally" connected by the AsA–GSH cycle, which ensures communication and coordination of the "vertical" modules. We suggest a "fence-like" mechanism (Fig. 2), in which information flows at different levels of molecular interaction. As it will be discussed below, the specificity of the AsA module resides in its involvement in the reactions catalyzed by 2-oxoglutarate-dependent dioxygenases (2-ODDs), resulting in hormone synthesis and gene expression, whereas GSH acts, among other non-antioxidant roles, as a specific activator/repressor of several proteins via glutathionylation (gtt), and as a substrate of S-transferases (GST). The "horizontal" cycle also provides connection with the NAD(P)H system, since this cofactor is required for the activity of GSH reductase which is an important component of the AsA–GSH cycle.

2 The Ascorbate Signalling Module

Ascorbic acid synthesis takes place in almost all higher eukaryotes, although with different biosynthetic pathways. Surprisingly, a comprehensive model of the main AsA biosynthetic pathway in plants was proposed not much more than a decade ago (Wheeler et al. 1998; Smirnoff et al. 2001). All genes of the Smirnoff–Wheeler pathway have been eventually identified and cloned (Conklin et al. 2006), but not much is known about the regulation of AsA biosynthesis. Available evidence shows that all steps of the plant pathway are cytosolic, excepting for the final one, catalysed by the mitochondrial enzyme L-galactono-1,4-lactone dehydrogenase (GalLDH). The peculiar location of GalLDH poses an interesting question concerning the regulation of AsA biosynthesis and its connection with energy metabolism. In fact,

Fig. 2 The "fence model" of the ascorbate–glutathione cycle. The ascorbate (AsA) and the glutathione (GSH) modules are represented as single "vertical" axes of the fence, connected by the horizontal AsA–GSH pathway. Outputs of the AsA module are the activities of 2-oxoglutarate-dependent dioxygenases (2-ODDs), some of which result in the production of plant hormones (ethylene, gibberellins). Specific outputs of the GSH module are the activity of GSH-S-transferases (GST) and glutathyolation (gtt) of selected proteins. Both AsA and GSH affect gene expression. The cycle also connects AsA and GSH with the NAD(P)H system

it has been demonstrated that oxidised cytochrome *c* is the electron acceptor in the reaction catalysed by GalLDH (Bartoli et al. 2000, 2006), and therefore AsA synthesis is at the same time regulated and regulator of respiratory electron transport (Millar et al. 2003).

As mentioned above, the actual function of AsA can be understood by observing the effects of AsA deprivation in organisms unable to synthesize it. All plants studied so far can produce their own ASC supply. In addition, recent studies showed the existence of alternative pathways of ASC biosynthesis (Agius et al. 2003; Lorence et al. 2004). It is therefore quite difficult to obtain "scurvy" plants to study AsA functions. Lycorine, an alkaloid described more than 3 decades ago by Arrigoni and co-workers (Arrigoni et al. 1975) was used in our and other labs as a tool to decrease AsA content in vivo (Liso et al. 1984; Cordoba-Pedregosa et al. 1996; Arrigoni et al. 1997a; De Tullio et al. 1998) and to inhibit the activity of GalLDH in vitro (Arrigoni et al. 1997b; Imai et al. 1998). Indeed, plants treated with lycorine showed reduced growth (Cordoba-Pedregosa et al. 1996; Arrigoni et al. 1997a).

The search for ozone-sensitive mutants led to the identification of some *Arabidopsis* mutants (called *vtc* for vitamin C) characterized by relatively low AsA content (Conklin et al. 2000). It has been demonstrated that the *vtc-1* mutant is defective in the gene coding for GDP-mannose pyrophosphorylase (mannose-1-phosphate guanylyltransferase), an enzyme catalysing an early step in AsA synthesis (Smirnoff et al. 2001), and other mutants have recently been characterized (Conklin et al. 2006). None of viable *vtc* mutants is completely devoid of AsA, whereas a null mutation in the VTC1 locus causes death of embryos before or during desiccation (Lukowitz et al. 2001), clearly indicating that at least a small amount of AsA is necessary for plant life.

Ascorbic acid is specifically required by many 2-oxoacid-dependent dioxygenases (2-ODDs) (Prescott and John 1996). This group of enzymes include the well-known peptidyl-prolyl-4-hydroxylase (P4H), which is responsible for the post-translational hydroxylation of proline residues incorporated in polypeptide chains in both animals (for the synthesis of collagen) and plants (for the synthesis of hydroxyproline rich glycoproteins, HRGPs) (Sommer-Knudsen et al. 1998). The inhibition of P4H in humans and consequent production of misfolded collagen is responsible for the scurvy syndrome (De Tullio 2004). In plants, P4H is especially important for the synthesis of extensin and other structural proteins of the extracellular matrix, but also of the interesting class of arabinogalactan proteins (AGPs), which are involved in cell proliferation (Serpe and Nothnagel 1994) and signalling (Schultz et al. 1998).

Although probably the best characterized for specific AsA requirement, P4H is not the only important plant 2-ODD requiring AsA as a co-substrate (Table 1). Indeed, this group includes enzymes involved in flavonoid biosynthesis and, very interestingly, enzymes acting in the synthesis of gibberellins, such as GA 13-hydroxylase, GA 20-oxidase, GA 3-β-hydroxylase and GA 2-β-hydroxylase (Hedden and Kamiya 1997). One very special 2-ODD is 1-aminocyclopropane-1-carboxylate oxidase (ACCO), the enzyme catalysing the last step of ethylene biosynthesis using ascorbic acid as a substrate instead of 2-oxoglutarate (Kende 1993; Rocklin et al. 1999). An effect of AsA on gene expression and selective mRNA stability is known since more than a decade. In animal systems, AsA was observed to stabilize collagen transcript and destabilize elastin transcript in smooth muscle

Table 1 Partial list of AsA-requiring enzymes

Enzyme	Function	References
Prolyl-4-hydroxylase	HRGP synthesis (AGP)	De Tullio (2004)
GA 20-oxidase	Gibberellin synthesis	Hedden and Kamiya (1997)
GA 13-hydroxylase	Gibberellin synthesis	Hedden and Kamiya (1997)
GA 3β-hydroxylase	Gibberellin synthesis	Hedden and Kamiya (1997)
GA 2β-hydroxylase	Gibberellin synthesis	Hedden and Kamiya (1997)
ACC oxidase	Ethylene synthesis	Rocklin et al. (1999)

2-oxoglutarate; GA, gibberellic acid; ACC, aminocyclopropane-1-carboxylate cells and skin fibroblasts (Davidson et al. 1997). The transcription of the 72-kDa type IV collagenase (matrix metalloproteinase-2) was down-regulated by AsA in cultured human amnion-derived cells (Pfeffer et al. 1998). Tyrosine hydroxylase transcription was enhanced by AsA treatment (Seitz et al. 1998), and the mRNA encoding for various forms of cytochrome P450 in liver microsomes from guinea pigs were also induced (Mori et al. 1997). In plants, the transcription of the maize *Hrgp* gene is induced by AsA (Garcia-Muniz et al. 1998), and microarray experiments showed changes in the expression of several defense-related genes in the AsA-deficient mutant *vtc-1* of *Arabidopsis* (Pastori et al. 2003; Kiddle et al. 2003).

Ascorbic acid utilization by 2-ODDS (and of course other forms of utilization in the reactions with ROS) results in one-electron AsA oxidation yielding the free radical monodehydroascorbate (MDH). In turn, MDH rapidly dismutates producing AsA and the double-oxidized form dehydroascorbate (DHA). A relatively high DHA amount has been observed only in the apoplastic space of plant cells, whereas the reduced form AsA is predominant in all other cellular compartments (Pignocchi and Foyer 2003). There is increasing evidence that DHA could have a key role in signalling since it is able to directly interact not only with GSH, but also with thiol-containing proteins with vicinal cysteine residues (Cys–X–X–Cys) (Fig. 3). This peculiar reactivity of DHA can change protein conformation and stability, thus regulating the activity of many signal-related and regulative proteins, including thioredoxins and transcription factors (Morell et al. 1997). For example, the involvement of DHA signalling in stomatal dynamics has been demonstrated (Fotopoulos et al. 2008), with a mechanism apparently requiring the involvement of protein phosphatases (Fan et al. 2009).

Signalling appears an important function of the AsA module. In particular, the AsA/DHA redox couple can be viewed as a reliable sensor of the surrounding environment, perceiving and integrating even small changes in external conditions, and traducing them into chemical signals which direct essential developmental processes, such as the organization of the root apical meristem (De Tullio et al. 2010).

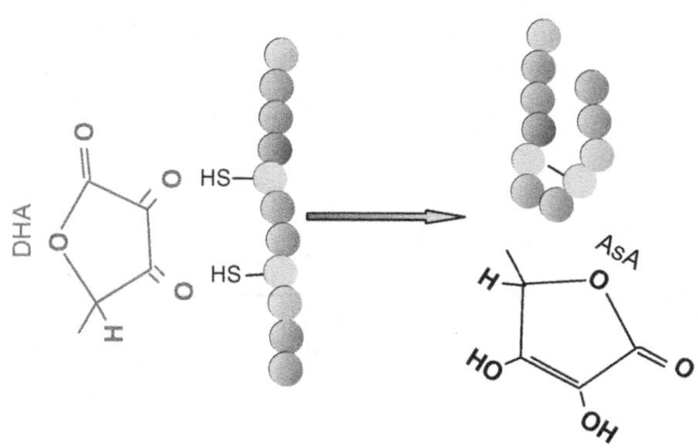

Fig. 3 Dehydroascorbate (DHA) affects protein folding by directly reacting with vicinal cysteine residues

3 The Glutathione Signalling Module

Synthesis of GSH occurs de novo from the amino acids glycine, cysteine and glutamic acid, and requires the consecutive action of two enzymes, γ-glutamylcysteine synthetase (γ-GCS) and GSH synthetase. Induction of γ-GCS expression appears environmentally regulated, occurring in response to diverse stimuli in a cell specific manner (Townsend et al. 2003). GSH represents the major pool of non protein reduced sulphur in the wide majority of prokaryotic and eukaryotic cells. Fahey and Sundquist (1991) found strong evidence for an evolutionary link between glutathione and aerobic eukaryotic metabolism; the findings indicate that glutathione evolved as a molecule that protects cells against oxygen toxicity. However, there is mounting evidence that the important role of GSH in plant development goes far beyond the strict boundaries of its antioxidant role. GSH not only provides reducing power needed for the conversion of DHA to AsA, but also for the conversion of ribonucleotides to deoxyribonucleotides and for a variety of thiol–disulfide interconversions; GSH is therefore important for the synthesis and repair of DNA, and for the folding of newly synthesized proteins (Sies 1999). Moreover, GSH can engage in thiol–disulphide exchange reactions that may be a key process in linking the regulation of gene expression to the redox state of cells or specific subcellular compartments (Schafer and Buettner 2001; Noctor et al. 2002).

The involvement of GSH in basic plant developmental processes has been ascertained. Compared with other tissues of higher plants, dry seeds contain higher amount of GSH (Klapheck et al. 1990). Changes in GSH content and GSH reductase activity during germination of some harbaceous plants have also been reported (Fahey et al. 1980; Kranner and Grill 1993). GSH is essential for cell cycle progression and root meristem activity, as demonstrated by the fact that roots fail to develop in the *Arabidopsis* mutant *ROOTMERISTEMLESS1* (*rml1*), which is GSH-deficient because of low activity of a key enzyme in GSH synthesis (Vernoux et al. 2000). Also embryogenesis and shoot development are regulated by GSH and GSSG (Stasolla 2010).

Glutathione deficiency may cause increased tendency to hemolysis, cataracts, and central nervous system abnormalities in humans (Meister 1994), and induces damage to skeletal muscle and to lung type 2 cell lamellar bodies in adult mice. Other structural effects of glutathione deficiency in mice include decreased numbers of microvilli on the alveolar surfaces of lung type 2 cells, swelling of mitochondria of lung capillary endothelial cells, blunting of the microvilli of lymphocytes, and marked mucosal damage in the jejunum and colon (Jain et al. 1992). Cancer, neurodegenerative diseases, cystic fibrosis, HIV, and ageing have been correlated to imbalances in GSH metabolism (Townsend et al. 2003).

Although this variety of effects of GSH deficiency in different organisms still lacks a detailed molecular explanation, it can hardly be explained just in terms of its reaction with ROS, i.e. its antioxidant function. GSH is often considered an important defence in stress conditions, and namely in plant responses to pollutants. Indeed, it is the precursor of phytochelatins, which bind supra-optimal concentrations of heavy metals (Grill et al. 1987, 1989), and is a substrate for the GSH S-transferases (GSTs), which catalyse the conjugation of GSH with potentially dangerous xenobiotics such as herbicides (Marrs 1996). However, GSTs they also have other important functions. For example, specific GSTs are inserted the anthocyanin synthesis pathway (Marrs 1996), and certain GSTs apparently function as flavonoid binding proteins as suggested for AN9, a GST required for efficient anthocyanin export from the cytosol in petunia (Mueller et al. 2000). Consistent with this is the observation that the anthocyanin content of *Arabidopsis* leaves correlated with GSH content in plants with modified capacity for GSH biosynthesis (Xiang et al. 2001). It has been reported that the expression of a human GST affects the expression of specific genes, by a mechanism still partly unknown (Castro-Caldasa et al. 2009), but the recent finding of a novel role of a GST in glutathionylation (see below) broadens the signalling potential of this class of enzymes (Townsend et al. 2009).

Beside its function as a substrate of GSTs, GSH is likely to exert a more general direct role, as a regulator of gene expression (Wingate et al. 1988; Baier and Dietz 1997). Protein glutathonylation is especially important for its implication in signalling. This post translational modification can modulate enzyme activity by modifying the structure of catalytic site cys residues or affect biological activity by competing with other thiol modifications (Sies 1999). Many proteins possess accessible cysteine residues, liable to undergo redox changes depending on variations of the intra- as well as extracellular conditions. GSH and its oxidized form GSSG are capable of affecting the redox status of such critical thiols in a number of proteins, including membrane channels and transporters, receptors, protein kinases and phosphatases and, above all, transcription factors (Arrigo 1999; Cross and Templeton 2004). In plants, more than 300 redox-target proteins with regulative functions have been identified so far (Dietz 2008), including Calvin cycle enzymes which are glutathionylated (Ito et al. 2003).

4 The Horizontal Connection

A major function of GSH in protection against oxidative stress is the re-reduction of AsA in the AsA–GSH cycle (Noctor and Foyer 1998). In this pathway, GSH acts as a recycled intermediate in the reduction of H_2O_2 using electrons derived ultimately, from H_2O, and was therefore renamed the "water–water cycle" by Asada (2003). Although known as a ROS-scavenging system, the AsA–GSH pathway could also be viewed as a connecting (horizontal) pathway regulating the AsA and GSH signalling modules briefly outlined above.

If all antioxidants were just a pool of undifferentiated reducing power, the redox potentials of all redox couples in a given cell compartment should be the same. Schafer and Buettner (2001) identified the redox state of the GSSG/GSH couple as a major indicator of physiological conditions in animal cells, and calculated half-cell reduction potentials ranging from −240 to −170 mV in proliferating and apoptotic cells, respectively. Attempts to measure intracellular redox potentials are made difficult by the lack of sensitive and reliable sensors, but recently redox potential values have been directly observed by a quantitative fluorescence-based method making use of redox-sensitive green fluorescent protein targeted to different plant cell compartments (Schwarzländer et al. 2008). Redox potentials of about −320 mV in the cytosol and −360 mV in the mitochondria of *Arabidopsis* leaf cells have been obtained, whereas slightly more oxidizing environments occur in plastids and peroxisomes in epidermal cells of tobacco leaves. The only oxidizing compartment was the lumen of the endoplasmic reticulum. However, specific interaction of redox-sensitive GFP with glutaredoxin in vitro strongly suggests that this method only allows calculation of the GSH redox potential (Schwarzländer et al. 2008).

Reliable methods for the calculations of the DHA/AsA redox potential are not presently available, so a direct comparison is not yet possible, but if GSH, and AsA are far from equilibrium with each other, the effects of each redox couple will be governed by the specific interactions of each molecule with proteins or other components. (the "vertical" modules), and in this case we will have an "AsA redox state" or "GSH redox state" but no single "cellular redox state" or "redox environment", as defined by Schafer and Buettner (2001). In other words, the two pairs would work independently ("vertically"), remaining connected by the horizontal module (AsA–GSH cycle).

5 Concluding Remarks: the AsA–GSH Cycle and the Evolution of the Metabolic Network

According to Theodosius Dobzhansky "Nothing in biology makes any sense except in the light of evolution" (Dobzhansky 1973). An evolutionary standpoint is presently unavoidable in many fields of biology, ranging from systematics to molecular biology, yet the situation is more complicated in biochemistry. The reasons for a less pronounced inclination of biochemists towards evolutionary aspects is probably to be found in the history of this discipline, which stems from chemistry and not from biology (Firn and Jones 2009). Another important limitation to an evolutionary approach is the difficulty in disentangling complex biochemical pathways strictly interconnected. However, the existing constellation of different pathways, some widespread among the entirety of living organisms, and some unique to a well-defined group, can hardly be explained in terms of an "ordered plan". On the contrary, it fits nicely with the concepts of variation and natural selection. It has been suggested that new pathways evolved by gene duplication and recruitment of enzyme activities from established pathways (Chapman and Ragan 1980). Additionally, the coordination of metabolic activities results from the co-evolution of substrates and their related enzymes: new metabolic functions derive from metabolite-driven recruitment of existing enzymatic mechanisms (Schmidt et al. 2003). Metabolites produced in a pathway can be used by other pathways, thus yielding a regulative loop. This seems to be the case for the AsA–GSH cycle. It is tempting to speculate that the two important molecules, standing "side to side" in two essential and independent pathways, started the "horizontal" collaboration which not only shares information between the two modules, but also ensures the well-known and celebrated antioxidant activity of the cycle.

References

Agius F, Gonzalez-Lamothe R, Caballero JL, Munoz-Blanco J, Botella MA, Valpuesta V (2003) Engineering increased vitamin C levels in plants by overexpression of a D-galacturonic acid reductase. Nat Biotechnol 21:177–181

Arrigo AP (1999) Gene expression and the thiol redox state. Free Radic Biol Med 27:936–44

Arrigoni O, Arrigoni-Liso R, Calabrese G (1975) Lycorine as an inhibitor of ascorbic acid biosynthesis. Nature 256:513–514

Arrigoni O, Calabrese G, De Gara L, Bitonti MB, Liso R (1997a) Correlation between changes in cell ascorbate and growth of *Lupinus albus* seedlings. J Plant Physiol 150:302–308

Arrigoni O, De Gara L, Paciolla C, Evidente A, de Pinto MC, Liso R (1997b) Lycorine: a powerful inhibitor of L-galactono-γ-lactone dehydrogenase. J Plant Physiol 150:362–364

Asada K (2003) The water-water cycle in chloroplasts. Annu Rev Plant Physiol Plant Mol Biol 50:601–639

Baier M, Dietz KJ (1997) The plant 2-cys peroxiredoxin BAS1 is a nuclear encoded chloroplast protein: its expression regulation, phylogenetic origin and implications for its specific physiological function in plants. Plant J 12:179–190

Bartoli CG, Pastori GM, Foyer CH (2000) Ascorbate biosynthesis in mitochondria is linked to the electron transport chain between complexes III and IV. Plant Physiol 123:335–344

Bartoli CG, Yu J, Gómez F, Fernández L, McIntosh L, Foyer CH (2006) Inter-relationships between light and respiration in the control of ascorbic acid synthesis and accumulation. J Exp Bot 37:1621–1631

Borsook H, Jeffreys CE (1936) Glutathione and ascorbic acid. Science 83:397–398

Castro-Caldasa M, Milagrea I, Rodriguesa E, Gama MJ (2009) Glutathione *S*-transferase pi regulates UV-induced JNK signaling in SH-SY5Y neuroblastoma cells. Neurosci Lett 451:241–245

Chapman DJ, Ragan MA (1980) Evolution of biochemical pathways: evidence from comparative biochemistry. Annu Rev Plant Physiol 31:639–678

Conklin PL, Saracco SA, Norris SR, Last RL (2000) Identification of ascorbic acid-deficient *Arabidopsis thaliana* mutants. Genetics 154:847–856

Conklin PL, Gatzek S, Wheeler GL, Dowdle J, Raymond MJ, Rolinski S, Isupov M, Littelchild JA, Smirnoff N (2006) Arabidopsis thaliana VTC4 encodes L-galactose-1-P phosphatase, a plant ascorbic acid biosynthetic enzyme. J Biol Chem 281:15662–156670

Cordoba-Pedregosa M, Gonzalez-Reyes JA, Canadillas M, Navas P, Cordoba F (1996) Role of apoplastic and cell-wall peroxidases on the stimulation of root elongation by ascorbate. Plant Physiol 112:1119–1125

Cross JV, Templeton DJ (2004) Oxidative stress inhibits MEKK1 by site specific glutathionylation in the ATP-binding domain. Biochem J 381:675–683

Davidson JM, LuValle PA, Zoia O, Quaglino D Jr, Giro M (1997) Ascorbate differentially regulates elastin and collagen biosynthesis in vascular smooth muscle cells and skin fibroblasts by a pre-translational mechanism. J Biol Chem 272:345–352

de Rey Pailhade MJ (1888) Sur un corps d'origine organique hydrogénant le soufre á froid. C R Acad Sci 106:1683–1684

De Tullio MC (2004) How does ascorbic acid prevent scurvy? A survey of the non-antioxidant functions of vitamin C. In: Asard H, May J, Smirnoff N (eds) Vitamin C, its functions and biochemistry in animals and plants. Bios Scientific Publishers, Oxford, UK, pp 159–72

De Tullio MC (2010) Antioxidants and redox regulation: changing notions in a changing world. Plant Physiol Biochem 48:289–291

De Tullio MC, De Gara L, Paciolla C, Arrigoni O (1998) Dehydroascorbate-reducing proteins in maize are induced by the ascorbate biosynthesis inhibitor lycorine. Plant Physiol Biochem 36:433–440

De Tullio MC, Jiang K, Feldman LJ (2010) Redox regulation of root apical meristem organization: connecting root development to its environment. Plant Physiol Biochem 48:328–336

Dietz KJ (2008) Redox signal integration: from stimulus to networks and genes. Physiol Plant 133:459–468

Dobzhansky T (1973) Nothing in biology makes any sense except in the light of evolution. Am Biol Teach 35:125–129

Fahey RC, Deena L, Di Stefano G, Meier P, Bryan RN (1980) Role of hydration state and thiol-sulphide status in the control of thermal stability and protein synthesis in wheat embryo. Plant Physiol 65:1062–1066

Fahey RC, Sundquist AR (1991) Evolution of glutathione metabolism. Adv Enzymol 64:1–53

Fan DP, Shi YC, Liu WQ (2009) Involvement of protein tyrosine phosphatase in signal pathway of dehydroascorbate. Plant Physiol Commun 45:843–846

Firn RD, Jones GG (2009) A Darwinian view of metabolism: molecular properties determine fitness. J Exp Bot 60:719–726

Fotopoulos V, De Tullio MC, Barnes J, Kanellis AK (2008) Altered stomatal dynamics in ascorbate oxidase over-expressing tobacco plants suggest a role for dehydroascorbate signalling. J Exp Bot 59:729–737

Foyer CH, Halliwell B (1976) The presence of glutathione and glutathione reductase in chloroplasts: a proposed role in ascorbic acid metabolism. Planta 133:21–25

Garcia-Muniz N, Martinez-Izquierdo JA, Puigdomenech P (1998) Induction of mRNA accumulation corresponding to a gene encoding a cell wall hydroxyproline-rich glycoprotein by fungal elicitors. Plant Mol Biol 38:623–632

Grill E, Winnacker EL, Zenk MH (1987) Phytochelatins, a class of heavy-metal-binding peptides from plants, are functionally analogous to metallothioneins. Proc Natl Acad Sci USA 84: 439–443

Grill E, Loffler S, Winnacker EL, Zenk MH (1989) Phytochelatins, the heavy-metal binding peptides of plants, ares ynthesized from glutathione by a specific c-glutamylcysteine dipeptidyl transpeptidase (phytochelatin synthase). Proc Natl Acad Sci USA 86:6838–6842

Groden D, Beck E (1979) H_2O_2 destruction by ascorbate dependent systems from chloroplasts. Biochim Biophys Acta 546:426–433

Hedden P, Kamiya Y (1997) Gibberellin biosynthesis: enzymes, genes and their regulation. Annu Rev Plant Physiol Plant Mol Biol 48:431–460

Hopkins FG (1929) On glutathione: a reinvestigation. J Biol Chem 84:269–320

Hopkins FG, Morgan EJ (1936) Some relations between ascorbic acid and glutathione. Biochem J 30:1446–1462

Imai T, Karita S, Shiratori G, Hattori M, Nunome T, Oba K, Hirai M (1998) L-galactono-gamma-lactone dehydrogenase from sweet potato: purification and cDNA sequence analysis. Plant Cell Physiol 39:1350–1358

Ito H, Iwabuchi M, Ogawa K (2003) The sugar-metabolic enzymes aldolase and triose-phosphate isomerase are targets of glutathionylation in *Arabidopsis thaliana*: Detection using biotinylated glutathione. Plant Cell Physiol 44:655–660

Jain A, Martenson J, Mehta T, Krauss AN, Auld PAM, Meister A (1992) Ascorbic acid prevents oxidative stress in glutathione-deficient mice: Effects on lung type 2 cell lamellar bodies, lung surfactant, and skeletal muscle. Proc Natl Acad Sci USA 89:5093–5097

Kende H (1993) Ethylene biosynthesis. Annu Rev Plant Physiol Plant Mol Biol 44:283–307

Kiddle G, Pastori GM, Bernard S, Pignocchi C, Antoniw J, Verrier PJ, Foyer CH (2003) Effects of leaf ascorbate content on defense and photosynthesis gene expression in *Arabidopsis thaliana*. Antiox Redox Sign 5:23–32

Klapheck S, Zimmer I, Cosse H (1990) Scavenging of hydrogen peroxide in the endosperm of *Ricinus communis* by ascorbate peroxidase. Plant Cell Physiol 31:1005–1013

Kosower EM, Kosower NS (1969) Lest I forget thee, glutathione.... Nature 224:117–120

Kranner I, Grill D (1993) Content of low molecular weight thiols during the imbibition of pea seeds. Physiol Plant 88:557–562

Liso R, Calabrese G, Bitonti MB, Arrigoni O (1984) Relationship between ascorbic acid and cell division. Exp Cell Res 150:314–320

Lorence A, Chevone BI, Mendes P, Nessler CL (2004) *myo*-Inositol oxygenase offers a possible entry point into plant ascorbate biosynthesis. Plant Physiol 134:1200–1205

Lukowitz W, Nickle TC, Meinke DW, Last RL, Conklin PL, Somerville CR (2001) *Arabidopsis cyt1* mutants are deficient in a mannose-1-phosphate guanylyltransferase and point to a requirement of N-linked glycosylation for cellulose biosynthesis. Proc Natl Acad Sci USA 98:2262–2267

Marrs K (1996) The functions and regulation of glutathione S transferases in plants. Annu Rev Plant Physiol Plant Mol Biol 47:127–158

Mårtensson JM, Meister A (1991) Glutathione deficiency decreases tissue ascorbate levels in newborn rats: ascorbate spares glutathione and protects. Proc Natl Acad Sci USA 88:4656–46560

Marx JL (1985) Oxygen free radicals linked to many diseases. Science 235:529–531

Meister A (1994) Glutathione, ascorbate, and cellular protection. Cancer Res Suppl 54:1969–1975

Millar AH, Mittova V, Kiddle G, Heazlewood JL, Bartoli CG, Theodoulou FL, Foyer CH (2003) Control of ascorbate synthesis by respiration and its implications for stress responses. Plant Physiol 133:443–447

Morell S, Follmann H, De Tullio M, Häberlein I (1997) Dehydroascorbate and dehydroascorbate reductase are phantom indicators of oxidative stress in plants. FEBS Lett 414:567–571

Mori T, Itoh S, Ohgiya S, Ishizaki K, Kamataki T (1997) Regulation of CYP1A and CYP3A mRNAs by ascorbic acid in guinea pigs. Arch Biochem Biophys 348:268–277

Mueller LA, Goodman CD, Silady RA, Walbot V (2000) AN9, a *Petunia* glutathione S-transferase required for anthocyanin sequestration, is a flavonoid binding protein. Plant Physiol 123:1561–1570

Nakano Y, Asada K (1980) Spinach chloroplasts scavenge hydrogen peroxide on illumination. Plant Cell Physiol 21:1295–1307

Noctor G, Foyer CH (1998) Ascorbate and glutathione: keeping active oxygen under control. Annu Rev Plant Physiol Plant Mol Biol 49:249–278

Noctor G, Gomez L, Vanacker H, Foyer CH (2002) Interactions between biosynthesis, compartmentation and transport in the control of glutathione homeostasis and signaling. J Exp Bot 53:1283–1304

Pastori GM, Kiddle G, Antoniw J, Bernard S, Veljovic-JovanovicS VPJ, Noctor G, Foyer CH (2003) Leaf vitamin C contents modulate plant defense transcripts and regulate genes that control development through hormone signaling. Plant Cell 15:939–951

Pfeffer F, Casanueva E, Kamar J, Guerra A, Perichart O, Vadillo-Ortega F (1998) Modulation of 72-kilodalton type IV collagenase (matrix metalloproteinase-2) by ascorbic acid in cultured human amnion-derived cells. Biol Reprod 59:326–329

Pignocchi C, Foyer CH (2003) Apoplastic ascorbate metabolism and its role in the regulation of cell signalling. Curr Opin Plant Biol 6:379–389

Prescott AG, John P (1996) Dioxygenases: molecular structure and role in plant metabolism. Annu Rev Plant Physiol Plant Mol Biol 47:245–271

Rocklin AM, Tierney DL, Kofman V, Brunhuber NM, Hoffman BM, Christoffersen RE, Reich NO, Lipscomb JD, Que L (1999) Role of the nonheme Fe(II) center in the biosynthesis of the plant hormone ethylene. Proc Natl Acad Sci USA 96:7905–7909

Sanchez-Fernandez R, Fricker M, Corben LB, White NS, Sheard N, Leaver CJ, Van Montagu M, Inzé D, May MJ (1997) Cell proliferation and hair tip growth in the *Arabidopsis* root are under mechanistichally different forms of redox control. Proc Natl Acad Sci USA 94:2745–2749

Schafer FQ, Buettner GR (2001) Redox environment of the cell as viewed through the redox state of the glutathione disulfide/glutathione couple. Free Radic Biol Med 30:1191–1212

Schmidt S, Sunyaev S, Bork P, Dandekar T (2003) Metabolites: a helping hand for pathway evolution? Trends Biochem Sci 28:336–341

Schultz C, Gilson F, Oxley D, Youl J, Bacic A (1998) GPI-anchors on arabinogalactan-proteins: implications for signalling in plants. Trends Plant Sci 3:426–431

Schwarzländer M, Fricker MD, Müller C, Marty L, Brach T, Novak J, Sweetlove LJ, Hell R, Meyer AJ (2008) Confocal imaging of glutathione redox potential in living plant cells. J Microsc 231:299–316

Seitz G, Gebhardt S, Beck JF, Bohm W, Lode HN, Niethammer D, Bruchelt G (1998) Ascorbic acid stimulates DOPA synthesis and tyrosine hydroxylase gene expression in the human neuroblastoma cell line SK-N-SH. Neurosci Lett 244:33–36

Serpe MD, Nothnagel EA (1994) Effects of Yariv phenylglicosides on *Rosa* cell suspensions: evidence for the involvement of arabinogalactan proteins in cell proliferation. Planta 193:542–550

Sies H (1999) Glutathione and its role in cellular functions. Free Radic Biol Med 27:916–21

Smirnoff N, Conklin P, Loewus FA (2001) Biosynthesis of ascorbic acid in plants: a renaissance. Annu Rev Plant Physiol Plant Mol Biol 52:437–467

Sommer-Knudsen J, Bacic A, Clarke AE (1998) Hydroxyproline-rich plant glycoproteins. Phytochemistry 47:483–497

Stasolla C (2010) Redox regulation of *in vitro* embryogenesis. Plant Physiol Biochem doi. doi:10.1016/j.plaphy.2009.10.007

Szent-Giörgyi A (1963) Lost in the twentieth century. Annu Rev Biochem 36:1–15

Townsend DM, Tew KD, Tapiero H (2003) The importance of glutathione in human disease. Biomed Pharmacother 57:145–155

Townsend DM, Manevich Y, He L, Hutchens S, Pazoles CJ, Tew KD (2009) Novel role for Glutathione S-Transferase π: regulator of protein S-glutathionylation following oxidative and nitrosative stress. J Biol Chem 284:436–445

Vernoux T, Wilson RC, Seeley KA, Reichheld JP, Muroy S, Brown S, Maughan SC, Cobbett CS, Van Montagu M, Inzé D, May MJ, Sung ZR (2000) The ROOT MERISTEMLESS1/CADMIUM SENSITIVE2 gene defines a glutathione-dependent pathway involved in initiation and maintenance of cell division during postembryonic root development. Plant Cell 12:97–110

Wheeler GL, Jones MA, Smirnoff N (1998) The biosynthetic pathway of vitamin C in higher plants. Nature 393:65–69

Wingate VMP, Lawton MA, Lamb CJ (1988) Glutathione causes a massive and selective induction of plant defense genes. Plant Physiol 87:206–210

Xiang C, Werner BL, Christensen EM, Oliver DJ (2001) the biological functions of glutathione revisited in Arabidopsis transgenic plants with altered glutathione levels. Plant Physiol 126:564–574

Chapter 10
Involvement of AsA/DHA and GSH/GSSG Ratios in Gene and Protein Expression and in the Activation of Defence Mechanisms Under Abiotic Stress Conditions

Vasileios Fotopoulos, Vasileios Ziogas, Georgia Tanou, and Athanassios Molassiotis

Abstract In a persistently changing environment there are many adverse abiotic stress conditions such as cold, heat, drought, salinity, heavy metal toxicity and oxygen deprivation, which remarkably influence plant growth and crop production. Plant cells produce oxygen radicals and their derivatives, so-called reactive oxygen species (ROS) during various processes associated with abiotic stress. Moreover, the generation of ROS is the main means for higher plants to transmit cellular signalling information concerning the changing environmental conditions. Therefore, plants have evolved inducible redox state-based sensing mechanisms that are activated or amplified in response to adverse environmental conditions. Ascorbate and glutathione, the key cellular redox buffers, are used for both detoxification of ROS and transmission of redox signals. In recent years, it has become clear that abiotic stress conditions induce changes in the reduction/oxidation (redox) state of signalling molecules, which in turn modulate gene and protein expression to increase plant acclimation to abiotic stress. This important redox state-related branch of science has given several clues in understanding the adaptive plant responses to different stressful regimes. In this chapter, an overview of the literature is briefly presented in terms of the main function of ascorbate and glutathione in plant cells. Further more, we describe how important forms of abiotic stress regulate the expression of genes and proteins involved in the ascorbate and glutathione redox sensing system.

Keywords Ascorbate • Glutathione • Redox state • Antioxidant enzymes • Gene expression • Abiotic stress • Reactive oxygen species

---

V. Fotopoulos
Department of Agricultural Sciences, Biotechnology and Food Science, Cyprus University of Technology, PC 3036 Limassol, Cyprus

V. Ziogas, G. Tanou, and A. Molassiotis (✉)
Faculty of Agriculture, Aristotle University of Thessaloniki, 54 124 Thessaloniki, Greece
e-mail: amolasio@agro.auth.gr

N.A. Anjum et al. (eds.), *Ascorbate-Glutathione Pathway and Stress Tolerance in Plants*,
DOI 10.1007/978-90-481-9404-9_10, © Springer Science+Business Media B.V. 2010

1 Introduction

One of the major outcomes of the aerobic life of plant cells is the formation of reactive oxygen species (ROS). The term ROS defines the partially reduced forms of atmospheric oxygen. In general they result from the photodynamic excitation of oxygen, a process that ends with the formation of singlet oxygen ($^1O_2$) or during the partial reduction of oxygen through the transfer of one, two, or three electrons to its molecule that has as a result the birth of superoxide radical ($O_2\cdot^-$), hydroxen peroxide (H_2O_2) and hydroxyl radical ($\cdot OH$), respectively (Mittler 2002).

During the life cycle of the cell, ROS are formed in chloroplasts, mitochondria, plasma membrane, apoplastic space and peroxisomes. In chloroplasts ROS are produced during reactions that participate in the mechanism of photosynthesis, enhancing the concept that ROS are an unavoidable part of the aerobic metabolism of the cell. In mitochondria, $O_2\cdot^-$ are produced in two segments of the electron transport chain: during respiration, in the flavoprotein NADH dehydrogenase (complex I) and in the ubiquinone zone. In the plasma membrane, the ROS-producing system is an NADPH oxidase. In the apoplastic space, cell wall-bound peroxidases have been shown to generate H_2O_2 and during stressful abiotic conditions ROS are formed in peroxisomes during photorespiration due to the formation of H_2O_2 by the enzymatic activity of glycolate oxidase (Corpas et al. 2001; Mittler 2002; Mittler et al. 2004).

Under stress-free growth conditions, the plant cell produces ROS, especially $O_2\cdot^-$ and H_2O_2, in proportions that are kept at a rather low level (Asada 1994; Polle 2001). These molecules, apart from being toxic to the cell, act as second messengers and key regulators of growth, development and defence pathways, strengthening the dual role of ROS in cell biology (Mittler 2004; Foyer and Noctor 2005a). However, plants are not always cultivated under ideal conditions; in many cases various abiotic stress factors occur and the cellular homeostasis is disrupted resulting in the elevation of the level of ROS formation, a situation that is characterized under the heading of 'oxidative stress'. These abiotic stresses may include drought, salinity, high light, chilling, heavy metals, heat shock, ozone stress and others (Mittler 2002; Chalapathi and Reddy 2008).

During extreme abiotic stress conditions, high levels of ROS may lead to cell death; that event is the result of oxidative processes such as membrane lipid peroxidation, DNA/RNA nicking, protein oxidation and enzyme inhibition (Tanou et al. 2009c). For that case, the cell is equipped with a battery of antioxidant strategies, including enzymatic and non-enzymatic molecules. The latter facilitate the cell to be detoxified during extreme abiotic stress conditions, and also to keep ROS at an optimum level thus allowing all the necessary signalling cascades to take place in order to make all the appropriate modifications in gene/protein expression and cell structures in response to environmental and developmental stimuli (Foyer and Noctor 2005a). Among the enzymatic and non-enzymatic antioxidant pathways, ascorbate and glutathione are molecules with a regulatory role that participate in the redox signalling of the plant cell under abiotic stress conditions (Noctor 2006; Anjum et al. 2008a; Meyer 2008; Khan et al. 2009; Szalai et al. 2009).

Reduced glutathione (GSH) is a low molecular weigh tripeptide thiol with the formula γ-glu-cys-gly. This multifunctional molecule is a vital part of the antioxidant armory of the plant cell against oxidative stress, and contributes to the cellular defence and protection (Potters et al. 2002). Its pivotal role as an antioxidant derives from the fact that GSH participates in the ascorbate–glutathione cycle as the reducing agent of dehydroascorbate (DHA) while it also possesses the ability to protect the integrity of the cellular plasma membrane by maintaining a-tocopherol and zeaxanthine in the reduced state as well as protecting proteins from denaturation caused by the oxidation of protein thiol groups (Garnczarska 2005; Paradiso et al., 2008). In addition, glutathione exerts its antioxidant ability through the direct ROS scavenging as well as by acting as the substrate for glutathione peroxidase (EC 1.11.1.9; GPx) and glutathione-*S*-transferases (EC 2.5.1.18; GST), enzymes that participate in ROS removal. Finally, glutathione is involved in the formation of phytochelatines and together with its oxidized form (GSSG) they consist the redox couple (GSH/GSSG) which plays a drastic role in the maintenance of the cellular homeostasis and signalling system in plants (Rauser 1995; Ha et al. 1999; Clemens 2006; Srivalli and Khanna-Chopra 2008), leading to the suggestion that the GSH/GSSG ratio, indicative of the cellular redox balance, may be involved in ROS perception (Shao et al. 2008).

The concentration of GSH is increased under stressful conditions, and such an increase gives the ability to the cell to counteract the oxidation of glutathione and provokes alterations in gene expression directly or through interplay with regulatory proteins and/or transcription factors (Pasqualini et al. 2001; Ruiz and Blumwald 2002; Freeman et al. 2004). This elevation in GSH concentration is of vital importance, because it induces the signal transduction and defence against ROS which is achieved through a pathway with various control points, which include orchestrated activation of genes encoding enzymes related with glutathione metabolism (Sečenji et al. 2010). Therefore, high levels of glutathione provide the cell with the ability to counteract the negative effects of oxidative stress syndrome (Srivalli and Khanna-Chopra 2008).

Ascorbate (AsA) is considered by many as one of the most powerful and crucial antioxidants in the plant cell. Its presence has been detected in a wide range of cellular compartments like the cytosol, chloroplast, vacuoles, mitochondria and extracellular matrix (apoplast) (Kollist et al. 2000; Dipierro et al. 2005; Foyer and Noctor 2005b; Cheng et al. 2007). Under non-stress conditions ascorbate is detected mostly in its reduced form (Garnczarska 2005). Ascorbate is involved in the protection of a wide range of cellular compartments against oxidative attacks due to its ultra crucial ability to function as a donor of electrons in a broad range of enzymatic and non-enzymatic reactions (Noctor and Foyer 1998). Ascorbate possesses the ability to directly scavenge ROS and assist to the detoxification from H_2O_2 through its reduction via the ascorbate–glutathione cycle (Veljovic-Jovanovic et al. 2001; Anjum et al. 2008a; Khan et al. 2009). Other antioxidant abilities of AsA are the regeneration of the oxidized form of a-tochopherol as well as its ability to act as a cofactor of violaxanthine de-epoxidase (Blokhina et al. 2003). Specific cytosolic APX isoforms are significantly induced under the effect of abiotic stress and redox perturbation (Kubo et al. 1995; Karpinski et al. 1997; Ranieri et al. 2000; Murgia et al. 2004).

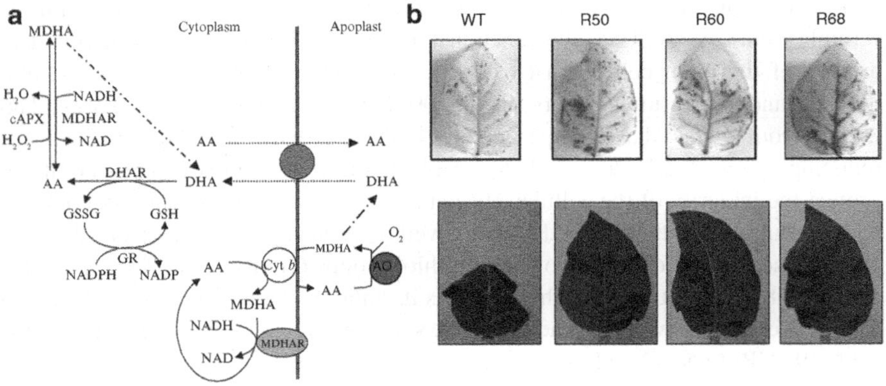

Fig. 1 (**a**) Schematic representation of the AsA–GSH cycle responsible for AsA recycling (Modified from Fotopoulos et al. 2006). Ascorbate oxidase which was targeted for over-expression in transgenic tobacco plants in the work by Fotopoulos et al. (2006, 2008) is *circled in red*. (**b**) Detached leaf assay and *in situ* localization of H_2O_2 in AO-over-expressing tobacco plants with reduced apoplastic AsA redox state (Taken from Fotopoulos et al. 2008). Fully-expanded leaves were detached from well-watered plants, immediately weighed, and held at room temperature for 1 h. Lower row displays images of representative leaves 1 h after detachment, while upper row displays leaves, (taken from plants grown under normal conditions, following incubation with 3,3′-diaminobenzidine (DAB) which polymerizes instantly (to form a reddish-brown complex which is stable in most solvents), as soon as it comes into contact with H_2O_2 in the presence of peroxidases. WT = wild-type, R50, R60 & R68 = transgenic tobacco lines over-expressing AO

Ascorbate and glutathione are linked through the ascorbate–glutathione cycle (outlined in Fig. 1a). During this cycle ascorbate is converted to the unstable radical monodehydroascorbate (MDHA) which rapidly disproportionates to yield dehydroascorbate (DHA) and AsA (Smirnoff, 2000). The latter is converted back to reduced ascorbate with the usage of reduced glutathione (GSH), acting as the electron donor in this reduction. During this process several enzymes coordinate their action in order to achieve the recycling of ascorbate, such as ascorbate peroxidase (EC 1.11.1.11; APX) and dehydroascorbate reductase (EC 1.8.5.1; DHAR) (Ushimaru et al. 1997). Efficient recycling of glutathione is ensured by glutathione reductase (EC 1.6.4.2; GR) activity (Noctor et al. 1998). Changes in turnover rates during the ascorbate–glutathione cycle may become manifested in altered redox ratios of AsA/DHA or GSH/GSSG (Tausz et al. 2004). Although DHA is reduced to ascorbate by DHAR, traces of DHA are always present in plant samples and the ratio of AsA/DHA is relatively lower compared to the ratio GSH/GSSG, especially under field conditions (Noctor et al. 1998).

During abiotic stress-driven oxidative stress, higher plants have the ability to sense, transduce and translate the ROS signals into specific cellular responses, a process that depends upon the presence of redox-sensitive proteins that possess the ability to accept reversible oxidation/reduction and may be functional or not according to the cellular redox state. ROS have the ability to oxidize redox-sensitive proteins directly or indirectly with the usage of molecules like glutathione. Cellular metabolism may be altered through the corresponding action of redox-sensitive metabolic enzymes, whereas redox sensitive proteins exert their function via downstream signalling components such as kinases, phosphatases and transcription factors (Shao et al. 2008). A number of regulatory and structural genes which are under the orchestration of thiol–disulfate status and ROS signalling have been identified in transgenic and wild-type plants, contributing to the clarification of the function of the redox network (Tausz et al. 2004). This network manipulates the levels or ROS during abiotic stress conditions by interpreting signals received from various cellular compartments, with the redox couple GSH/GSSG having an exceptional role in its fine tuning (Szalai et al. 2009).

Although ascorbate and glutathione are united together through redox flux and coordinate their action during the metabolism of ROS, each one has specific roles in the regulation of the cellular status (Noctor 2006). Additionally, several studies conducted in a number of plant species under abiotic stress conditions have elucidated the fact that a high ratio of GSH/GSSG and/or AsA/DHA sustained by increased GSH and AsA or diminution of GSSG and DHA, may be the key element for efficient protection against abiotic stress-induced accumulation of ROS (Szalai et al. 2009). In this chapter, we present an up-to-date analysis of existing studies that reveal aspects of how ascorbate/glutathione-related responses are modulated by various envinromental stresses in plants.

2 Drought Stress

Drought is perhaps the most common abiotic stress limiting crop productivity world-wide. Different drought levels caused an oxidation of the glutathione pool in barley and pine leaves (Smirnoff 1993; Tausz et al. 2001), indicating that the maintenance of GSH/GSSG ratio appears to function as an important component of the plant's antioxidant defence system under drought stress. However, links between glutathione redox state and the drought tolerance are less clear. Lascano et al. (2001) found no clear differences between four different drought-tolerant wheat varieties following one month of drought exposure, but the two more resistant cultivars responded with an increase in total glutathione during the rewatering period. In the work of Loggini et al. (1999) two wheat cultivars were compared with different drought tolerance and the researchers found that both cultivars showed a higher (more reduced) GSH/GSSG ratio after a month of drought. Also, an increase in glutathione-related enzyme activities was only observed in the more susceptible cultivar. Moreover, Herbinger et al. (2002) reported an induction in total glutathione in wheat flag leaves exposed to drought stress, particularly evidenced in the more susceptible cultivar. In addition, in the work of Khanna-Chorpa and Selote (2007), susceptibility of Moti leaves during severe water stress was evident from the failure in H_2O_2 management and by a significant reduction in AsA/DHA and GSH/GSSG ratios. The high H_2O_2 level and/or oxidation of the ascorbate pool might have an inhibitory effect on antioxidant enzymes, particularly APX and GR, in the susceptible and in non-acclimated Moti leaves (Shigeoka et al. 2002). In the study of Sharma and Dubey (2005) a decline in the concentration of total ascorbate as well as decline in AsA/DHA ratio was observed under drought stress in rice. Furthermore, the previous authors found that mild drought stress induced a significant increase in glutathione concentration which can be explained on the basis of significant increase in GR and DHAR activities. Results in this study further indicate enhanced activities of all the enzymes of the ascorbate–glutathione cycle, signifying a potential role of these enzymes in providing antioxidant defence under drought stress conditions.

It is known that an increase in AsA and an induction of ascorbate–glutathione cycle enzymes during water stress minimized the oxidative damage, but decrease in AsA content intensified oxidative processes during severe water stress conditions (Sgherri and Navari-Izzo 1995). AsA showed a reduction under drought stress in maize and wheat, suggesting its vital involvement in the oxidative response (Nayyar and Gupta 2006). Lascano et al. (2001) stated that lesser oxidative damage in the tolerant wheat cultivar during osmotic stress is due to higher AsA and induction of AsA–GSH cycle enzymes. Furthermore, Al-Ghamdi (2009) found that susceptibility of wheat leaves during severe water stress was evident from the failure in H_2O_2 management and by drastic oxidation of ascorbate–glutathione pool and significant reduction in AsA/DHA ratio. Finally, APX activity increased under drought stress in *Euphorbia escula* (Davis and Swanson 2001), *Zea mays* (Jiang and Zhang 2002), soybean (Riekert van Heerden and Kruger 2002), wheat (Dalmia and Sawhney, 2004), and *Populus accutifolius* (Turkan et al. 2005).

Several reports demonstrate the importance of the ascorbate–glutathione cycle in the regulation of the plant's response to drought stress, while numerous approaches have been made to produce transgenic crops tolerant to drought via the genetic manipulation of key enzymes in the ROS detoxifying and ascorbate recycling pathways. Examples include tall fescue and potato plants over-expressing SOD and APX (Tang et al. 2006; Lee et al. 2007 – for a comprehensive review see Cruz de Carvalho 2008).

Latest findings by Sečenji et al. (2010) who compared the responses of two wheat genotypes with differing capacity to withstand reduced water supply revealed major differences in ascorbate metabolism: both ascorbate oxidation and transcription levels of enzymes processing ascorbate were changed. Relative transcript levels of APX, monodehydroascorbate reductase (MDHAR), DHAR and GR isoenzymes showed different transcriptional changes in the two genotypes. Specifically, expression levels of two cytosolic APX isoenzymes and a thylakoid-bound variant increased significantly in the drought tolerant wheat cultivar while a cytosolic and a stromal APX coding transcript were found to be higher in the drought sensitive cultivar after a 4-week-long water deficit stress. In addition, mRNA levels of two cytosolic MDHAR isoforms were induced in the sensitive genotype, whereas only one was induced in the tolerant cultivar. An up-regulated chloroplastic DHAR was detected only in the sensitive cultivar. However, increased expression levels of a cytosolic and a chloroplastic GR were detected only in the tolerant wheat cultivar. Such a pattern of gene expression regulation following imposition of water stress was accompanied by a significantly lower AsA redox state in leaves of the sensitive cultivar compared with the tolerant one, indicating that more robust transcription of ascorbate-based detoxification machinery may prevent an adverse shift of the cellular redox balance.

In a study by Chen and Gallie (2004), plants with an increased guard cell AsA redox state were generated by increasing DHAR expression, and these exhibited a reduction in the level of guard cell H_2O_2. In addition, a higher percentage of open stomata, an increase in total open stomatal area, increased stomatal conductance, and increased transpiration were observed. Guard cells with an increase in AsA redox state were less responsive to H_2O_2 or abscisic acid signalling, and the plants exhibited greater water loss under drought conditions, whereas suppressing DHAR expression conferred increased drought tolerance, thus suggesting that DHAR serves to maintain a basal level of AsA recycling in guard cells. The protective role of DHAR against oxidative stress was further supported by Eltayeb et al. (2006), who developed transgenic tobacco plants over-expressing cytosolic DHAR gene from *Arabidopsis thaliana*. Transgenic plants exhibited 2.3–3.1 fold higher DHAR activity and 1.9–2.1 fold higher level of reduced AsA compared with non-transformed control plants, resulting in enhanced tolerance to drought stress in terms of higher net photosynthesis.

Control of stomatal aperture is of paramount importance for plant adaptation to the surrounding environment. Work by Fotopoulos et al. (2008) reported on several parameters related to stomatal dynamics and performance in transgenic tobacco plants (*Nicotiana tabacum* cv. Xanthi) over-expressing cucumber ascorbate oxidase (AO), a cell wall-localized enzyme of uncertain biological function that oxidizes AsA to MDHA acid which dismutates yielding AsA and DHA. Transgenic tobacco AO-overexpressing plants exhibited significantly lowered AsA redox state in the apoplast. In comparison to wild type plants, leaves of AO over-expressing plants exhibited reduced stomatal conductance (due to partial stomatal closure), higher water content, and reduced rates of water loss on detachment. Transgenic plants also exhibited elevated levels of hydrogen peroxide and a decline in hydrogen peroxide-scavenging enzyme activity under normal growth conditions (see Fig. 1b). It should be noted that these transgenic tobacco plants also displayed enhanced sensitivity to various oxidative stress-promoting chemical agents, while RNA blot analyses suggested a correlation between such a response with a general suppression of the plant's antioxidative metabolism as demonstrated by lower expression levels of AsA recycling genes (Fotopoulos et al. 2006). Treatment of epidermal strips with either 1 mM DHA or 100 mM hydrogen peroxide resulted in rapid stomatal closure in wild type plants, but not in AO-over-expressing plants, therefore suggesting that signal perception and/or transduction associated with stomatal closure is altered by AO over-expression. These data support a specific role for cell wall-localized AsA in the perception of environmental cues, and suggest that DHA acts as a regulator of stomatal dynamics.

3 Salinity Stress

Salinity is a major abiotic stress that plants experience and greatly affects agricultural productivity. Due to changes in the osmotic balance, high salinity can cause water loss in cells and induce drought stress in plant tissues and results in ionic toxicity which can affect the physiological and biochemical function of the plant cell. Transgenic plants with low or high levels of AsA and GSH also exhibit an altered response to NaCl-induced oxidative damage (Huang et al. 2005; Yadav et al. 2005), demonstrating the crucial importance of a tightly orchestrated redox buffering capacity under salinity.

There is considerable evidence that GSH plays a protective role in salinity tolerance by maintaining the redox state (Gossett et al. 1996; Shalata et al. 2001). Salt stress induces increases in both GSH and γ-ECS activity (Ruiz and Blumwald 2002), which is known to be a key factor controlling the amount of GSH in leaves (Noctor et al. 1998). Additionally, Mittova et al. (2004) provided evidence that salt-dependent up-regulation of γ-ECS activity in tomato occurs at the level of transcription. In the study of Chaparzadeh et al. (2004), the increased foliar GR activity was accompanied by a decrease in GSH/GSSG ratio, suggesting that a predominant GSH oxidation took place under salinity. Furthermore, increases in GR activity during salt stress were reported in pea (Hernandez et al. 1993, 1995, 2000), cantaloupe (Fahmy et al. 1998), citrus (Gueta-Dahan et al. 1997), soybean (Comba et al. 1998), rice (Dionisio-Sese and Tobita 1998; Lin and Kao 2000; Vaidyanathan et al. 2003; Demiral and Turkan 2005; Tsai et al. 2005), tomato (Shalata et al. 2001; Molina et al. 2002; Mittova et al. 2003), *Arabidopsis thaliana* (Huang et al. 2005), wheat (Sairam et al. 2005), *Vigna radiate* (Sumithra et al. 2006), *Setaria italica* (Sreenivasulu et al. 2000), and *Helianthus annuus* (Davenport et al. 2003). On the other hand, over-expression of GR in plants leads to an increase in the antioxidant capacity and in the resistance to oxidative stress (Kocsy et al 2001).

It has been previously reported that salt stress leads to a decrease in the ascorbate–glutathione cycle components in salt sensitive cultivars and to an increase in salt tolerant ones (Hernández et al. 2001; Shalata et al. 2001; Mittova et al. 2003, 2004). Salt-induced oxidative damage led to necrotic lesions in the minor veins of pea leaves, as oxidative stress was higher in the apoplasts (Hernández et al. 2001). The GSH/GSSG ratio declined under salt stress, and there was no GR activity in the apoplasts of the sensitive pea cultivar Lincoln. Continuous exposure to salt stress in rice seedlings made them more tolerant when the GSG/GSSG levels returned to normal values after an initial decline (Fadzilla et al. 1997). The results of Mittova et al. (2003) indicate that salt tolerance is linked to the ability to up-regulate enzymes of GSH synthesis and utilization and that this is absent from the salt sensitive species. A comparison of the total amounts of GSH and the GSH/GSSG ratios in the tissues of both species under salinity indicates that the dramatic decrease in the GSH/GSSG ratio in salt-stressed *Solanum lycopersicum* roots is due to other factors such as NADPH deficits, rather than limitations on GR activity. A recent analysis in citrus plants revealed that pre-exposure to sodium nitroprusside (SNP), a donor of nitric oxide (NO), prior to salinity resulted in higher (less oxidized) glutathione redox compared to NaCl-treated as well as control plants (Tanou et al. 2009b) providing a link between glutathione and nitric oxide during the establishment of salt tolerance. Additionally, proteome-wide analysis revealed the involvement of many ascorbate/glutathione-related proteins in the acclimation of citrus plants to high salinity (Tanou et al. 2009a).

Some researchers have reported that salt stress leads to a decrease in ascorbate content in salt-sensitive cultivars (Hernández et al. 2001; Shalata et al. 2001; Mittova et al. 2003, 2004). The biosynthetic capacity of ascorbate is impaired under stress conditions because the ascorbate pool is generally determined by its rates of not only regeneration but also synthesis (Song et al. 2005). It is also reported that regeneration of ascorbate under salinity is insufficient or that ascorbate synthesis is lower than ascorbate catabolism (Shalata et al. 2001; Amor et al. 2006). However, in some plants acclimated to salinity a significant increase in total ascorbate was found (Shalata et al. 2001). More important than the total AsA content is the AsA/DHA ratio that, in the case of roots in high and leaves in low salinity conditions, was found to be comparable to the values observed in salt-tolerant cotton plants and calli (Gossett et al. 1996). The low AsA/DHA ratio in leaves at high salinity might be an indication of APX participation in ROS scavenging. A proper increase of AsA, during H_2O_2 increase in conditions of high salinity, may be important for maintaining APX activity, being that APX is inactivated when ascorbate concentration drops dramatically (Asada 1999). Under salinity, AsA is mainly regenerated from MDHA (Shalata and Tal 1998; Mittova et al. 2000) or DHA (Meneguzzo et al. 1999). In marigold plants under salinity stress (*C. officinalis*), the decreasing trends of both MDHAR and DHAR activities may suggest that a non-enzymatic disproportionation of MDHA to AsA and DHA or a reduction of MDHA by reduced ferredoxin and b-type cytochrome (Noctor and Foyer 1998) could participate in AsA regeneration.

In view of the known links between oxidative stress and cell death, it is surprising that few studies have addressed the role of salt stress and cell death responses in plants. The results of a recent study by Tanou et al. (2009c) illustrate how the oxidation of ascorbate/glutathione pool participates in the oxyradical-mediated necrotic death-like destruction in salt-stressed leaves. The model of ascorbate/glutathione-mediated in salt dependent-oxidative cell death, based on components identified in strawberry response leaves, is outlined in Fig. 2.

Several approaches have been made in order to produce transgenic plants with an acquired tolerance to salinity stress by manipulating gene expression levels of various antioxidant enzymes involved in the ascorbate–glutathione cycle such as DHAR and APX (for a comprehensive review see Ashraf 2009). Enhanced expression of DHAR was achieved by the expression of rice DHAR in transgenic Arabidopsis plants (Ushimaru et al. 2006). This study further demonstrated that transgenic Arabidopsis plants showed enhanced tolerance to salt despite the fact that there had been a slight increase in DHAR activity and total ascorbate in the transgenic plants. The protective role of DHAR against oxidative stress was further supported by Eltayeb et al. (2006), who developed transgenic tobacco plants over-expressing cytosolic DHAR gene from *Arabidopsis thaliana*. Transgenic plants

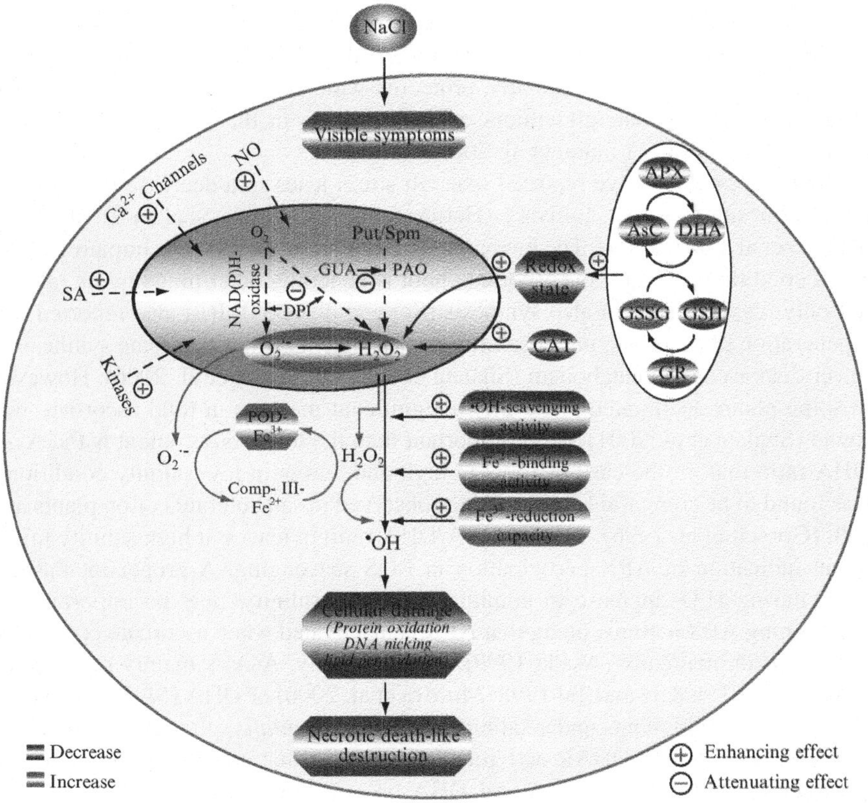

Fig. 2 Model schematizing some of the relationships among cellular redox state, ROS production and necrotic death-like destruction in leaves of strawberry plants exposed to salinity. NaCl ensures activation of membrane-bound NAD(P)H-oxidase produces H_2O_2 and $O_2\cdot^-$. H_2O_2 also generated via the polyamine oxidase (PAO)-mediated catabolism of high Put and Spm. The low cell redox state, as indicated by the accumulation of the oxidized forms of ascorbate (DHA) and glutathione (GSSG), together with the inhibition in the activities of some H_2O_2-scavenging enzymes ensures H_2O_2 generation. Nitric oxide (NO), salicylic acid (SA), protein kinase and Ca^{2+} channel activity promote NaCl-dependent H_2O_2 production. NaCl-induced high Fe^{3+}-reduction capacity, and low Fe^{2+}-binding and ·OH scavenging activities provoke the conversion of H_2O_2 to ·OH via the Fe^{2+}-catalyzed Fenton reaction chemistry and favor the cellular oxidative damage and necrotic death-like destruction. It is possible that apoplastic peroxidase (POD), transformed into the Compound III by $O_2\cdot^-$, catalyzes also the generation of ·OH in the presence of H_2O_2 (Reprint from Tanou et al. 2009c)

displayed enhanced DHAR activity and higher levels of reduced AsA compared with non-transformed control plants, maintaining redox state of AsA and ultimately resulting in enhanced tolerance to salt stress. Similarly, differential tolerance to salinity was achieved by over-expressing/suppressing the expression of ascorbate oxidase in tobacco and Arabidopsis (Yamamoto et al. 2005). AO activities in the transgenic tobacco plants expressing the gene in sense and antisense orientations were, respectively, about 16- and 0.2-fold of those in the wild type. At high salinity conditions, the percentage germination and photosynthetic activity were higher in antisense plants. In addition, the redox state of apoplastic ascorbate in sense plants was very low even under normal growth conditions, whereas upon salt stress, the redox state of symplastic and apoplastic ascorbate decreased among the three types of plants, but was lowest in the sense plants.

Acquired tolerance to salinity has also been demonstrated via genetic manipulation of other gene targets not directly linked to the ascorbate–glutathione pathway. Yadav et al. (2005) developed transgenic tobacco plants over-expressing glyoxalase pathway enzymes that suppress accumulation of methylglyoxal (MG) in plants under salt stress. The transgenic plants showed enhanced activity of various glutathione-related antioxidant enzymes under both control and saline conditions. Additionally, these plants maintained high contents of reduced glutathione and overall increased glutathione redox state under salt stress, thus suggesting that hindrance in an increase in MG coupled with maintaining higher reduced glutathione levels can be considerably achieved by the over-expression of glyoxalase pathway enzymes for developing salt tolerant plants. Latest findings by G. Tanou et al. (unpublished data) provide significant evidence that NaCl application in citrus leaves results in the oxidation of the leaf ascorbate redox state, while the glutathione redox state remained unaltered. Antioxidant enzyme activity assays revealed that an array of redox-related proteins is induced as a result of high salinity stress, whereas transcript regulation was more variable.

4 Chilling-Low Temperature Stress

Chilling impairs all photosynthetic components provoking, for example, reduction of stomatal conductance, changes in pigment complexes and losses of photochemical efficiency, modifications in the biophysical properties of thylakoid lipids and restriction of electron transport, as well as decreases in enzyme activity and protein metabolism (Ensminger et al. 2006). The level of damage is dependent on the plant species, the developmental and phenological stage, the organ and tissues, and the degree of stress. ROS are also involved in freezing stress, participating in lipid peroxidation, protein destruction, and collapses of the antioxidant defence systems (Polle 1997). Thus, the improvement of chilling stress tolerance is often related to enhancement of activities of antioxidant systems in plants.

It has been documented that in order to prevent hydrogen peroxide accumulation to toxic levels during chilling stress, a high capacity of the AsA–GSH cycle is necessary (Kocsy et al. 2001). In the work of Zhang et al. (2008) a significant increase in ascorbate and glutathione concentration was observed during chilling stress. Higher glutathione concentrations in chilled maize plants are the result of an induction of key enzymes of glutathione synthesis, as well as sulphate reduction, which also increases cysteine levels (Kopriva et al. 2001). Increases in GSH levels and/or GR activity during chilling stress have been observed in many plant species such as tomato (Walker and McKersie 1993), sorghum (Badiani et al. 1997), wheat

(Kocsy et al. 2001), jack pine (Zhao and Blumwald 1998), and poplar (Foyer et al. 1995). A long duration chilling stress experiment in maize showed increased activities of various antioxidant enzymes, including GR (Hodges et al. 1997). In line, chilling stress and cold acclimation studies in rice showed an increase in GR activity (Oidaira et al. 2000; Kuk et al. 2003). In cotton, the increase of the activities of glutathione–ascorbate cycle enzymes in chloroplasts by genetic manipulation increased resistance to chilling-related photo-oxidative stress under laboratory conditions (Payton et al. 2001). When transgenic cotton overproducing GR was field-grown, there was no significant difference with wild plants as GR activity doubled in wild-type cotton during slow chilling exposure in the field (Logan et al. 2003). Genetic transformation studies for freezing tolerance in maize showed the up-regulation of three genes, including GSTs, under both normal and cold-acclimated conditions (Wang 2005). GSTs are involved in the oxidative signalling pathway and contribute to the genetic acclimation towards freezing tolerance in maize. Proteome analysis of chilling stress in rice showed the up-regulation of cysteine synthase (Yan et al. 2006). This enzyme is responsible for the final step in cysteine biosynthesis, a key-limiting step in GSH production. Thus, GSH and GR are important for resistance to chilling stress.

During chilling and cold acclimation, the maintenance of a high GSH/GSSG ratio is very important in order to ensure that GSH can function appropriately in the AsA–GSH cycle and other physiological processes. During chilling, the GSH content and the GSH/GSSG ratio were higher in tolerant genotypes of tomato compared with sensitive ones (Walker and McKersie 1993). Correspondingly, the GSH/GSSG ratio was generally higher in chilling-tolerant maize genotypes than in sensitive ones at 11°C (Hodges et al. 1996), demonstrating that the maintenance of a high GSH/GSSG ratio contributes to improved chilling tolerance. The higher GSH/GSSG ratio in the freezing-tolerant genotypes may keep the sulphhydryl groups of the proteins in reduced form, thus decreasing the possibility of intermolecular disulphide bridge formation when the plants are exposed to freezing temperatures (Levitt 1962).

In the works of Huang and Guo (2005) and Dai et al (2009), conducted using tolerant and sensitive rice and barley cultivars, respectively, the AsA content had a different pattern among the two cultivars in response to chilling. The tolerant one showed elevated amounts of AsA in contrast to lower levels in the sensitive one. In the same study the tolerant cultivar exhibited higher activities of APX and GR, an observation that is in accordance with the findings of Tao et al (1998) who used pine trees as a research model. These elevated activities of APX and GR facilitate the cell to cope with the oxidative stress due to chilling conditions. In addition, work by Guo et al. (2006) who tested four rice cultivars under chilling condition concluded that the chilling tolerance was well correlated with the enhanced antioxidant capacity of the cultivars, a capacity that was attributed to the higher AsA content and enhanced activity of antioxidant enzymes like APX and GR.

Freezing injury has been shown to involve the participation of ROS. Antioxidant enzymes can protect plant cells from oxidative stress imposed by freezing injury; therefore, cold acclimation may involve an increase in the expression of antioxidant enzymes. In a work carried out by Baek and Skinner (2003), quantitative RT-PCR was used to measure the expression levels of a wide range of antioxidant enzymes during cold acclimation in near-isogenic lines (NILs) of wheat, differing in the Vrn1-Fr1 chromosome region that conditions winter versus spring wheat growth habit. The expression levels of several antioxidant enzyme transcripts were induced (Mn-SOD, MDAR, t-APX, DHAR, GPX, and GR), suppressed (CAT), or remained relatively constant (FeSOD and Cu/Zn–SOD). The Vrn1-Fr1 region appeared to have a role in regulating the expression level of some of the antioxidant enzyme genes because CAT, t-APX and Mn-SOD transcripts were expressed at significantly higher levels in the winter wheat NIL than the spring wheat NIL after 4 weeks' cold acclimation. More recently, Fortunato et al. (2010) performed the characterization of the antioxidant system of *Coffea* sp. genotypes with different cold acclimation abilities using an integrated biochemical and molecular approach. Cold-tolerant variety Icatu showed the greatest ability to control oxidative stress, as reflected by the enhancement of several antioxidant components (Cu/Zn–SOD and APX activities; ascorbate, α-tocopherol and chlorogenic acids contents) and lower reactive oxygen species contents (H_2O_2 and ·OH). Gene expression studies showed that GR and DHAR might also be involved in the cold acclimation ability of the cold-tolerant variety. The difference in the triggering of antioxidant components supports the hypothesis of their importance to cold tolerance in *Coffea* sp. and could provide a useful probe to identify tolerant genotypes.

5 Heat Stress

Although there have been several reports on oxidative stress and the response of antioxidant defence mechanisms in heat-stressed plants (Dat et al. 1998; Anderson and Padhye 2004), there have been fewer reports focusing on glutathione homeostasis. Treatment of maize roots to heat shock temperatures of 40°C resulted in decrease of cysteine levels and increase in GSH levels (Nieto-Sotelo and Ho 1986). There was an increase in the GSH-synthesizing capacity in maize root cells, which was related to the cell's capacity to cope with heat stress conditions. Accumulation of GSH has also been observed in heat-stressed tomato seedlings (Rivero et al. 2004). In wheat, it was established that heat stress induced accumulation of GSH levels and increased the activity of the enzymes involved in GSH synthesis and the GSH/GSSG ratio (Kocsy et al. 2001). Heat stress increased GSH levels during grain development in the flag leaf of two wheat genotypes with contrasting behavior under heat stress (Chauhan 2005). The GSH content was also higher in a heat-tolerant cabbage genotype compared to a sensitive one after heat stress.

There is increasing evidence for considerable interlinking between ascorbate redox state-related responses and heat stress. In the study of Ma et al. (2008), the contents of total ascorbate, AsA, total glutathione and GSH in apple leaves was increased during imposition of high temperature (40°C) for 2h. The high thermal activation of GR observed in apple leaves after exposure to 40°C suggests that GR also play an important role in the adjustment of metabolism to high temperature. APX and GR play an important role in the protection of plants from high temperature stress by preventing the oxidation of enzymes and membranes (Almeselmani et al. 2006). Furthermore, the study of Ma et al. (2008) demonstrated that, gene expression of APX, DHAR and GR in apple leaves in the high temperature (40°C) treatment was increased, compared with the control (28°C). In addition, cultivation of Arabidopsis plants at elevated but non-stress temperatures led to the increase of APX enzymatic activity and of foliar concentration of AsA (Panchuk et al. 2002). This suggests that the activation of AsA-dependent antioxidation system may be a pre-adaptive reaction to an enhanced production of ROS under severe heat stress. Song et al. (2005) stated in their study that the redox state of AsA is believed to play a pivotal role in influencing APX isoenzymes activities. Double Arabidopsis mutant *tylapx* and *capx1* mutant lacking thylakoid ascorbate peroxidase (tylapx) and cytosolic ascorbate peroxidase1 (capx1) showed enhanced tolerance to heat stress (Muller et al. 2007).

Not an extremely wide body of evidence exists linking the gene expression involved in the regulation of the plant's response to heat stress with ascorbate–glutathione redox states. Work by Larkindale et al. (2005) investigated the importance of different processes to heat stress tolerance. Plants tested were reactive oxygen metabolism mutants with lowered ascorbate levels (*vtc1*, *vtc2*), and these were more defective in basal (heating to 45°C) than acquired thermotolerance (pre-exposure to 38°C, followed by acclimation to room temperature and re-heating at 45°C), especially under high light. All mutants accumulated wild-type levels of heat shock protein 101 and small heat shock proteins, which are typical markers detected under stressful conditions.

6 Heavy Metal Stress

Heavy metal pollution of soils and waters is a major environmental problem. Some of the heavy metals are essential for the plant growth when they are present in normal levels; however, when they are present in excess, they cause toxic effects on plant growth, ultimately resulting in decreased yields or even plant death. Heavy metals are also known to induce free radicals in plants and, consequently, oxidative damage (Dietz et al. 1999).

Cadmium (Cd) is a non-redox metal unable to produce ROS via Fenton and/or Haber–Weiss reactions. However, several lines of evidence have revealed that oxidative stress is a major component of Cd phyto-toxicity (Piqueras et al. 1999; Sandalio et al. 2001; Romero-Puertas et al. 2004; Cho and Seo 2005; Hsu and Kao 2007; Anjum et al. 2008a, b, c, d; Khan et al. 2009).

Phytochelatins (PCs) comprise one of the mechanisms involved in the chelation of heavy metals by a family of peptide ligands. Many trace metals in an environment are known to induce PC production by plants and Cd has been found to be the most effective inducer of phytochelatins (Nishikawa et al. 2006). Phytochelatins form a family of peptides with a structure based on repetitions (2–11 times) of the γ-Glu-Cys dipeptide followed by C-terminal glycine (Gly), and they are structurally related to glutathione (GSH), which is a substrate for their synthesis (Rauser 1995; Ha et al. 1999; Clemens 2006). Thus, one of the mechanisms by which plants can withstand Cd toxicity is by maintaining high levels of phytochelatin or its precursor, GSH, which functions as a heavy metal ligand (Cánovas et al. 2004). Upon heavy metal exposure, GSH concentrations drop as a consequence of initiated PCs biosynthesis. This causes oxidative stress and in turn short-term toxicity (Schützendübel and Polle 2002; Nocito et al. 2006).

Cadmium was shown to induce a significant increase in the mRNA expression level of genes involved in GSH synthesis (*gsh1* and *gsh2*) and phytochelatin synthase (*pcs1*) in leaves of *Arabidopsis thaliana* (Semane et al. 2007). The authors observed a significant decrease of reduced GSH in Cd-treated plants, while Cd treatment increased the accumulation of GSSG, keeping the GSH/GSSG ratio lower than in control plants. The accumulation of GSSG was accompanied by suppressed GR mRNA levels, while the activity of GR was significantly enhanced. In addition, the authors observed a general increase of ROS-scavenging enzymes such as APX, CAT or SOD, indicating that the plants respond to Cd stress by activation of the AsA–GSH defence network at both transcriptional and enzymatic level. Studies of transgenic *Brassica juncea* plants, in which the expression of the GSH biosynthetic pathway enzymes was increased, have shown that PC biosynthesis and Cd tolerance have been correlated with over-expression of GSH (Zhu et al. 1999). Zhang and Ge (2008) found a close relationship between Cd level and GSH content as well as GST activity, suggesting that these two parameters of antioxidant defence system may be used as biomarkers of Cd-induced stress. Accumulated evidence suggests that GR plays an important role in the detoxification of Cd-induced ROS, possibly via the ascorbate–glutathione cycle. Increased GR activity in the roots exposed to Cd was reported in plants, including *Phaseolus vulgaris* (Chaoui et al. 1997), potato (Stroinski et al. 1999), radish (Vitoria et al. 2001), soybean (Ferreira et al. 2002), sugarcane (Fornazier et al. 2002), *Arabidopsis thaliana* (Skorzynska-Polit et al. 2003, 2004) and alfalfa (Sobrino-Plata et al. 2009). Moreover, GR activity was enhanced in shoots and roots of alfalfa plants exposed to Cd (Sobrino-Plata et al. 2009).

Notably, apart from its involvement in PC biosynthesis, GSH may contribute in several other ways to heavy metal tolerance. It may sequester toxic metal ions in the cytosol, and such complexes may activate PC synthase, transfer metal ions to newly synthesized PCs, or transport them to the vacuole (Howden et al. 1995; May et al. 1998; Xiang et al. 2001). Interestingly, recent findings by Wojas et al. (2008) suggest that the ability of high rate PC synthesis in transgenic tobacco plants over-expressing an *A. thaliana* PC synthase is insufficient to cope with the metal load if the functionality of the antioxidant system is simultaneously hampered. The authors demonstrated that transgenic plants were Cd-hypersensitive compared with wild-type plants, as manifested by strong depletion of GSH and higher oxidative stress.

Ascorbate concentrations were elevated following Cd exposure indicating that this antioxidant compound is necessary for redox cellular homeostasis under Cd stress (Sobrino-Plata et al. 2009). When AsA biosynthesis was enhanced by feeding plants with its last biosynthetic precursor, l-galactono-γ-lactone (GalL), Cd uptake was not affected (Paradiso et al. 2008). Chao et al. (2010) demonstrated that Cd toxicity of rice seedlings was accompanied by a decrease in the contents of AsA and AsA + DHA and in the ratios of AsA/DHA in leaves. Conversely, pre-treatment with AsA resulted in an increase in the contents of AsA and GSH, the ratios of AsA/DHA and GSH/ GSSG, and the activities of APX and GR in the leaves of rice seedlings, while several transcripts encoding APX and GR isoenzymes (*OsAPX2*-7 and *OsGR1*) were induced in rice leaves following AsA pre-treatment. Moreover, Hatata and Abdel-Aal (2008) found that AsA was markedly decreased to a very low level at high concentration (100 μM) of Cd stress, with a corresponding increase in the level of reduced ascorbate (DHA) indicating that DHA content was significantly enhanced probably through suppressed glutathione-dependent DHAR activity and/or due to a decrease in AsA synthesis. In addition, APX activity increased and a new basic root peroxidase isoform was found in Cd-treated alfalfa plants (Sobrino-Plata et al. 2009).

The role of AsA as an efficient scavenger for oxidative compounds is well-known (Polle and Rennenberg 1993). Furthermore, the effectiveness of AsA–GSH-regenerating enzyme system comprising MDHAR, DHAR and GR, and the maintenance of AsA, DHA, GSH and GSSG pools may contribute to controlling Cd-caused oxidative stress in plants (Paradiso et al. 2008; Anjum et al. 2010). The cellular concentration of AsA is, in fact, determined by the rate of its synthesis and decay. DHA is rapidly hydrolyzed into 2,3-diketogulanic acid if not reduced by DHAR. Anjum et al. noticed an enhancement in DHAR activity in Cd-exposed plants which could not maintain the AsA pool. Hence, they suggested that AsA generated by DHAR was utilized by some other metabolic function(s). It is also important to mention here that parallel to enhancement in AsA-regenerating enzyme activities, the activity of APX, an H_2O_2 scavenging enzyme, consumes/uses AsA as a reductant (Willekens et al. 1995). Besides, Anjum et al. (2010) reported Cd-induced increase in DHA with a corresponding increase in MDHAR activity and confirmed that this metabolite was chiefly formed by enzymatic action and not by non-enzymatic disproportionation which is in coincidence with results of Paradiso et al. (2008). In addition, Anjum et al. (2010) reported Cd-induced decrease in GSH pool in Cd-treated moongbean cultivars and suggested that the depletion of GSH pool due to Cd stress in spite of higher GR activity may indicate the mechanism of antioxidant defense through enhanced oxidation of GSH to GSSG by DHAR thus yielding AsA which was later utilized by APX for the detoxification of H_2O_2. In fact, GSH functions as an antioxidant by scavenging ROS, resulting in the oxidation of GSH to GSSG. It is well-established that not only the pool of GSH but also GSH/GSSG ratio is important to maintain the redox status of the cell (May et al. 1998; Paradiso et al. 2008). In the study of Anjum et al. (2010), the pool of GSH and also the ratio of GSH/GSSG (more, mainly due to Cd-induced decline in GSH pool) were significantly reduced. Furthermore, the reduced GSH/GSSG redox state of glutathione under Cd stress also indicated that maximum metabolic load was exerted to maintain redox buffer status of the cells, suggesting a leading role of GSH in an adaptive response to Cd stress and the maintenance of redox status in physiological conditions to a greater extent in Cd-tolerant moongbean cv. Pusa 9531 than in Cd-susceptible cv. PS 16. Similar results have been reported earlier by Anjum et al. (2008a, c) in *Brassica campestris* and *Vigna radiata*.

Copper (Cu) is an essential micronutrient required by all living organisms. Nevertheless, the reactive nature of ionic Cu makes it a toxic metal if not properly handled by the cell. One of the main reasons for these disturbances is due to the generation of hydroxyl radicals by free copper, which readily oxidize disulfide bonds within proteins, destroying their secondary structure, and also causing catastrophic damage to lipids and nucleic acids (Hanna and Mason 1992). Like Cd, PCs were proved to play an important role in Cu toxicity. Cobbett and Goldsbrough (2002) found that Cu is also a strong activator of PC biosynthesis both *in vivo* and *in vitro* and can form stable complexes with PCs. However, GSH does not appear to be directly involved in Cu detoxification and tolerance in *A. thaliana* since its role as a substrate for PC synthesis in Cu-stressed plants is not fully deciphered (Wójcik et al. 2009). In earlier reports, a reduction in cellular GSH levels as well as GR activity was observed in plants exposed to toxic levels of Cu (Gallego et al. 1996; Mazhoudi et al. 1997; Patra and Panda 1998; Tewari et al. 2006). In the study of Drazkiewicz et al. (2003), the GSH/GSSG ratio played an important role in a short-time exposure of plants to Cu (after 1 and 3 days) as well as GSH after 7 days. Though the concentration of total ascorbate increased with the increasing supply of Cu, the ratio of the redox couple (DHA/AsA) increased in Cu-deficient or Cu-excessive mulberry plants (Tewari et al. 2006). The latter authors conclude that the disturbed redox state of the cellular environment due to increased DHA/AsA ratio could have caused accelerated senescence and poor growth of the Cu-excess plants. In addition, proteomic analysis of Cu-treated Arabidopsis seedlings (Smith et al. 2004) revealed a regulatory role towards enzymes such as GSTs and GPx which contribute to the redox poise of the cell and, thus reflect the modified redox state of the cells induced by Cu.

Aluminium (Al) is a major constituent of soil and, consequently, plants often grow in soil environments in which the roots are potentially exposed to high concentrations of aluminium. The possible connection between Al toxicity and oxidative stress had been previously suggested by Cakmak and Horst (1991) following the finding that the Al-induced inhibition of root elongation was correlated with enhanced lipid peroxidation. Oxidative stress is evidently involved in the development of Al-induced toxic symptoms. Thus, Al-tolerant lines of wheat developed by *in vitro* microspore selection produced lower amounts of ROS in response to Al and grew better in comparison with the Al-sensitive genotype. The Al tolerance could tentatively be ascribed to the higher activity of GST observed in the tolerant lines (Darkó et al. 2004). Oxidative stress caused by Al toxicity is an early symptom that can trigger cellular redox state as well as root growth inhibition in Macaca (Al-sensitive) and SMIC148-A (Al-tolerant) potato clones (Tabaldi et al. 2009). Furthermore, the increased level of hydrogen peroxide in pumpkin (*Cucurbita pepo*) roots treated with aluminium sulphate was matched by both increased APX activity and ascorbate free radical reductase (AFRR) activity, while DHAR and GR did not change (Dipierro et al. 2005). Sharma and Dubey (2007) stated that the enhanced activities of enzymes of ascorbate–glutathione cycle MDHAR, DHAR and GR observed in Al^{3+} stressed rice seedlings appear to be due to the need of maintaining a favourable redox state, by maintaining a sufficient amount of reduced ascorbate and reduced glutathione and to overcome the possible problems of oxidation. Several approaches have been made in order to produce transgenic plants with an acquired tolerance to heavy metal stress including Al by manipulating gene expression levels of various antioxidant enzymes involved in the ascorbate–glutathione cycle, as well as genes involved in phytochelatin biosynthesis (for a comprehensive review see Sharma and Dietz 2009). Basu et al. (2001) demonstrated the importance of oxidative stress in Al tolerance by producing transgenic oilseed rape plants over-expressing a mitochondrial bread wheat Mn-SOD. Transgenic plants showed 1.5–2.5-fold greater SOD activity than wild type plants, resulting in reduced malondialdehyde accumulation and overall growth inhibition in response to Al. Yin et al. (2010) investigated the role of MDAR and DHAR in AsA regeneration during Al stress using transgenic tobacco plants over-expressing Arabidopsis cytosolic MDHAR or DHAR. Transgenic DHAR-over-expressing (but not MDHAR-over-expressing) plants showed better root growth than wild type plants after exposure to toxic Al level accompanied by lower hydrogen peroxide content, less lipid peroxidation and lower level of oxidative DNA damage. Furthermore, DHAR-over-expressing plants maintained a higher AsA level both with and without Al exposure when compared with wild type plants, in contrast with MDAR-over-expressing plants which maintained a higher AsA level only without Al exposure. These findings allowed the authors to suggest that the over-expression of DHAR, but not of MDAR, confers Al tolerance, and that maintenance of a high AsA level is essential towards acquiring Al tolerance.

Although nickel (Ni) is an essential element for plants, it is strongly phytotoxic at high concentrations (Eskew et al. 1983; Eskew et al. 1984). Nickel tolerance and hyperaccumulation in *Thlaspi* species is linked to the constitutive ability to accumulate higher concentrations of GSH, Cys and O-acetyl-L-serine (Freeman et al. 2004). The elevated GSH concentrations in *T. goesingense* were driven by constitutively elevated activities of serine acetyl transferase (SAT), which provides the C skeleton for Cys synthesis. The causality was proven in a transgenic approach: the overproduction of *T. goesingense* SAT in the non-accumulator *A. thaliana* led to the accumulation of O-acetyl-L-serine, Cys and GSH and coincided with strongly enhanced resistance to Ni-induced growth inhibition and oxidative stress (Freeman et al. 2004). On the other hand, the ascorbate recycling seems to be a mechanism of great importance in controlling the cellular redox state and the phenotypic performance of plants exposed to Ni. Exposure to 0.25 mM $NiCl_2 \times 6H_2O$ for 5 days resulted in toxicity symptoms, such as formation of reddish-brown mottled spots on the leaf blade in soybean seedlings (Saeidi-Sar et al. 2007). Exogenous addition of 1 mM AsA totally reversed these Ni-associated toxic symptoms. Also, the Ni-stressed seedlings exposed to AsA exhibited an improved growth as compared to Ni-treated plants while AsA considerably reduced root-to-shoot translocation of Ni (Saeidi-Sar et al. 2007). Contrary to DHAR and GR, of which the activities were unaffected by Ni, MDHAR was stimulated by Ni treatment in maize (*Zea mays* cv. LG 23/01) plants; it appears, thus, that the regeneration of ascorbate was accomplished through an activation of MDHAR following toxic Ni exposure (Baccouch et al. 2001).

7 Light Stress

In the natural environment, plants are frequently exposed to fluctuating light intensities and often absorb more light energy than can be consumed by photosynthetic metabolism and thus require that excess excitation energy be dissipated. Failure to dissipate excitation energy results in over-reduction of the photosynthetic chain components that direct linear electron flux from water to NADPH (Baker et al. 2007). Part of the absorbed light energy is dissipated as heat in the light-harvesting complexes of photosystem II (PSII) through non-photochemical quenching (Müller et al. 2001). Additional dissipation of excitation energy is also achieved by photochemical quenching (Baker et al. 2007) and reflects that action of processes such as the reduction of molecular oxygen at photosystem I by the Mehler reaction (Ort and Becker 2002). In this reaction, the photoreduction of molecular oxygen via PSI leads to the formation of $O_2^{\cdot -}$, which in turn becomes rapidly dismutated to H_2O_2, either spontaneously or enzymatically via SOD; H_2O_2 is then converted into water by APX ('water–water' cycle; Asada 1999). A soluble APX (sAPX) is present in the chloroplastic stroma, whereas another APX (tAPX) is anchored to the thylakoid membrane via a C-terminal transmembrane domain, its catalytic site facing the stroma (Shigeoka et al. 2002; Ishikawa and Shigeoka 2008). In recent years, it has become apparent that APX function is a key element in controlling many aspects of light stress in plants. The expression of both *Apx1* and *Apx2*, paralogous genes encoding cAPX in Arabidopsis, was found to be induced by H_2O_2 accumulation and redox changes in photosynthetic electron transport system through a plastoquinone pool within 15–30 min upon a low-light to high-light shift (Karpinski et al. 1997); this was found to be part of a systemic response to excess excitation energy (Karpinski et al. 1999), providing supporting evidence on the time frames in which acclimation responses occur. Studies were carried out using transgenic tobacco over-expressing tAPX or catalase in chloroplasts that showed a high degree of tolerance to photo-oxidative stress through a significant reduction in excess accumulation of H_2O_2 (Miyagawa et al. 2000; Yabuta et al. 2002).

Recently, Maruta et al. (2010) demostrated that both chloroplastic APXs, but particularly tAPX, are important for photoprotection and gene regulation under photooxidative stress in Arabidopsis leaves. Additionally, the cAPX expression under light stress has been examined and the available data revealed that its regulation is directly associated with the heat shock transcription factors (HSFs). For example, transgenic Arabidopsis plants expressing a dominant-negative form of Hsf21 (encoded by HSFA4a), which lacks an activation domain, showed inhibition of

APX1 mRNA accumulation under H_2O_2-producing light stress conditions (250 mmol m^{-2} s^{-1}) (Davletova et al. 2005). In view of the physiological significance of APX against oxidative stress, transgenic plants over-expressing APX isoenzymes have been generated to improve stress tolerance, as in the case of other antioxidant enzymes. For example, transgenic tobacco and Arabidopsis plants over-expressing tAPX in chloroplasts showed remarkable tolerance against photo-oxidative stress following exposure to strong light and paraquat treatment (Yabuta et al. 2002; Murgia et al. 2004).

On the basis of the evidence previously presented it is not suprising that light signals influence leaf ascorbate accumulation (Gatzek et al. 2002). Leaves exposed to light contain more ascorbate than leaves in the shade (Grace and Logan 1996), and ascorbate levels in the leaves show a diurnal rhythm (Dutilleul et al. 2003). The AsA pool size decreased, in apple leaves and fruit peel, after the whole trees were shaded with shade-net for 20 days (Li et al. 2009). Also, Mieda et al. (2004) reported that spinach leaves grown for 3 days under dark condition showed a significant decrease of the mRNA expression and enzyme activity of L-galactose dehydrogenase and L-galactono-1,4-lactone dehydrogenase. Bartoli et al. (2006) provided proof that through an interplay of light and respiratory controls, the regulation of the activity of L-galactono-1,4-lactone dehydrogenase (L-GalLDH), an enzyme that catalyzes the last reaction of the AsA biosynthetic pathway, determines the extent of leaf AsA accumulation. Growth at low light has been found to lower the abundance of transcripts encoding enzymes involved in AsA synthesis such as L-GalLDH and GDP-mannose pyrosphorylase (Tabata et al. 2002). Comparative proteomics analysis revealed differential accumulation of a number of thylakoid-associated proteins between ascorbate-deficient mutant *vtc2-2* and wild type Arabidopsis plants after transition to high light (1,000 μmol m^{-2} s^{-1}). These proteins included Fe–SODs and Cu/Zn–SOD, HSP70, PsbS, and a chloroplast-localized glyoxalate I, all of which have been associated with stress–response functions. The differential accumulation of these proteins shows that the ascorbate deficiency does have a significance, albeit impacts on the chloroplast stress response under high light stress (Giacomelli et al. 2006). Furthermore, it was reported that Arabidopsis leaves grown under low-light condition also contained low activities of DHAR and MDHAR compared with those grown in high-light condition (Bartoli et al. 2006), further indicating that the rate of ascorbate turnover is tightly controlled by light conditions.

Findings by Chen and Gallie (2008) implicate the regulation of the plant's response to high light by the ascorbate redox state following a genetic engineering approach. The authors constructed transgenic tobacco plants over-expressing and suppressing DHAR expression, thus leading to an increase and decrease of ascorbate redox state respectively. Suppression of DHAR expression resulted in reduced amounts of chlorophyll and xanthophyll pigments, quantum yield of photosystem II, and CO_2 assimilation, whereas the level of reactive oxygen species increased. Significant photoinhibition was also observed following exposure to high light. Direct feeding with AsA reversed effects observed in DHAR-suppressed leaves. In contrast, DHAR over-expression increased the pool size of xanthophyll and chlorophyll pigments as well as the rate of CO_2 assimilation, particularly at high light intensities, whereas the level of reactive oxygen species was reduced. Leaves of DHAR over-expressing plants experienced less photoinhibition than wild-type plants following exposure to high light, indicating the protective role of enhanced DHAR activity during high light stress.

Although significant evidence points to a central role of glutathione redox state-related responses in response to high light, the pattern of these responses is not uniform. Since the reduction of GSSG by GR is dependent on NADPH, which is mainly provided by the photosynthetic electron transport chain, a light-dependence of the GSH/GSSG ratio might be expected. Moreover, some studies reported a more reduced glutathione pool in the light, but in other cases the GSH pool became more reduced during a dark period (Tausz et al. 1999). Burritt and MacKenzie (2003) found that the pools of both glutathione and ascorbate in leaves of begonia plants grown under light were greater than those in shade. A different hypothesis concerning the role of light stress on GSH was addressed. For example, it was stated that GSSG might act as a molecule that inactivates light-modulated stromal enzymes such as fructose 1,6-bisphosphatase through an oxidation of their –SH groups (Alscher 1989; Vivekanandan and Edwards 1987). According to Baena-Gonzalez et al. (2001), high light intensity causes an elevation of the GSH/GSSG ratio, an effect that has as a consequence the inactivation of the plastid transcription kinase (PTK), an enzyme that was found to respond to changes in thiol/disulfide redox state mediated by glutathione (Link et al. 1997), and this inactivation causes changes in the phosphorylation of RNA polymerase subunits leading to an elevation of transcriptional activity. Similarly, Ogrzewalla et al. (2002) reported on a redox-sensitive plastid kinase that associates with the transcriptional complex and links transcriptional activity of chloroplast genes to the glutathione system (for excellent reviews on redox-regulated chloroplast gene expression in plants see Barnes and Mayfield 2003; Link 2003).

It is interesting to note that UV-B radiation was also shown to induce changes in glutathione metabolism and gene expression in pea plants (Kalbin et al. 1997). GSH and GSSG levels remained unchanged after exposure to the lowest UV-B dose rates; however, increased irradiation gave rise to 60- and 4.5-fold increases in GSSG and GSH respectively. Also, *cab* (Chlorophyll-a/b-binding protein) transcript levels decreased and *CHS* (chalcone synthase) and *PAL* (phenylalanine ammonia-lyase) transcript levels increased after shorter UV-B exposures to the higher dose rate of UV-B, and after exposure to an intermediate dose rate.

8 Low Oxygen Stress

Plants in the natural environment often experience limited oxygen availability due to transient or continuous flooding (reviewed by Drew 1997). Data on antioxidant levels and the activity of antioxidant-regenerating enzymes are somewhat contradictory, as both increases and decreases in tissue antioxidant capacity have been reported. An investigation on MDHAR and DHAR activities, and AsA and GSH contents in various species exposed to anoxia revealed an increase in MDHAR and/or DHAR in the anoxia-tolerant plants (Wollenweber-Ratzer and Crawford 1994). GSH decreased significantly during the post-anoxic period, while AsA showed increased values in the tolerant species. Induction of enzymes involved in the ascorbate–glutathione cycle (APX, MDHAR, DHAR and GR) has been demonstrated for anaerobically germinated rice seedlings and roots of wheat (*Triticum aestivum*) seedlings (Albrecht and Wiedenroth 1994; Ushimaru et al. 1997). Anoxia caused a significant inhibition of MDHAR, DHAR and GR activities (Biemelt et al. 1998) whereas inhibition of GR and APX was observed in corn leaves under prolonged flooding (Yan et al. 1996). Recent data have shown that the leaf and root cell ascorbate/glutathione systems react differentially to changes in oxygen concentration (Skutnik and Rychter 2009).

Immediately after the end of an anaerobic exposure, a rapid oxidation of ascorbate and glutathione pool took place in barley leaves, suggesting that these low-mass antioxidants are the "first line" of ROS detoxification under low oxygen stress (Skutnik and Rychter 2009). Conversely, ascorbate and glutathione pools became even more reduced in roots. Also, the imposition of anoxia and subsequent reoxygenation caused a decrease both in the content of ascorbate and in its reduction state in the roots of cereals and the rhizomes of *Iris* spp. (Blokhina et al. 2000). The highly reducing conditions prevailing under anoxia were also reflected in roots of wheat seedlings by increasing levels of AsA and GSH, leading to increased reduction states; however, the onset of re-aeration of plants caused enhanced oxidation of the reduced fractions, resulting in decreased AsA/DHA and GSH/GSSG ratios (Biemelt et al. 1998). Garnczarska (2005) observed increased content of total ascorbate in hydroponically grown lupine roots following oxygen stress, whereas total glutathione level decreased. However, a significant increase in the reduced forms of both metabolites was found directly after hypoxia.

Hypoxia triggers the induction of multiple transcripts responsible for ROS formation and utilization (e.g. genes associated with ascorbate and glutathione metabolism). The production of ROS is an integral part of the response to anoxic/hypoxic conditions which encompasses several levels of metabolic regulation to sustain redox signalling and to prevent oxidative damage. Major examples of mRNAs directly related to ROS handling and utilization include SOD, MDHAR, APX and several GSH-related enzymes involved in H_2O_2 detoxification. Modern microarray analyses have allowed the profiling of gene transcripts that are regulated by hypoxic conditions: Klok et al. (2002) showed that some peroxidases are down-regulated in Arabidopsis root cultures, while Lasanthi-Kudahettige et al. (2007) demonstrated catalase to be suppressed or unaffected in anoxic rice coleoptiles under hypoxia. The authors went on to suggest that low catalase transcript abundance acts as a signalling factor and is of paramount importance for H_2O_2 build-up. Further microarray-based analyses examining transient expression patterns revealed that these may also reflect a temporal necessity for ROS accumulation for signalling purposes. In particular, transcripts coding for APX were induced transiently in Arabidopsis as early as 0.5 h after low O_2 treatment, followed by MDHAR and ATP4a peroxidase (Branco-Price et al. 2005; Liu et al. 2005).

9 Ozone Stress

The air pollutant, ozone (O_3), is a molecule that possesses the ability to cause extended tissue damage resulting in chlorosis, water-logging, premature loss of chlorophyll, leaf abscission, decrease of photosynthesis and accelerated senescence in asymptomatic leaves in a variety of plant species (Reich and Amundson 1985; Krupa and Manning 1988; Heath 1994; Heggestad 1991; Strohm et al. 1999). The ingress of ozone seems to depend on the number and size of stomata in different plant species (Conklin and Last 1995). Once ozone enters the stomata, it reacts instantaneously with the apoplastic cell structures and generates secondary oxyproducts like H_2O_2 and $O_2{\cdot}^-$ (Pell and Dann 1991; Rao 1992; Baier et al. 2005).

Several differences in the effects of ozone on ascorbate/glutathione-related antioxidant mechanism have been reported in the literature for different species and also for the same species under treatment with different ozone concentrations. It has been suggested that changes in glutathione metabolism play a role in limiting damage to oxidative stress conditions induced by ozone (reviewed by Hausladen and Alscher 1993). In poplar, after an initial decline in GSH content, an overall increase in GSH, GSSG, and total glutathione content was observed in leaves in response to ozone fumigation (Sen Gupta et al. 1991). In an ozone sensitive cultivar of *Phaseolus vulgaris,* GSH levels were found to be lower than in the ozone-tolerant cultivars (Guri 1983), illustrating the importance of the capacity for reduction of the glutathione pool in ozone tolerance mechanism in plants. Futhermore, it has been shown that ozone exposure leads to up-regulation of sulphur assimilation in response to decreases in GSH/GSSG, via activation of adenosine 5′A-phosphosulphate reductase (Bick et al. 2001). On the other hand, Noctor et al. (1998) demonstrated that neither foliar glutathione accumulation nor the redox state of the GSH pool correlated with ozone sensitivity. The absence of an effect of GSH on ozone tolerance may be due to the compartmentalization of this antioxidant. Ozone enters the leaf through open stomata and produces ROS in the apoplastic fluid which therefore represents the first line of defence against ozone damage. In many plant species, however, the apoplast contains little or no glutathione (Polle et al. 1990). Therefore, increases in total foliar glutathione contents would be of little avail in improving defence against ozone. In this sense, it is proposed that increases in GSH synthesis or the GSH pool in the cytosol or chloroplast may not be sufficient for ozone stress tolerance, and enhanced tolerance may be achieved by increasing antioxidant capacity in the apoplast, which is likely to be the first target of ozone stress (Schützendübel and Polle 2002).

Available evidence suggests that increasing AsA levels via enhanced recycling has been implicated to be the first line of defence to ozone (Ranieri et al. 1999; Plochl et al. 2000; Ranieri et al. 2000, 2001; Moldau and Bichele 2002; Ranieri et al. 2003). In the work of Mahalingam et al. (2006) the inability of the ozone-treated plants to increase their AsA content to a level that is seen in the control plants suggests that AsA regenerative systems are adversely affected by ozone or ozone-derived ROS. Low AsA decreases the threshold for sensing stress and can trigger cell death and systemic acquired resistance (SAR), as seen in AsA-deficient mutants (Pastori et al. 2003; Foyer and Noctor 2005a). However, in others works it was suggested that apoplastic AsA alone could not explain differential ozone tolerance in *Trifolium* clones and soybean leaves (D'Haese et al. 2005; Cheng et al. 2007). In support to this hypothesis, Kollist et al. (2000) provide experimental evidence that the direct removal of O_3 by apoplastic AsA is limited in various plant species because mesophyll cell walls are thin and the effective path length for the reaction is short. Padu et al. (2005) found that the cell wall/plasmalemma/cytosol system in birch (*Betula pendula*) had sufficient capacity to maintain AsA redox state in the apoplast, without necessity to restrict O_3 uptake by stomatal closure. Finally, Nali et al (2004) found that the content of reduced glutathione, but not that of reduced ascorbate, was significantly increased in the ozone-treated (110 ppb O_3, 5 h day^{-1}) leaves of strawberry and phillyrea, thus making it unlikely for AsA to have an involvement in the differential O_3-sensitivity exhibited by these species. Similarly, total ascorbate contents, APX, MDHAR and DHAR activities were not affected by acute ozone exposure (Noctor et al. 1998).

Several attempts have been made to produce transgenic plants tolerant to ozone stress by manipulating components of the AsA recycling pathway. The most common "target" for genetic engineering has been DHAR which is well-documented as a protective antioxidant (see previous chapters). Chen and Gallie (2005) carried out ozone exposure experiments with tobacco plants over-expressing DHAR which demonstrated increased endogenous levels of AsA. Interestingly, DHAR-over-expressing plants had a lower oxidative load, a lower level of oxidative-related enzyme activities, and a higher level of photosynthetic activity following exposure to ozone compared with control plants. Conversely, reducing the size of the AsA pool size through suppression of DHAR expression had the opposite effect. Similar findings were reported by Eltayeb et al. (2006) who constructed DHAR-over-expressing tobacco plants with higher levels of reduced AsA compared with non-transformed control and who observed enhanced tolerance to ozone in terms of higher net photosynthesis. The importance of apoplastic AsA for ozone tolerance was further supported by Yoshida et al. (2006) who characterized an Arabidopsis mutant with a deficient cytosolic DHAR. The mutant completely lacked cytosolic DHAR activity and demonstrated significantly lower amounts of apoplastic AsA, resulting in high ozone sensitivity. Notably, similar tolerance to ozone stress was observed following over-expression of MDHAR in transgenic tobacco plants, which exhibited higher levels of reduced AsA compared to non-transformed control plants (Eltayeb et al. 2007).

Ozone induced alteration of ascorbate/glutathione-related enzymatic anti-oxygenic activities as well as gene expression levels have been reported in various plants, indicating antioxidant systems linking to ozone tolerance (Creissen et al. 1994; Kangasjarvi et al. 1994; Willekens et al. 1994). Increase in GR activity during ozone stress was reported in pea (Madamanchi et al. 1992; Edwards et al. 1994), *Spinacia oleracea* (Tanaka et al. 1988), *Triticum aestivum* (Rao et al. 1995), and *Arabidopsis thaliana* (Kubo et al. 1995; Rao et al. 1996). Edwards et al. (1994) reported that ozone exposure induced two isoforms of GR in pea plants with no significant changes either in the GR protein or in the mRNA transcripts encoding

GR. Exposure to O_3 caused a sharp decline in chloroplastic GR mRNA levels in both tobacco cv. Bel B and Bel W3, known for their differential sensitivity to ozone (Pasqualini et al. 2001). Additionally, it is reported that the increased expression of APX activity was observed in the ozone-exposed Arabidopsis, while other antioxidant enzymes, such as MDHAR, DHAR and GR, were unaffected (Kubo et al. 1995). Analogous gene expression patterns were observed following exposure of *Arabidopsis thaliana* plants to ozone, in which transcript levels of genes coding for cytosolic APX, Cu/ZnSOD and GST increased while FeSOD and GR were suppressed (Conklin and Last 1995). Interestingly, over-expression of APX in the chloroplast of transgenic tobacco plants did not protect them from ozone stress (Torgethausen et al. 1997). Furthermore Price et al. (1990) observed a marked increase in GST activity in barley tissues as a result of ozone fumigation that was correlated with increased lipid peroxidation. Moreover, the relatively large increase in GST mRNA levels observed in ozone-treated plants suggests that GST may be an appropriate gene for detailed studies on ozone-mediated gene activation (Price et al. 1990).

Ranieri et al. (2000) found an enhancement in the apoplastic and symplastic activity of APX in sunflower plants exposed to 150 nL L^{-1} of O_3 (4 h day^{-1} for 4 days), while stromal and thylakoid-bound chloroplastic APX activity was unaffected. PAGE analysis revealed different apoplastic and cytoplasmic APX isoenzyme patterns between control and O_3-treated plants. Also, immunoreaction with a cytosolic APX antibody revealed stonger apoplastic and symplastic signal in the O_3-treated plants than in the control ones (Ranieri et al. 2000). APX was linked with NC-R (resistant) white clover clones treated with ozone (60 ppb for 56 days, 5 h day^{-1}) (Nali et al. 2005). An early decrease in APX activity was observed in tomato genotypes exposed to O_3, which, however, was recovered in the post-fumigation period. The activity of APX and MDHAR proved to increase in two clover species exposed to ozone (Scebba et al. 2003). Meanwhile, Sanmartin et al. (2003) showed that expression of cucumber ascorbate oxidase (AO) in the cell wall/apoplast of tobacco (*Nicotiana tabacum* L. cv. Xanthi) leaves resulted in the oxidation of apoplast AsA and shifts in the redox state of AsA and GSH, without affecting total AsA (AsA + DHA) and GSH (GSH + GSSG) contents. These effects were associated with an increased sensitivity of transformed lines to O_3-induced oxidative stress and clearly reveal the role of AO in the regulation of AsA in redox signalling under ozone-induced oxidative stress (Sanmartin et al. 2003).

An alternative pathway of protection against ozone-induced damage is achieved with the activity of alternative oxidase (AOX). Plant mitochondrial AOX transfers electrons from the ubiquinione pool to oxygen without energy conservation and prevents the formation of ROS when the ubiquinone pool is over-reduced. Pasqualini et al. (2007) constructed transgenic tobacco plants presenting differing AOX activity levels. Exposure of plants to acute ozone fumigation resulted severe damage in transgenic plants with increased AOX activity, contrary to transgenic plants with reduced AOX activity which remained undamaged. Further molecular and enzymatic analyses revealed a differential response between wild type and transgenic plants with increased AOX activity, in which gene expression and enzymatic activity profiles of GPX and APX were suppressed in the latter, suggesting that AOX over-expression leads to increased O_3 sensitivity via disturbances of ascorbate/glutathione regulation.

10 Conclusions and Perspectives

In a constantly changing environment, the plant has to adapt primarily by quickly sensing all factors that could impose stressful conditions towards its function and development. It is therefore not surprising that a complex signalling system has evolved to provide protection against the many challenges of a harmful environment. This important role in orchestrating these diverse signalling pathways has been attributed to cellular redox regulation, which acts as a sensitive mechanism able to perceive even small changes in environmental conditions, and co-ordinate appropriate responses. Taking advantage of the latest advances in genomic, proteomic and transcriptomic technologies, researchers are focusing on the production and use of transgenic plants with altered levels of AsA/GSH content as well as of antioxidant enzymes and other protective metabolites, leading to enhanced antioxidant protection and ultimately to increased agricultural productivity. On the basis of the up-to-date findings outlined in this chapter, it is safe to conclude that redox regulation and antioxidants are one of the most promising areas of research for several years to come.

References

Albrecht G, Wiedenroth EM (1994) Protection against activated oxygen following re-aeration of hypoxically pre-treated wheat roots. The response of the glutathione system. J Exp Bot 45:449–455

Al-Ghamdi AA (2009) Evaluation of Oxidative Stress Tolerance in Two Wheat (Triticum aestivum) Cultivars in Response to Drought. Intern J Agric Biol, 11:7–12

Almeselmani M, Deshmukh PS, Sairam RK, Kushwaha SR, Singh TP (2006) Protective role of antioxidant enzymes under high temperature stress. Plant Sci 171:382–388

Alscher RG (1989) Biosynthesis and antioxidant function of glutathione in plants. Physiol Plant 77:457–464

Amor NB, Jimenez A, Megdiche W, Lundqvist M, Sevilla F, Abdelly C (2006) Response of antioxidant systems to NaCl stress in the halophyte Cakile maritima. Physiol Plant 126:446–457

Anderson JA, Padhye SR (2004) Protein aggregation, radical scavenging capacity, and stability of hydrogen peroxide defense systems in heat-stressed Vinca and sweet pea leaves. J Am Soc Horti Sci 129:54–59

Anjum NA, Umar S, Ahmad A, Iqbal M, Khan NA (2008a) Sulphur protects mustard (*Brasasica campestris* L.) from cadmium toxicity by improving leaf ascorbate and glutathione. Plant Growth Regul 54:271–279

Anjum NA, Umar S, Ahmad A, Iqbal M, Khan NA (2008b) Ontogenic variation in response of *Brassica campestris* L. to cadmium toxicity. J Plant Interact 3:189–198

Anjum NA, Umar S, Ahmad A, Iqbal M (2008c) Responses of components of antioxidant system in moongbean [*Vigna radiata* (L.) Wilczek] genotypes to cadmium stress. Commun Soil Sci Plant Anal 39:2469–2483

Anjum NA, Umar S, Singh S, Nazar R, Khan NA (2008d) Sulfur assimilation and cadmium tolerance in plants. In: Khan NA et al (eds) Sulfur assimilation and abiotic stress in plants. Springer-Verlag, Berlin, Heidelberg, pp 271–302

Anjum NA, Umar S, Iqbal M, Khan NA (2010) Cadmium causes oxidative stress in moongbean [*Vigna radiata* (L.) Wilczek] by affecting antioxidant enzyme systems and ascorbate-glutathione cycle metabolism. Russ J Plant Physiol 58:00–00

Asada K (1994) Production and action of active oxygen species in photosynthetic tissues. In: Foyer CH, Mullineaux PM (eds) Causes of photooxidative stress and amelioration of defence systems in plants. CRC Press, Boca Raton, FL, pp 77–104

Asada K (1999) The water-water cycle in chloroplasts: scavenging of active oxygens and dissipation of excess photons. Annu Rev Plant Physiol Plant Mol Biol 50:601–639

Ashraf M (2009) Biotechnological approach of improving plant salt tolerance using antioxidants as markers. Biotech Adv 27:84–93

Baccouch S, Chaoui A, El Ferjani E (2001) Nickel toxicity induces oxidative damage in *Zea mays* roots. J Plant Nutr 24:1085–1097

Badiani M, Paolacci AR, Fusari A, Dovidio R, Scandalios JG, Porceddu E, Sermanni GG (1997) Non-optimal growth temperatures and antioxidants in the leaves of Sorghum bicolor (L.) Moench. II. Short-term acclimation. J Plant Physiol 151:409–421

Baek KH, Skinner DZ (2003) Alteration of antioxidant enzyme gene expression during cold acclimation of near-isogenic wheat lines. Plant Sci 165:221–1227

Baena-Gonzalez E, Baginsky S, Mulo P, Summer H, Aro E-M, Link G (2001) Chloroplast transcription at different light intensities. Glutathione-mediated phosphorylation of the major RNA polymerase involved in redox regulated organellar gene expression. Plant Physiol 127:1044–1052

Baier M, Kandlbinder A, Golldack D, Dietz KJ (2005) Oxidative stress and ozone: perception, signalling and response. Plant Cell Environ 28:1012–1020

Baker NR, Harbinson J, Kramer DM (2007) Determining the limitations and regulation of photosynthetic energy transduction in leaves. Plant Cell Environ 30:1107–1125

Barnes D, Mayfield SP (2003) Redox control of posttranslational processes in the chloroplast. Antioxid Redox Signal 5:89–94

Bartoli C, Yu J, Gómez F, Fernández L, McIntosh L, Foyer CH (2006) Inter-relationships between light and respiration in the control of ascorbic acid synthesis and accumulation in *Arabidopsis thaliana* leaves. J Exp Bot 57:1621–1631

Basu U, Good AG, Taylor GJ (2001) Transgenic *Brassica napus* plants overexpressing aluminium-induced mitochondrial manganese superoxide dismutase cDNA are resistant to aluminium. Plant Cell Environ 24:1269–1278

Bick JA, Setterdahl AT, Knaff DB, Chen Y, Pitcher LH, Zilinskas BA, Leustek T (2001) Regulation of the plant-type 5 -adenylylsulfate reductase by oxidative stress. Biochem 40:9040–9048

Biemelt S, Keetman U, Albrecht G (1998) Re-aeration following hypoxia or anoxia leads to activation of the antioxidant defence system in roots of wheat seedlings. Plant Physiol 116:651–658

Blokhina OB, Virolainen E, Fagerstedt KV, Hoikkala A, Wähälä K, Chirkova TV (2000) Antioxidant status of anoxia-tolerant and -intolerant plant species under anoxia and reaeration. Physiol Plant 109:396–403

Blokhina O, Virolainen E, Fagerstedt K (2003) Antioxidant, oxidative damage and oxygen deprivation stress: a review. Ann Bot 91:179–194

Branco-Price C, Kawaguchi R, Ferreira RB, Bailey-Serres J (2005) Genome-wide analysis of transcript abundance and translation in *Arabidopsis* seedlings subjected to oxygen deprivation. Ann Bot 96:647–660

Burritt D, Mackenzie S (2003) Antioxidant metabolism during acclimation of Begonia x erythrophylla to high light levels. Ann Bot 91:783–794

Cakmak I, Horst WJ (1991) Effect of aluminum on lipid peroxidation, superoxide dismutase, catalase, and peroxidase activities in root tips of soybean (*Glycine max*). Physiol Plant 83:463–468

Cánovas D, Vooijs R, Schat H, De Lorenzo V (2004) The role of thiol species in the hypertolerance of *Aspergillus* sp. P37 to arsenic. J Biol Chem 279:51234–51240

Chalapathi Rao ASV, Reddy A (2008) Glutathione reductase: a putative redox regulatory system in plant cells. In: Khan NA et al (eds) Sulfur assimilation and abiotic stress in plants. Springer-Verlag, Berlin, Heidelberg, pp 111–147

Chao YY, Hong CY, Kao CH (2010) The decline in ascorbic acid content is associated with cadmium toxicity of rice seedlings. Plant Physiol Biochem. doi:10.1016/j.plaphy.2010.01.009

Chaoui A, Mazhoudi S, Ghorbal MH, Ferjani E (1997) Cadmium an Zinc induction of lipid peroxidation and effects on antioxidant enzyme activities in bean (Phaseolus vulgaris L). Plant Sci 127:139–147

Chaparzadeh N, Lucia D'Amico M, Khavari-Nejad RA, Izzo R, Navari-Izzo F, (2004) Antioxidative responses of Calendula officinalis under salinity conditions. Plant Physiol Biochem 42: 695–701

Chauhan S (2005) Physiological and molecular basis of heat tolerance with emphasis on oxidative stress metabolism in wheat. PhD Thesis. HNB Garhwal University, Srinagar, Uttaranchal, India

Chen Z, Gallie DR (2004) The ascorbic acid redox state controls guard cell signalling and stomatal movement. Plant Cell 16:1143–1162

Chen Z, Gallie DR (2005) Increasing tolerance to ozone by elevating foliar ascorbic acid confers greater protection against ozone than increasing avoidance. Plant Physiol 138:1673–1689

Chen Z, Gallie DR (2008) Dehydroascorbate reductase affects non-photochemical quenching and photosynthetic performance. J Biol Chem 283:21347–21361

Cheng FY, Burkey KO, Robinson JM, Booker FL (2007) Leaf extracellular ascorbate in relation to O_3 tolerance of two soybean cultivars. Environ Pollut 150:355–362

Cho UH, Seo NH (2005) Oxidative stress in *Arabidopsis thaliana* exposed to cadmium is due to hydrogen peroxide accumulation. Plant Sci 168:113–120

Clemens S (2006) Evolution and function of phytochelatin synthase. J Plant Physiol 163:319–332

Cobbett C, Goldsbrough P (2002) Phytochelatins and metallothioneins: roles in heavy metal detoxification and homeostasis. Annu Rev Plant Biol 53:59–182

Comba ME, Benavides MP, Tomaro ML (1998) Effect of salt stress on antioxidant defense system in soybean root nodules. Austr J Plant Physio 25:665–671

Conklin PL, Last RL (1995) Differential accumulation of antioxidant mRNAs in *Arabidopsis thaliana* exposed to ozone. Plant Physiol 109:203–212

Corpas F, Barroso J, del Río L (2001) Peroxisomes as a source of reactive oxygen species and nitric oxide signal molecules in plant cells. Trends Plant Sci 6:145–150

Creissen GP, Edwards EA, Mullineaux PM (1994) Glutathione reductase and ascorbate peroxidase. In: Foyer CH, Mullineaux PM (eds) Causes of photooxidative stress and amelioration of defence systems in plants. CRC Press, Boca Raton, FL, pp 343–364

Cruz de Carvalho MH (2008) Drought stress and reactive oxygen species. Production, scavenging and signalling. Plant Signal Behav 3:156–165

Dai F, Huang Y, Zhou M, Zhang G (2009) The influence of cold acclimation on antioxidant enzymes and antioxidants in sensitive and tolerant barley cultivars. Biol Plantar 53:257–262

Dalmia A, Sawhney V (2004) Antioxidant defence mechanism under drought stress in wheat seedlings. Physiol Mol Biol – Plants 10:109–114

Darkó E, Ambrus H, Stefanovits-Bányai E, Fodor J, Bakos F, Barnabás B (2004) Aluminium toxicity, Al tolerance and oxidative stress in an Al-sensitive wheat genotype and in Al-tolerant lines developed by in vitro microspore selection. Plant Sci 166:583–591

Dat JF, Lopez-Delgado H, Foyer CH, Scott IM (1998) Parallel changes in H_2O_2 and catalse during thermotolerance induced by salicylic acid or heat acclimation in mustard seedlings. Plant Physiol 116:1351–1357

Davenport SB, Gallego SM, Benavides MP, Tomaro ML (2003) Behaviour of antioxidant defense system in the adaptive response to salt stress in *Helianthus annuus* L. cells. Plant Growth Regul 40:81–88

Davis DG, Swanson HR (2001) Activity of stress-related enzymes in the perennial weed leafy spurge (*Euphorbia esula* L.). Environ Exp Bot 46:95–108

Davletova S, Rizhsky L, Liang H, Shengqiang Z, Oliver DJ, Coutu J, Shulaev V, Schlauch K, Mittler R (2005) Cytosolic ascorbate peroxidase 1 is a central component of the reactive oxygen gene network of Arabidopsis. Plant Cell 17:268–281

Demiral T, Turkan I (2005) Comparative lipid peroxidation, antioxidant defence systems and praline content in roots of two rice cultivars differing in salt tolerance. Environ Exp Bot 53:247–257

D'Haese D, Vandermeiren K, Asard H Horemans N (2005) Other factors than apoplastic ascorbate contribute to the differential ozone tolerance of two clones of Trifolium repens L. Plant Cell Environ 28:623–632

Dietz KJ, Baier M, Kramer U (1999) Free radicals and reactive oxygen species as mediators of heavy metal toxicity in plants. In: Prasad MNV, Hagemeyer J (ed) Heavy metal stress in plants: from molecules to ecosystems. Springer-Verlag, Berlin, pp 73–79

Dionisio-Sese ML, Tobita S (1998) Antioxidant responses of rice seedlings to salinity stress. Plant Sci 135:1–9

Dipierro N, Mondelli D, Paciolla C, Brunetti G, Dipierro S (2005) Changes in the ascorbate system in the response of pumpkin (*Cucurbita pepo* L.) roots to aluminium stress. J Plant Physiol 162:529–536

Drazkiewicz M, Skorzynska-Polit E, Krupa Z (2003) Response of the ascorbate–glutathione cycle to excess copper in Arabidopsis thaliana (L.). Plant Sci 164:195–202

Drew MC (1997) Oxygen deficiency and root metabolism: injury and acclimation under hypoxia and anoxia. Annu Rev Plant Physiol Plant Mol Biol 48:223–250

Dutilleul C, Garmier M, Noctor G, Mathieu CP, Foyer CH, de Paepe R (2003) Leaf mitochondria modulate whole cell redox homeostasis, set antioxidant capacity, and determine stress resistance through altered signalling and diurnal regulation. Plant Cell 15:1212–1226

Edwards EA, Enard C, Creissen GP, Mullineaux PM (1994) Synthesis and properties of glutathione reductase in stressed peas. Planta 192:137–143

Eltayeb AE, Kawano N, Badawi GH, Kaminaka H, Sanekata T, Morishima I, Shibahara T, Inanaga S, Tanaka K (2006) Enhanced tolerance to ozone and drought stresses in transgenic tobacco overexpressing dehydroascorbate reductase in cytosol. Physiol Plant 127:57–65

Eltayeb AE, Kawano N, Badawi GH, Kaminaka H, Sanekata T, Shibahara T, Inanaga S, Tanaka K (2007) Overexpression of monodehydroascorbate reductase in transgenic tobacco confers enhanced tolerance to ozone, salt and polyethylene glycol stresses. Planta 225:1255–1264

Ensminger I, Busch F, Huner NPA (2006) Photostasis and cold acclimation: sensing low temperature through photosynthesis. Physiol Plant 126:28–44

Eskew DL, Welch RM, Norvell WA (1983) Nickel: an essential micronutrient for legumes and possibly all higher plants. Science 222:621–623

Eskew DL, Welch RM, Norvell WA (1984) Nickel in higher plants: further evidence for an essential role. Plant Physiol 76:671–693

Fadzilla NM, Robert P, Finch RP, Burdon RH (1997) Salinity oxidative stress and antioxidant response in shoot cultures of rice. J Exp Bot 48: 325–331

Fahmy AS, Mohamed TM, Mohamed SA, Saker MM (1998) Effect of salt stress on antioxidant activities in cell suspension cultures of cantaloupe (Cucumis melo). Egyp J Physiol Sci 22: 315–326

Ferreira RR, Fornazier RF, Vitoria AP, Lea PJ, Azevedo RA (2002) Changes in antioxidant enzyme activities in soybean under cadmium stress. J Plant Nutr 25:327–342

Fornazier RF, Ferreira RR, Vitoria AP, Molina SMG, Lea PJ, Azevedo RA (2002) Effects of cadmium on antioxidant enzyme activities in sugarcane. Biol Plant 41:91–97

Fortunato A, Lidon F, Batista-Santos AP, Leitão E, Pais I, Ribeiro A, Ramalho JC (2010) Biochemical and molecular characterization of the antioxidant system of *Coffea* sp. under cold conditions in genotypes with contrasting tolerance. J Plant Physiol 167:333–342

Fotopoulos V, Sanmartin M, Kanellis AK (2006) Effect of ascorbate oxidase over-expression on ascorbate recycling gene expression in response to agents imposing oxidative stress. J Exp Bot 57:3933–3943

Fotopoulos V, De Tullio MC, Barnes J, Kanellis AK (2008) Altered stomatal dynamics in ascorbate oxidase over-expressing tobacco plants suggest a role for dehydroascorbate signalling. J Exp Bot 59:729–737

Foyer C, Noctor G (2005a) Oxidant and antioxidant signalling in plants: a re-evaluation of the concept of oxidative stress in a physiological context. Plant Cell Environ 28:1056–1071

Foyer CH, Noctor G (2005b) Redox homeostasis and antioxidant signalling: a metabolic interface between stress perception and physiological responses. Plant Cell 17:1866–1875

Foyer CH, Souruau N, Perret S, Lelandais M and Kunert KJ, Provust C, Jouanin L (1995) Overexpression of glutathione reductase but not glutathione synthetase leads to increases in antioxidant capacity and resistance to photoinhibition in poplar trees. Plant Physiol 109:1047–1057

Freeman JL, Persans MW, Nieman K, Albrecht C, Peer W, Pickering IJ, Salt DE (2004) Increased glutathione biosynthesis plays a role in nickel tolerance in *Thlaspi nickel* hyper accumulators W inside box sign. Plant Cell 16:2176–2191

Gallego SM, Benavides MP, Tomaro ML (1996) Effect of heavy metal ion excess on sunflower leaves: evidence for involvement of oxidative stress. Plant Science, 121:151–159

Garnczarska M (2005) Response of the ascorbate-glutathione cycle to re-aeration following hypoxia in lupine roots. Plant Physiol Biochem 43:583–590

Gatzek S, Wheeler GL, Smirnoff N (2002) Antisense suppression of L-galactose dehydrogenase in *Arabidopsis thaliana* provides evidence for its role in ascorbate synthesis and reveals light modulated L-galactose synthesis. Plant J 30:541–553

Giacomelli L, Rudella A, van Wijk KJ (2006) High light response of the thylakoid proteome in *Arabidopsis* wild-type and the ascorbate-deficient mutant *vtc2-2*. A comparative proteomics study. Plant Physiol 141:685–701

Gossett DR, Banks SW, Millhollon EP, Lucas MC (1996) Antioxidant response to NaCl stress in a control and a NaCl-tolerant cotton cell line grown in the presence of paraquat, buthionine sulfoximine and exogenous glutathione. Plant Physiol 112:803–809

Grace SC, Logan BA (1996) Acclimation of foliar antioxidant systems to growth irradiance in three broad-leaved evergreen species. Plant Physiol 112:1631–1640

Gueta-Dahan Y, Yaniv Z, Zilinkas BA, Ben-Hayyim G (1997) Salt and oxidative stress: similar and specific responses and their relation to salt tolerance in citrus. Planta 203:460–469

Guo Z, Ou W, Lu S, Zhong Q (2006) Differential responses of antioxidant system to chilling and drought in four rice cultivars differing in sensitivity. Plant Physiol Biochem 44:828–836

Guri A (1983) Variation in glutathione and ascorbic acid content among selected cultivars of *Phaseolus vulgaris* prior to and after exposure to ozone. Can J Plant Sci 63:733–737

Ha SB, Smith AP, Howden R, Dietrich WH, Bugg S, O'Connell MJ, Goldsbrough PB, Cobbet CS (1999) Phytochelatin synthase genes from *Arabidopsis* and yeast *Schizosaccharomyces pombe*. Plant Cell 11:1153–1164

Hanna PM, Mason RP (1992) Direct evidence for inhibition of free radical formation from Cu(I) and hydrogen peroxide by glutathione and other potential ligands using the EPR spin-trapping technique. Arch Biochem Biophys 295:205–213

Hatata MM, Abdel-Aal EA (2008) Oxidative Stress and Antioxidant Defence Mechanisms in Response to Cadmium Treatments, American-Eurasian J Agric Environ Sci 4:655–669

Hausladen A, Alscher RG (1993) Glutathione. In: Alscher RG, Hess J (eds) Antioxidants in higher plants. CRC Press, Boca Raton, FL, pp 1–30

Heggestad HE (1991) Origin of Bel-W3, Bel-C and Bel-B tobacco varieties and their use as indicators of ozone. Environ Pollut 74:264–291

Heath RL (1994) Alterations of plant metabolism by ozone exposure. In RG Alscher, AR Wellburn, (eds) Plant Response to the Gaseous Environment: Molecular, Metabolic and Physiological Aspects. Chapman and Hall, London, pp 121–145

Herbinger K, Tausz M, Wonisch A, Soja G, Sorger A, Grill D (2002) Complex interactive effects of drought and ozone stress on the antioxidant defence systems of two wheat cultivars. Plant Physiol and Biochem 40:691–696

Hernandez JA, Corpas FJ, Gomez M, Del Rio LA, Sevilla F (1993) Salt-induced oxidative stress mediated by activated oxygen species in pea leaf mitochondria. Physiol Planta 89:103–110

Hernandez JA, Olmos E, Corpas FJ, Sevilla F, del Rio LA (1995) Salt-induced oxidative stress in chloroplasts of pea plants. Plant Sci 105:151–167

Hernandez JA, Jimenez A, Mullineaux PM, Sevilla F (2000) Tolerance of pea (Pisum sativum L.) to long term salt stress is associated with induction of antioxidant defenses. Plant Cell & Environ 23:853–862

Hodges DM, Andrews CJ, Johnson DA, Hamilton RI (1997) Antioxidant enzyme responses to chilling stress in differentially sensitive inbred maize lines. J Exp Bot 48:1105–1113

Hodges DM, Andrews CJ, Johnson DA, Hamilton RI (1996) Antioxidant compound responses to chilling stress in differentially sensitive inbred maize lines. Physiol Plant 98:685–692

Howden R, Goldsbrough PB, Andersen CR, Cobbett CS (1995) Cadmium-sensitive *cad1* mutants of *Arabidopsis thaliana* are phytochelatin deficient. Plant Physiol 107:1059–1066

Hsu YT, Kao CH (2007) Toxicity in leaves of rice exposed to cadmium is due to hydrogen peroxide accumulation. Plant Soil 298:231–241

Huang M, Guo Z (2005) Responses of antioxidant system to chilling stress in two rice cultivars differing in sensitivity. Biol Plant 49:81–84

Huang C, He W, Guo J, Chang X, Su P, Zhang L (2005) Increased sensitivity to salt stress in an ascorbate-deficient *Arabidopsis* mutant. J Exp Bot 56:3041–3049

Ishikawa T, Shigeoka S (2008) Recent advances in ascorbate biosynthesis and the physiological significance of ascorbate peroxidase in photosynthesizing organisms. Biosci Biotechnol Biochem 72:1143–1154

Jiang MY, Zhang JH (2002) Water stress induced abscisic acid accumulation triggers the increased generation of reactive oxygen species and up regulates the activities of antioxidant enzymes in maize leaves. J Exp Bot 53:2401–2410

Kalbin G, Ohlsson AB, Berglund T, Rydström J, Strid A (1997) UV-B radiation causes changes in intracellular levels of nicotinamide, trigonelline and glutathione in *Pisum sativum* leaves. Phyton – Ann Rei Botan 37:115–124

Kangasjarvi J, Talvinen J, Utriainen M, Karjalainen R (1994) Plant defence systems induced by ozone. Plant Cell Environ 17:783–794

Karpinski S, Escobar C, Karpinska B, Creissen G, Mullineaux PM (1997) Photosynthetic electron transport regulates the expression of cytosolic ascorbate peroxidase genes in Arabidopsis during excess light stress. Plant Cell 9:627–640

Karpinski S, Reynolds H, Karpinska B, Wingsle G, Creissen G, Mullineaux P (1999) Systemic signalling and acclimation in response to excess excitation energy in *Arabidopsis*. Science 284:654–657

Khan NA, Anjum NA, Nazar R, Iqbal N (2009) Increased activity of ATP-sulfurylase and increased contents of cysteine and glutathione reduce high cadmium-induced oxidative stress in mustard cultivar with high photosynthetic potential. Russ J Plant Physiol 56:743–750

Klok EJ, Wilson IW, Wilson D, Chapman SC, Ewing RM, Somerville SC, Peacock WJ, Dolferus R, Dennis ES (2002) Expression profile analysis of the low oxygen response in Arabidopsis root cultures. Plant Cell 14:2481–2494

Kocsy G, Galiba G, Brunold C (2001) Role of glutathione in adaptation and signalling during chilling and cold acclimation in plants. Physiol Plant 113:158–164

Kollist H, Moldau H, Mortensen L, Rasmussen SK, Jorgensen LB (2000) Ozone flux to plasmalemma in barley and wheat is controlled rather by stomata than by direct reaction of ozone with apoplastic ascorbate. J Plant Physiol 156:645–651

Kopriva S, Jones S, Koprivova A, Suter M, von Ballmos P, Brander K, Fluckiger H (2001) Influence of chilling stress on the intercellular distribution of assimilatory sulfate reduction and thiols in Zea mays. Plant Biol 3:24–31

Krupa SV, Manning WJ (1988) Atmospheric ozone: formation and effects on vegetation. Environ Pollut 50:101–137

Kubo A, Saji H, Tanaka K, Kondo N (1995) Expression of Arabidopsis cytosolic ascorbate peroxidase gene in response to ozone or sulphur dioxide. Plant Mol Biol 29:479–489

Kuk YI, Shin JS, Burgos NR, Hwang TE, Han O, Cho BH, Jung S, Guh JO (2003) Antioxidative enzymes offer protection from chilling damage in rice plants. Crop Sci 43:2109–2117

Larkindale J, Hall JD, Knight MR, Vierling E (2005) Heat stress phenotypes of Arabidopsis mutants implicate multiple signalling pathways in the acquisition of thermotolerance. Plant Physiol 138:882–897

Lasanthi-Kudahettige R, Magneschi L, Loreti E, Gonzali S, Licausi F, Novi G, Beretta O, Vitulli F, Alpi A, Perata P (2007) Transcript profiling of the anoxic rice coleoptile. Plant Physiol 144:218–231

Lascano HR, Antonicelli GE, Luna CM, Melchiorre MN, Gomez LD, Racca RW, Trippi VS, Casano LM (2001) Antioxidant system response of different wheat cultivars under drought: field and in vitro studies. Austral J Plant Physiol 28:1095–1102

Lee SH, Ahsan N, Lee KW, Kim DH, Lee DG, Kwak SS, Kwon SY, Kim TH, Lee BH (2007) Simultaneous overexpression of both CuZn superoxide dismutase and ascorbate peroxidase in transgenic tall fescue plants confers increased tolerance to a wide range of abiotic stresses. J Plant Physiol 164:1626–1638

Levitt J (1962) A sulfhydryl-disulfide hypothesis of frost injury and resistance in plants. J Theor Biol 3:355–391

Li M, Ma F, Shang P, Zhang M, Hou C, Liang D (2009) Influence of light on ascorbate formation and metabolism in apple fruits. Planta 230:39–51

Lin CC, Kao CH (2000) Effect of NaCl stress on H_2O_2 metabolism in rice leaves. Plant Growth Regul 30:151–155

Link G, Tiller K, Baginsky S (1997) Glutathione, a regulator of chloroplast transcription. In KK Hatzios, (eds) Regulation of Enzymatic Systems Detoxifying Xenobiotics in Plants. Kluwer Academic Publishers, Dordrecht, The Netherlands, pp 125–137

Link G (2003) Redox regulation of chloroplast transcription. Antioxid Redox Signal 5:79–87

Liu F, Van Toai T, Moy LP, Bock G, Linford LD, Quackenbush J (2005) Global transcription profiling reveals comprehensive insights into hypoxic response in *Arabidopsis*. Plant Physiol 137:1115–1129

Logan BA, Monteiro G, Kornyeyev D, Payton P, Allen RD, Holaday AS (2003). Transgenic overproduction of gluatathione reductase does not protect cotton, Gossypium hirsutum (Malvaceae) from photoinhibition during growth under chilling conditions. Am J Bot 90:1400–1403

Loggini B, Scartazza A, Brugnoli E, Navari-Izzo F (1999) Antioxidant defence system, pigment composition, and photosynthetic efficiency in two wheat cultivars subjected to drought. Plant Physiol 119:1091–1099

Ma YH, Ma FW, Zhang JK, Li MJ, Wang YH, Liang D (2008) Effects of high temperature on activities and gene expression of enzymes involved in ascorbate-glutathione cycle in apple leaves. Plant Sci 175:761–766

Madamanchi NR, Anderson JV, Alscher RG, Cramer CL, Hess JL (1992) Purification of multiple forms of glutathione reductase from pea *(Pisum sativum* L.) seedlings and enzyme levels in ozone fumigated pea leaves. Plant Physiol 100:138–145

Mahalingam R, Jambunathan N, Gunjan SK, Faustin E, Weng H, Ayoubi P (2006) Analysis of oxidative signalling induced by ozone in *Arabidopsis thaliana*. Plant Cell Environ 29:1357–1371

Maruta T, Tanouchi A, Tamoi M, Yabuta Y, Yoshimura K, Ishikawa T, Shigeoka S (2010) Arabidopsis chlAPXs contribute not only to photoprotection, but also to the regulation of H_2O_2-responsive genes in response to photooxidative stress. Plant Cell Physiol 51:190–200

May MJ, Vernoux T, Leaver C, van Montagu M, Inzé D (1998) Glutathione homeostasis in plants: implications for environmental sensing and plant development. J Exp Bot 49:649–667

Mazhoudi S, Chaoui A, Ghorsab MW, Ferjani EE (1997) Response of antioxidant enzymes to excess copper in tomato (Lycopersicon esculentum Mill.). Plant Sci 127:129–137

Meneguzzo S, Navari-Izzo F, Izzo R (1999) Antioxidant responses of shoots and roots of wheat to increasing NaCl concentrations. J Plant Physiol 155:274–280

Meyer A (2008) The integration of glutathione homeostasis and redox signalling. J Plant Physiol 165:1390–1403

Mieda T, Yabuta Y, Rapolu M, Motoki T, Takeda T, Yoshimura K, Ishikawa T, Shigeoka S (2004) Feedback inhibition of spinach l-galactose dehydrogenase by l-ascorbate. Plant Cell Physiol 45:1271–1279

Mittler R (2002) Oxidative stress, antioxidants and stress tolerance. Trends Plant Sci 9:405–410

Mittler R, Vanderauwera S, Gollery M, van Breusegem F (2004) Reactive oxygen gene network of plants. Trends Plant Sci 9:490–498

Mittova V, Guy M, Tal M, Volokita M (2004) Salinity up-regulates the antioxidative system in root mitochondria and peroxisomes of the wild salt-tolerant tomato species Lycopersicon pennellii. J Exp Bot 55:1105–1113

Mittova V, Volokita M, Guy M, Tal M (2000) Activities of SOD and the ascorbate–glutathione cycle enzymes in subcellular compartments in leaves and roots of the cultivated tomato and its wild salt-tolerant relative *Lycopersicon pennellii*. Physiol Plant 110:42–51

Mittova V, Theodoulou FL, Kiddle G, Gómez L, Volokita M, Tal M, Foyer CH, Guy M (2003) Coordinate induction of glutathione biosynthesis and glutathione-metabolizing enzymes is correlated with salt tolerance in tomato. FEBS Lett 554:417–421

Miyagawa Y, Tamoi M, Shigeoka S (2000) Evaluation of the defence system in chloroplasts to photooxidative stress caused by paraquat using transgenic tobacco plants expressing catalase from *Escherichia coli*. Plant Cell Physiol 41:311–320

Moldau H, Bichele I (2002) Plasmalemma protection by the apoplast as assessed from above-zero ozone concentrations in leaf intercellular air spaces. Planta 214:484–487

Molina A, Bueno P, Marín MC, Rosales MPR, Belver A, Venema K, Donaire JP (2002) Involvement of endogenous salicylic acid content, lipoxygenase and antioxidant enzyme activities in the response of tomato cell suspension cultures to NaCl. New Phytolog 156:409–415

Muller IM, Jensen PE, Hansson A (2007) Oxidative modifications to cellular components in plants. Ann Rev Plant Biol 58:459–481

Müller P, Li XP, Niyogi KK (2001) Non-photochemical quenching. A response to excess light energy. Plant Physiol 125:1558–1566

Murgia I, Tarantino D, Vannini C, Bracale M, Carravieri S, Soave C (2004) *Arabidopsis thaliana* plants overexpressing thylakoidal ascorbate peroxidase show increased resistance to paraquat-induced photooxidative stress and to nitric oxide-induced cell death. Plant J 38:940–953

Nali C, Paoletti E, Marabottini R, Della Rocca G, Lorenzini G, Paolacci AR, Ciaffi M, Badiani M (2004) Ecophysiological and biochemical strategies of response to ozone in Mediterranean evergreen broadleaf species. Atmos Environ 38:2247–2257

Nali C, Pucciariello C, Mills G, Lorenzini G (2005) On the different sensitivity of white clover clones to ozone: physiological and biochemical parameters in a multivariate approach. Water Air Soil Poll 164:137–153

Nayyar H, Gupta D (2006) Differential sensitivity of C3 and C4 plants to water deficit stress: association with oxidative stress and antioxidants. Environ Exp Bot 58:106–113

Nieto-Sotelo J, Ho T-HD (1986) Effect of heat shock on the metabolism of glutathione in maize roots. Plant Physiol 82:1031–1035

Nishikawa K, Onodera A, Tominaga N (2006) Phytochelatins do not correlate with the level of Cd accumulation in *Chlamydomonas* spp. Chemosphere 63:1553–1559

Nocito FF, Lancilli C, Crema B, Fourcroy P, Davidian JC, Sacchi GA (2006) Heavy metal stress and sulfate uptake in maize roots. Plant Physiol 141:1138–1148

Noctor G (2006) Metabolic signalling in defence and stress: the central roles of soluble redox couples. Plant Cell Environ 29:409–425

Noctor G, Foyer CH (1998) Ascorbate and glutathione: keeping active oxygen under control. Annu Rev Plant Physiol Plant Mol Biol 49:249–279

Noctor G, Arisi ACM, Jouanin L, Kunert KJ, Rennenberg H, Foyer CH (1998) Glutathione: biosynthesis, metabolism and relationship to stress tolerance explored in transformed plants. J Exp Bot 49:623–647

Ogrzewalla K, Piotrowski M, Reinbothe S, Link G (2002) The plastid transcription kinase from mustard (*Sinapis alba* L.). A nuclear-encoded CK2-type chloroplast enzyme with redox-sensitive function. European. J Biochem 269:3329–3337

Oidaira H, Sano S, Koshiba T, Ushimaru T (2000) Enhancement of antioxidative enzyme activities in chilled rice seedlings. J Plant Physiol 156:811–813

Ort DR, Becker NR (2002) A photoprotective role for O_2 as an alternative electron sink in photosynthesis? Curr Opin Plant Biol 5:193–198

Padu E, Kollist H, Tulva I, Oksanen E, Moldau H (2005) Components of apoplastic ascorbate use in *Betula pendula* leaves exposed to CO_2 and O_3 enrichment. New Phytol 165:131–143

Panchuk J, Volkov R, Schoffl F (2002) Heat Stress- and Heat Shock Transcription Factor-Dependent Expression and Activity of Ascorbate Peroxidase in Arabidopsis. Plant Physiol 129:838–853

Paradiso A, Berardino R, De Pinto MC, Sanità Di Toppi L, Storelli MM, Tommasi F, De Gara L (2008) Increase in ascorbate-glutathione metabolism as local and precocious systemic responses induced by cadmium in durum wheat plants. Plant Cell Physiol 49:362–374

Pasqualini S, Batini P, Ederli L, Porceddu C, Piccioni C, De Marchis F, Antonielli M (2001) Effects of short-term fumigation on tobacco plants: response of the scavenging system and expression of the glutathione reductase. Plant Cell Environ 24:245–252

Pasqualini S, Paolocci F, Borgogni A, Morettini R, Ederli L (2007) The overexpression of an alternative oxidase gene triggers ozone sensitivity in tobacco plants. Plant Cell Environ 30:1545–1556

Pastori GM, Kiddle G, Antoniw J, Bernard S, Veljovic-Jovanovic S, Verrier PJ, Noctor G, Foyer CH (2003) Leaf vitamin C contents modulate plant defence transcripts and regulate genes that control development through hormone signalling. Plant Cell 15:939–951

Patra J, Panda BB (1998) A comparison of biochemical responses to oxidative and metal stress in seedlings of barley, Hordeum vulgare L. Environ Pollut 101:99–105

Payton P, Webb R, Kornyeyev D, Allen R, Holaday A (2001) Protecting cotton photosynthesis during moderate chilling at high light intensity by increasing chloroplastic antioxidant enzyme activity. J Exp Bot 52:2345–2354

Pell EJ, Dann MS (1991) Multiple stress-induced foliar senescence and implications for whole-plant longevity. In: Monney HA, Winner WE, Pell EJ (eds) Responses of plants to multiple stresses. Academic, San Diego, CA, pp 389–403

Piqueras A, Olmos E, Martińez-Solano JR, Hellín E (1999) Cd-induced oxidative burst in tobacco BY2 cells: time course, subcellular location and antioxidant response. Free Radic Res 31:33–38

Plochl M, Lyons T, Ollerenshaw J, Barnes J (2000) Simulating ozone detoxification in the leaf apoplast through the direct reaction with ascorbate. Planta 210:454–467

Polle A (1997) Defence against photo-oxidative damage in plants. In JG Scandalios, (eds) Oxidative Stress and the Molecular Biology of Antioxidant Defenses. Cold Spring Harbor Laboratory Press, Cold Spring Harbor, NY, pp 623–666

Polle A (2001) Dissecting the superoxide dismutase–ascorbate peroxidase–glutathione pathway in chloroplasts by metabolic modeling. Computer simulations as a step towards flux analysis. Plant Physiol 126:445–462

Polle A, Rennenberg H (1993) Significance of antioxidants in plant adaptation to environmental stress. In: Fowden L, Mansfield T, Stoddart J (eds) Plant adaptation to environmental stress. Chapman & Hall, London, pp 264–273

Polle A, Chakrabarti K, Schurmann W, Rennenberg H (1990) Composition and properties of hydrogen peroxide decomposing systems in extracellular and total extracts from needles of norway spruce (*Picea abies* L. Karst). Plant Physiol 94:312–319

Potters G, De Gara L, Asard H, Horemans N (2002) Ascorbate and glutathione: guardians of the cell cycle, partners in crime? Plant Physiol Biochem 40:537–548

Price A, Lucas PW, Lea PJ (1990) Age dependent damage and glutathione metabolism in ozone fumigated barley: a leaf section approach. J Exp Bot 41:1309–1317

Ranieri A, Castagna A, Padu E, Moldau H, Rahi M, Soldatini GF (1999) The decay of O_3 through direct reaction with cell wall ascorbate is not sufficient to explain the different degrees of O_3-sensitivity in two poplar clones. J Plant Physiol 150:250–255

Ranieri A, Castagna A, Soldatini GF (2000) Differential stimulation of ascorbate peroxidase isoforms by ozone exposure in sunflower plants. J Plant Physiol 156:266–271

Ranieri A, Giuntini D, Ferraro F, Nali C, Baldan B, Lorenzini G, Soldatini GF (2001) Chronic ozone fumigation induces alterations in thylakoid functionality and composition in two poplar clones. Plant Physiol Biochem 39:999–1008

Ranieri A, Castagna A, Pacini J, Baldan B, Mensuali Sodi A, Soldatini GF (2003) Early production and scavenging of hydrogen peroxide in the apoplast of sunflowers plants exposed to ozone. J Exp Bot 54:2529–2540

Rao MV (1992) Cellular detoxifying mechanisms determine age dependent injury in tropical plants exposed to SO_2. J Plant Physiol 140:733–740

Rao MV, Hale B, Ormrod DP (1995) Amelioration of ozone induced oxidative damage in wheat plants grown under high carbon dioxide. Role of antioxidant enzymes. Plant Physiol 109:421–432

Rao MV, Paliyath G, Ormrod DP (1996) Ultraviolet-B and ozone-induced biochemical changes in antioxidant enzymes of *Arabidopsis thaliana*. Plant Physiol 110:125–136

Rauser WE (1995) Phytochelatins and related peptides. Structure, biosynthesis and function. Plant Physiol 109:1141–1149

Reich RB, Amundson RG (1985) Ambient levels of ozone reduce net photosynthesis in tree and crop species. Science 230:566–570

Riekert van Heerden PD, Kruger GHJ (2002) Separately and simultaneously induced dark chilling and drought stress effects on photosynthesis, proline accumulation and antioxidant metabolism in soybean. J Plant Physiol 159:1077–1086

Rivero RM, Ruiz JM, Romero L (2004) Oxidative metabolism in tomato plants subjected to heat stress. J Horti Sci Biotech 79:560–564

Romero-Puertas MC, Rodriguez-Serrano M, Corpas FJ, Gómez M, del Rio LA, Sandalio LM (2004) Cadmium-induced subcellular accumulation of and H_2O_2 in pea leaves. Plant Cell Environ 27:1122–1134

Ruiz JM, Blumwald E (2002) Salinity-induced glutathione synthesis in *Brassica napus*. Planta 214:965–969

Saeidi-Sar S, Khavari-Nejad RA, Fahimi H, Ghorbanli M, Majd A (2007) Interactive effects of gibberellin A_3 and ascorbic acid on lipid peroxidation and antioxidant enzyme activities in glycine max seedlings under nickel stress. Russ J Plant Physiol 54:74–79

Sairam RK, Srivastava GC, Agarwal S, Meena RC (2005) Differences in antioxidant activity in response to salinity stress in tolerant and susceptible wheat genotypes. Biol Plant 49:85–91

Sandalio LM, Dalurzo HC, Gómez M, Romero-Puertas MC, del Rio LA (2001) Cadmium-induced changes in the growth and oxidative metabolism of pea plant. J Exp Bot 52:2115–2126

Sanmartin M, Drogoudi PD, Lyons T, Pateraki I, Barnes J, Kanellis AK (2003) Over-expression of ascorbate oxidase in the apoplast of transgenic tobacco results in altered ascorbate and glutathione redox states and increased sensitivity to ozone. Planta 216:918–928

Scebba F, Pucciarelli I, Soldatini GF, Ranieri A (2003) O_3-induced changes in the antioxidant systems and their relationship to different degrees of susceptibility of two clover species. Plant Sci 165:583–593

Schützendübel A, Polle A (2002) Plant responses to abiotic stresses: heavy metal-induced oxidative stress and protection by mycorrhization. J Exp Bot 53:1351–1365

Sečenji M, Hideg E, Bebes A, Györgyey J (2010) Transcriptional differences in gene families of the ascorbate–glutathione cycle in wheat during mild water deficit. Plant Cell Rep 29:37–50

Semane B, Cuypers A, Smeets K, Van Belleghem F, Horemans N, Schat H, Vangronsveld J (2007) Cadmium responses in *Arabidopsis thaliana*: glutathione metabolism and antioxidant defence system. Physiol Plant 129:519–528

Sen Gupta A, Alsher RG, McCune D (1991) Response of photosynthesis and cellular antioxidants to ozone in *Populus* leaves. Plant Physiol 96:650–655

Sgherri CLM, Navari-Izzo F (1995) Sunflower seedlings subjected to increasing water deficit stress: oxidative stress and defence mechanisms. Physiol Plant 93:25–30

Shalata A, Tal M (1998) The effect of salt stress on lipid peroxidation and antioxidants in the leaf of the cultivated tomato and its wild salt tolerant relative *Lycopersicon pennellii*. Physiol Plant 104:169–174

Shalata A, Mittova V, Volokita M, Guy M, Tal M (2001) Response of the cultivated tomato and its wild salt-tolerant relative *Lycopersicon pennellii* to salt-dependent oxidative stress: the root antioxidant system. Physiol Plant 112:487–494

Shao HB, Chu LY, Lu ZH, Kang CM (2008) Primary antioxidant free radical scavenging and redox signalling pathways in higher plant cells. Int J Biol Sci 4:8–14

Sharma SS, Dietz KJ (2009) The relationship between metal toxicity and cellular redox imbalance. Trends Plant Sci 14:43–50

Sharma P, Dubey RS (2005) Drought induces oxidative stress and enhances the activities of antioxidant enzymes in growing rice seedlings. Plant Growth Regul 46:209–221

Sharma P, Dubey RS (2007) Involvement of oxidative stress and role of antioxidative defense system in growing rice seedlings exposed to toxic concentrations of aluminum. Plant Cell Rep 26:2027–2038

Shigeoka S, Ishikawa T, Tamoi M, Miyagawa Y, Takeda T, Yabuta Y, Yoshimura K (2002) Regulation and function of ascorbate peroxidase isoenzymes. J Exp Bot 53:1305–1319

Skorzynska-Polit E, Drazkiewicz M, Krupa Z (2003/4) The activity of the antioxidative system in cadmium-treated *Arabidopsis thaliana*. Biol Plant 47:71–78

Skutnik M, Rychter AM (2009) Differential response of antioxidant systems in leaves and roots of barley subjected to anoxia and post-anoxia. J Plant Physiol 166:926–937

Smirnoff N (1993) The role of active oxygen in the response of plants to water deficit and desiccation. New Phytol 125:27–58

Smirnoff N (2000) Ascorbic acid: metabolism and functions of a multi-facetted molecule. Curr Opin Plant Biol 3:229–235

Smith AP, De Ridder BP, Guo WJ, Seeley EH, Regnier FE, Goldsbrough PB (2004) Proteomic analysis of *Arabidopsis* glutathione S-transferases from benoxacor- and copper-treated seedlings. J Biol Chem 279:26098–26104

Sobrino-Plata J, Ortega-Villasante C, Flores-Cáceres ML, Escobar C, Del Campo FF, Hernández LE (2009) Differential alterations of antioxidant defences as bioindicators of mercury and cadmium toxicity in alfalfa. Chemosphere 77:946–954

Song XS, Hu WH, Mao WH, Ogweno JO, Zhou YH, Yu JQ (2005) Response of ascorbate peroxidase isoenzymes and ascorbate regeneration system to abiotic stresses in *Cucumis sativus* L. Plant Physiol Biochem 43:1082–1088

Sreenivasulu N, Grimma B, Wobusa U, Weschkea W (2000) Differential response of antioxidant compounds to salinity stress in salt-tolerant and salt-sensitive seedlings of foxtail millet (*Setaria italica*). Physiol Plant 109:435–442

Srivalli S, Khanna-Chopra R (2008) Role of glutathione in abiotic stress tolerance. In: Khan NA et al (eds) Sulfur assimilation and abiotic stress in plants. Springer-Verlag, Berlin, Heidelberg, pp 207–225

Strohm M, Eiblmeier M, Langebartels C, Jouanin L, Polle A, Sandermann H, Rennenberg H (1999) Responses of transgenic poplar (*Populus tremula x P.alba*) overexpressing glutathione reductase to acute ozone stress: visible injury and leaf gas exchange. J Exp Bot 50:365–374

Stroinski A, Kubis J, Zielezinska M (1999) Effect of cadmium on glutathione reductase in potato tubers. Acta Physiol Planta 21:201–207

Sumithra K, Jutur PP, Dalton CB, Reddy AR (2006) Salinity-induced changes in two cultivars of Vigna radiata: responses of antioxidative and proline metabolism. J Plant Growth Regul 50:11–22

Szalai G, Kellos T, Galiba G, Kocsy G (2009) Glutathione as an antioxidant and regulatory molecule in plants under abiotic stress conditions. J Plant Growth Regul 28:66–80

Tabaldi LA, Cargnelutti D, Gonçalves JF, Pereira LB, Castro GY, Maldaner J, Rauber R, Rossato LV, Bisognin DA, Schetinger MR, Nicoloso FT (2009) Oxidative stress is an early symptom triggered by aluminum in Al-sensitive potato plantlets. Chemosphere 76:1402–1409

Tabata K, Takaoka T, Esaka M (2002) Gene expression of ascorbic acid-related enzymes in tobacco. Phytochemistry 61:631–635

Tanaka K, Saji H, Kondo N (1988) Immunological properties of spinach glutathione reductase and inductive biosynthesis of the enzyme with ozone. Plant Cell Physiol 29:637–642

Tang L, Kwon SY, Kim SH, Kim JS, Cho KY, Sung CK (2006) Enhanced tolerance of transgenic potato plants expressing both superoxide dismutase and ascorbate peroxidase in chloroplasts against oxidative stress and high temperature. Plant Cell Rep 25:1380–1386

Tanou G, Job C, Rajjou L, Arc E, Belghazi M, Diamantidis G, Molassiotis A, Job D (2009a) Proteomics reveals the overlapping roles of hydrogen peroxide and nitric oxide in the acclimation of citrus plants to salinity. Plant J 60:795–804

Tanou G, Molassiotis A, Gr D (2009b) Hydrogen peroxide- and nitric oxide-induced systemic antioxidant prime-like activity under NaCl-stress and stress-free conditions in citrus plants. J Plant Physiol 166:1904–1913

Tanou G, Molassiotis A, Diamantidis G (2009c) Induction of reactive oxygen species and necrotic death-like destruction in strawberry leaves by salinity. Environ Exp Bot 65:270–281

Tao DL, Oquist G, Wingsle G (1998) Active oxygen scavengers during cold acclimation of Scots pine seedlings in relation to freezing tolerance. Cryobiology 37:38–45

Tausz M, Bytnerowicz A, Arbaugh MJ, Weidner W, Grill D (1999) Antioxidants and protective pigments of *Pinus ponderosa* needles at gradients of natural stresses and ozone in the San Bernardino Mountains in California. Free Radic Res 31:113–120

Tausz M, Bytnerowicz A, Arbaugh MJ, Wonisch A, Grill D (2001) Biochemical response patterns in *Pinus ponderosa* trees at field plots in the San Bernardino Mountains (Southern California). Tree Physiol 21:329–336

Tausz M, Sircelj H, Grill D (2004) The glutathione system as a stress marker in plant ecophysiology: is a stress-response concept valid? J Exp Bot 55:1955–1962

Tewari RK, Parma PK, Sharma N (2006) Antioxidant responses to enhanced generation of superoxide anion radical and hydrogen peroxide in the copper-stressed mulberry plants. Planta 223:1145–1153

Tsai YC, Hong CY, Liu LF, Kao CH (2005) Expression of ascorbate peroxidase and glutathione reductase in roots of rice seedlings in response to NaCl and H_2O_2. J Plant Physiol 162:291–299

Torsethaugen G, Pitcher LH, Zilinskas BA, Pell EJ (1997) Overproduction of ascorbate peroxidase in the tobacco chloroplast does not provide protection against ozone. Plant Physiol 114:529–537

Turkan I, Bor M, Ozdemir F, Koca H (2005) Differential responses of lipid peroxidation and antioxidants in the leaves of drought tolerant *Populus acutifolius* Gray and drought sensitive *P. vulgaris* L. subjected to polyethylene glycol mediated water stress. Plant Sci 168:223–231

Ushimaru T, Maki Y, Sano S, Koshiba K, Asada K, Tsuji H (1997) Induction of enzymes involved in the ascorbate-dependent antioxidant system, namely ascorbate peroxidase, mono dehydroascorbate reductase and dehydroascorbate reductase, after exposure to air of rice (*Oryza sativa*) seedlings germinated under water. Plant Cell Physiol 38:541–549

Ushimaru T, Nakagawa T, Fujioka Y, Daicho K, Naito M, Yamauchi Y, Nonaka H, Amako K, Yamawaki K, Murata N (2006) Transgenic *Arabidopsis* plants expressing the rice dehydroascorbate reductase gene are resistant to salt stress. J Plant Physiol 163:1179–1184

Vaidyanathan H, Sivakumar P, Chakrabarty R, Thomas G (2003) Scavenging of reactive oxygen species in NaCl-stressed rice (Oryza sativa L.) differential response in salt-tolerant and sensitive varieties. Plant Sci 165:1411–1418

Veljovic-Jovanovic DS, Pignocchi C, Noctor G, Foyer HC (2001) Low ascorbic acid in the *vtc-1* mutant of *Arabidopsis* is associated with decreased growth and intracellular redistribution of the antioxidant system. Plant Physiol 127:426–435

Vitoria AP, Lea PJ, Azevedo RA (2001) Antioxidant enzymes responses to cadmium in radish tissues. Phytochem 57:701–710

Vivekanandan M, Edwards GE (1987) Activation of NADP-malate dehydrogenase in C3 plants by reduced glutathione. Photosynth Res 14:113–124

Walker MA, McKersie BD (1993) Role of the ascorbate-glutathione antioxidant system in chilling resistance of tomato. J Plant Physiol 141:234–239

Wang K (2005) Genetic acclimaton for freezing tolerance. ISB news report

Willekens HD, Van Camp WV, Montagu M, Inze D, Langerbelts C, Sandermann H (1994) Ozone, sulfur dioxide and UV-B radiation have similar effects on mRNA accumulation of antioxidant genes in *Nicotiana plumbaginifolia* L. Plant Physiol 106:1007–1014

Willekens HD, Inze M, Montagu MV, Camp WV (1995) Catalase in plants. Mol Breed 1:207–228

Wojas S, Clemens S, Hennig J, Skłodowska A, Kopera E, Schat H, Bal W, Antosiewicz DM (2008) Overexpression of phytochelatin synthase in tobacco: distinctive effects of AtPCS1 and CePCS genes on plant response to cadmium. J Exp Bot 59:2205–2219

Wójcik M, Pawlikowska-Pawlega B, Tukiendorf BA (2009) Physiological and ultrastructural changes in *Arabidopsis thaliana* as affected by changed GSH level and Cu excess. Russ J Plant Physiol 56:906–916

Wollenweber-Ratzer B, Crawford RMM (1994) Enzymatic defence against post-anoxic injury in higher plants. Proc R Soc Edinb 102:381–390

Xiang C, Werner BL, Christensen EM, Oliver DJ (2001) The biological functions of glutathione revisited in *Arabidopsis* transgenic plants with altered glutathione levels. Plant Physiol 126:564–574

Yabuta Y, Motoki T, Yoshimura K, Takeda T, Ishikawa T, Shigeoka S (2002) Thylakoid membrane-bound ascorbate peroxidase is a limiting factor of antioxidant systems under photooxidative stress. Plant J 32:915–925

Yadav SK, Singla-Pareek SL, Reddy MK, Sopory SK (2005) Transgenic tobacco plants overexpressing glyoxalase enzymes resist an increase in methylglyoxal and maintain higher reduced glutathione levels under salinity stress. FEBS Lett 579:6265–6271

Yamamoto A, Bhuiyan MNH, Waditee R, Tanaka Y, Esaka M, Oba K, Jagendorf AT, Takabe T (2005) Suppressed expression of the apoplastic ascorbate oxidase gene increases salt tolerance in tobacco and *Arabidopsis* plants. J Exp Bot 56:1785–1796

Yan B, Dai Q, Liu X, Huang S, Wang Z (1996) Flooding-induced membrane damage, lipid oxidation and activated oxygen generation in corn leaves. Plant Soil 179:261–268

Yan SP, Zhang QY, Tang ZC, Su WA, Sun WN (2006) Comparative proteomic analysis provides new insights into chilling stress responses in rice. Molecular and Cell Proteomics 5:484–496

Yin L, Wang S, Eltayeb AE, Uddin MI, Yamamoto Y, Tsuji W, Takeuchi Y, Tanaka K (2010) Overexpression of dehydroascorbate reductase, but not monodehydroascorbate reductase, confers tolerance to aluminum stress in transgenic tobacco. Planta 231:609–621

Yoshida S, Tamaoki M, Shikano T, Nakajima N, Ogawa D, Ioki M, Aono M, Kubo A, Kamada H, Inoue Y, Saji H (2006) Cytosolic dehydroascorbate reductase is important for ozone tolerance in *Arabidopsis thaliana*. Plant Cell Physiol 47:304–308

Zhang C, Ge Y (2008) Response of Glutathione and Glutathione S-transferase in Rice Seedlings Exposed to Cadmium Stress. Rice Sci 15:73–76

Zhang Y, Lou Y, Hao J, Chen Q, Tang HR (2008) Chilling acclimation induced changes in the distribution of H_2O_2 and antioxidant system of strawberry leaves. Agric J 3:286–291

Zhao S, Blumwald E (1998) Changes in oxidationreduction state and antioxidant enzymes in the roots of jack pine seedlings during cold acclimation. Physiol Plant 104:134–142

Zhu OYL, Pilon-Smits EAH, Jouanin L, Terry N (1999) Overexpression of glutathione synthetase in Indian mustard enhances cadmium accumulation and tolerance. Plant Physiol 11901:73–79

Chapter 11
Ascorbate–Glutathione Cycle: Enzymatic and Non-enzymatic Integrated Mechanisms and Its Biomolecular Regulation

Juan Pablo Martínez and Héctor Araya

Abstract In plant physiology, low level of oxidative stress produces a favourable effect on the metabolism of a plant. However, when antioxydant defenses are overpassed by the action of oxidative compounds it will produce metabolic alteration that would end in cell death. The increase of oxidative stress is conditioned by different types of abiotic (salinity, drought, heavy metal) and biotic (fungus and insects) stresses. The main objective of this chapter will be to evaluate the role of ascorbate–glutathione cycle in the defense mechanisms and antioxidant capacity in plants under different stress conditions, integrating non enzymatic (ascorbate and glutathione) and enzymatic pathways and its biomolecular regulation. This chapter will describe the integration of both non-enzymatic and enzymatic mechanisms specially those concerning with the regeneration of active compounds. Enzymatic pathway preserves the active form of ascorbate and glutathione. Furthemore, we will also describe the regulation of ascorbate-glutation cycle developed by alterated redox status, which can be produced produced by abiotic and biotic stress.

In addition, in the discussion we will emphasize the agronomic and genomic approaches on how to improve the plant tolerance against the biotic and abiotic stresses, so that a higher antioxidant level can be obtained, which will benefit the healthy quality of vegetable foods.

J.P. Martínez (✉)
Institute of Agriculture Research (INIA), Chorrillos 86, La Cruz, Regi6n de Valparaíso, Chile
and
Regional Center for the Study of Healthy Foods (CREAS), Blanco 1623, Oficina 1402, Edificio Torres Mar del Sur 2, Valparaíso
e-mail: jpmartinez@inia.cl

H. Araya
Faculty of Medicine, Departament of Nutrition, University of Chile, Independencia 1027, Santiago, Chile
and
Faculty of Pharmacy, Department of Food and Nutrition, University of Valparaíso, Avda Gran Bretaña 1111, Valparaíso, Regi6n de Valparaíso, Chile
and
Regional Center for the Study of Healthy Foods (CREAS), Blanco 1623, Oficina 1402, Edificio Torres Mar del Sur 2-, Valparaíso

N.A. Anjum et al. (eds.), *Ascorbate-Glutathione Pathway and Stress Tolerance in Plants*,
DOI 10.1007/978-90-481-9404-9_11, © Springer Science+Business Media B.V. 2010

Keywords Glutathione • Ascorbate • Antioxidants

1 Introduction

Reactive oxygen species (ROS) are partially reduced forms of atmospheric oxygen (O_2) and they typically result from the transfer of one, two or three electrons to O_2 to form a superoxide radical (O_2^-), hydrogen peroxide (H_2O_2) or hydroxyl radical (HO^-), respectively. In contrast to atmospheric oxygen, ROS are able of unrestricted oxidation of various cellular components and can lead to the oxidative destruction of cells (Mittler 2002). Under optimal conditions, the constitutive antioxidants in plant cells control that basal production of ROS is maintaining their concentrations at levels compatible with the specific metabolic requirements of cellular comparments. However, in natural environments, plants are exposed to many abiotic and biotic stresses that cause an over-production of ROS leading to an imbalance between ROS production and defence mechanisms, a process referred to as oxidative stress (Kuzniak and Sklodowska 2005). This imbalance becomes a common trait of all environmental stresses and is regarded as the major cause of oxidative damage found in cells under stressful conditions. Since ROS are highly reactive and may cause oxidative damage to lipids, proteins, nucleic acids, and affect cell membrane properties, their accumulation may lead to the oxidative destruction cells. Therefore, plant tolerance to abiotic and biotic stresses has to be supported by an integraded non enzymatic and enzymatic antioxidant systems. This response assures the development of efficient ROS-scaveging mechanisms. This adaptation mechanism depends on the transduction signals and expression of genes associated to the enzymatic defence and levels of antioxidant molecules.

2 Production of Reactive Oxygen Species (ROS) in Plants by Abiotic and Biotic Stresses

It is well-known that oxidative stress is increased when plants are directly exposed to biotic and abiotic stresses, such as insect and pathogen attack, salinity, drought, high light irradiance, high or low temperature, ozone, wounding, flooding, herbicides, mineral nutrients deficiency or mineral ion toxicity (Kuzniak and Sklodowska 2005; Wang et al. 2009). However, plants are equipped with an array of non-enzymic scavengers and antioxidant enzymes acting in concert to alleviate cellular damage under oxidative stress conditions (Foyer and Noctor 2005a, b). Although oxygen is essential for life and is produced under normal growth conditions and their concentration remains low (Polle 2001), its reduction results in the production of ROS such as superoxide, H_2O_2 and the hydroxyl radical that can easily perturb a variety of metabolic processes within a plant. The central role of redox regulation in plant networks contributes to this defence that is contributing to plant survival. There are several evidences that plants with high levels of antioxidants either constitutive or induced, show greater tolerance to the oxidative damage caused by ROS. Antioxidants are classified by their mechanism of action in (Gomez et al. 2007): (i) Primary antioxidants, which are able of stabilizing or deactivating free radicals by donating hydrogen or electrons to them before they attack cells (Kaliora et al. 2006), (ii) synergistic antioxidants, which are classified as oxygen scavengers and chelators (redox activity), (iii) secondary antioxidants, which prevent oxidation by decomposing lipid peroxides into stable end products. For example, in tomato, the properties of carotenes (CARs), polyphenolic compounds, vitamins and powerful antioxidant capacity have been attracting much attention. According to Minoggio et al. (2003), antioxidant content and behaviour base is related with the cultivar and agronomic culture. Different cultivars of cherry tomato have shown exceptionally high lycopene (LYC) content and significant differences on beta-carotene, alpha-tocopherol, vitamin C (ascorbic acid and dehydroascorbic acid) (Raffo et al. 2006; Lenucci et al. 2006), total phenolic and flavonoid contents (Lenucci et al. 2006). Typically breeders are continuously improving their breeding lines, either by making new crosses with their own material or by using the cultivars of their competitors, which is allowed by breeders' law (Bai and Lindhout 2007). Vitamin C is the major non enzymatic antioxidant in plants, while glutathione (GSH), vitamin E, and CARs; polyamines and flavonoids also are at high concentration in stressed plants providing protection from free radical injury. Ascorbate–glutathione cycle is the major antioxidant pathway in plastids, where ROS are generated during normal biochemical processes that include photosynthetic transfer of electron and when stress occurs, they increase its levels several times. While increasing ROS they are acting as a key regulator of the plant network antioxidant systems. Regulation of the contents of antioxidants and antioxidants enzymes constitutes an important mechanism for avoiding oxidative stress. Therefore, plant resistance to stresses is closely associated with the adequate functioning of the antioxidant system (Mittler 2002).

2.1 Biotic and Abiotic Stress

The effect of environmental conditions such as drought, salinity, heat, heavy metals and excess of light on plants are known as abiotic and biotic stresses. Drought and salinity are one of the major abiotic stresses affecting plant growth and productivity. Desertification and salinization are rapidly increasing on a global scale and currently affect more than 10% of arable land, which results in a decline of the average yields of major crops greater than 50% (Wang et al. 2009). Therefore, understanding the mechanisms of plant tolerance to drought and high salinity is a crucial environmental research topic. Generally, exposure to drought or salt stresses triggers many common reactions in plants that lead to cellular dehydration with concomitant osmotic change, removal of water from cytoplasm into the extracellular space that result in a decrease of cytosolic and vacuolar volumes (Wang et al. 2009). Another consequence of drought and salinity stress in plant is the excessive generation of reactive oxygen species (ROS intermediates) such as superoxide, H_2O_2 and hydroxyl radical (Mittler 2002; Bartels and Sunkar 2005).

The excess of ROS production during drought and salinity stress results from impaired electron transport processes in chloroplast and mitochondria as well as from pathway such as photorespiration (Vaidyanathan et al 2003; Wang et al. 2009). The production of ROS which can damage DNA, proteins, chlorophyll and membrane function, is a by-product of oxidative metabolism in chloroplasts, mitochondria and peroxisomes. Under saline conditions, it is not surprising that oxidative stress is also an important component of the salinity damage to plants. Several researchers, working with different plant species demonstrate that enzymes change with salt stress, suggesting the resistance to oxidative stress may, at least in part, be involved in salt stress tolerance (Mittova et al. 2002a; Azevedo Neto et al. 2006). For example, Chookhampaeng et al. (2008) observed enhanced activities of superoxide dismutase (SOD), ascorbate peroxidase (APX), catalase (CAT) and glutathione reductase (GR) to defend against the ROS. Mittova et al. (2002b) found in leaves of salt-stressed wild tomato (*Lycopersicon pennellii*), that the activity of APX was most enhanced (4.7 folds) and POD (1.9 folds). Harinasut et al. (2003) reported the enhanced activity of antioxidative enzymes (APX and SOD) in leaves of mulberry grown under salt stress (150 mM NaCl) conditions. Other environmental stress is the low and high temperatures which follow the same model described for salinity and drought. Although ROS were thought to be involved in light-associated chilling stress, molecular and biochemical evidence was presented to indicate that low temperature also imposes oxidative stress in dark-grown maize seedlings during chilling treatment (Prasad et al. 1994).

The model for insect and pathogen attack operates in different form. In this case, ROS play a central role in the defense of plants against pathogen attack. During this response by plant cells via the enhanced enzymatic activity of plasma-membrane-bound peroxidases and amine oxidases in the apoplast (Hammond-Kosack and Jones 1996; Grant and Loake 2000). H_2O_2 produced during this response is thought to diffuse into cells and, together with salicylic acid (SA) and nitric oxide (NO) are able to active many of the plant defenses, including programmed cell death (PCD) (Dangl et al. 1996). Vadassery et al. (2009) observed that MDAR2 and DHAR5 are crucial for producing sufficient ASC to maintain the interaction between *Priformospora indica* and *Arabidopsis* in a mutualistic state.

According to Mittler (2002), the role ROS play during PCD appears to be opposite to the role they play during abiotic stresses, during which ROS induced ROS-scavening mechanisms such as APX and CAT that decrease the steady state level of ROS in cells. The differences in the function of ROS between abiotic and biotic stresses might result from action of NO and hormones such as SA.

3 Plant Defense Mechanisms: Role of Ascorbate–Glutathione Cycle

Redox alteration is induced by biotic and abiotic stresses (Wormuth et al. 2007) is accompanied by a oxidative burst mediated by NADPH oxidases (Mittler et al. 2004) and a generation of H_2O_2 occurrs by the action of SOD on the superoxide radical at almost diffusion-limited rates to produce H_2O_2 (Scandalios 1997). Root mitochondrias, like those of leaves (Moller 2001; del Rio et al. 2002) contain MnSOD (Malecka et al. 2001). H_2O_2 produced by SOD and some oxidases, is scavenged by peroxidases, especially APX, CAT and glutathione peroxidase (GPX). GPX is a key enzyme of antioxidant network in plants (Chang et al. 2009). In tomato, Mittova et al. (2002a) reports an increased activity of total SOD, APX, monodehydroascorbate reductase (MDHAR), glutathione-S-transferase (GST), phospholipid hydroperoxide glutathione peroxidase (PHGPX) and several isoforms of non-specific peroxidases (POD) found in chloroplasts of salt-treated wild tomato (*Lycopersicon pennellii*) plants. CAT activity has been reported in roots (Shalata et al. 2001), and it has been mainly found in leaf peroxisomes where it functions mainly removing H_2O_2 formed in photorespiration or in b-oxidation of fatty acids in the glyoxysomes (Dat et al. 2000) and in the process of oxidative stress. Kerdnaimongkol and Woodson (1999) observed that inhibition of CAT by antisense RNA increases susceptibility to oxidative stress in transgenic tomato plants, indicating an important role for CAT in oxidative stress defence system. However, Seong et al. (2007) showed enhanced resistance to salinity and oxidative stress due to tomato plants overexpressing CaKR1. These overexpressed transgenic plants produced lower levels of free oxygen radicals, such as O_2^- and H_2O_2.

3.1 Enzymes of the Ascorbate–Glutathione Cycle

The most important antioxidative enzymes of the ascorbate–glutathion cycle are a series of coupled redox reactions involving a set of four enzymes, ascorbate peroxidase (APX, EC 1.11.1.11), monodehydroascorbate reductase (MDHAR, EC 1.6.5.4), glutathione-dependent dehydroascorbate reductase (DHAR, EC 1.8.5.1), and glutathione reductase (GR, EC 1.6.4.2) (Chew et al. 2003). This cycle is an important antioxidant protection system against H_2O_2 generated in different cell compartments. Its occurrence has been reported in chloroplast, mitochondria, peroxisomes, and cytoplasm (Palma et al. 2006). This cycle operates mainly in chloroplasts of plants in order to remove the large amounts of H_2O_2 generated during photosynthetic operations in the light through thylakoid electron transport chain components with electrochemical potentials capable of produce a direct reduction of O_2 (Asada and Takahashi 1987). The specific activity of these enzymes in plastids and extensive work on its operation clearly indicates that this is the major H_2O_2 metabolizing in these photosynthetic organelles (Noctor and Foyer 1998).

3.2 Non-enzymatic Compounds

The main non-enzymatic antioxidant molecules participating in the cycle are ascorbate and glutathione (Noctor and Foyer 1998; Halliwell and Gutteridge 2000). Vitamin C is one of the most extensively non enzymatic anti-oxidant compounds studied and has been detected in the majority of plant cell types, organelles and apoplast (Smirnoff 2000; Borland et al. 2006). It is synthesized in the mitochondria and is transported to the other cell components through a proton electrochemical gradient or through facilitated diffusion. There is compelling genetic evidence that the biosynthesis of AsA proceeds via a D-mannose/L-galactose pathway and is the most significant source of AsA in plants (Ishikawa and Shigeoka 2008). Further, vitamin C also has been implicated in regulation of cell elongation (Yabuta et al. 2004). Vitamin C concentration has been reported to increase with leaf age, being high when chloroplasts are fully developed and chlorophyll levels are highest. Although it has been determined that D-glucose is the precursor of vitamin C, the synthetic pathway has not been totally understood (Foyer and Noctor 2005a, b). Vitamin C has effects on many physiological processes including the regulation of growth, differentiation and metabolism of plants. A fundamental role of vitamin C in plant defence system is related with metabolic processes depleting H_2O_2 and other toxic derivatives of oxygen. Acting essentially as a reductant and scavenging many types of free radicals. AsA is a important reducing sustrate for H_2O_2 catabolism in photosynthetic organism donating electron to reduce H_2O_2 to water with the generation of monodehydroascorbate (MDHA). Beside, reacts non-enzymatically with superoxide (O_2^-), H_2O_2, and singlet oxygen ($^1O_2$). Also particepates in the regeneration of the active form of tocopherol acting coordination with glutathione or participating in the synthesis of zeaxanthin in the xanthophyll cycle.

Therefore, vitamin C influences many enzyme activities, minimizing the oxidative damage through synergic function with other antioxidants (Foyer and Noctor 2005a, b). In the ascorbate–glutathione cycle, two molecules of vitamin C are utilized by APX to reduce each molecule of H_2O_2 to water (H_2O) with concomitant generation of MDHA. MDHA is a radical with a short life time, rapidly transformed into dehydroascorbate and ascorbic acid. The electron donor is usually NADPH and catalyzed by MDHAR or ferredoxin in water–water cycle in the chloroplasts (Gapper and Dolan 2006). Vitamin C can also directly scavenge $^1O_2$, super oxide (O_2^-) and HO^- and regenerate tocopherol from tocopheroxyl radicals, thus providing membrane protection (Li and Jin 2007). Vitamin C also plays a role as a co-factor of violaxanthin de-epoxidase, thus sustaining dissipation of excess excitation energy (Li and Jin 2007; Pourcel et al. 2007). Antioxidants such as vitamin C and GSH are envolved in the neutralization of secondary products of ROS reactions (Foyer and Noctor 2003; Shao et al. 2008).

Glutathione (GSH) is a tripeptide (α-glutamyl cysteinylglycine), detected virtually in all cell compartments such as cytosol, chloroplasts, endoplasmic reticulum, vacuoles and mitochondria (Millar et al. 2003), being the major source of non-protein thiols in most plant cells. The chemical reactivity of the thiol group of GSH makes it particularly suitable to serve a broad range of biochemical functions in all organisms. The nucleophilic nature of the thiol group is also important in the formation of mercaptide bonds with metals and for reacting with selected electrophiles. This reactivity along with the relative stability and high water solubility of GSH makes it an ideal biochemical to protect plants against stress including oxidative stress, heavy metals and certain exogenous and endogenous organic chemicals

(Millar et al. 2003; Foyer and Noctor 2005a, b). GSH takes part in the control of H_2O_2 levels (Foyer and Noctor 2005a, b). The change in the ratio of GSH to oxidized (GSSG) form during the degradation of H_2O_2 is important in certain redox signalling pathways (Millar et al. 2003). It has been suggested that the GSH/GSSG ratio, indicative of the cellular redox balance, may be involved in ROS perception (Millar et al. 2003; Foyer and Noctor 2005a, b). A dual signalling mechanism involving GSH and reduced glutathione/oxidised glutathione (GSH/GSSG) ratio may operate as an intracellular signalling mechanism for the onset of DNA replication in response to stress and GSH also can be involved in transport of signals out of the cell (Russo et al. 2008). GSH acts as an antioxidant and is involved directly in the reduction of most active oxygen radicals generated due to stress, helping to withstand oxidative stress in transgenic lines of tobacco as reported by Foyer and Noctor (2005a, b). In response to added Cu to the roots a higher requirement for glutathione was observed to counteract ROS formation with and increasig GR activity in roots. Aerobic organism have veloped strategies to utilise differen biological signals that control various genetic stress responses (Russo et al. 2008).

4 Non-enzymatic and Enzymatic Integration in the Ascorbate and Glutathione Cycle

Antioxidant enzyme systems and non enzymatic antioxidative compounds are regulated by ROS produced in the cells by means of a complex system which occur in different organelles which can also act harmonically. At least four enzymes of the scavenning system have targeted to both the chloroplasts and the mitochondria (Valero et al. 2009). There are the following enzymes:

4.1 Ascorbate Peroxidase and Monodehydrascorbate Reductase

Ascorbate peroxidase (APX) comprises a family of haemperoxidases isoenzymes located in several organelles. This enzyme catalyses the reduction of H_2O_2 to H_2O and has high specifity and affinity for ascorbate (AsA) as reductant. APX uses two molecules of ASC to reduce H_2O_2 to H_2O and production of two molecules of monodehydroascorbate (MDHA) (Table 1). APX have a much higher affinity for H_2O_2 than CAT. The high affinity of H_2O_2, in conjunction with findings of ascorbate-glutathion cycle in almost all cellular compartments, suggests that this cycle plays a crucial role in controlling the levels of ROS in these compartments. Additionally, APX migh also be responsible for the fine modulation of H_2O_2 for signaling. On the basis of its localization within the plant cell, different forms of APX have been found in chloroplast, cytosol, mitocondria, peroxisomes, glyoxysomes and apoplast (Shigeoka et al. 2002; Ashraf 2009). APX and MDHAR activities increased when sun exposed apple peels were studied (Ma and Cheng 2003).

Table 1 Major reactive oxygen species-scavenging enzymes of ascorbate-glutathione cycle, enzyme code and reactions catalyzed

Enzymatic antioxidants	Enzyme code	Reactions catalyzed
Ascorbate peroxidase (APX)	EC 1.11.1.11	H_2O_2 + 2 ASC → 2H2O + 2MDHA
Monodehydroascorbate reductase (MDHAR)	EC 1.6.5.4	NADPH + 2MDHA → ASC + $NAD(P)^+$
Dehydroascorbate reductase (DHAR)	EC 1.8.5.1	DHA + 2GSH → ASC + GSSG
Glutathione reductase (GR)	EC 1.6.4.2	NADPH + GSSG → 2GSH + $NAD(P)^+$

4.2 Dehydrascorbate Reductase and Glutathione Reductase

Dehydrascorbate reductase (DHAR) activity increased when sun exposed apple peels were studied (Ma and Cheng 2003). DHAR reduces DHA to ASC, using reduced glutathione (GSH) as the reducing substrate. This reaction generates reduced glutathione (GSSG), which is turn re-reduced to GSH by NADPH, a reaction catalized by GR. The removal of H_2O_2 through the series reactions is known as ascorbate–glutathione cycle or Halliell–Asada pathway (McKersie 1994). ASC and GSH are not consumed in this pathway but participate in the cycle transfert of reducing equivalents, which permit the reduction of H_2O_2 to H_2O, with NADPH as the donor of reducing equivalents (Table 1).

5 Metabolic Regulation of the Ascorbate–Glutathione Cycle

Modulation of the ROS-antioxidant interaction plays a part in many stresses, as well as other response to the environment, and in the regulation of plant development. Although oxygen is essential for life and is produced under normal growth conditions and their concentration remains low (Polle 2001), its reduction results in the production of ROS such as H_2O_2, superoxide (O_2^-) and the hydroxyl radical (HO^-) that can easily perturb a variety of metabolic processes within a plant. The central role of redox regulation in plant networks contributes to this defence that is contributing to plant survival. Mitochondria and chloroplast in addition to their action in the respiration of the plant and in the photosyntetic process are recognized as important sites for the generation, dispensation and removal of a number of intracellular signaling effects including H_2O_2. These compounds alters the redox status of the cell being the GSH/GSSG ratio the key signal that iniciates the transduction signal to induce the metabolic regulation of different enzyme patways being the ascorbate–glutathione cycle the most important in to maintain a sustentable GSH/GSSG ratio. Thus, the rate at which H_2O_2 is emitted is considered as an important centinel of the plant function and modulator of the overall cellular redox environment.

5.1 Modulation of the Aascorbate–Glutathione Cycle

Signaling mediated by ROS involves heterotrimeric G-proteins (Joo et al. 2005) and protein phosphorylation regulated by MAP kinases and protein Tyr phophatase (Kovtun et al. 2000; Gupta and Luan 2003; Rentel et al. 2004; Foyer and Noctor 2005a, b). The biochemical and structural basis of kinase pathway activation by ROS remains to be established in plants, but thiol oxidation likely plays a key role. The best-characterized redox signal transduction system in plants is the stromal ferredoxin–thioredoxin system, which functions in the regulation of photosynthetic carbon metabolism. Signal transmission involves disulfide–thiol conversion in target enzymes and is probably achieved by a light-induced decrease in the thioredoxin redox potential. Thiol groups are likely important in other types of redox signal transduction, including ROS sensing by receptor kinases, such as ETR1 (Desikan et al. 2005). Thiol-based regulation may be important in plant acclimation to environmental changes, particularly where redox interactions play a key role in the orchestration of the abiotic stress response. Thermodynamics (redox potential of oxidizable thiols) and kinetics (ability to compete with the antioxidative system) are key considerations in assessing the functional importance of putative thiol based ROS sensor in plants. Recent studies have identified several components involved in the signal transduction pathway of plants that senses ROS. These include the mitoge-activated protein (MAP) kinase kinase kinases AtANP1 and NtNPK1, and the MAP kinases AtMPK3/6 and Ntp46MAPK (Kovtun et al. 2000; Samuel et al. 2000). Loss of AtMPK6 renders *Arabidopsis* more susceptible to oxidative stress (Miles et al. 2009). In addtion, calmodulin has been implicated in ROS signaling (Desikan et al. 2001). A hypothetical model depicting some of the players involved in this pathway. H_2O_2 is sensed by a sensor that might be a two-component histidine kinase, as in yeast. Calmodulin and a MAP-kinase cascade are then activated, resulting in the activation or suppression of several transcription factors. These regulate the response of plants to oxidative stress (Desikan et al. 2001). This model is simplified and is likely to change as research advances our understanding of this pathway (Fig. 1).

Reactive oxygen species act as signals that mediate the systemic activation of gene expression in response to pathogen attack (Alvarez et al. 1998), wounding (Orozco-Cardenas and Ryan 1999) and high light (Mullineaux and Karpinski 2002). They were suggested to act in conjunction with a compound that travels systemically and activates their production in distal parts of the plant, where they mediate the induction of gene expression (Orozco-Cardenas and Ryan 1999). The involvement of ROS in the regulation of stomatal closure (Pei et al. 2000) and in other cellular responses involving auxin (Kovtun et al. 2000). This might suggest that more signaling pathways involving ROS as inducers of systemic signals await discovery. It is unlikely that ROS can travel systemically because they are highly reactive and would be scavenged along the way by the many antioxidative mechanisms and antioxidants present in the apoplast. However, it is possible that a wave of activity similar to the "oxidative burst" is activated in cells along the systemic path and in distal tissues, resulting in the accumulation of ROS. Future studies

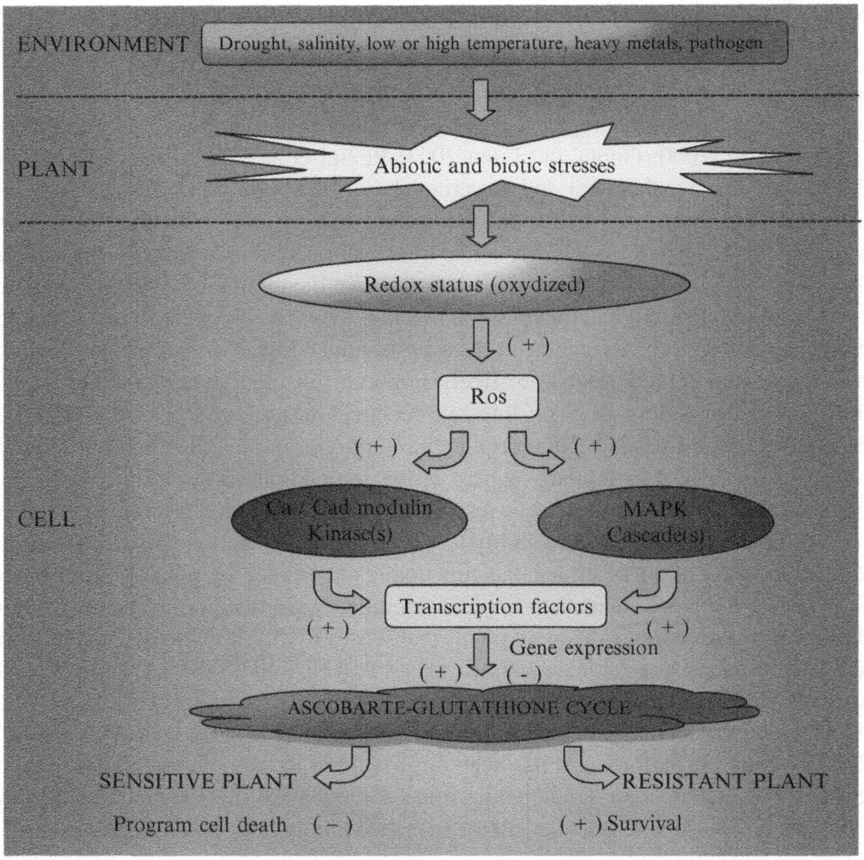

Fig. 1 Enviromental stress factors, ROS production and the signal regulation of the ascorbate–glutathione cycle

using plants with altered levels of ROS-scavenging resolve this question. According to Foyer and Noctor (2005a, b), the current knowlege of redox controls in the apoplast and cytoplasm predicts that: (1) the plasma membrane is an important site for perception and transduction of environmental change through redox signals, (2) apoplastic redox changes facilitate interactions between receptor proteins containing oxidizable thiols that are sited in or near the membrane surface, (3) because the apoplastic redox gradient is present across the plasma membrane; this gradient triggers or elaborates membrane channel activity involving calcium release and aquaporin or peroxiporin function.

Reactive oxygen species is detected by cellular receptors or sensors. The regulation of gene expression by the different transcription factor results of various defence pathways, such as reactive oxygen intermediate scavenging and heat-shock proteins (HSPs), and in the suppression of some ROS-producing mechanisms and photosynthesis. Furthermore, Chew et al. (2003) mentioned that many diverse functions may be linked by a single gene in the genome, which helps to coordinate the function and biogenesis of mitochondria and plastides without the need for additional pathways to facilitate intra-organellar communication (Larkin et al. 2003; Strand et al. 2003). So if any independent regulation needs to occur it must be via a different mechanism at the level of protein import or post-translational modification. Diverses authors have discussed the need for signaling pathways to coordinate antioxidant defense in different cellular organelles in plants. But, it appears that at least in the context of the mitochondrial and the chloroplast ascorbate–glutathione cycle, this coordination occurs by dual targeting rather retrograde signaling.

5.2 Gene Expression of Ascorbate–Glutathione Cycle Enzymes

Eight types of genes encoding APX have been described in *Arabidopsis* (Table 2): three cytosolic (apx1, apx2, apx6), two chloroplastic (stromal, sapx; thylakoid, tapx), and three microsomal (apx3, apx4, apx5). Gene expression is conditioned by different enviromental stresses such as ozone, high light, extremes of temperatures, drought, heat stress and their interactions. Currently, APX activities increase along with activities of other antioxidant enzymes like CAT, SOD, and GR in response

Table 2 Genes encoding the enzymes of arcobate glutathione cycle, its localization in the cell and on the chromosomes of *Arabidopsis thaliana*

Gene	Enzyme	Localization	Chromosome	Reference
apx1	APX1	cytosolic	1	Koussevitzky et al. 2008
apx2	APX2	cytosolic	3	Rossel et al. 2006
apx3	APX3	microsomal	4	Navendra et al. 2006
apx4	APX4	microsomal	4	Schubert et al. 2002
apx5	APX5	microsomal	4	Chew et al. 2003
sapx6	sAPX	stromal	4	Shigeoka et al. 2002
tapx	tAPX	thylakoid	1	Miller et al. 2007
mdar1	MDAR1[a]	cytosolic	3	Eltayeb et al. 2007
mard4	MARD4	peroxisome	4	Eastmond 2007
atmdar2	ATMDAR2	peroxisome	5	Lisenbee et al. 2005
gr1	GR1	cytosolic	3	Xiang and Olivier 1998
gr2	GR2	plastidic	3	Peltier et al. 2006
dhar1	DHAR1	-	1	Elter et al. 2007
dhar1	DHAR2	cytosolic	1	Yoshida et al. 2006
dhar2	DHAR3	chloroplast	5	Dixon et al. 2002

[a]MDAR (or MDHAR)

to various environmental stress factors, suggesting that the components of ROS-scavenging are co-regulated (Shigeoka et al. 2002). Drought-resistant maize shows greater induction of APX activity than sensitive plants, in addition to a significant increase in GR activity (Pastori and Trippi 1992). Koussevitzky et al. (2008) showed that *Arabidopsis thaliana* plants subjected to a combination of drought and heat stress, APX1 and mRNA accumulated during the stress combination. When exposed to heat stress combined with drought, on APX1-deficient mutant (apx1) accumulate more H_2O_2 and was significantly more sensitive to the stress combination than wild type. In contrast, mutants deficient in thylakoid or stromal/mitochondrial APXs were not more sensitive to the stress combination than apx1 or wild type. This author suggests that cytosolic APX1 plays a key rol in the acclimation of plants to a combination of drought and heat stress. The heat shock response, for example, can be completely inhibited by effective removal of H_2O_2 because the expression of genes such as that encoding APX1 is modulated by ROS signals (Davletova et al. 2005). The analysis of the protein level and activity of APX indicates that when pea is recupering from drought stress is bing regulted by at level of protein synthesis.

Monodehydroascorbate reductase (MDHAR or MDAR) is crucial for ASC regeneration and esential for maintaining a reduced pool of AsA (Table 1). MDAR gene encodes a member of the monodehydroascorbate reductase gene family. For example, MDAR4 gene encodes a peroxisome membrane bound MDAR, which removes toxic H_2O_2 (Table 2). Transgenic plants overexpressing *Arabidopsis thaliana* MDAR gene (AtMDAR1) showed enhanced stress tolerance in term of significantly higher net photosynthesis rates under ozone, salt and osmotic stress and greater Photosystem II (PSII) effective quantum yield inder ozone and salt stresses (Eltayeb et al. 2007). In the same study, theses transgenic plants exhibited significantly lower H_2O_2 level when tested under salt stress.

Glutathione reductases (GR) are enzymes that catalyse the oxydised form of glutathione to reduced glutathione (GSH). GHS is synthesized in both the chloroplast and the cytosol of plant leaves, where it is maintained almost exclusively in the reduced form by the action of GR (Table 1). When GSH is oxidized as part of its antioxidant activity, it forms glutathione disulfide (GSSG), the oxidized form of GSH. The GRs reduce GSSG back to GSH by using reducing equivalent from NADPH. Two genes encoding GR have been identified in *Arabidopsis thaliana*; one gr 2, encodes a plastidic isoform, and the other, gr1, encodes a cytosolic enzyme (Xiang and Olivier 1998) (Table 2). Genes encoding GR have also described in others species such as *Populus trichocarpa* (Ptrc2) *Ricinus communis* (RCOM-1020350), *Nicotiana tabacum* (gr), and *Oryza sativa* (Os02g0813500). Ding et al. (2009) studied the possible mechanisms of GR in protecting against oxidative stress in *Nicotiana tabacum*. These researchers obtained transgenic tabacco (*Nicotiana tabacum*) plants with 30–70% decreased GR activity by using a gene encoding tabacco chloroplastic GR for RNAi construct. Oxidative stress was induced by application of methyl viologen vivo on wild type and transgenic plants. This study demonstrated that transgenic plants exhibited enhanced sensitivity to oxidative stress in relation to wild type. Secenji et al. (2010) compared the responses of two wheat (*Triticum aestivum*) genotypes, the drought-tolerant Plainsman V and the drought-sensitive Cappelle Desprez, to reduced amounts of irrigation water, they found differences in ascorbate metabolism: both ascorbate oxidation and transcription levels of enzymes processing ascorbate were changed. These authors found an up-regulated chloroplastic DHAR (chlDHAR) mRNA only in the sensitive Cappelle Desprez. However, increased expression levels of a cytosolic GR (cGR) and a chloroplastic GR (chlGR) were detected only in the tolerant Plainsman V. After 4 weeks of reduced irrigation, a significantly lower ascorbate/dehydroascorbate ratio was detected in leaves of the sensitive Cappelle Desprez than in the tolerant Plainsman V. This study indicate that more robust transcription of ascorbate-based detoxification machinery may prevent an adverse shift of the cellular redox balance.

Dehydroascorbate reductase (DHAR) is a key component of the ascorbate recycling system. DHAR gene encodes members of the dehydroascorbate reductase gene family. Three functional DHAR genes are encoded in the Arabidopsis genome. DHAR1 gene (Table 2), critical for a mutualistic symbiosis between the host *Arabidopsis* and the root colonizing fungus *Piriformospora indica*. These genes are found in othes plant species such as *Zea mays*, *Ricinus communis* and *Vitis vinifera*. Yoshida et al. (2006) observed that in *Arabidopsis* normal plant, ozone exposure increased the expression of the cytosolic DHAR (cytDHAR) gene alone. However, in *Arabidopsis* mutant with a deficient cytDHAR, completely lacked cytDHAR activity, was highly ozone sensitive. The amounts of total ASC and GSH were similar in both lines, but the amount of apoplastic ascorbate in the mutant was 61.5% lower. This study showed that the apoplastic ascorbate, which is generated through the reduction of DHA by cytDHAR, is important for ozone tolerance.

6 Strategies to Improve the Antioxidant Levels in Crops

Antioxidant defence in context of photosynthesis is a condition sine qua non for plant survival and has been explored in many studies. Especially in the field, oxidative damage is a major factor in limiting plant productivity. The detailed comparison of sensitive and tolerant genotypes and species, and a systematic understanding of antioxidant defense promise new ways for efficient breeding and targeted genetic engineering of plants with improved perfomance. These agronomical and biotechnological actions can increase cropped areas, thus increasing the ecological importance of these resistant crops.

Plants are directly exposed to abiotic stresses which are known to induce oxidative stress, being salinity the abiotic stress that has been studied most. This is due to the fact that vast area of the available land on the globe is affected by salinity (Ashraf 2009). The question arises as to how a plant controls and speeds up its rate of RSO production and ROS scavenging when it is exposed to a abiotic stress. This defence network is activated by a change in the cellular redox state which transmit the redox information into transcriptional regulation vitamin C is the major antioxidant compound in plants, while GSH, vitamin E, and CARs; polyamines and flavonoids also provide protection from free radical injury. Ascorbate–glutathione cycle is the major antioxidant pathway in plastids, where ROS are generated during normal biochemical processes that include photosynthetic transfer of electron. The increase of the antioxidant compounds is an indicator of the healthy value of these crops. This can contribute to low the risk of non transmisible illness such as diabetes, obesity, cardiovascular desease, cancer. The increasing demand for natural antioxidants, in parallel with the introduction of new technologies to meet higher quality standards, justifies the search for new sources of natural antioxidants. Consequently, there is an increasing interest for the development of food crops with enhanced levels of antioxidants in vegetables such as tomatoes. This major crop is an excellent candidate because their fruit could increase its levels of antioxidants either by conventional plant breeding, by genetic manipulation (Fraser and Bramley 2004) or by management practices. For example, in tomato several studies have demonstrated that plants submitted to salinity or water stress augment the fruit's antioxidant capacity and widen its range of health benefit properties. Firstly, conventional breeding has extensively contributed to the improvement of vegetable yields, quality, post-harvest conditions, and resistance to biotic and abiotic stresses. The art of vegetable breeding basically consists on identifying and combining the specific traits. For exemple, fresh tomato fruits are offered in a wide range of flavours, tastes, shapes and sizes, from the small cherry tomato to very large beef tomatoes. Current efforts in vegetable breeding are focused to discover and exploiting genes for the most important traits in plant germplasm (Bai and Lindhout 2007). Breeders commonly start their work from naturally occurring re-combinations to develop cultivars that combine favourable traits. New traits are rarely introduced from wild germplasm requiring many generations to remove the deleterious genes that go along with the introduced genes due to linkage drag. However, there are many constraints in conventional breeding, which can only be overcome by advancements made in modern biotechnology. In the last decade various traits such as biotic stress resistance, quality and storage life have been successfully engineered into vegetable crops and some of them have been commercialized. Currently, agronomical research in vegetable is focused on identifying new varieties with high yield and fruits of good organoleptic quality and high antioxidant levels either as nutrients or non-nutrient chemicals. The incorporation of wild food plant varieties to improve the traditional process would be interesting to introduce desirable characteristics into a new variety, as the increasing of the antioxidant capacity of the plant. The selection of genotypes in vegetables and fruit trees such as tomato having high antioxidant capacity must be part of long term programmes directed to achieve this objective. Experiments with plant wild genotypes would be an advantage to unravel genes of resistance to high light, low or high temperature, drought and saline stress conditions. Secondly, over last years significant progress has been made to manipulate vegetable crops for abiotic stress tolerance, quality improvement and pharmaceutical and industrial applications. Although the progress in commercialization of transgenic vegetable crops has been relatively slow, transgenic vegetables engineered for nutraceutical and pharmaceutical use will contribute significantly to add value in agriculture products in the near future (Dalal et al. 2006). Undoubtedly, during the last few decades, genetic engineering approach has gained a considerable ground because progress in terms of improving a trait in a crop could be achieved within the shortest possible time period. Furthermore, tolerance to a multitude of enviromental stresses has been correlated well with enhanced activities of antioxidant genes provides an opportunity to develop plants with enhanced tolerance to abiotic and biotic stress. The improvement of saline and drought resistance in different plant species is possible through the genetically engineered overexpressions of specific enzymes. Such studies provide sound evidence that expression of antioxidant componds could be used as a prospective criteria for breeding for salt tolerance indifferent crops (Ashraf 2009). Biotechnical manipulations have been used to adopt a transgenic approach generating plant transformants in which sequences encoding key biosynthetic enzymes, either alone or in combination, or encoding regulatory elements are ectopically expressed. For example, these approaches resulted in a number of modified tomato lines with increased levels of flavonols (Verhoeyen et al. 2002). Finally, in the last decade various traits such as biotic stress resistance, quality and storage life have been successfully engineered into vegetable crops and some of them have been commercialized. Although most consumers do not seem comfortable with genetic engineering, the idea of increasing health benefits in daily consumed vegetables is promising and alluring. As many of the health compounds belong to the plant defence system, controlled stress to plants will probably increase the concentration of desirable components (Krauss et al. 2006). For example, in the case of cropped tomato, it is expected that the effect of salinity on a reduction of productivity will be evident compared to wild tomato. Since fruit quality will be improved by increasing both the level of antioxidant compounds and the enzymatic activity of the antioxidant system, screening for salinity tolerant genotypes will be more efficient if physiological traits related to saline stress are identified.

On the other hand, large and increasing proportions of the world's irrigated land are severely affected by drought and salinity. While the exact area affected is not known, it is estimated that approximately 25% of the world's irrigated land is damaged by salinization (Flowers and Flowers 2005). Naturally occurring salt-affected soils cover about a billion hectares; these areas, which are largely coastal salt marshes or inland deserts, are not as important to agriculture as the areas of secondary salinization brought about through irrigation and forest clearance. There is currently nearly 275 million hectares of irrigated land (273 Mha in 2001; http://apps.fao.org, January 2004) of which about 20% is salt affected (Ghassemi et al. 1995); perhaps half of all irrigation schemes are subject to salinization. This is particularly unfortunate, since irrigated land is the most productive in world's agriculture. Dry land salinity is also linked to rising water tables brought about by increased deep drainage of rainfall following forest clearance and a change from deep-rooted perennial plants to a shallow-rooted annual crop (Flowers and Flowers 2005). Climate change is expected to cause a net increase in the proportion of land classed as semi-arid and saline (Yeo 1998) by a rainfall reduction, as it would be existing in the northern of Chile or in the southern of Spain. While salt-affected soils occur extensively under natural conditions, salt problems of greatest importance to agriculture arise when previously productive cultivated soil becomes salinized as a result of irrigation (so-called secondary salinization). Human activities have increased the extent of salt-affected areas considerably by the redistribution of water (hence salt) through irrigation. Another important factor to consider is the increasing of salinity as a result of the waste water from the big city, which contains high sales concentrations. Therefore, this proposal tries to unravel the response of the crop under different stresses, considering the importance of drought and saline soils, environmental tendencies and global climate change around the world.

Both traits yield and fruit quality (essentially health compound levels and sensory characteristics) are commonly opposite when stress conditions are applied and the same fact is observed in plant productivity. Consequently, it is necessary to study the yield stress conditions in order to obtain a less reduction on yield while maintaining fruit quality and antioxidant levels. It is suggested that the research should be also focused on manipulation of environmental stresses according to plant responses and greenhouse environments providing a more precise and effective application of stress treatment such as water stress, salinity, solar radiation intensity among others stresses.

Acknowledgements The authors are grateful to Dr. Mei Lie Tan for stimulating discussions and comments on this manuscript. This work was supported by the FONDECYT (grants project N° 1090405), National Commision for Scientific and Technological Research (CONICYT), Ministery of Education, Chile.

References

Alvarez ME, Pennell RI, Meijer PJ, Ishikawa A, Dixon RA, Lamb C (1998) Reactive oxygen intermediates mediate a systemic signal network in the establishment of plant immunity. Cell 92:773–784

Asada K, Takahashi M (1987) Production and scavenging of active oxygen radicals in photosynthesis. In: Kyle DJ, Osmond CB, Arntzen CJ (eds) Photoinhibition, vol 9. Elsevier, Amsterdam, pp 227–288

Ashraf M (2009) Biotechnological approach of improving salt tolerance using antioxidants as a markers. Biotech Adv 27:84–93

Azevedo Neto AD, Prisco JT, Enéas-Filho J, Braga de Abreu CE, Gomes-Filho E (2006) Effect of salt stress on antioxidative enzymes and lipid peroxidation in leaves and roots of salt-tolerant and salt-sensitive maize genotypes. Environ Exp Bot 56:87–94

Bai Y, Lindhout P (2007) Domestication and breeding of tomatoes: what have we gained and what can we gain in the future? Ann Bot 100:1085–1094

Bartels D, Sunkar R (2005) Drought and salt tolerance in plants. Crit Rev Plant Sci 24:23–58

Borland A, Elliott S, Patterson S (2006) Are the metabolic components of crassulacean acid metabolism up-regulated in responses to an increase in oxidative burden? J Exp Bot 57:319–328

Chang CCC, Slesak I, Jordá L, Sotnikov A, Melzer M, Miszalski Z, Mullineaux PM, Parker JE, Karpinska B, Karpinski S (2009) Arabidopsis chloroplastic glutathine peroxidases play a role in cross talk between photooxidative stress and immune responses. Plant Physiol 150:670–683

Chew O, Whelan J, Millar AH (2003) Molecular definition of the ascorbate-glutathione cycle in Arabidopsis mitochondria reveals dual targeting of antioxidant defenses in plants. J Biol Chem 278:46869–46877

Chookhampaeng S, Pattanagul W, Theerakulpisut P (2008) Effect od salinity on growth, activity of antioxidant enzymes and sucrose content in tomato (*Lycopersicon esculentum* Mill.) at the reproductive stage. SciAsia 34:69–75

Dalal M, Dani RG, Kumar PA (2006) Current trends in the genetic engineering of vegetable crops. Sci Hort 107:215–225

Dat J, Vandenabeele S, Vranova E, Van Montagu M, Inzé D, Van Breusegem F (2000) Dual action of the active oxygen species during plant stress responses. Cell Mol Life Sci 57:779–795

Dangl JL, Dietrich RA, Richberg MH (1996) Death don't have no mercy: cell death programs in plant-microbe interactions. Plant Cell 8:1793–1807

Davletova S, Rizhsky L, Liang H, Shengqiang Z, Oliver DJ, Coutu J, Shulaev V, Schlauch K, Mittler R (2005) Cytosolic ascorbate peroxidase 1 is a central component of the reactive oxygen gene network of *Arabidopsis*. Plant Cell 141:341–345

del Rio LA, Corpas FJ, Sandalio LM, Palma JM, Gomez M, Barroso JB (2002) Reactive oxygen species, antioxidant systems and nitric oxide in peroxisomes. J Exp Bot 53:1255–1272

Desikan R, Mackerness S, Hancock J, Neill SJ (2001) Regulation of the Arabidopsis transcriptosome by oxidative stress. Plant Physiol 127:159–172

Desikan R, Hancock JT, Bright J, Harrison J, Weir I, Hooley R, Neill SJ (2005) A role for ETR1 in hydrogen peroxide signaling in stomatal guard cells. Plant Physiol 137:831–834

Ding S, Lu Q, Zhan Y, Yang Z, Wen X, Zhang L, Lu C (2009) Enhanced sensitivity to oxidative stress in transgenic tobacco plants with decreased glutathione reductase activity leads to a decrease in ascorbate pool and ascorbate redox state. Plant Mol Bio 69:577–592

Dixon DP, Davis BG, Edwards R (2002) Functional divergence in the glutathione transferase superfamily in plants. Identification of two classes with putative functions in redox homeostasis in Arabidopsis thaliana. J Biol Chem 277:30859–30869

Eastmond PJ (2007) MONODEHYROASCORBATE REDUCTASE4 is required for seed storage oil hydrolysis and postgerminative growth in Arabidopsis. Plant Cell 19:1376–87

Elter A, Hartel A, Sieben C, Hertel B, Fischer-Schliebs E, Lüttge U, Moroni A, Thiel G (2007) A plant homolog of animal chloride intracellular channels (CLICs) generates an ion conductance in heterologous systems. J Biol Chem 282:8786–87892

Eltayeb AE, Kawano N, Badawi GH, Kaminaka H, Sankata T, Shibahara T, Inanaga S, Tanaka K (2007) Overexpression of monodehydroascorbate reductase in transgenic tobacco confers enhanced tolerance to ozone, salt and polyethylene glycol stresses. Planta 225:1255–1264

Flowers TJ, Flowers SA (2005) Why does salinity pose such a difficult problem for plant breeders? Agric Water Manag 78:15–24

Foyer CH, Noctor G (2005a) Redox homeostasis and antioxidant signaling: a metabolic interface between stress perception and physiological responses. Plant Cell 17:1866–1875

Foyer CH, Noctor G (2005b) Oxidant and antioxidant signaling in plants: a re-evaluation of the concept of oxidative stress in a physiological context. Plant Cell Environ 28:1056–1071

Foyer CH, Noctor G (2003) Redox sensing and signaling associated with reactive oxygen in chloroplasts, peroxisomes and mitochondria. Physiol Plant 119:355–364

Fraser PD, Bramley PM (2004) The biosynthesis and nutritional uses of carotenoids. Prog Lipid Res 43:228–265

Gapper C, Dolan L (2006) Control of plant development by reactive oxygen species. Plant Physiol 141:341–345

Ghassemi F, Jakeman AJ, Nix HA (1995) Salinisation of land and water resources. human causes, extent management and case studies. University of New South Wales, Sydney, p 526

Gomez M, Arráez D, Segura A, Fernández A (2007) Analytical determination of antioxidants in tomato: typical components of the Mediterranean diet. J Sep Sci 30:452–461

Grant JJ, Loake GJ (2000) Role of reactive oxygen intermediates and cognate redox signaling in disease resistance. Plant Physiol 126:21–29

Gupta R, Luan S (2003) Redox control of protein tyrosine phosphatases and mitogen-activated protein kinases in plants. Plant Physiol 132:1149–1152
Halliwell B, Gutteridge JMC (2000) Free radicals in biology and medicine. Oxford University Press, Oxford, UK
Hammond-Kosack KE, Jones JDG (1996) Resistance gene-dependent plant defense responses. Plant Cell 8:1773–1791
Harinasut P, Poonsopa D, Roengmongkol K, Charoensataporn R (2003) Salinity effects on antioxidant enzymes in Mulberry cultivar. SciAsia 29:109–113
Ishikawa T, Shigeoka S (2008) Recent advances in ascorbate biosynthesis and the physiological significance of ascorbate peroxidase in photosynthesizing organisms. Biosci Biotechnol Biochem 72:1143–1154
Joo JH, Wang S, Chen JG, Jones AM, Fedoroff NV (2005) Differnt signaling and cell death roles of heterotrimeric G protein a and b subunits in the Arabidopsis oxidative stress response to ozone. Plant Cell 17:957–979
Kaliora AC, Dedoussis GVZ, Schmidt H (2006) Dietary antioxidants in preventing atherogenesis. Atherosclerosis 187:1–17
Kerdnaimongkol KR, Woodson RW (1999) Inhibition of catalase by antisense RNA increases susceptibility to oxidative stress and chilling injury in transgenic tomato plants. J Am Soc Hort Sci 124:330–444
Koussevitzky S, Suzuki N, Huntington S, Armijo L, Sha W, Cortes D, Shulaev V, Mittler R (2008) Ascorbate peroxidase 1 plays a key role in the rsponse of Arabidopsis thaliana to stress combination. J Biol Chem 283:34197–34203
Kovtun Y, Chiu WL, Tena G, Sheen J (2000) Functional analysis of oxidative stress-activated mitogen-activated protein kinase cascade in plants. Proc Natl Acad Sci USA 97:2940–2945
Krauss S, Schnitzler WH, Grassmann J, Woitke M (2006) The influence of different electrical conductivity values in a simplified recirculating soil less system on inner and outer fruit quality characteristics of tomato. J Agric Food Chem 54:441–448
Kuzniak E, Sklodowska M (2005) Compartment-specific role of the ascorbate-glutathione cycle in the response of tomato leaf cells to Botrytis cinerea infection. J Exp Bot 56:921–933
Larkin RM, Alonso JM, Ecker JR, Chory J (2003) GUN4, a regulator of chlorophyll synthesis and intracellular signaling. Science 299:902–906
Lenucci MS, Cadinu D, Taurino M, Piro G, Dalessandro G (2006) Antioxidant composition in cherry and high-pigment tomato cultivars. J Agric Food Chem 54:2606–2613
Li JM, Jin H (2007) Regulation of brassinosteroid signaling. Trends Plant Sci 12:37–41
Lisenbee CS, Lingard MJ, Trelease RN (2005) Arabidopsis peroxisomes possess functionally redundant membrane and matrix isoforms of monodehydroascorbate reductase. Plant J 43:900–91
Ma F, Cheng L (2003) The Sun-exposed peel of apple fruit has higher xanthophyll cycle-dependent thermal dissipation and antioxidants of the ascorbate-glutathione pathway than the shade peel. Plant Sci 165:819–827
Malecka A, Jarmuszkiewicz W, Tomaszewska B (2001) Antioxidative defense to lead stress in subcellular compartments of pea root cells. Acta Biochim Pol 48:687–698
McKersie BD (1994) Oxidative stress. In: McKersie BD, Leshem YY (eds) Stress and stress coping in cultivated plants. Kluwer, Dordrecht, pp 15–55
Miles GP, Samuel MA, Ranish JA, Sperrazzo GM, Ellis BE (2009) Quantitative proteomics identifies oxidant-induced, AtMPK6-dependent changes in *Arabidopsis thaliana* protein profiles. Plant Signal Behav 4:497–505
Millar AH, Mittova V, Kiddle G (2003) Control of ascorbate synthesis by respiration and its implications for stress responses. Plant Physiol 133:443–447
Miller G, Suzuki N, Rizhsky L, Hegie A, Koussevitzky S, Mittler R (2007) Double mutants deficient in cytosolic and thylakoid ascorbate peroxidase reveal a complex mode of interaction between reactive oxygen species, plant development, and response to abiotic stresses. Plant Physiol 144:1777–1785
Minoggio M, Bramati L, Simonetti P, Gardana C, Lemoli L, Santangelo E, Mauri L, Spigno P, Soressi GP, Pietta PG (2003) Polyphenol pattern and antioxidant activity of different tomato lines and cultivars. Ann Nutr Metab 47:64–69

Mittler R (2002) Oxidative stress, antioxidants and stress tolerance. Trends Plant Sci 7:405–410

Mittler R, Vanderauwera S, Gollery M, Van Breusegem FV (2004) Reactive oxygen gene network of plants. Trends Plant Sci 9:490–498

Mittova V, Guy M, Tal M, Volokita M (2002a) Response of the cultavted tomato and its wild salt-tolerant relative *Lycopersicon pennelli* to salt-dependent oxidative stress: increased activities of antioxidants enzymes in root plastids. Free Radic Res 36:195–202

Mittova V, Tal M, Volokita M, Guy M (2002b) Salt stress induces up-regulation of an efficient chloroplast antioxidant system in the salt-tolerant wild tomato species *Lycopersicon pennellii* but not in the cultivated species. Physiol Plant 115:393–400

Moller IM (2001) Plant mitochondria and oxidative stress: electron transport, NADPH turnover, and metabolism of reactive oxygen species. Ann Rev Plant Physiol Plant Mol Bio 52:561–591

Mullineaux P, Karpinski S (2002) Signal transduction in response to excess light: getting out of the chloroplast. Curr Opin Plant Biol 5:43–48

Narendra S, Venkataramani S, Shen G, Wang J, Pasapula V, Lin Y, Kornyeyev D, Holaday AS, Zhang H (2006) The Arabidopsis ascorbate peroxidase 3 is a peroxisomal membrane-bound antioxidant enzyme and is dispensable for Arabidopsis growth and development. J Exp Bot 57:3033–3042

Noctor G, Foyer CH (1998) Ascorbate and glutathione: keeping active oxygen under control. Ann Rev Plant Physiol Plant Mol Biol 49:249–279

Orozco-Cardenas M, Ryan CA (1999) Hydrogen peroxide is generated systemically in plant leaves by wounding and systemin via the octadecanoid pathway. PNAS 96:6553–6557

Palma JM, Jiménez A, Sandalio LM, Corpas FJ, Lundqvist M, Gómez M, Sevilla F, Río LA (2006) Antioxidative enzymes from chloroplasts, mitochondria, and peroxisomes during leaf senescence of nodulated pea plants. J Exp Bot 57:1747–1758

Pastori GM, Trippi VS (1992) Oxidative stress induces high rate of glutathione reductase synthesis in a drought-resistant maize strain. Plant Cell Physiol 33:957–961

Pei ZM, Murata Y, Benning G, Thomine S, Klüsener B, Allen GJ, Grill E, Schroeder JI (2000) Calcium channels activated by hydrogen peroxide mediate abscisic acid signalling in guard cells. Nature 406:731–734

Peltier JB, Cai Y, Sun Q, Zabrouskov V, Giacomelli L, Andrea Rudella A, Ytterberg AJ, Rutschow H, van Wijk KJ (2006) The oligomeric stromal proteome of Arabidopsis thaliana chloroplasts. MCP 5:114–133

Polle A (2001) Dissecting the superoxidismutase-ascorbate-glutathione pathway by metabolic modelling: computer analysis as a step towards flux analysis. Plant Physiol 126:445–462

Pourcel L, Routaboul JM, Cheynier V (2007) Flavonoid oxidation in plants: from biochemical properties to physiological functions. Trends Plant Sci 12:29–36

Prasad TK, Anderson MD, Martin BA, Stewart CR (1994) Evidence for chilling-induced oxidative stress in maize seedlings and a regulatory role for hydrogen peroxide. Plant Cell 6:65–74

Raffo A, La Malfa G, Fogliano V, Maiani G, Quaglia G (2006) Seasonal variations in antioxidant components of cherry tomatoes (*Lycopersicon esculentum* cv. Naomi F1). J Food Compos Anal 19:11–19

Rentel MC, Lecourrieux D, Ouaked F, Usher SL, Peterson L, Okamato H, Knight H, Peck SC, Grierson CS, Hirt H, Knight MR (2004) OXI1 Kinase is necessary for oxidative burst-mediated signal-ling in *Arabidopsis*. Nature 427:858–861

Rossel JB, Walter PB, Hendrickson L, Chow WS, Poole A, Mullineaux PM, Pogson BJ (2006) A mutation affecting ASCORBATE PEROXIDASE 2 gene expression reveals a link between responses to high Light and drought tolerante. Plant, Cell and Environment 29:269–281

Russo M, Sgherri C, Izzo R, Navari-Izzo F (2008) *Brassica napus* subjected to copper excess: phospolpases C and D and glutathione system in signalling. Env Exp Bot 62:238–246

Samuel MA, Miles GP, Ellis BE (2000) Ozone treatment rapidly activates MAP kinase signalling in plants. Plant J 22:367–376

Scandalios JG (1997) Molecular genetics of superoxide dismutase in plants. In: Scandalios JG (ed) Oxidative stress and the molecular biology of antioxidant defenses. Cold Spring Harbor Laboratory Press, New York, pp 527–568

Schubert M, Petersson UA, Haas BJ, Funk C, Schröder WP, Kieselbach T (2002). Proteome Map of the Chloroplast Lumen of Arabidopsis thaliana. J Biol Chem 277: 8354–8365

Secenji M, Hideg E, Bebes A, Györgyey (2010) Transcriptional differences in gene families of the ascorbate-glutathione cycle in wheat during mild water deficit. Plant Cell Rep 29:37–50

Seong ES, Cho HS, Choi D, Joung YH, Lim CK, Hur JH, Wang MH (2007) Tomato plants overexpressing CaKR1 enhanced tolerance to salt and oxidative stress. Biochem Biophys Res Commun 363:983–988

Shalata A, Mittova V, Volokita M, Guy M, Tal M (2001) Response of the cultivated tomato and its wild salt-tolerant relative *Lycopersicon pennellii* to salt-dependent oxidative stress: the root antioxidative system. Physiol Plant 112:487–494

Shao HB, Chu LY, Lu ZH, Kang CM (2008) Primary antioxidant free radical scavenging and redox signaling pathways in higher plant cells. Int J Biol Sci 4:8–14

Shigeoka S, Ishikawa T, Tamoi M, Miyagawa Y, Tkeda T, Yabuta Y, Yoshimura K (2002) Regulation and function of ascorbate peroxidase isoenzymes. J Exp Bot 53:1305–1319

Smirnoff N (2000) Ascorbic acid: metabolism and functions of a multifacetted molecule. Curr Opin Plant Biol 3:229–235

Strand A, Asami T, Alonso J, Ecker JR, Chory J (2003) Chloroplast to nucleus communication triggered by accumulation of Mg-protoporphyrinIX. Nature 421:79–83

Vadassery J, Tripathi S, Prasad R, Varma A, Oelmüller R (2009) Monodehydroascorbate reductase 2 and dehydroascorbate reductase 5 are crucial for a mutualistic interaction between Piriformospora indica and Arabidopsis. J Plant Physiol 166:1263–1274

Vaidyanathan H, Sivakumar P, Chakrabarty R, Thomas G (2003) Sacvenging of reactive oxygen species in NaCl-stressed rice (Oryza sativa L.)- differencial response in salt-tolerant and sensitive varieties. Plant Sci 165:1411–1418

Valero E, González-Sánchez MI, Maciá H, García-Carmona F (2009) Computer simulation of the dynamic behavior of the glutathione-ascorbate redox cycle in chloroplasts. Plant Physiol 149:1958–1969

Verhoeyen ME, Bovy A, Collins G, Muir S, Robinson S, de Vos CHR, Colliver S (2002) Increasing antioxidant levels in tomatoes through modification of the flavonoid biosynthetic pathway. J Exp Bot 53:2099–2106

Wang WB, Kim YH, Lee HS, Kim KY, Deng XP, Kwak SS (2009) Analysis of antioxidant enzyme activity during germination of alfalfa under salt and drought stresses. Plant Physiol Biochem 47:570–577

Wormuth D, Heiber I, Shaikali J, Kandlbinder K, Baier M, Dietz KJ (2007) Redox regulation and antioxidative defence in *Arabidopsis* leaves viewed from a systems biology perspective. J Biotech 129:229–248

Yabuta Y, Maruta T, Yoshimura K (2004) Two distinct redox signaling pathways for cytosolic APX induction under photooxdidative stress. Plant Cell Physiol 45:1586–594

Yeo A (1998) Predicting the interaction between the effects of salinity and climate change on crop plants. Sci Hort 78:159–174

Yoshida S, Tamaoki M, Shikano T, Nakajima N, Ogawa D, Ioki M, Aono M, Kubo A, Kamada H, Inoue Y, Saji H (2006) Cytosolic dehydroascorbate reductase is important for ozone tolerance in *Arabidopsis thaliana*. Plant Cell Physiol 47:304–308

Xiang C, Olivier DJ (1998) Glutathione metabolic genes coordinately respond to heavy metals and jasmonic acid in *Arabidopsis*. Plan Cell 10:1539–1550

Chapter 12
Coordinate Role of Ascorbate–Glutathione in Response to Abiotic Stresses

Imran Haider Shamsi, Sisi Jiang, Nazim Hussain, Xianyong Lin, and Lixi Jiang

Abstract Glutathione (GSH), the tripeptide (nonprotein) thiol (γ-glutamyl cysteinyl glycine) and ascorbic acid (AsA, vitamin C) are the most prominent and worth functional low molecular weight soluble antioxidants in plant cells. Environmental fluctuations can lead the plants/crops to abiotic stress conditions. Thus a prompt and effective response, involving many genes and biochemical–molecular mechanism to cope with these conditions is inevitable. In response to these stresses, the ratio of reduced glutathione (GSH) and oxidized form of glutathione (GSSG) tends to decrease due to the oxidation of GSH during the detoxification of reactive oxygen species and changes in its metabolism, which leads to the activation of various defence mechanisms through a redox signalling pathway, including several oxidants, antioxidants, and stress hormones. Glutathione along with AsA plays a pivotal role in protecting cell function. They detoxify H_2O_2 in the AsA–GSH cycle and are involved in cellular redox regulation and buffering. The ascorbate and glutathione (AsA/DHA and GSH/GSSG) redox pairs are often found to be coupled in plants favours net electron flow from reduced glutathione to dehydroascorbate (DHA). The antioxidation property of ascorbate and glutathione plays a key role in the redox signal transduction process. The plausible reasons could be: (a) signal transduction is influenced as ascorbate and glutathione regulate the cellular H_2O_2. (b) As these metabolites are responsible for regulating gene expression, so the compartment-specific variations in AsA/DHA and GSH/GSSG ratios may have substantial significance for redox signalling. The aim of present chapter is: firstly, to enlighten the pivotal role of AsA and GSH in plant metabolism and tolerance to abiotic stresses such as oxidative, drought, salinity, heat, and cold stress. Secondly, to emphasize the importance of identification and analysis of AsA–GSH genes responsible for multiple stress tolerance in plant species.

I.H. Shamsi, N. Hussain, and L. Jiang (✉)
Institute of Crop Science, College of Agriculture and Biotechnology, Zhejiang University, 268 Kaixuan Road, Hangzhou 310029, PR of China
e-mail: jianglx@zju.edu.cn

S. Jiang and X. Lin
Institute of Agricultural Chemistry, College of Environmental and Resource Sciences, 288 Kaixuan Roas, Hangzhou 310029, PR of China

N.A. Anjum et al. (eds.), *Ascorbate-Glutathione Pathway and Stress Tolerance in Plants*, DOI 10.1007/978-90-481-9404-9_12, © Springer Science+Business Media B.V. 2010

Keywords Ascorbate • Glutathione • Coordination • Role • Abiotic stress • Response

1 Introduction

1.1 Glutathione

Glutathione (GSH) is a multifunctional metabolite present in all organisms. In plants it is present in the cytosol, plastids and mitochondria. GSH being one of the components of the AsA–GSH cycle, plays a pivotal role as antioxidant in prevention of cells against oxidative damage under different abiotic or biotic stresses by equilibrating the redox status (Noctor and Foyer 1998; Gullner and Kómíves 2001). It is depicted from the research studies that its concentration is in millimole in various plant tissues (Creissen et al. 1999; Meyer and Fricker 2002; Noctor et al. 2002).

Reactivity of GSH depends on the thiol (–SH) group, and hence glutathione and homo-glutathiones may not show significant differences in their redox chemistry (Hausladen and Alscher, 1993). GSH has an oxidation reduction potential of −0.23 V that allows it to act as an effective electron acceptor and donor for numerous biological reactions (Xiang et al. 2001). The nucleophilic nature of the thiol group is also important in the formation of mercaptide bonds with metals and for reacting with select electrophiles (Xiang et al. 2001). This reactivity, along with the relative stability and high water solubility of GSH makes it an ideal bio-molecule to protect plants against environmental and biotic stresses.

1.1.1 Glutathione Synthesis

Cysteine and γ-glutamyl cysteine (γEC) are the precursors of glutathione biosynthesis. Synthesis of GSH comprises of two ATP dependent steps, catalysed by γ-glutamyl cysteine synthetase (γ-ECS) and glutathione synthetase (GS). The sites for this reaction are both photosynthetic as well as non-photosynthetic tissues, i.e. chloroplasts and cytosol of plant cells (Noctor and Foyer 1998). The accumulation of GSH in plants depends upon the activity of γ-ECS and the availability of cysteine (Noctor et al. 2002). Cystine, γ-glutamylcystine (ESSE) and glutathione disulphide (GSSG) are the oxidised forms of these reduced compounds are also present in the cell. GSSG in the presence of a specific enzyme, glutathione reductase (GR, EC 1.6.4.2) and by using NADPH as electron donor regenerates the reduced form of GSH.

1.2 Ascorbate

Ascorbate (an ion of ascorbic acid) is required for a range of essential metabolic reactions in all animals and plants. The functions of ascorbic acid as a crucial antioxidant and ascorbate peroxidase (APX) are well documented (Nijs and Kelley 1991; Buettner and Jurkiewicz 1996). Ascorbic acid, in addition to a significant role in photoprotection and the regulation of photosynthesis (Foyer and Harbinson 1994; Forti and Elli 1995), also plays an important role in preserving the activities of enzymes that contain prosthetic transition metal ions (Padh 1990). Although ascorbate peroxidases (APX) and antioxidant property of ascorbate has well been documented, yet many other functions of ascorbate are to be explored. Recent research advances reveal the modification of gene transcription and plant behavior in response to environmental changes (Desikan et al. 2001; Neill et al. 2002) and the plausible reason is the fluctuations in the amount of ascorbate (Veljovic-Jovanovic et al. 2001; Pastori et al. 2003). As a powerful secondary antioxidant, it reduces the oxidized form of α-tocopherol, hence links GSH to the dominant free radical scavenger in membranes (Padh 1990; Hess 1994).

Other functions are; the control of mitosis and cell growth (Bielawski and Joy 1986), gene transcription and mediation of plant responses to the environment (Anderson et al. 1992; Beeor-Tzahar et al. 1995). Modifications in amount of ascorbate bring changes in gene expression (Bowler et al. 1992; Buettner and Jurkiewicz 1996). Deficiency in ascorbate contents leads to changes in plant development through hormonal signaling pathways (Bowler et al. 1992). It also contributes in plant defense network by activating genes normally associated with pathogenesis resistance rather than antioxidative enzymes (Bowler et al. 1992).

1.2.1 Ascorbate Synthesis

All plants can synthesize ascorbic acid (AsA) (Burns 1957). In plants, ascorbate can accumulate to millimolar concentrations in both photosynthetic and non-photosynthetic tissues (Foyer and Noctor 2005a). Of the total percentage of ascorbate, about 30–40% is present in the chloroplast and the concentration in stroma has been reported as high as 50 mM (Foyer and Noctor 2005a). Mature leaves contain highest concentration of ascorbate, as the chloroplasts are fully developed and the chlorophyll level is highest. The site of its biosynthesis is the mitochondria, from where it is transported to the other cell components through a proton-electrochemical gradient or through facilitated diffusion. Although it has been determined that D-glucose is the precursor of Leaf-AA, the synthetic pathway has not been totally understood (Foyer and Noctor 2003, 2005b).

2 Cellular Redox Homeostasis and Signaling

Glutathione as the main thiol buffer having a tight co-operation with ascorbate, plays a central role in cellular redox regulation (Pastori and Foyer 2002; Foyer and Noctor 2003). GSH in its reduced and oxidised forms can act as signal-transducing molecules to regulate cell cycle and cell growth (Potters et al. 2002).

Reactivity of GSH depends on the thiol (–SH) group (Hausladen and Alscher 1993). Alteration in the GSH pool and its redox status could result in redox signals modulating gene expression, including antioxidant defence genes. Having oxidation reduction potential of –0.23 V, GSH acts as an effective electron acceptor and donor for numerous biological reactions. Thiol group is nucleophilic in nature and plays vital role in the formation of mercaptide bonds with metals and in reacting with selective electrophiles (Xiang et al. 2001). Reactivity of such nature, along with the relative stability and high water solubility, are the properties which make GSH an ideal bio-molecule to protect plants against abiotic and biotic stresses, including oxidative stress, heavy metals and certain exogenous and endogenous organic chemicals (Foyer and Noctor 2005a; Millar et al. 2003).

Glutathione and AsA co-operation is the key for the cellular redox homeostasis in the antioxidative AsA–GSH cycle as well as in redox regulation of signalling pathways, gene expression and plant metabolism. Fundamental role of ascorbate and glutathione redox pairs coupling in plant responses to abiotic and biotic stress has been documented (Foyer and Noctor 2003).

3 Ascorbate–Glutathione Cycle

The AsA–GSH cycle is a metabolic pathway that detoxifies hydrogen peroxide (H_2O_2) generated during photosynthetic operations in the light through thylakoid electron transport chain components with electrochemical potentials capable of direct reduction of O_2 (Noctor and Foyer 1998). In plants, this cycle operates in the cytosol, mitochondria, plastids and peroxisomes. (Jimenez et al. 1998; Meyer 2009). Photosynthetic operation and broad spectrum of stress conditions are responsible for the induction of transcripts of the cytosolic and chloroplast enzymes of the Asc–glutathione cycle in plants (reviewed in Kiddle et al. 2003).

This cycle operates in chloroplasts of plants. Four enzymes, ascorbate peroxidase (APX), monodehydroascorbate reductase (MDHAR), glutathione-dependent dehydroascorbate reductase (DHAR), and glutathione reductase (GR) catalyze this cycle along with antioxidant metabolites including glutathione and ascorbate and NADPH. The specific activity and extensive functional operation of these enzymes in plastids illustrate its importance as a major H_2O_2 metabolizing pathway in these photosynthetic organelles (Meyer 2009). The primary step of this cycle is the reduction of H_2O_2 to water by ascorbate peroxidase (APX), whereas ascorbate is used as the electron donor. This step yields to the monodehydroascorbate radical (MDHA). Being a radical, MDHA it can be converted (reduced) to Asc by MDHAR (Jimenez et al. 1998) or disproportionate to ascorbate and dehydroascorbate (DHA) non-enzymatically. The reduction of dehydroascorbate to ascorbate by the enzyme DHAR is at expense of glutathione (GSH), results in the production of oxidized glutathione (GSSG). The final step of this cycle is the reduction of GSSG by GR using NADPH as electron donor. Hence it can be concluded that ascorbate and glutathione are not consumed and the net electron flow is from NADPH to H_2O_2 (Asada and Takahashi 1978).

4 Crop (Plant), Abiotic Stresses and Response of AsA–GSH Genes

4.1 General Stress Response

Ever since the start of plants or crops domestication, abiotic and biotic stresses proved to be a powerful force influencing the evolution of plant populations in the wild (Hoffmann and Parsons 1997), as well as a challenging factor limiting economic yield in commercially valuable species (Boyer 1982; Blum 1988), hence leads to dynamic scientific research studies in the field of crop sciences to understand and evolve a wide range of stress-resistance and tolerance mechanisms in plants/crops (Singh et al. 2002; Mahalingam et al. 2003; Chen and Zhu 2004; Bohnert et al. 2006). As these stresses bring modifications in the redox status of the cell, so the adaptability of a plant to cope with these stresses depends on its capability to respond to the increased oxidative load (Pastori and Foyer 2002). To evaluate the genetic response of plant species to the stress environment, extensive studies from various aspects including; cellular physiology (Singh et al. 2002; Mahalingam et al. 2003; Chen and Zhu 2004), evolutionary biology (Hoffmann and Parsons 1991, 1997), and molecular biology and biotechnology (Holmberg and Bülow 1998; Kasuga et al. 1999; Wang et al. 2003; Pellegrineschi et al. 2004; Denby and Gehring 2005; Vinocur and Altman 2005) have been carried out. The involvement of various genes in the complex plant response to abiotic stresses is well reviewed and grouped in three major categories by Wang et al. (2003) viz. (a) signaling cascades and transcriptional control genes, such as MyC, MAP kinases and SOS kinase (Shinozaki and Yamaguchi-Shinozaki 1997; Munnik et al. 1999; Zhu 2001), phospholipases (Chapman 1998; Frank et al. 2000), and transcriptional factors such as HSF, and the CBF/DREB and ABF/ ABAE families (Stockinger et al. 1997; Schöffl et al. 1998; Choi et al. 2000, Shinozaki and Yamaguchi-Shinozaki 2000); (b) membranes and proteins protecting genes, such as heatshock proteins (Hsps) and chaperones, late embryogenesis abundant (LEA) proteins (Vierling 1991; Ingram and Bartels 1996; Bray et al. 2000), osmoprotectants, and free-radical scavengers (Bohnert and Sheveleva 1998); (c) water and ion uptake and transport functioning genes such as aquaporins and ion transporters (Maurel 1997; Serrano et al. 1999; Tyerman et al. 1999; Zimmermann and Sentenac 1999; Blumwald 2000). Among the abiotic stresses, drought, salinity, light, heat, pollutants and oxidative alterations (Inzé and Van Montagu 1995) are the most fatal ones to plants life, accompanied by the formation of reactive oxygen species (ROS) such as O_2, H_2O_2, and OH (Price et al. 1989; Moran et al. 1994; Mittler 2002).

4.1.1 Reactive Oxygen Species in Response to Abiotic Stresses

The most prominent outcomes of the studies on the response of plants to environmental stresses depicted that these stresses cause increase ROS levels in plants, and the resistance or susceptibility of the plant has been related to the proportion and functional efficiency of network of low molecular weight antioxidants, and ROS-scavenging enzymes (Karpinski et al. 1997; Asada 1999) produced inside plant body against these stresses. It is also reported that the accumulation of ROS in plants by themselves cause the activation of defence gene expression in response to biotic and abiotic stimuli/stresses (Karpinski et al. 1999; Grant and Loake 2000; Fryer et al. 2003; op den Camp et al. 2003). Enzymes (catalase, superoxide dismutase (SOD), ascorbate peroxidase (APX) and glutathione reductase) as well as non-enzyme molecules (ascorbate, glutathione, carotenoids, and anthocyanins) are important antioxidants for scavenging ROS. Some additional compounds, such as osmolytes, proteins (peroxiredoxin) and amphiphilic molecules (tocopherol) can also perform the same function (Bowler et al. 1992; Noctor and Foyer 1998). Among these antioxidant species, glutathione–ascorbate and their co-ordinate response at gene level can be elaborated as a crucial index for adaptive stress responses. Realizing the importance, glutathione has been traditionally used as a marker for various stresses (Navari-Izzo et al. 1997; Mittova et al. 2003a; Srivalli et al. 2003).

4.2 Drought Stress

During drought stress, enzymatic and non-enzymatic plant antioxidants under continuous oxidative conditions regulate the defence mechanism to counter the phenomenon of oxidative stress in plants (Shao et al. 2007). In response to water stress, the availability of CO_2 for carbon assimilation is also decreased as the stomata are closed, hence results in the non-availability of electron acceptor (NADP) leading to the generation of free radicals (Smirnoff 1993; Asada 2000; Sairam and Saxena 2000; Reddy et al. 2004). Antioxidants such as ascorbic acid and glutathione are involved in the neutralization of secondary products of ROS reactions (Hare et al. 1998; Foyer and Noctor 2003). Scientific research depicted that resistance of plant against drought can be related to accumulation of AsA and glutathione concentrations and enzymes related to these. AsA has the ability to protect or regenerating oxidized carotenoids or tocopherols (Shao et al. 2005c, 2006). It has been reported that under drought stress significant increase in antioxidant AsA concentration in turf grass occurred (Shao et al. 2005a) while in contrary another study showed reduced AsA concentration under drought stress in maize and wheat, suggesting its vital involvement in deciding the oxidative response (Shao et al. 2005b). Reviewing the work on glutathione elucidate that during drought stress, high levels of GSH are achieved through high activity of glutathione reductase (GR) (Gamble and Burke 1984). However, this trend is not always the same. Another factor i.e. the GSH/GSSG ratio is likely to be influenced by the duration and severity of stress and the plant species (Smirnoff 1993).

4.3 Salt Stress

Among the abiotic stresses, salinity also adversely affect the crop productivity, hence, is one of the most thought provoking factor for agro-biologist to cope with for enhancing crop productivity. Osmotic inhibition and ionic toxicity are the two major processes, which can affect the physiological and biochemical functions of the plant cell as a result of salt stress (Chinnusamy et al. 2005; Sumithra et al. 2006). The response of plants to salinity stress is documented as mulitigenic in nature as the adaptation to high salt levels involves osmotic adjustment, toxic ions' compartmentation and oxidative stress tolerance. Great similarities have been found in the symptoms caused by salinity and drought stress. Great fluctuations in the activity of GR concentration during salinity stress have been reported in various crops, e.g., enhanced activity of GR was observed in foxtail millet (Sreenivasulu et al. 2000), in pea (Hernandez et al. 1993, 1995, 2000), cantaloupe (Fahmy et al. 1998), citrus (Gueta-Dahan et al. 1997), soybean (Comba et al. 1998), rice (Dionisio-Sese and Tobita 1998; Lin and Kao 2000; Vaidyanathan et al. 2003; Demiral and Turkan 2005; Tsai et al. 2005), *Cicer arietinum* (Kukreja et al. 2005), tomato (Shalata et al. 2001; Molina et al. 2002; Mittova et al. 2003b), *Arabidopsis thaliana* (Huang et al. 2005), wheat (Sairam et al. 2005), *Vigna radiata* (Sumithra et al. 2006), *Setaria italica* (Sreenivasulu et al. 2000), and *Helianthus annuus* (Davenport et al. 2003). Compare to salt-tolerant transgenic over expressing a vacuolar Na^+/H^+ antiporter, a threefold increase in glutathione and cysteine was recorded in wild type *Brassica* plants (Ruiz and Blumwald 2002). Formation of necrotic lesions in the minor veins of pea leaves was observed to be caused by salt-induced higher oxidative stress in the apoplasts (Hernández et al. 2001). Similarly, changes were also observed in the GSH/GSSH ratio under salt stress, i.e. decrease in GSH/GSSH ratio and no GR activity in the apoplasts. Rice seedlings became more tolerant when the GSG/GSSG levels returned to normal values after an initial decline when exposed to continuous salt stress (Fadzilla et al. 1997). Glutathione peroxidase (GPX) can reduce alkyl and lipid hydroperoxides to protect the cells from lipid peroxidation. Increase in the levels of *GPX-1* and *GPX-2* transcripts in Synechocystis PCC 6803 during salt stress highlighted their importance in protecting the cellular machinery against oxidative damage (Gaber et al. 2004). The over expression of cDNA which encodes an enzyme with both glutathione S-transferase (GST) and GPX in transgenic tobacco plants resulted in improved seed germination and seedling growth under salt or low temperature stress (Roxas et al. 1997).

In addition to enhanced GST/GPX activities, higher level of GSH and AsA were also recorded in the transgenic seedlings when compared with wild-type plants (Roxas et al. 2000).

4.4 *Photo-Oxidative Stress*

Photo-oxidative stress can alter plants physiological functions by diminishing photosynthetic metabolism, increasing photo-reduction of molecular oxygen (O_2), photorespiration, and driving away the excitation energy at photosystem II (Asada 1999; Ort and Baker 2002). GSH induces several defense genes (Dron et al. 1988; Wingate et al. 1988) and also regulates plastid gene transcription (Link et al. 1997). Ball et al (2004) reported two genes *GSH1* (coding for plastidial g-ECS) and *GSH2* (a putative cytosolic glutathione synthetase), identified in *Arabidopsis thaliana* and many other plant species (Rawlins et al. 1995, Cobbett et al. 1998; Noctor et al. 2002). Fluctuations in the expression of *GSH1* in Arabidopsis resulted in the increased or decreased glutathione level (Cobbett et al. 1998; Xiang and Oliver 1998; Vernoux et al. 2000). Similarly under normal conditions *ascorbate peroxidase2* (*APX2*), a vital component of encoding the antioxidant network, shows very low level of expression. In contrary, enhanced expression of gene in bundle sheath tissue can be observed under excess light or wounding, as it caused photo-oxidative stress (Karpinski et al. 1997, 1999; Fryer et al. 2003; Chang et al. 2004).

5 Conclusion

To make plants/crop species to reach their potential growth and productivity, adaptation to stress environment and enhanced tolerance mechanism are inevitable. Exploration of enzymatic, non-enzymatic compounds, structural and functional genes involve in the stress tolerance mechanism such as SODs having key role in antioxidant defenses is potentially of greater interest as they are involved in the synthesis and metabolism of AsA and GSH. Recent advances in technologies, especially in the field of molecular biology have enabled scientists to have a deeper understanding of the key roles of these components. Some worth achievements through induced tolerance by gene transformation has been obtained; however, lots of responsible genes and genetically complex mechanisms of abiotic stress tolerance with their potential detrimental side effects are yet to be explored. Emphasis should be given to target the detoxification pathways including the involvement of antioxidants such as GSH and AsA, to obtain plants with wide range stress tolerance traits. Further advancement is needed in research at molecular level to get better understanding of the GSH and AsA levels and their gene expression (enhanced or depressed) under abiotic stress environment.

Acknowledgments The work of our lab was supported by the Agricultural Ministry of China (the program code: 2009ZX08009-076B), Natural Science Foundation of China (the program code: 30671339 and 30971700), National High-Tech R&D Program (the program code: 2006AA10A113).

References

Asada K (1999) The water-water cycle in chloroplasts: scavenging of active oxygen species and dissipation of excess photons. Annu Rev Plant Physiol Mol Biol 50:601–639

Anderson JV, Chevone BI, Hess JL (1992) Seasonal variation in the antioxidant system of eastern white pine needles: evidence for thermal dependence. Plant Physiol 98:501–508

Asada K (2000) The water-water cycle as alternative photon and electron sinks. Phil Trans R Soc Lond B 355:1419–1431

Asada K, Takahashi M (1978) Production and scavenging of activated oxygen in photosynthesis. In: Kyle DJ, Osmond CB, Arntzen CJ (eds) Photoinhibition. Elsevier Science Publishers, Amsterdam, The Netherlands, pp 227–287

Ball L, Accotto G-P, Bechtold U, Creissen G, Funck D, Jimenez A, Kular B, Leyland N, Mejia-Carranza J, Reynolds H, Karpinski S, Mullineaux PM (2004) Evidence for a direct link between glutathione biosynthesis and stress defense gene expression in *Arabidopsis*. Plant Cell 16:2448–2462

Beeor-Tzahar T, Ben-Hayyim G, Holland D, Faltin Z, Eshdat Y (1995) A stress-associated citrus protein is a distinct plant phospholipid hydroperoxide glutathione peroxidase. FEBS Lett 366:151–155

Bielawski W, Joy KW (1986) Reduced and oxidised glutathione and glutathione reductase activity in tissues of *Pisum sativum*. Planta 169:267–272

Blum A (1988) Plant Breeding for Stress Environment. CRC Press, Boca Raton, FL

Blumwald E (2000) Sodium transport and salt tolerance in plants. Curr Opin Cell Biol 12:431–434

Bohnert HJ, Sheveleva E (1998) Plant stress adaptations – making metabolism move. Curr Opin Plant Biol 1:267–274

Bohnert HJ, Qingqiu G, Pinghua L, Ma S (2006) Unraveling abiotic stress tolerance mechanisms – getting genomics going. Curr Opin Plant Biol 9:180–188

Bowler C, Van Montagu M, Inźe D (1992) Superoxide dismutase and stress tolerance. Annu Rev Plant Physiol Plant Mol Biol 43:83–116

Boyer JS (1982) Plant productivity and environment. Science 218:443–448

Bray EA, Bailey-Serres J, Weretilnyk E (2000) Responses to abiotic stresses. In: Gruissem W, Buchannan B, Jones R (eds) Biochemistry and molecular biology of plants. American Society of Plant Physiologists, Rockville, MD, pp 1158–1249

Buettner GR, Jurkiewicz BA (1996) Chemistry and biochemistry of ascorbic acid. In: Cadenas E, Packer L (eds) Handbook of antioxidants. Marcel Dekker, New York, pp 91–115

Burns JJ (1957) Missing step in man, monkey and guinea pig required for the biosynthesis of L-ascorbic acid. Nature 180:553

Chang CCC, Ball L, Fryer MJ, Baker NR, Karpinski S, Mullineaux PM (2004) Induction of ASCORBATE PEROXIDASE 2 expression in wounded Arabidopsis leaves does not involve known wound-signalling pathways but is associated with changes in photosynthesis. Plant J 38:499–511

Chapman D (1998) Phospholipase activity during plant growth anddevelopment and in response to environmental stress. Trends Plant Sci 3:419–426

Chen WJ, Zhu T (2004) Networks of transcription factors with roles in environmental stress response. Trends Plant Sci 9:591–596

Chinnusamy V, Jagendorf A, Zhu JK (2005) Understanding and improving salt tolerance in plants. Crop Sci 45:437–448

Choi HI, Hong JH, Ha J, Kang JY, Kim SY (2000) ABFs, a family of ABA-responsive element binding factors. J Biol Chem 275:1723–1730

Cobbett CS, May MJ, Howden R, Rolls B (1998) The glutathione deficient, cadmium-sensitive mutant, cad2-1 of *Arabidopsis thaliana* is deficient in g-glutamylcysteine synthetase. Plant J 16:73–78

Comba ME, Benavides MP, Tomaro ML (1998) Effect of salt stress on antioxidant defense system in soybean root nodules. Aus J Plant Physiol 25:665–671

Creissen GP, Firmin J, Fryer M, Kular B, Leyland N, Reynolds H, Pastori G, Wellburn F, Baker N, Wellburn A, Mullineaux PM (1999) Elevated glutathione biosynthetic capacity in the chloroplasts of transgenic tobacco plants paradoxically causes increased oxidative stress. Plant Cell 11:1277–1291

Davenport SB, Gallego SM, Benavides MP, Tomaro ML (2003) Behaviour of antioxidant defense system in the adaptive response to salt stress in *Helianthus annuus* L. cells. Plant Growth Regul 40:81–88

Demiral T, Turkan (2005) Comparative lipid peroxidation, antioxidant defense systems and proline content in roots of two rice cultivars differing in salt tolerance. Environ Exp Bot 53:247–257

Denby K, Gehring C (2005) Engineering drought and salinity tolerance in plants: lessons from genome-wide expression profiling in *Arabidopsis*. Trends Biotechnol 23:547–552

Desikan R, Mackerness SA-H, Hancock JT, Neill SJ (2001) Regulation of the *Arabidopsis* transcriptome by oxidative stress. Plant Physiol 127:159–172

Dionisio-Sese ML, Tobita S (1998) Antioxidant responses of rice seedlings to salinity stress. Plant Sci 135:1–9

Dron M, Clouse SD, Dixon RA, Lawton MA, Lamb CJ (1988) Glutathione and fungal elicitor regulation of a plant defense gene promoter in electroporated protoplasts. Proc Natl Acad Sci USA 85:6738–6742

Fadzilla NM, Finch RP, Burdon RH (1997) Salinity, oxidative stress and antioxidant responses in shoot cultures of rice. J Exp Bot 48:325–331

Fahmy AS, Mohamed TM, Mohamed SA, Saker MM (1998) Effect of salt stress on antioxidant activities in cell suspension cultures of cantaloupe (*Cucumis melo*). Egypt J Physiol Sci 22:315–326

Forti G, Elli G (1995) The function of ascorbic acid in photosynthetic phosphorylation. Plant Physiol 109:1207–1211

Foyer CH, Harbinson J (1994) Oxygen metabolism and the regulation of photosynthetic electron transport. In: Foyer CH, Mullineaux P (eds) Causes of photooxidative stresses and amelioration of defense systems in plants. CRC Press, Boca Raton, FL, pp 1–42

Foyer CH, Noctor G (2003) Redox sensing and signaling associated with reactive oxygen in chloroplasts, peroxisomes and mitochondria. Physiol Plant 119:355–364

Foyer CH, Noctor G (2005a) Oxidant and antioxidant signaling in plants a re-evaluation of the concept of oxidative stress in a physiological context. Plant Cell Environ 28:1056–1071

Foyer CH, Noctor G (2005b) Redox homeostis and antioxidant signaling: a metabolic interface between stress perception and physiological responses. Plant Cell 17:1866–1875

Frank W, Munnik T, Kerkmann K, Salamini F, Bartels D (2000) Water deficit triggers phospholipase D activity in the resurrection plant *Craterostigma plantagineum*. Plant Cell 12:111–124

Fryer MJ, Ball L, Oxborough K, Karpinski S, Mullineaux PM, Baker NR (2003) Control of ascorbate peroxidase 2 expression by hydrogen peroxide and leaf water status during excess light stress reveals a functional organisation of Arabidopsis leaves. Plant J 33:691–705

Gaber A, Yoshimura K, Tamoi M, Takeda T, Nakano Y, Shigeoka S (2004) Induction and functional analysis of two reduced nicotinamide adenine dinucleotide phosphate-dependent glutathione peroxidase-like proteins in *Synechocystis* PCC 6803 during the progression of oxidative stress. Plant Physiol 136:2855–2861

Gamble PE, Burke JJ (1984) Effect of water stress on the chloroplastic antioxidant system. I. Alterations in glutathione reductase activity. Plant Physiol 76:615–621

Grant JJ, Loake GJ (2000) Role of reactive oxygen intermediates and cognate redox signaling in disease resistance. Plant Physiol 124:21–29

Gueta-Dahan Y, Yaniv Z, Zilinkas BA, Ben-Hayyim G (1997) Salt and oxidative stress: similar and specific responses and their relation to salt tolerance in citrus. Planta 203:460–469
Gullner G, Kómíves T (2001) The role of glutathione and glutathione-related enzymes in plant-pathogen interactions. In: Grill D, Tausz M, De Kok LJ (eds) Significance of glutathione to plant adaptation to the environment. Kluwer, Dordrecht, The Netherlands, pp 207–239
Hare PD, Cress WA, Van SJ (1998) Dissecting the roles of osmolyte accumulation during stress. Plant Cell Environ 21:535–553
Hausladen A, Alscher RG (1993) Glutathione. In: Alscher RG (ed) Antioxidants in higher plants. CRC Press, Boca Raton, FL, pp 1–30
Hernandez JA, Corpas FJ, Gomez M, Del Rio LA, Sevilla F (1993) Salt-induced oxidative stress mediated by activated oxygen species in pea leaf mitochondria. Physiol Plant 89:103–110
Hernandez JA, Olmos E, Corpas FJ, Sevilla F, del Rio LA (1995) Salt-induced oxidative stress in chloroplasts of pea plants. Plant Sci 105:151–167
Hernandez JA, Jimenez A, Mullineaux PM, Sevilla F (2000) Tolerance of pea (*Pisum sativum* L.) to long term salt stress is associated with induction of antioxidant defenses. Plant Cell Environ 23:853–862
Hernández JA, Ferrer MA, Jime'nez A, Barcelo' AR, Sevilla F (2001) Antioxidant systems and O_2-/H_2O_2 production in the apoplast of pea leaves. Its relation with salt-induced necrotic lesions in minor veins. Plant Physiol 127:817–831
Hess JL (1994) Free radical scavenging. In: Alscher RG, Wellburn AR (eds) Plant responses to the gaseous environment. Chapman & Hall, London, pp 165–180
Hoffmann AA, Parsons PA (1991) Evolutionary genetics and environmental stress. Oxford University Press, Oxford
Hoffmann AA, Parsons PA (1997) Extreme environmental change and evolution. Cambridge University Press, Cambridge, UK
Holmberg N, Bülow L (1998) Improving stress tolerance in plants by gene transfer. Trends Plant Sci 3:61–66
Huang C, He W, Guo J, Chang X, Su P, Zhang L (2005) Increased sensitivity to salt stress in an ascorbate-deficient *Arabidopsis* mutant. J Exp Bot 56:3041–3049
Ingram J, Bartels D (1996) The molecular basis of dehydration tolerance in plants. Annu Rev Plant Biol 47:377–403
Jimenez A, Hernandez JA, Pastori G, del Rio LA, Sevilla F (Dec 1998) Role of the ascorbate-glutathione cycle of mitochondria and peroxisomes in the senescence of pea leaves. Plant Physiol 118:1327–1335
Karpinski S, Escobar C, Karpinska B, Creissen G, Mullineaux PM (1997) Photosynthetic electron transport regulates the expression of cytosolic ascorbate peroxidase genes in Arabidopsis during excess light stress. Plant Cell 9:627–640
Karpinski S, Reynolds H, Karpinska B, Wingsle G, Creissen G, Mullineaux P (1999) Systemic signaling and acclimation in response to excess excitation energy in *Arabidopsis*. Science 284:654–657
Kasuga M, Liu Q, Miura S, Yamaguchi-Shinozaki K, Shinozaki K (1999) Improving plant drought, salt, and freezing tolerance by gene transfer of a single stress-inducible transcription factor. Nat Biotechnol 17:287–291
Kiddle G, Pastori GM, Bernard S, Pignocchi C, Antoniw J, Verrier PJ, Foyer CH (2003) Effects of leaf ascorbate content on defense and photosynthesis gene expression in *Arabidopsis thaliana*. Antioxid Redox Signal 5:23–32
Kukreja S, Nandwal AS, Kumar N, Sharma SK, Unvi V, Sharma PK (2005) Plant water status, H_2O_2 scavenging enzymes, ethylene evolution and membrane integrity of *Cicer arietinum* roots as affected by salinity. Biol Plant 49:305–308
Lin CC, Kao CH (2000) Effect of NaCl stress on H_2O_2 metabolism in rice leaves. Plant Growth Regul 30:151–155
Link G, Tiller K, Baginsky S (1997) Glutathione, a regulator of chloroplast transcription. In: Hatzios KK (ed) Regulation of enzymatic systems detoxifying xenobiotics in plants. Kluwer, Dordrecht, The Netherlands, pp 125–137
lnzé D, Van Montagu M (1995) Oxidative stress in plants. Curr Opin Biotech 6:153–158

Mahalingam R, Gomez-Buitrago A, Eckardt N, Shah N, Guevara-Garcia A (2003) Characterizing the stress/defense transcriptome of *Arabidopsis*. Genome Biol 4:R20

Maurel C (1997) Aquaporins and water permeability of plant membranes. Annu Rev Plant Biol 48:399–429

Meyer A (2009) The integration of glutathione homeostasis and redox signaling. J Plant Physiol 165:1390–1403

Meyer AJ, Fricker MD (2002) Control of demand-driven biosynthesis of glutathione in green *Arabidopsis* suspension culture cells. Plant Physiol 130:1927–1937

Millar AH, Mittova V, Kiddle G (2003) Control of ascorbate synthesis by respiration and its implications for stress responses. Plant Physiol 133:443–447

Mittler R (2002) Oxidative stress, antioxidants and stress tolerance. Trends Plant Sci 7:405–410

Mittova V, Theodoulou FL, Kiddle G, Gomez L, Volokita M, Tal M, Foyer CH, Guy M (2003a) Coordinate induction of glutathione biosynthesis and glutathione-metabolizing enzymes is correlated with salt tolerance in tomato. FEBS Lett 554:417–421

Mittova V, Tal M, Volokita M, Guy M (2003b) Up-regulation of the leaf mitochondrial and peroxisomal antioxidative systems in response to salt-induced oxidative stress in the wild salt-tolerant tomato species *Lycopersicon pennellii*. Plant Cell Environ 26:845–856

Molina A, Bueno P, Marín MC, Rosales MPR, Belver A, Venema K, Donaire JP (2002) Involvement of endogenous salicylic acid content, lipoxygenase and antioxidant enzyme activities in the response of tomato cell suspension cultures to NaCl. New Phytol 156:409–415

Moran JF, Becana M, Iturbe-Ormaetxe I, Frechilla S, Klucas RV, Aparicio-Tejo P (1994) Drought induces oxidative stress in pea plants. Planta 194:346–352

Munnik T, Ligterink W, Meskiene I, Calderini O, Beyerly J, Musgrave A, Hirt H (1999) Distinct osmo-sensing protein kinase pathways are involved in signaling moderate and severe hyper-osmotic stress. Plant J 20:381–388

Navari-Izzo F, Meneguzzo S, Loggini B, Vazzana C, Sgherri CLM (1997) The role of the glutathione system during dehydration of *Boea hygroscopica*. Physiol Plant 99:23–30

Neill SJ, Desikan R, Hancock JT (2002) Hydrogen peroxide signalling. Curr Opin Plant Biol 5:388–395

Nijs D, Kelley PM (1991) Vitamins C and E donate single hydrogen atoms *in vivo*. FEBS Lett 284:147–151

Noctor G, Foyer CH (1998) Ascorbate and glutathione: keeping active oxygen under control. Annu Rev Plant Physiol Plant Mol Biol 49:249–279

Noctor G, Gomez L, Vanacker H, Foyer CH (2002) Interactions between biosynthesis, compartmentation and transport in the control of glutathione homeostasis and signalling. J Exp Bot 53:1283–1304

Op den Camp RGL, Przbyla D, Ochsenbein C, Laloi C, Kim C, Danon A, Wagner D, Hideg E, Gobel C, Feussner I, Nater M, Apel K (2003) Rapid induction of distinct stress responses after the release of singlet oxygen in Arabidopsis. Plant Cell 15:2320–2332

Ort DR, Baker NR (2002) A photoprotective role for O_2 as an alternative electron sink in photosynthesis. Curr Opin Plant Biol 5:193–198

Padh H (1990) Cellular functions of ascorbic acid. Biochem Cell Biol 68:1166–1173

Pastori GM, Foyer CH (2002) Common components, networks, and pathways of cross-tolerance to stress. The central role of "redox" and abscisic acid-mediated controls. Plant Physiol 129:460–468

Pastori GM, Kiddle G, Antoniw J, Bernard S, Veljovic-Jovanovic S, Verrier PJ, Noctor G, Foyer CH (2003) Leaf vitamin C contents modulate plant defense transcripts and regulate genes controlling development through hormone signaling. Plant Cell 15:939–951

Pellegrineschi A, Reynolds M, Pacheco M, Brito RM, Almeraya R (2004) Stress-induced expression in wheat of the *Arabidopsis thaliana* DREB1A gene delays water stress symptoms under greenhouse conditions. Genome 47:493–500

Potters G, De Gara L, Asard H, Horemans N (2002) Ascorbate and glutathione: guardians of the cell cycle, partners in crime? Plant Physiol Biochem 40:537–548

Price AM, Atherton NM, Hendry GAF (1989) Plants under drought-stress generate activated oxygen. Free Radic Res Commun 8:61–66

Rawlins MR, Leaver CJ, May MJ (1995) Characterisation of a cDNA encoding Arabidopsis glutathione synthetase. FEBS Lett 376:81–86

Reddy AR, Chaitanya KV, Vivekanandan M (2004) Drought-induced responses of photosynthesis and antioxidant metabolism in higher plants. J Plant Physiol 161:1189–1202

Roxas VP, Smith RK, Allen ER, Allen RD (1997) Over expression of glutathione S-transferase/glutathione peroxidase enhances the growth of transgenic tobacco seedlings during stress. Nat Biotech 15:988–991

Roxas VP, Lodhi SA, Garrett DK, Mahan JR, Allen RD (2000) Stress tolerance in transgenic tobacco seedlings that overexpress glutathione S-transferase/glutathione peroxidase. Plant Cell Rep 41:1229–1234

Ruiz JM, Blumwald E (2002) Salinity-induced glutathione synthesis in *Brassica napus*. Planta 214:965–969

Sairam PK, Saxena DC (2000) Oxidative stress and antioxidants in wheat genotypes: possible mechanism of water stress tolerance. J Agron Crop Sci 184:55–61

Sairam RK, Srivastava GC, Agarwal S, Meena RC (2005) Differences in antioxidant activity in response to salinity stress in tolerant and susceptible wheat genotypes. Biol Plant 49:85–91

Schöffl F, Prandl R, Reindl A (1998) Regulation of the heat-shock response. Plant Physiol 117:1135–1141

Serrano R, Mulet JM, Rios G, Marquez JA, de Larrinoa IF, Leube MP, Mendizabal I, Pascual-Ahuir A, Proft M, Ros R, Montesinos C (1999) A glimpse of the mechanisms of ion homeostasis during salt stress. J Exp Bot 50:1023–1036

Shalata A, Mittova V, Volokita M, Guy M, Tal M (2001) Response of the cultivated tomato and its wild salt-tolerant relative *Lycopersicon pennellii* to salt-dependent oxidative stress: the root antioxidative system. Physiol Plant 112:487–494

Shao HB, Liang ZS, Shao MA (2005a) Adaptation of higher plants to stresses and stress signal transduction. Acta Ecol Sin 25:1871–1882

Shao HB, Liang ZS, Shao MA (2005b) Dynamic changes of anti-oxidative enzymes of 10 wheat genotypes at soil water deficits. Biointerfaces 42:187–195

Shao HB, Liang ZS, Shao MA, Sun Q, Hu ZM (2005c) Investigation on dynamic changes of photosynthetic characteristics of 10 wheat (*Triticum aestivum L.*) genotypes during two vegetative-growth stages at water deficits. Biointerfaces 43:221–227

Shao HB, Chu LY, Zhao CX, Guo QJ, Liu XA, Ribaut JM (2006) Plant gene regulatory net work system under abiotic stress. Acta Biol Sezeged 50:1–9

Shao HB, Jiang SY, Li FM, Chu LY, Zhao CX, Shao MA, Zhao XN, Li F (2007) Some advances in plant stress physiology and their implications in the systems biology era. Biointerfaces 54:33–36

Shinozaki K, Yamaguchi-Shinozaki K (1997) Gene expression and signal transduction in water-stress response. Plant Physiol 115:327–334

Shinozaki K, Yamaguchi-Shinozaki K (2000) Molecular responses to dehydration and low temperature: differences and cross-talk between two stress signalling pathways. Curr Opin Plant Biol 3:217–223

Singh KB, Foley RC, Onate-Sanchez L (2002) Transcription factors in plant defense and stress responses. Curr Opin Plant Biol 5:430–436

Smirnoff N (1993) The role of active oxygen in the response of plants to water deficit and desiccation. New Phytol 125:27–58

Sreenivasulu N, Grimma B, Wobusa U, Weschkea W (2000) Differential response of antioxidant compounds to salinity stress in salt-tolerant and salt-sensitive seedlings of foxtail millet (*Setaria italica*). Physiol Plant 109:435–442

Srivalli B, Sharma G, Khanna-Chopra R (2003) Antioxidative defense system in an upland rice cultivar subjected to increasing intensity of water stress followed by recovery. Physiol Plant 119:503–512

Stockinger EJ, Gilmour SJ, Thomashow MF (1997) Arabidopsis thaliana CBF1 encodes an AP2 domain-containing transcriptional activator that binds to the C-repeat/DRE, a cis-acting DNA regulatory element that stimulates transcription in response to low temperature and water deficit. Proc Natl Acad Sci USA 94:1035–1040

Sumithra K, Jutur PP, Dalton CB, Reddy AR (2006) Salinity-induced changes in two cultivars of *Vigna radiata*: responses of antioxidative and proline metabolism. Plant Growth Regul 50:11–22

Tsai YC, Hong CY, Liu LF, Kao CH (2005) Expression of ascorbate peroxidaes and glutathione reductase in roots of rice seedlings in respons to NaCl and H_2O_2. J Plant Physiol 162:291–299

Tyerman SD, Bohnert HJ, Maurel C, Steudle E, Smith JAC (1999) Plant aquaporins: their molecular biology, biophysics and significance for plant water relations. J Exp Bot 50:1055–1071

Vaidyanathan H, Sivakumar P, Chakrabarty R, Thomas G (2003) Scavenging of reactive oxygen species in NaCl-stressed rice (*Oryza sativa* L.) – differential responses in salt tolerant and sensitive varieties. Plant Sci 165:1411–1418

Veljovic-Jovanovic SD, Pignocchi C, Noctor G, Foyer CH (2001) Low ascorbic acid in the vtc1 mutant of Arabidopsis is associated with decreased growth and intracellular redistribution of the antioxidant system. Plant Physiol 127:426–435

Vernoux T, Wilson RC, Seeley KA, Reicheld JP, Muroy S, Brown S, Maughan SC, Cobbett CS, Van Montagu M, Inzé D, May MJ, Sung ZR (2000) The ROOT MERISTEMLESS/CADMIUM SENSITIVE 2 gene defines a glutathione dependent pathway involved in initiation and maintenance of cell division during postembryonic root development. Plant Cell 12:97–110

Vierling E (1991) The roles of heat-shock proteins in plants. Annu Rev Plant Biol 42:579–620

Vinocur B, Altman A (2005) Recent advances in engineering plant tolerance to abiotic stress: achievements and limitations. Curr Opin Biotechnol 16:123–132

Wang W, Vinocur B, Altman A (2003) Plant responses to drought, salinity and extreme temperatures: towards genetic engineering for stress tolerance. Planta 218:1–14

Wingate VPM, Lawton MA, Lamb CJ (1988) Glutathione causes a massive and selective induction of plant defense genes. Plant Physiol 87:206–210

Xiang C, Oliver DJ (1998) Glutathione metabolic genes co-ordinately respond to heavy metals and jasmonic acid in *Arabidopsis*. Plant Cell 10:1539–1550

Xiang C, Werner BL, Christensen EM, Oliver DJ (2001) The biological functions of glutathione revisited in *Arabidopsis* transgenic plants with altered glutathione levels. Plant Physiol 126:564–574

Zhu JK (2001) Cell signaling under salt, water and cold stresses. Curr Opin Plant Biol 4:401–406

Zimmermann S, Sentenac H (1999) Plant ion channels: from molecular structures to physiological functions. Curr Opin Plant Biol 2:477–482

Chapter 13
Regulation of Genes Encoding Chloroplast Antioxidant Enzymes in Comparison to Regulation of the Extra-plastidic Antioxidant Defense System

Margarete Baier, Nicola T. Pitsch, Marina Mellenthin, and Wei Guo

Abstract Within a plant cell, chloroplasts produce reactive oxygen species (ROS) at the highest rates. Singlet oxygen, superoxide anions and peroxides accumulate when the light intensity exceeds the capacity of the photosynthetic electron transport chain and metabolic electron consumption. Embedded in the heterotrophic cell, chloroplasts have a high-potential antioxidant defense system. The enzymatic components link to and substitute non-enzymatic protection by ascorbate and glutathione. Recent analysis demonstrated that the expressional regulation of the enzymatic components of the chloroplast antioxidant defense system is widely independent from regulation of extracellular antioxidant enzymes. It includes specific transcriptional regulation by differentially regulated transcription factors, post-transcriptional regulation by alternative splicing, complementary RNA and microRNAs and regulated protein-import into chloroplasts. Most chloroplast antioxidant enzymes show a high background expression activity, while it is low for most cytosolic components. Instead of the strong induction kinetic reported for cytosolic APx1 and APx2, most genes for chloroplast enzymes are two-directionally regulated in response to environmental signals. In this review an overview over the present state of knowledge on regulation of genes for chloroplast and extra-plastid antioxidant enzymes is provided based on the detailed analysis of model genes and comparison of transcriptome data.

Keywords Antioxidant enzymes • Chloroplast • Regulation • Gene expression • Peroxidases • Reactive oxygen species

---

M. Baier (✉)
Plant Science, Heinrich-Heine-University, Universitaetsstrasse 1, 40225 Düsseldorf, Germany
and
Plant Physiology, Free University Berlin, Königin-Luise-Strasse 12-16, 14195 Berlin, Germany
e-mail: margarete.baier@uni-duesseldorf.de

N.T. Pitsch, M. Mellenthin, and W. Guo
Plant Science, Heinrich-Heine-University, Universitaetsstrasse 1, 40225 Düsseldorf, Germany

N.A. Anjum et al. (eds.), *Ascorbate-Glutathione Pathway and Stress Tolerance in Plants*,
DOI 10.1007/978-90-481-9404-9_13, © Springer Science+Business Media B.V. 2010

1 Reactive Oxygen in Plants

Plants produce reactive oxygen species in high amounts when exposed to stress-full conditions or developmental stages such as pathogen attack (Yakimova et al. 2009), at the onset of apoptosis (Gao et al. 2008), early seedling development (Pena-Ahumada et al. 2006) and senescence (Bhattacharjee 2005). Besides these extreme ROS-accumulating stages, plant cells and, especially mesophyll cells, are constantly exposed to low doses of reactive oxygen species (ROS). ROS are generated in the environment and in metabolic reactions. In the light, especially in UV-light, highly reactive ozone accumulates in the atmosphere (Vaida 2005), enters plants via the stomata, react with the plant surfaces and induce secondary redox-reactions (Kangasjarvi et al. 2005). In addition, plant cells produce ROS in the apoplasm in response to environmental stimuli, such as pathogens, and in response to developmental signals. Activation of apoplast peroxidases (Murakami et al. 2007) and plasmamembrane NADPH oxidase (Torres et al. 2005) results in peroxides and superoxide anions. These ROS can accumulate locally to millimolar concentrations. They are antipathogenic and support hardening of cell walls by stimulating radical induced polymerization of phenolics (Zarra et al. 1999).

Inside cells, photorespiratorily generated H_2O_2 can escape from peroxisomes (Queval et al. 2007). Furthermore, H_2O_2 is produced in many enzyme reactions, such as sulfide and polyamine oxidation (Cona et al. 2003; Hansch et al. 2006). Chloroplasts generate superoxide (O_2^-), singlet oxygen ($^1O_2$) and H_2O_2 to a much greater extent by uncontrolled energy and electron transfer from the photosynthetic electron transport chain to oxygen (Foyer et al. 1994; Baker et al. 2000).

In principle, all aerobic organisms face oxidative stress by exogenous and endogenous ROS. In plants, compared to other organisms, the risks for ROS intoxication is dramatically increased due to photosynthesis: (i) In thylakoid membranes, photosensitive chlorophylls are densely packed. When excited by light, they can transfer excitation energy to oxygen. In this reaction singlet oxygen ($^1O_2$) is formed. $^1O_2$ can take up electrons more easily than triplet oxygen and, consequently, is highly reactive with many biomolecules. In presence of water, singlet oxygen leads to formation of hydroxyl radicals and superoxide ions (Elstner 1990). (ii) Reduced photosystem I is the most electronegative biological compound (Gou et al. 2006). With its redox potential of −354 to −403 mV it can reduce almost all biomolecules, including oxygen (Gou et al. 2006). The generated superoxide anions (Mehler 1951; Asada 2000) are still active enough to react with most biomolecules, including lipids and amino acids (Elstner 1990). Alternatively, it spontaneously and enzymatically disproportionates into oxygen and H_2O_2 in water. Finally, even H_2O_2 is still sufficiently reactive to peroxidize, epoxidize or oxidize metabolites and structural components of the cell (Baier and Dietz 1999).

Asada (2000) estimated that in higher plants 20–30% of the electrons transported in the photosynthetic electron transport chain are consumed by ROS turn-over. Under unfavorable environmental conditions, such as high light, low and high temperatures, wounding and in response to pathogen attacks, the relative ratio of electron consumption can dramatically increase. Under severe stress, even up to 100% of electrons transported in the photosynthetic electron transport chain can be lost to ROS formation and subsequent detoxification of superoxide radicals, singlet oxygen and hydrogen peroxide (Asada 2000).

2 Complexity of the Chloroplast and Extra-plastidic Antioxidant Systems

2.1 Complexity and Subcellular Localization of the Antioxidant System

To protect cells from the destructive potential of oxidative stress, all cellular compartments are equipped with low molecular weight antioxidants and antioxidant enzymes. Various components are connected in respect of redox control. In all cellular compartments, they form combinatorial redox-sensitive antioxidant networks.

The chloroplast antioxidant system controls the cellular redox poise. A recent study (Robert et al. 2009) demonstrated that accumulation of even apoplast superoxides is regulated by chloroplast redox signals. Redox imbalances are propagated over the chloroplast envelopes to the plasma membrane, where ROS-generation is activated. Vice versa, the chloroplast H_2O_2 detoxification system collapses if the cytosolic peroxidase capacity is limited (Pnueli et al. 2003; Mittler et al. 2006). These examples demonstrate that the compartment-specific defense systems are closely regulated.

Antioxidants control the propagation of redox imbalances within and between compartments. According to Halliwell (1994), all metabolites which can protect others from oxidation are antioxidants. In this series of reviews, special focus is given to ascorbate and glutathione, which are the major low molecular weight antioxidants in plants (Noctor and Foyer 1998). They accumulate up to millimolar concentrations inside cells (Baier et al. 2000). Under most developmental and environmental conditions their redox state is by 95–97% reduced (Baier et al. 2000). During development (Pena-Ahumada et al. 2006) and upon stress (Noctor and Foyer 1998) the redox state shifts to more oxidized values, if the enzymatic antioxidant regeneration system is overwhelmed.

Antioxidant enzymes, such as ascorbate and glutathione peroxidases (Eshdat et al. 1997; Asada 2000), accelerate the cellular redox poise control by catalyzing ascorbate and glutathione/thioredoxin-dependent ROS detoxification (Raven et al. 2004; Navrot et al. 2006). In parallel, peroxiredoxins reduce H_2O_2 and alkyl hydroperoxides to H_2O and alkyl alcohols. These antioxidant enzymes (Baier and Dietz 1996; Stacy et al. 1996; Horling et al. 2002) act either by an intramolecular or, like glutathione peroxidases, by an intermolecular thiol–disulfide mechanism (Dietz et al. 2002). Peroxide detoxification takes place at a conserved, catalytic cysteine residue (König et al. 2003). In contrast, ascorbate peroxidases and catalases coordinate a heme prosthetic group in the active site (Hiner et al. 2002). The redox state of the central iron ion controls the reactivity with peroxides. In the first reaction step of ascorbate peroxidases and catalases, H_2O_2 interacts with the Fe-centre. By oxidation, an oxyferryl iron is formed. Subsequently, the catalytic site is regenerated by reaction with a second H_2O_2 and release of H_2O and O_2 in catalases (Fita and Rossmann 1985). In ascorbate peroxidases, regeneration of the Fe-centre takes place by reaction with two molecules of ascorbate, which are oxidized to monodehydroascorbate (Miyake and Asada 1996). Cyanobacteria and fungi express hybrid enzymes having both ascorbate peroxidase and catalase functions (Obinger et al. 1999; Passardi et al. 2007). In higher plants, the heme-enzymes are mono-functional and, thus, have either catalase or peroxidase activity (Passardi et al. 2007).

The antioxidant enzymes and enzymes involved in biosynthesis of ascorbate and glutathione are all nuclear encoded, irrespective of their subcellular localization. Translation and protein sorting takes place in the cytosol. The organellar isoforms are targeted to chloroplasts and mitochondria by N-terminal targeting signals in most plants (Baier and Dietz 1997; Jespersen et al. 1997). Sorting of microsomal and apoplastic proteins is translationally regulated at the rough endoplasmatic reticulum and depends on intrinsic peptide signals and intracellular vesicle trafficking (Jürgens 2004). In addition the isoenzymes are distributed between membranes and soluble compartments by transmembrane anchors or membrane attachment (Jespersen et al. 1997; König et al. 2002).

The antioxidant enzymes most directly linked to ascorbate and glutathione pathways are ascorbate peroxidases (APx; E.C. 1.11.1.11), glutathione peroxidases (GPx; E.C. 1.11.1.9 and E.C. 1.11.1.12), peroxiredoxins (1CP, 2CP, PrxII and PrxQ; E.C. 1.11.1.15), glutathione reductases (GR; E.C. 1.8.1.7) and mono- and dehydroascorbate reductases (MDHAR and DHAR; E.C. 1.6.5.4 and 1.8.5.1)). Together with the other antioxidant enzymes, such as superoxide dismutases (FeSOD/Fsd, MnSOD/Msd and CuZnSOD/Csd; E.C. 1.15.1.1) and catalase (Cat; E.C. 1.11.1.6) they influence the redox-poise of ascorbate and glutathione (Aono et al. 1997; Baier et al. 2000; Chen et al. 2003; Kangasjarvi et al. 2008). Their cellular distribution is summarized in Table 1.

Table 1 Cellular localization of antioxidant enzymes in higher plants

	Cytosol	Chloroplast	Mitochondria	Microbodies	Apoplasm	Nucleus
APx	x	x	x	x	x	
GPx	x	x	x	x		
MDHAR	x	x	x	x	x	
DHAR	x	x	x			
GR	x	x				
1CP						x
2CP		x				
PRXII	x	x	x			
PrxQ		x				
FeSOD		x				
MnSOD			x			
CuZnSOD	x	x		x		
Cat				x		

2.2 Isoenzyme Patterns and Subcellular Localization

2.2.1 Ascorbate Peroxidases

Ascorbate peroxidases (APx) are the most efficient peroxidases in plant cells with, e.g. v_{max}-values of 23.4±4.2 mmol H_2O_2 min^{-1} mg $protein^{-1}$ (Lu et al. 2009). Based on interspecies sequence comparisons, organellar ascorbate peroxidases are supposed to have evolved from the same single ancestor gene as catalase-peroxidases. The ancestral gene already existed prior to separation of archea and eubacteria (Zamocky et al. 2000). In plants, APx diversified into organellar, microbody/apoplastic and cytosolic ascorbate peroxidases (Passardi et al. 2007). Due to a specific protein loop close to the heme binding site, chloroplast APx are more sensitive to inactivation by H_2O_2 and, therefore, prone to inactivation upon oxidative stress (Kitajima 2008) while the cytosolic isoforms are more stable under oxidative stress.

Recent bioinformatic comparison of the amino acid sequences of APx from various plants and exon-intron structures in the chlorophytic green algae *Chlamydomonas reinhardtii*, the moss *Physcomitrella patens*, the lycophyte *Selaginella moellendorffii* and the seed plant *Arabidopsis thaliana* (N.T. Pitsch, B. Witsch, M. Baier (2010) submitted) revealed that organellar and extra-organellar APx may have separated early in plant evolution. We concluded that the ancestral gene of organellar APx duplicated creating genes encoding stromal and thylakoid-bound isoforms prior to separation of lycophytes and seed plants (N.T. Pitsch, B. Witsch, M. Baier (2010) submitted). In some plants, APx genes were secondarily lost during evolution or further amplified in specific plant lineages. Studies with Arabidopsis knockout lines demonstrated that plants can survive loss of sAPx or tAPx under moderate environmental stress conditions (Kangasjarvi et al. 2008). However, limitations in ascorbate peroxidase capacity can be assumed to be limiting under natural conditions. In some plants, such as tobacco, spinach and pumpkin, the APx protection system has been stabilized by alternative splicing (Jespersen et al. 1997; Mano et al. 1997; Yoshimura et al. 1999). In *Physcomitrella patens*, loss of the ancestral APx gene was stabilized by evolutionary adaptation of a novel APx gene of retrotransposonal origin (N.T. Pitsch, B. Witsch, M. Baier (2010) submitted).

2.2.2 Peroxiredoxins

Peroxiredoxins are non-heme peroxidases. Plants, like animals, fungi and many prokaryotes, encode four types of peroxiredoxins: 1-Cys peroxiredoxins (1CP; At1g48130), 2-Cys peroxiredoxins (2CP; At3g11630 and At5g06290), type-II-peroxiredoxins (Prx-II; At1g65990; At1g65980; At1g65970; At1g60740; At3g52960; At3g06050) and atypical peroxiredoxins, designated peroxiredoxin-Q (PrxQ; At3g26060). 1-Cys peroxiredoxins are nuclear proteins (Chew et al. 2003). 2-Cys peroxiredoxins and PrxQ are post-translationally targeted to chloroplast

(Baier 1997; Dietz et al. 2002). 2CP are stromal proteins (Baier and Dietz 1997), while PrxQ are localized in the thylakoid lumen (Petersson et al. 2006). The various isoforms of PrxII are distributed between cytosol, mitochondria and chloroplasts (Dietz et al. 2002). In Arabidopsis PrxIIF is post-translationally targeted to mitochondria (Finkemeier et al. 2005), PrxIIE to chloroplasts and PrxIIB and PrxIIC are cytosolic proteins (Horling et al. 2003).

The peroxiredoxins reduce H_2O_2 and alkyl hydroperoxides by a thiol–disulfide mechanisms (Dietz et al. 2002; König et al. 2003). For regeneration, the various types of small redox active proteins are used, such as thioredoxins, glutaredoxins and NTR-C (König et al. 2002; Rouhier et al. 2002; Kirchsteiger et al. 2009). Besides antioxidant function, 2-Cys peroxiredoxins were recently been shown to act as molecular chaperones, which prevent proteins from denaturation (Kim et al. 2009a).

2.2.3 Glutathione Peroxidases

Glutathione peroxidases have, like peroxiredoxins (König et al. 2002), a broad substrate spectrum. They reduce H_2O_2 and alkylhydroperoxides, such as lipid peroxides, using reduced glutathione as co-substrate (Eshdat et al. 1997). In comparison to the four types of glutathione peroxidases found in animal cells (Toppo et al. 2008; Lu and Holmgren 2009), only the monomeric and presumably more ancient non-selenium type has maintained in higher plants (Eshdat et al. 1997), while the chlorophytic green alga *Chlamydomonas reinhardtii* encodes one non-selenium and two selenium-type GPx (Dayer et al. 2008).

Since thioredoxins are the preferred electron donors of plant GPx, and not glutathione or glutharedoxin, Navrot et al. (2006)) suggested to rename glutathione peroxidases and to designate the enzyme family as a fifth type of peroxiredoxins. Since glutathione peroxidases lack characteristic sequence features of peroxiredoxins, we suggest to designate them thiol peroxidases (considering that the name thioredoxin peroxidase is already in use as a synonym for thioredoxin-regenerative peroxiredoxins). The Km-values of heterologously expressed *Arabidopsis* GPx with *E. coli* thioredoxin are 14.0–25.4 μM for H_2O_2 and 2.2–4.0 μM for cumenehydroperoxide (Iqbal et al. 2006). v_{max} ranges between 346 and 433 nmol min^{-1} mg^{-1} protein for H_2O_2 and 60 and 233 nmol min^{-1} mg^{-1} protein for cumenehydroperoxide.

In *Arabidopsis thaliana*, two of seven isoforms, GPx1 and GPx7, are post-translationally targeted to chloroplasts (Rodriguez Milla et al. 2003), where they provide antioxidant protection and regulate salicylic and ROS triggered plant immune responses (Chang et al. 2009). The other GPx isoforms are localized in the cytosol, mitochondria and the endoplasmatic reticulum (Rodriguez Milla et al. 2003). Arabidopsis GPx3 may link redox signaling with ABA signaling since it physically interacts with the redox-sensitive protein phosphatases ABI1 and ABI2, which control ABA-signal transduction (Rodriguez Milla et al. 2003).

2.2.4 Enzymes of the Foyer–Halliwell-Cycle

Glutathione reductase (GR), and mono- and dehydroascorbate reductases (MDAR and DHAR) form the Foyer–Halliwell-cycle (Foyer and Halliwell 1977). The cycle couples ascorbate and glutathione and controls regeneration of the low molecular weight antioxidants. Although first described for chloroplasts (Foyer and Halliwell 1977), its enzymes are localized in most cellular compartments. According to Meister (1994) and Noctor et al. (2000) the reaction mechanism is proposed to be common for all kinds of organisms.

Arabidopsis thaliana encodes one cytosolic (At3g24170) and one organellar isoform of glutathione reductase (GR; At3g54660), five isoforms for MDAR, of which two are organellar (At1g63940; At3g27829) and three cytosolic (At3g09940; At3g52880; At5g03630), and one cytsolic (At5g36270) and two organellar isoforms of DHAR (At1g75270; At5g16710). Together they provide redox coupling of ascorbate and glutathione in chloroplasts, mitochondria and the cytosol (Foyer and Halliwell 1977; Jimenez et al. 1998; Noctor et al. 2000, 2007). Comparison of ascorbate and glutathione interchange in transgenic lines with either decreased catalase or lower 2-Cys peroxiredoxin levels (Baier et al. 2000; Willekens et al. 1997), suggests lower relative DHAR activity (and, therefore, less redox coupling) in chloroplasts than in the cytosol (Noctor et al. 2000). According to Morell et al. (1997), in presence of high MDAR activity, even total absence of DHAR may be tolerable for plant cells, suggesting that DHAR is only a supportive enzyme. However, over-expression of DHAR increased the ascorbate reduction state in maize and tobacco (Chen et al. 2003) demonstrating that DHAR activity is actually limiting in plant cells.

Glutathione reductase activity can also be assumed to be restrictive in plants. Over-expression results in increased protection against oxidative stress and higher reduction states for ascorbate and glutathione (Aono et al. 1997). Analysis of transgenics with increased glutathione amounts demonstrated that unchanged GR activities can cope with higher glutathione concentrations (Noctor et al. 1996; Arisi et al. 1997). Thus, limitation depends not on the glutathione concentration, but the redox state of glutathione. Various authors report induction of glutathione reductase expression upon stress (Stevens et al. 1997; Lee et al. 1998). Over-expression of GR (Aono et al. 1997) demonstrated that induction may not be sufficient to protect plants under severe oxidative stress and may rather provide a physiological window to control ROS- or thiol-regulated stress signaling.

2.2.5 Superoxide Dismutases

Superoxide dismutases (SOD) disproportionate superoxide anions into H_2O_2 and O_2. Plants express three SOD isoenzyme families, namely CuZn-, Mn- and Fe-superoxide dismutases, which have different transition metals in their catalytic sites. *Arabidopsis thaliana* encodes three Csd (At1g08830; At2g28190, At5g18100), three Fsd (At4g25100, At5g51100, At5g23310) and two Msd (At3g10920, At3g56350). Csd1 is cytosolic, Csd2 and at least Fsd2 (At5g51100) and Fsd3 (At5g23310) are chloroplastic, Csd3 is peroxisomal and the Msd-isoenzymes are mitochondrial (Kliebenstein et al. 1998). Localization of Fsd1 is not finally clarified. Based on co-purification with chloroplast proteins, originally chloroplast localization was assumed (Kliebenstein et al. 1998). However, Myouga et al. (2008) recently showed with FSD1–GFP fusion proteins that Fsd1 very likely is a cytoplasmic protein. Since the experimental evidence for cytosolic localization is stronger than for chloroplast localization, here we categorize Fsd1 as a cytosolic protein.

Various SODs, e.g. Csd1, Csd2, Fsd2 and Fsd3, were shown to be induced by oxidative stress (Kliebenstein et al. 1998; Myouga et al. 2008). Nevertheless, the role of SOD in protection from oxidative stress symptoms was still unclear. Over-expression of SOD in transgenics and correlation of SOD expression with stress resistance gave inconclusive results (Bowler et al. 1989, 1991; Pitcher et al. 1991; Mckersie et al. 1993; VanCamp et al. 1996; Van Breusegem et al. 1999; Alscher et al. 2002). While some authors reported increased protection to stress, others described higher susceptibility to oxidative stressors (Pitcher et al. 1991). The different observations have been explained (besides differences in growth conditions and stress application) by limited peroxidase activity to detoxify accumulating H_2O_2 (VanCamp et al. 1996; Noctor and Foyer 1998), by differences in the subcellular localization (Pitcher et al. 1991; Pitcher and Zilinskas 1996) and by the catalytic properties of the over-expressed enzymes (VanCamp et al. 1996). In addition, combinatorial feed-back and post-transcriptional control, such as described in Chapter 4, may mask part of the antioxidant function in transgenic lines with increased and decreased SOD expression and may even explain, why plants lacking detectable Csd1, Csd2 and Fsd1 activity do not show differences in their performance under photooxidative stress conditions (Cohu et al. 2009).

2.3 *Dual-Targeting and Alternative Splicing Increase the Complexity of the Antioxidant System*

Comparison of gene complexities and transcript patterns demonstrated that in various plants the complexity of antioxidant enzymes is bigger than indicated by the gene numbers. For example, in spinach, tobacco and pumpkin (Jespersen et al. 1997; Mano et al. 1997; Yoshimura et al. 1999), soluble and thylakoid-bound ascorbate peroxidases (sAPX: stromal ascorbate peroxidase; tAPx: thylakoid ascorbate peroxidase) result from alternative splicing of the same gene. The 5′-ends of the resulting transcripts encode alternatively an early stop codon or a C-terminal transmembrane helix which anchors the protein in the thylakoid membrane. By this mechanism, a single gene encodes two differentially subcellular localized enzymes.

Furthermore, recent proteomic approaches, in vivo protein labeling studies and bioinformatic analysis revealed that monodehydroascorbate reductase (MDHAR), stromal ascorbate peroxidase (sAPx) and glutathione reductase are dually-targeted to mitochondria and chloroplasts in *Arabidopsis* (Obara et al. 2002; Chew et al.

2003; Teixeira et al. 2006). There are also strong indications for dual-targeting in rice (Obara et al. 2002; Chew et al. 2003; Morgante et al. 2009) suggesting that the mechanism is conserved at least in angiosperms.

In case of MDHAR5 (At1g63940), alternative targeting is controlled by distinct transcription initiation sites (Obara et al. 2002) (Fig. 1). The longer transcript encodes the mitochondrial isoform, the shorter one a chloroplast MDHAR. Since only the transit prepetides are affected, the mature chloroplast and mitochondrial isoforms are identical. Transcription initiation for chloroplast MDHAR takes place in close proximity of I-boxes (GATA/GATAAG) in an intron of the heteronuclear mitochondrial MDHAR transcript (Obara et al. 2002). I-boxes are known to provide light-responsiveness and strong expression activity to many nuclear genes for chloroplast proteins, such as genes for chlorophyll-a/b-binding proteins and RUBISCO small subunit (rbcS) (Donald and Cashmore 1990; Martinez-Hernandez et al. 2002). Thus, it is tempting to assume that the alternative transcription initiation sites of the MDHAR gene are linked to differential transcriptional activity.

For Arabidopsis GPx6, alternative transcription start sites probably control the availability of the transit peptide and direct the expression of cytosolic and mitochondrial isoforms (Rodriguez Milla et al. 2003). In case of poplar CuZn–SOD, alternative splicing of exon 6 (of seven exons) further increases the gene complexity (Srivastava et al. 2009). The resulting hipI-SODC1b protein (GeneBank FJ393059) is 23 amino acids longer than hip1-SODC1s (GenBank FJ393058). Expression analysis demonstrated that alternative splicing is tissue and development specific. hip1-SODC-1s was found in phloem, cambium and xylem cells, while the hip1-SODC1b transcript, encoding a presumably less stable isoform, was not detected in phloem cells and developing xylem (Srivastava et al. 2009).

```
              TS-mitochondria
1     TTTGTTATAGAGCCAACACTACAGAGAGACTCACACACTTGTTTCAATAA

                                       TL-mitochondria
51    TTAAATCTGAATTTGGCTCTTCTTATAAACTAATGTCTGCAGgtcttCTT

      TS-chloroplast
101   ATCtctctcactcaccaccatcttcttcctcgattgtcaaaccctagat

                                                TL-chloroplast
151   cgaaatCTTATCtctctaatctgttgttacagTTCGTAGAGTCATGGCGT
```

Fig. 1 Transcription of the chloroplast isoforms starts in the intron of the longer heteronuclear transcript (nRNA) for mitochondrial MDHAR according to Obara et al. (2002). Place analysis revealed two I-boxes close to the transcription start site of the chloroplast isoforms. TS: transcription start site; TL: translation start site; *gray*: GATA/GATAAG-boxes = I-boxes; *italics*: intron

3 Expressional Regulation of Antioxidant Enyzmes

3.1 *Comparison of Gene Regulation: Plastid and Extra-plastid Antioxidants*

The analysis of transgenic lines with decreased expression of antioxidant enzymes (Baier et al. 2000; Kangasjarvi et al. 2008; Shaikhali et al. 2008) and mutants with defects in transcriptional regulation of antioxidant enzymes (Heiber et al. 2007) demonstrated that the cellular antioxidant system is a combinatorial system. Plastidic and extraplastidic antioxidant enzymes act in concert and limitations in one component can be compensated by others.

Consequently, it is not surprising that the transcript levels of antioxidant enzymes differentially respond to most stress treatments (Fig. 2). From the genes encoding chloroplast antioxidant enzymes, array experiments performed by the AtGenExpress Consortium (Kilian et al. 2007) showed that, for example, Csd2 (At2g28190), was selectively induced by UV-B light, while the DHAR2 (At1g75270) transcript level was elevated in response to 10 μM methyl viologen (oxidative stress data set), UV-B and osmotic stress. Much higher regulatory amplitudes were observed for cytosolic antioxidant enzymes, such as APx2 (synonym APx1b; At3g09640), Csd1 (At1g08830) and PrxIIC (At1g65970). With more than 600-fold increases in transcript abundance, APx2 was the strongest regulated gene encoding an antioxidant enzyme in this comparative analysis. Its signal transduction pathway has been intensively studied and is described below (Section 3.2).

Generally, the genes for cytosolic antioxidant enzymes are more responsive to stress treatments. 24 h after onset of stress, the average transcript level regulation factor was 2.05-fold higher for genes encoding cytosolic antioxidant enzymes than for chloroplast antioxidant enzymes (Table 2). An exception within the group of genes encoding chloroplast antioxidant enzymes is GPx7 (At4g31870). The transcript level is transiently 6.22-fold increased after 12 h of cold treatment and 10.59-fold after 3 h UV-B stress (data not shown). Afterwards, the transcript levels decrease to almost pre-stress levels.

The strongest differences in long-term (24 h) regulation of chloroplast and extra-plastidic antioxidants take place in response to osmotic and salt stress (Fig. 3). These treatments cause a decrease of the transcript levels of almost all genes for chloroplast antioxidant enzymes, but an increase of various transcript levels for cytosolic antioxidant enzymes (Fig. 2). In contrast, the mean transcript levels are balanced within the first 6 h in response to cold, heat, methyl viologen (labeled "oxidative stress") and wounding, but peak after 12 h. Upon cold, the transient increase depends on strong induction of GPx1 (At2g25080) and GPx7 (At4g31870) and slighter increases in GR2 (At3g54660), Fsd1 (At4g25100) and sAPx (At4g08390) transcript levels. Fsd1 and sAPx transcript levels are increased after 12 h upon all treatments (Fig. 4) indicating a common slow stress response, which is independent from continuously suppression of salt and osmotic stress signaling during 24 h of treatment (Fig. 3).

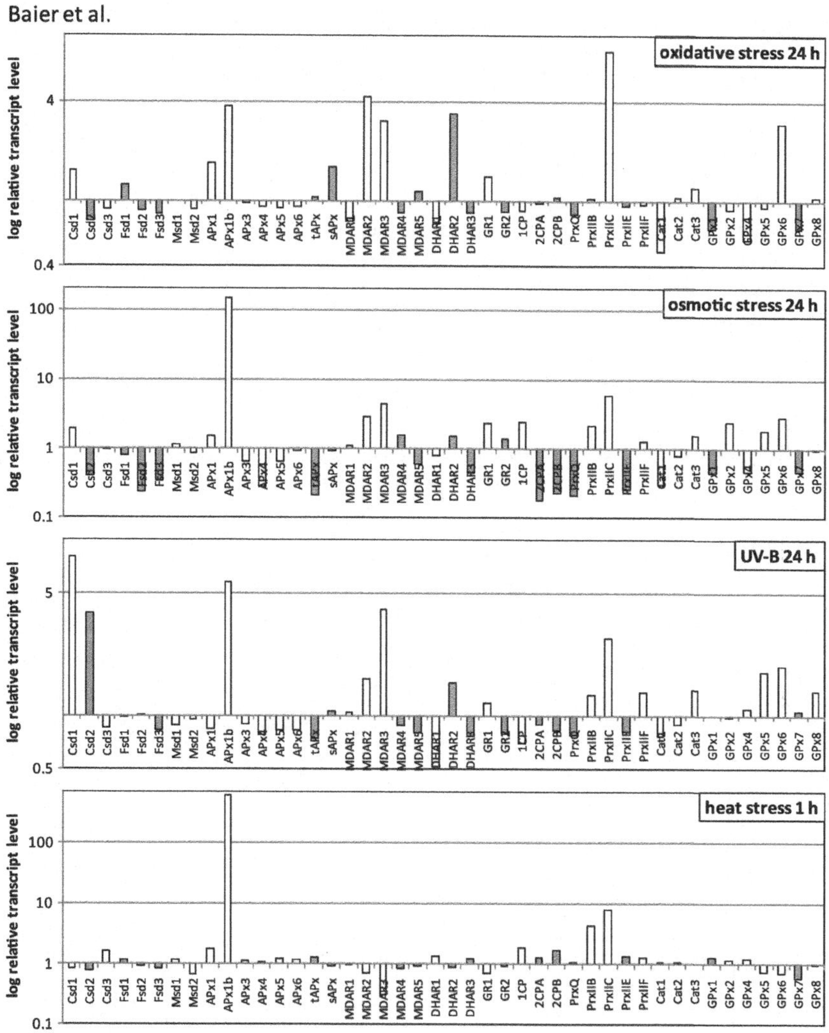

Fig. 2 Transcript abundance regulation of genes encoding antioxidant enzymes. The signals for chloroplast targeted enzymes are labeled in *gray*, for extra-plastidic enzymes in *white*. Data were collected from AtGenExpress (Kilian et al. 2007)

Recently, Soitamo et al. (2008) studied the effects of cold and light on Arabidopsis gene expression. 50% of the light and cold-responsive genes were chloroplast targeted. Consistent with other array analysis (NASC arrays described above, Gadjev et al. 2006; Pastori et al. 2003), responses of genes encoding antioxidant enzymes were comparably low. Only for GPx1 (At2g25080) correlation with ROS formation was observed. Over the years, our own work taught us that it is challenging to induce a strong increase of the transcript levels encoding some antioxidant enzymes. For example, 2-Cys peroxiredoxin transcript levels are only slightly

Table 2 Comparison of regulatory amplitudes of genes encoding chloroplast and cytosolic antioxidant enzymes after 24 h of treatment. The means were calculated from data provided in the AtGenExpress data collections via the NASC array web page (http://affymetrix.arabidopsis.info/narrays)

	Chloroplast	Cytosolic	Chloroplast/Cytocol
Cold stress	0.81	1.46	0.55
Heat stress	1.12	2.26	0.49
Osmotic stress	0.60	7.05	0.09
Salt stress	0.76	6.84	0.11
Drought stress	0.84	1.29	0.67
Oxidative stress	1.09	1.61	0.67
UV-B stress	1.10	1.69	0.65
Wounding stress	0.94	1.41	0.67

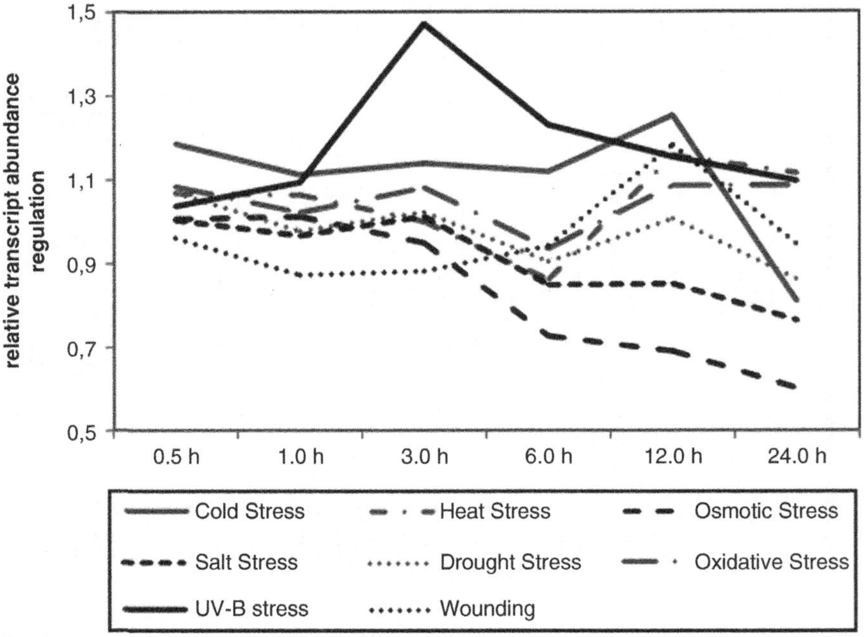

Fig. 3 Regulation of transcript abundance of genes encoding chloroplast antioxidant enzymes 0.5–24 h after onset of stress. The means were calculated from array data provided by the AtGenExpress Consortium (Kilian et al. 2007)

enhanced upon oxidative stress (Baier and Dietz 1997; Baier et al. 2004). In contrast, they are strongly suppressed in response to antioxidants (Horling et al. 2003; Baier et al. 2004). Like for various other genes, e.g. those encoding the chloroplast ascorbate peroxidases, the regulatory amplitudes are low in response to most treatments (Baier et al. 2000; Pena-Ahumada et al. 2006; Heiber et al. 2007; Shaikhali et al. 2008). Therefore, they are ignored in most array experiments, which usually list only the prominently regulated genes. On top of strong expression, even a 1.2-fold

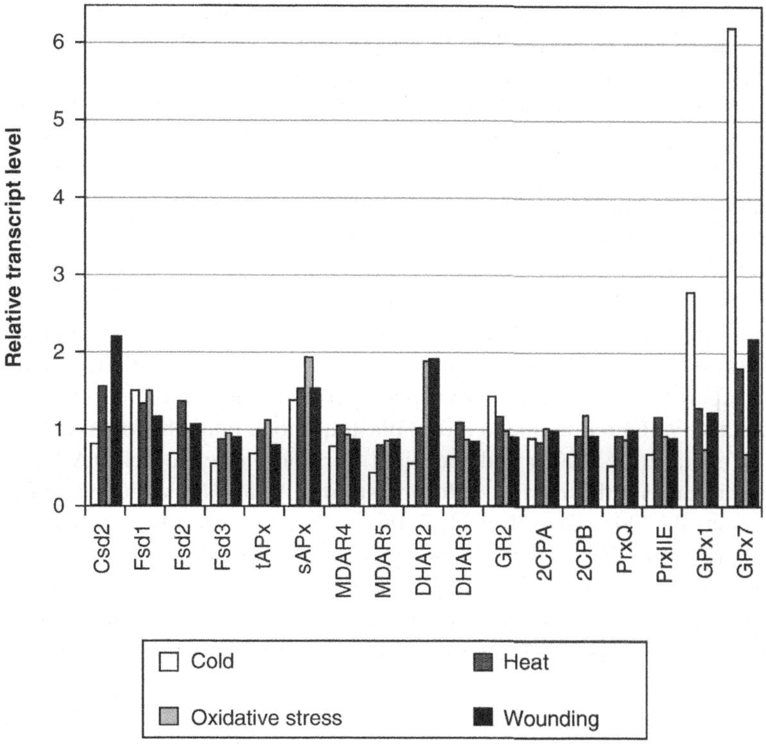

Fig. 4 Relative transcript levels of genes encoding chloroplast antioxidant enzymes 12 h after onset of cold, heat methylviologen (oxidative stress data set) and wounding stress (compared to untreated plant material). Data were taken from AtGenExpress (Kilian et al. 2007). The regulation is shown in comparison to untreated samples (level = 1)

induction, such as recently quantified by qPCR analysis (I.Heiber and M. Baier, in preparation), can result in a significant increase in the antioxidant capacity of chloroplasts.

3.2 Regulation of Cytosolic Antioxidant Enzymes

Like many pathogenesis-related genes (Van Verk et al. 2009), some genes for cytosolic antioxidant enzymes are several-fold induced upon stress (Fig. 2). APx1 and APx2 (synonym APx1b) are strongly regulated genes and well investigated ones:

3.2.1 Regulation of Cytosolic Ascorbate Peroxidases

In redox-signal transduction analysis APx1 (At1g07890) and APx2 (At3g09640) serve as model genes for strongly induced cytosolic antioxidant defenses. APx1 transcripts are usually expressed at low levels under non-stress conditions, while APx2 transcription is barely detectable. Upon stress, e.g. high light, wounding, heat and mild water deficit the genes are induced within minutes (Karpinski et al. 1997, 1999; Fryer et al. 2002). Correlation of promoter activity with ROS accumulation (Fryer et al. 2003) suggests responsiveness to redox signals. Mittler et al. (2004) suggested three possible sensing mechanisms: (i) ROS could be sensed via yet unidentified receptor proteins, (ii) via redox-sensitive transcription factors, or (iii) by ROS-specific inhibition of phosphatases (Mittler 2002; Neill et al. 2002; Vranova et al. 2002; Apel and Hirt 2004). Besides this, hydrogen peroxide and glutathione mediated redox shifts induce the accumulation of calcium ions (Price et al. 1994; Gomez et al. 2004) and are involved in the regulation of cytosolic APx (Karpinski et al. 1999; Ball et al. 2004; Rossel et al. 2006). Apart from glutathione, hydrogen peroxide, and calcium ions, photosynthetic activity, the redox-status of the plastoquinone pool (Karpinski et al. 1997, 1999; Yabuta et al. 2004) and the plant hormone abscisic acid (ABA), whose precursors are synthesized in chloroplasts in a redox/stress dependent manner (Pastori et al. 2003; Rossel et al. 2006), were suggested to regulate cytosolic APx expression. Until today it is not precisely known how ROS are sensed and their signal transmitted. According to recent literature, the first steps of ROS-signal-transduction could involve mitogen activated protein kinase (MAPKs) cascades and/or heat shock factors (HSFs) (for review see: Mittler et al. 2004, 2008).

3.2.2 Regulation via MAPK-Cascades

MAPK-signaling pathways participate in the regulation of cell growth, death, the cell cycle and stress responses (for review see: Jonak et al. 2002). A typical MAPK-cascade comprises at least a MAP kinase kinase kinase (MAPKKK) followed by a MAP kinase kinase (MAPKK) and finally a MAP kinase (MAPK). It transmits the signal by phosphorylation of the target. Sometimes also a MAP kinase kinase kinase kinase (MAPKKKK) is involved. Usually, the cascade is activated by an upstream receptor sensing a primary signal, e.g. ROS.

MAPKKK are serine/threonine kinases. They activate MAPKKs by phosphorylation of two serine/threonine residues. MAPKKs are threonine/tyrosine kinases, which phosphorylate MAPKs, which are again serine/threonine kinases and phosphorylate various substrates such as transcription factors and protein kinases (for review see: Jonak et al. 2002).

Arabidopsis thaliana encodes 60 MAPKKKs, 10 MAPKKs, and 20 MAPKs (Ichimura et al. 2002). Concerning the regulation of plant defense, two MAPKKKs have been identified: ANP1 and MEKK1 (Kovtun et al. 2000; Rossel et al. 2007; Pitzschke et al. 2009). Both are considered to be induced upon peroxide accumulation (Kovtun et al. 2000; Ichimura et al. 2006) and to regulate antioxidant defense enzymes, such as APx1, APx2 and Fsd1 (Fig. 5).

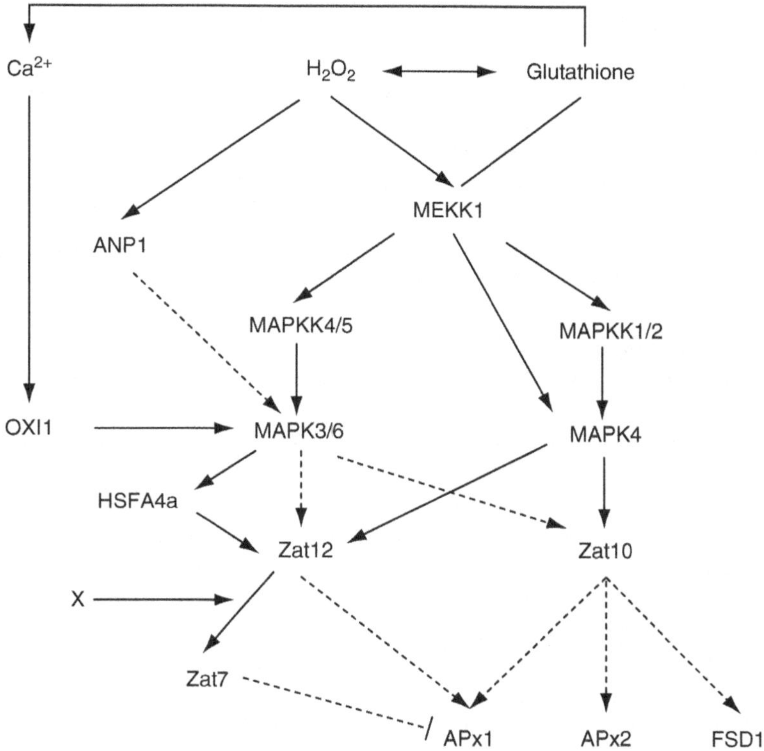

Fig. 5 Transcription factors and kinases involved in APx1 and APx2 regulation. Proven interactions are marked with *plain lines*, while postulated pathways are *dotted*

MEKK1 is suggested to regulate the APx genes via MAPKK1, MAPKK2, and MAPK4 (Ichimura et al. 1998; Ichimura et al. 2006; Qiu et al. 2008). Additionally, MEKK1 can activate MAPK4 in an MAPKK1 and MAPKK2 independent pathway (Pitzschke et al. 2009). MAPK4 subsequently activates the zinc finger transcription factor Zat10 (Pitzschke et al. 2009), which will be described below. Besides this, MEKK1 can activate APx1 and APx2 expression by phosphorylating MAPKK4 and MAPKK5, which in turn activate MAPK3 and MAPK6 (Asai et al. 2002). These MAPKs are involved in regulation of stress response genes and HSFs. Alternatively, MAPK3 and MAPK6 can be activated by MEKK1 by an presumably H_2O_2 independent mechanism (Pitzschke et al. 2009).

MAPK3 and MAPK6 generally play an essential role in plant defense. In response, Ca^{2+}-signals, which can be induced by redox imbalances (Price et al. 1994; Gomez et al. 2004), 14-3-3-mediated signals (Otterhag et al. 2006) and via the phosphoinositide-dependent kinase 1 (PDK1) (Anthony et al. 2004), OXI1 and other serine/threonine kinases (for NDPK2 see below), can also activate the MAPKs (Rentel et al. 2004). However, in contrast to MAPK4, until today no direct link between MAPK3 and MAPK6 and the expression of APx1 or APx2 has been established (Fig. 5).

Regulation via HSFs

HSFs are putative alternative redox sensors. They are considered to participate in sensing stresses and signal transduction of various environmental stresses inducing ROS production (for review see: Miller and Mittler 2006). Their mode of regulation involves DNA binding to heat stress elements (HSE: 5′-nGAAnnTTCnnGAAn-3′ or 5′-nTTCnnGAAnnTTCn-3′) (Wu 1995) via a highly conserved N-terminal binding domain. HSEs can be found in promoters of various genes involved in stress responses including APx1 and APx2 (Storozhenko et al. 1998; Schramm et al. 2006).

HSFs are subgrouped into class A and class B. While class A HSFs are considered to have an activating function and could act as hydrogen peroxide sensors (Miller and Mittler 2006), class B HSFs may be repressors (Czarnecka-Verner et al. 2000, 2004).

Among class A HSFs only few are known to play roles in the regulation of Arabidopsis APx1 and APx2. In *Arabidopsis*, APx1-knockout plants, the steady state transcript levels of HSFA4a (synonym: HSF21) and HSFA8 were elevated if compared to wildtype plants. Additionally, increased concentrations of H_2O_2 were maintained (Pnueli et al. 2003; Davletova et al. 2005). From these data, Davletova et al. (2005) concluded that HSFA4a is induced by a H_2O_2-controlled signal. HSFA4a regulates the expression of the zinc finger transcription factor Zat12 which is involved in the expression of APx1 (Rizhsky et al. 2004; Davletova et al. 2005). Besides this, Davletova et al. (2005) suggested HSFA4a may transmit a redox signal to Zat12, since the promoter of ZAT12 has a heat shock element (HSE) (Rizhsky et al. 2004). Consistently, HSFA4a dominant negative Arabidopsis lines remained unaltered APx1 transcript levels in response to light stress (Davletova et al. 2005).

Baniwal et al. (2007) shed light on a negative regulator of HSFA4a. By using tobacco mesophyll protoplasts they found HSFA5 interacting with HSFA4a and thereby repressing its activity. The transcript levels of HSFA5 are usually lower than those of HSFA4a. Upon oxidative stress the HSFA5 transcript amount slightly increases. It should be taken into account that three molecules of HSFA4a are necessary to form the functional trimer and merely one HSFA5 molecule is sufficient to repress HSFA4a function. Consequently, a small increase in the abundance of HSFA5 could be adequate to have a strong repressive effect.

Besides H_2O_2 induction of HSFA4a (Davletova et al. 2005), inducibility still has to be verified for other HSFs involved in cytosolic APx regulation: HSF1a (synonym: HSF3), HSFA2, and HSF5 (Panchuk et al. 2002; Nishizawa et al. 2006; Schramm et al. 2006). Each of these HSF is known to participate in the regulation of Arabidopsis cytosolic APx. In transgenic lines overexpressing HSF1a, the HSF was shown to induce APx1 and APx2 expression (Panchuk et al. 2002), while HSFA2 is involved in regulation of APx2 (Panchuk et al. 2002; Nishizawa et al. 2006) and HSF5 participates in the induction of APx1 (Davletova et al. 2005) (Fig. 6). Recently, HSE-based interlinking of oxidative stress and heat stress responses were also postulated for *Pennisetum glaucum* based on HSF-binding to a single HSE in the APx1 promoter (Reddy et al. 2009). From all these data, HSFs are good candidates for stress-regulators of cytosolic APx in Arabidopsis and other

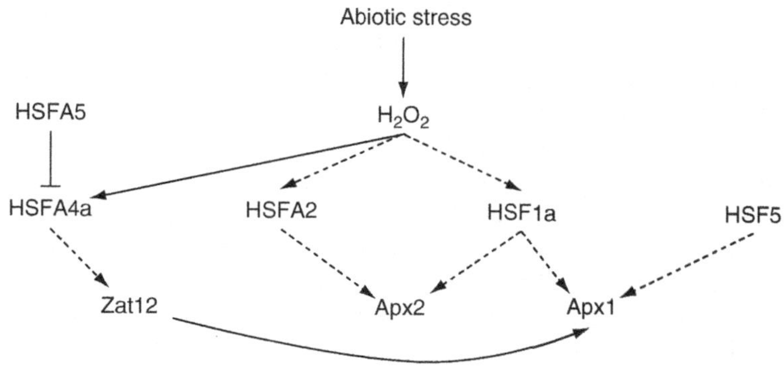

Fig. 6 Heat shock factors involved in APx1 and APx2 regulation. Proven interactions are marked with *plain lines*, while postulated pathways are *dotted*

plants. In the future, the interaction of the postulated factors with the putative HSE in APx1 and APx2 promoters needs to be verified.

Regulation via ZAT Factors

ZAT factors are zinc finger transcription factors. They can be sub-grouped into C2H2, C2C2, C2HC, C2C2C2C2, and C2HCC2C2 types (for review see: Ciftci-Yilmaz and Mittler 2008). ZAT7, ZAT10, and ZAT12 are considered to be involved in the regulation of Arabidopsis cytosolic APx1 and APx2 (Rizhsky et al. 2004; Davletova et al. 2005; Mittler et al. 2006; Ciftci-Yilmaz et al. 2007). They are C2H2 type ZAT factors, which contain a DNA binding motif of approximately 30 amino acids with two conserved cysteine and two conserved histidine residues. Besides this, the QALGGH motif is essential for DNA binding (Kubo et al. 1998).

ZAT7, ZAT10, and ZAT12 are involved in APx regulation (Fig. 5). Like some other ZAT-factors, they have an ethylene-responsive-element-binding-factor (ERF)-associated amphiphilic repression domain (EAR) (Englbrecht et al. 2004), which is discussed to mediate transcriptional control (Ohta et al. 2001; Kazan 2006).

ZAT10 recognizes AGTnnnACT or ACTnnnnGGT motifs in promoter sequences (Sakamoto et al. 2004) under control of MAPK4 (Pitzschke et al. 2009). ZAT10 over-expression in Arabidopsis resulted in an increase of the transcript levels of APx1 and APx2 as well as Fsd1 during high light treatment (Mittler et al. 2006). Consistent with ZAT10 regulation, in ZAT10 knockout lines APx2 and Fsd1 were induced to a much weaker extent (Mittler et al. 2006). ZAT10 over-expressing Arabidopsis are more tolerant to photoinhibitory light and exogenous hydrogen peroxide application due to an increased expression of genes encoding antioxidant enzymes (Rossel et al. 2006). Unlike Mittler et al. (2006); Rossel et al. (2006) observed that a change in transcript levels of Zat10 was accompanied by altered levels of APx2 and Fsd1 but not of APx1 and Csd2. From these results, it is still in question whether ZAT10 is a regulator of both, APx1 and APx2, or only of APx2.

Like Zat10, Zat12 is induced upon cold, high light treatment, and oxidative stress (Davletova et al. 2005; Mittler et al. 2006; Vanderauwera et al. 2005). Presumably, MAPK4 is also involved in the regulation of ZAT12 expression (Pitzschke et al. 2009). Besides this, ZAT12 transcription is regulated by HSF4a (Rizhsky et al. 2004; Davletova et al. 2005). Zat12 is essential for the expression of the *Arabidopsis* cytosolic APx1 but also for the defense related WRKY25 (Rizhsky et al. 2004; Davletova et al. 2005). Constitutive expression of ZAT12 resulted in higher tolerance to high light and osmotic and oxidative stress, while the sensitivity was increased when Zat12 was knocked out or suppressed by expression of an antisense construct (Iida et al. 2000; Rizhsky et al. 2004). In Arabidopsis ZAT12 over-expressors APx1 transcript levels were not elevated indicating a rather complex regulatory mechanism (Rizhsky et al. 2004). Since APx1 was yet still high-light-inducible in ZAT12 knockout plants (Rizhsky et al. 2004), a high-light-pathway could exist which is unaffected by the absence of ZAT12.

ZAT12 regulates not only the expression of APx1 and WRKY25, but also of ZAT7 (Rizhsky et al. 2004). The H_2O_2 triggered induction of ZAT7 was abolished in Arabidopsis ZAT12 knockout plants. However, when Zat12 was overexpressed, no induction of ZAT7 occurred under non-stress conditions. It is tempting to assume that ZAT12-mediated ZAT7 induction depends on priming by other stress signals (Rizhsky et al. 2004). The transcript level of ZAT7 was elevated in Arabidopsis APx1-knockout plants, probably in response to the oxidative stress caused by the lack of APx1 activity (Ciftci-Yilmaz and Mittler 2008). Besides this, Ciftci-Yilmaz et al. (2007) proposed that ZAT7 suppresses APx1 expression. When ZAT7 was constitutively expressed, it conferred an enhanced tolerance to salinity and temperature stress, which was dependent on the presence of the EAR motif within the transcription factor (Pnueli et al. 2003; Ciftci-Yilmaz and Mittler 2008). Over-expression of ZAT7 in *Arabidopsis thaliana* increased the transcript levels of alternative oxidase (AOX1) and pathogen defense related WRKY70 (Pnueli et al. 2003; Ciftci-Yilmaz and Mittler 2008) providing additional stress protection.

ZAT12 regulates the expression of WRKY25 and ZAT7 that of WRKY70 (Rizhsky et al. 2004; Davletova et al. 2005; Ciftci-Yilmaz and Mittler 2008). Both WRKYs are considered to regulate defense related genes in response to biotic as well as abiotic stresses (Li et al. 2004; Zheng et al. 2007). They have redox-sensitive zinc finger DNA binding domains in which two cysteine and two histidine residues coordinate a zinc atom (for review see: Eulgem et al. 2000; Eulgem and Somssich 2007). In addition, they comprise the WRKY motif at their N-terminus (WRKYGQK). Their target motifs are W-boxes [(T)(T)TGAC(C/T)] (Rushton et al. 1996), which are frequently found in promoters of stress-regulated genes (Scarpeci et al. 2008).

ZAT factors coordinate directly or via regulation of other transcription factors, such as WRKY, plant defense against biotic and abiotic stressors. They can either directly induce the expression of the desired gene or they repress an expressional repressor of the target gene (Ohta et al. 2001; Englbrecht et al. 2004; Davletova et al. 2005; Vogel et al. 2005). Alternatively, they influence the expression of target genes by impacting on the regulator of the target gene (Rizhsky et al. 2004). With APx1 and APx2, and presumably Fsd1, three targets important in antioxidant protection have been identified (Mittler et al. 2006).

3.2.3 Regulation of Defense Genes via the NPR1-TGA1-as-1-System

As-1 is a well known promoter element, which responds to oxidative stress and salicylic acid (Garreton et al. 2002). In as-1, two TGACG motifs bind dimeric TGA transcription factors (Xiang et al. 1997; Johnson et al. 2001). They are basic leucine zippers (bZIP). Various members of the TGA-family are regulated by Non-repressor of PR genes (NPR1) (Fan and Dong 2002; Despres et al. 2003; Shearer et al. 2009), which triggers systemic aquired resistance (SAR) (Cao et al. 1994). NPR1 comprises two protein-protein interaction motifs called BTB/POZ (Aravind and Koonin 1999) and ankyrin repeats (Cao et al. 1997). The NPR1 protein can shuttle between the cytosol, where it is involved in the crosstalk between salicylic acid- and jasmonate-dependent defense signaling, and the nucleus (Despres et al. 2000; Spoel et al. 2003). In the nucleus, it interacts with TGA transcription factors, such as TGA1 and TGA7 (Shearer et al. 2009). The interaction with NPR1 stimulates binding of the TGA factors to the as-1 motifs (Despres et al. 2000). As-1 was originally described in bacterial and viral promoters and later found to bind TGA transcription factors in plants (Lam and Lam 1995). Similar elements designated as asf-1 were reported for the promoters of Arabidopsis chloroplast and cytosolic APx (Srivastava et al. 2009), suggesting similar regulation.

TGA are known to regulate PR-genes (Lam and Lam 1995; Johnson et al. 2003; Van Verk et al. 2009). TGA1 is phosphorylated by MAPK10, which may influence its activity (Popescu et al. 2009). In the cytosol, NPR1 exists as uninduced oligomers in which the monomers are linked through disulfide bridges (Mou et al. 2003). Upon pathogen attack, these oligomers are reduced. Subsequently, the monomers can migrate into the nucleus. The ratio of NPR1 oligo- and monomers can be adjusted by the ratio of reduced and oxidized glutathione in the medium. Since the redox state of the glutathione pool shifts to more oxidized levels soon after pathogen attack (Noctor et al. 2002), it is unlikely that NPR1 localization is directly regulated by the thiol redox signature. Laloi et al. (2004) suggested that specific thioredoxins, such as cytosolic thioredoxin Trxh5, might influence the activity of NPR1. Since the reduction of two cysteine residues in the C-terminus of TGA1 (C260 and C266) is a prerequisite for the interaction with NPR1, similar mechanisms may prepare TGA1 for its interaction with its regulator (Despres et al. 2003). Laloi et al. (2004) additionally suggested that Trxh5 is controlled by WRKY6, which in turn may be induced by MEKK1 (Pitzschke et al. 2009) indicating that NPR1 dynamics are under expressional control.

3.2.4 Regulation of Catalase Genes by ARE and ABRE

Most higher plants encode three catalase genes (summarized in: Iwamoto et al. 2000), of which two are regulated by circadian rhythms (Iwamoto et al. 2004). Many cDNA array studies (e.g. Fig. 2) report only slight transcript abundance regulation, while regulation is described in RT-PCR and Northern-blot based studies. Cat1 (synonyme: CatA) is stress-induced. It responds to H_2O_2 and ABA (Guan et al. 2000). As shown in ABA-deficient embryos, ABA-signalling takes place by two independent pathways and DNA motifs. The ABRE2 motif is activated by CBF1 (Cat1-binding factor 1), the ABRE2 motif by CBF2. H_2O_2, which was transiently induced in ABA-treated plant material (Guan et al. 2000), is supposed to activate the ABA-response indirectly via an ARE (antioxidant response element: TGACTCA) localized upstream of ABRE1 and ABRE2.

Similar ARE were observed in the promoters of all three catalase genes of maize (Polidoros and Scandalios 1999). They respond, like the maize superoxide dismutase genes SOD3 and SOD4, to various types of ROS, including H_2O_2 (Rushmore et al. 1991). Signal transduction is supposed to take place via redox-sensitive MAPK-cascades (Scandalios 2005) and finally results in binding of a regulatory protein (ARE-binding protein; ARE-BP) to ARE. Comparison of the promoter structure of the three maize catalase promoters demonstrates distinct positioning of AREs and combination with other putative stress-responsive promoter elements. While the ARE is in position −470 in the Cat1 promoter, the Cat2-promoter shows signatures for a strong ARE in position −56 and a weak one at the transcription initiation site (+1). In the Cat 3 promoter ARE is localized in position −145 relative to the transcription initiation site (Guan et al. 1996; Scandalios 2005). Although many scientists refer to the prediction of ARE in plant promoters (Guan et al. 1996; Polidoros and Scandalios 1999; Scandalios 2005), to date the function of the AREs seem not to have been confirmed by promoter deletions analysis or reporter gene studies and the nature of the ARE-binding protein has not been identified. For Arabidopsis, Zimmermann et al. (2006) relate H_2O_2-regulation of catalase genes to the transcription factor WRKY53 during senescence. It may be triggered by MEKK1 (Miao et al. 2007) or a novel DNA-binding protein with HPT-type kinase function (Miao et al. 2008).

3.2.5 Peroxiredoxin Regulation: PrxIIB and PrxIIC

Figure 2, like many other array and RT-PCR based transcript abundance studies (Horling et al. 2003) reports strong increases of PrxIIB (At1g65980) and PrxIIC (At1g65970) transcript levels upon osmotic, UV-B and heat stress. On chromosome 1 of *Arabidopsis thaliana*, the two peroxiredoxins are arranged next to each other. PrxIIB is encoded upstream of PrxIIC. The exon–intron structure is conserved indicating recent gene duplication. In PrxIIB, the two introns are only longer than in PrxIIC. While PrxIIB is stronger expressed than PrxIIC (Fig. 2; Horling et al. 2003), the regulation amplitude of PrxIIC upon stress is higher than that of

PrxIIB (Fig. 2). In a comparions of peroxiredoxin gene expression, Horling et al. (2003) reported induction of PrxIIC by H_2O_2 and t-butyl hydroperoxides and in response to dimaide treatment, which acts as a pro-oxidant by blocking thiol groups. Unlike the strongly ascorbate-suppressive chloroplast peroxiredoxins 2CPA, 2CPB, PrxQ and PrxIIE, the transcript level of PrxIIC is increased in response to ascorbate application (Horling et al. 2003). PrxIIC also shows atypical regulation in response to osmotic stress: While the transcript levels of most peroxiredoxin genes decrease, PrxIIC transcripts accumulate (Horling et al. 2002).

The 5′-end of PrxIIC overlaps with the 5′-end of a gene encoding glutamate decarboxylase GAD2 (At1g65960). The genes are expressed in the same tissues at the same time of development (information taken from array data summarized in the Arabidopsis eFP browser). According to array based comparison of transcript abundance using the Arabidopsis eFP browser (Winter et al. 2007) PrxIIC is much stronger expressed than GAD2 in seeds, in pollen, following infection with various pathogenes and UV-B illumination (data not shown). Transcript abundance regulation may take place by antisense interference of the 5′-ends, since PrxIIC transcript accumulation is accompanied with a decrease of GAD2 transcript levels in seeds and upon pathogen attack. A similar post-transcriptional regulation mechanism has been proven for SRO5 and P5CDH (Borsani et al. 2005) (Fig. 7). Two types of siRNAs were found to be involved in transcript abundance regulation. The siRNAs are generated by the antisense overlapping gene pair of Δ^1-pyrroline-5-carboxylate dehydrogenase (P5CDH) (At5g62530), a gene related to stress response, and SRO5 (At5g62520), which is a gene of unknown function, but similar to radical induced cell death-1 (RCD1), which is involved in controlling ROS-induced cell death (Ahlfors et al. 2004; Borsani et al. 2005; Babajani et al. 2009) and presumably participates in regulation of 2CPA via interaction with the transcription factor Rap2.4a (J. Shaikhali et al., unpublished data).

The SRO5 mRNA is induced by osmotic and oxidative stress (Borsani et al. 2005). It interacts with the constitutively expressed P5CDH mRNA. From the overlapping double-stranded RNA (dsRNA) a 24-nt nat-siRNA is generated by DCL2, RDR6, SGS3 and NRPD1A (Borsani et al. 2005). This 24-nt nat-siRNA induces the cleavage of the P5CDH transcripts. Catalyzed by DCL1 this leads to formation of a 21-nt nat-siRNA. These two types of siRNA guide the cleavage the P5CDH mRNA. Limitations in P5CDH result in accumulation of pyrroline-5-carboxylate (P5C) and, finally, proline. The latter functions as a cytosolic osmoprotectant. Consistently, its biosynthesis increases osmotic stress tolerance. However, increasing proline is also accompanied by accumulation of ROS (Borsani et al. 2005). SRO5 protein, therefore, mediates ROS detoxification and can balance the dangers caused by ROS induction due to reduced P5CDH activity (Borsani et al. 2005). In a similar manner than SRO5 and P5CDH, a pair of *cis*-antisense GAD2 – PrxIIC transcripts may regulate PrxIIC expression.

3.2.6 Regulation of Cytosolic MDHAR

MDHAR2 and MDHAR3 are besides APx1, APx2, PrxIIB and PrxIIC, dynamically regulated upon stress (Fig. 2). So far very little is known about the mechanisms of

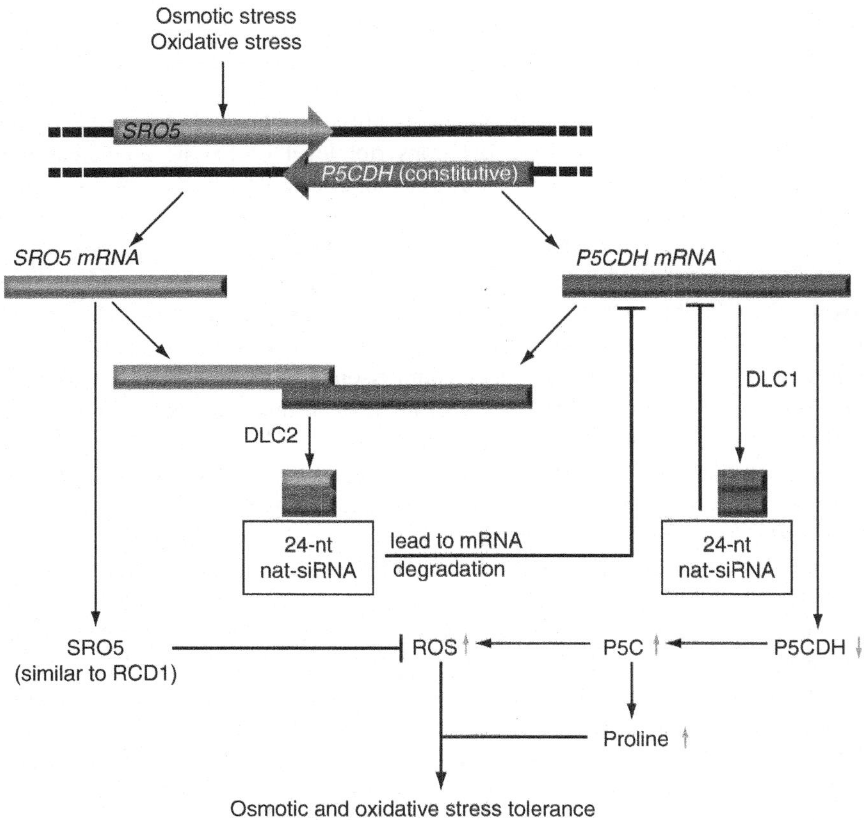

Fig. 7 An example for regulation of transcript abundance by antisense constructs. Upon oxidative and osmotic stress expression of SRO5 is induced. The 5′-end of SRO5 interacts with the 5′-end of the constitutively expressed P5CDH transcript. DLC2 recognizes the RNA hybrid and generates a 24-nt siRNA, which leads to P5CDH RNA degradation

MDAR2 and MDHAR3 regulation. According to array data depicted in the eFP browser (Winter et al. 2007), MDHAR2 transcript levels are up-regulated upon senescence, while MDHAR3 is almost constitutively expressed during development. Recently, based on analysis of T-DNA insertion lines Vadassery et al. (2009) reported that MDHAR2, like a glutathione-S-transferase (At1g19570; annotated as dehydroascorbate reductase in Vadassery et al. 2009), plays a crucial role for the mutalistic interaction between the endophytic function *Piriformospora indica* and *Arabidopsis thaliana*. Consistent with a protective function upon biotic stress, the MDHAR2 transcript level is up-regulated in wild-type plants treated with *P. indica*. The regulation is specific for MDHAR2, since none of the other four MDHAR-genes or DHAR1, DHAR2 or DHAR3 responded.

Like for PrxIID, the MDHAR2 transcript levels were decreased 48 h aluminium application (Kumari et al. 2008). In contrast the transcript levels of organellar

MDHAR was increased after 6 h and that of cytosolic MHDAR5 after 48 h. In the long-term response also the transcript levels of APx2, cytosolic glutathione reductase and Fsd1 were elevated, suggesting a global induction of genes encoding cytosolic antioxidant enzymes in response to Al (Kumari et al. 2008)

3.3 Regulation of Genes Encoding Chloroplast Antioxidants

Compared to regulation of cytosolic antioxidant defenses, for which APx1 and APx2 regulation provide detailed information on signaling pathways and suggests an overlap with regulation of the well studied pathogenesis response pathways, analysis of the transcriptional regulation of genes encoding chloroplast antioxidant enzymes has just started.

3.3.1 Redox-Regulation of 2-Cys Peroxiredoxin-A (2CPA)

The 2CPA gene is used as a model gene for transcriptional regulation of a nuclear gene encoding a chloroplast antioxidant enzyme. Transcript abundance analysis demonstrated that the transcript level is strongly sensitive to antioxidants. In the liverwort *Riccia fluitans*, the monocot barley and the dicot Arabidopsis the transcript levels decreased in response to application of ascorbate (Horling et al. 2001, 2003; Baier and Dietz 1997) and glutathione (Baier and Dietz 1997; Horling et al. 2001) and the glutathione precursor (Queval et al. 2009) cysteine (Baier and Dietz 1997).

In the 16 years, since we discovered the first plant homologue of 2CP (that time designated thiol-specific antioxidant) (Baier 1993; Baier and Dietz 1996), own data (especially: Baier and Dietz 1996, 1997; Baier 1997; Horling et al. 2003; Pena-Ahumada et al. 2006) and in recent years the accumulating array data from various labs all over the world (e.g.: Gadjev et al. 2006; Queval et al. 2007; Soitamo et al. 2008; Vandenbroucke et al. 2008), demonstrate that the transcript levels of 2CPA are - unlike many genes encoding extra-plastidic protecting enzymes, e.g. APx1, APx2 (Mittler et al. 2006) and many PR genes (Choi et al. 2007) – hardly inducible by commonly used oxidative stressors, such as methylviologen and ozone (Baier and Dietz 1997; Baier 1997).

2CPA expression is strongly linked to chloroplast development (Baier and Dietz 1996; Baier et al. 2004; Pena-Ahumada et al. 2006). Recent transcript abundance studies in sAPx and tAPx knockout lines demonstrated induction of peroxiredoxins in response to insufficient chloroplast antioxidant protection (Kangasjarvi et al. 2008). In addition, in one of our first expression studies (Baier and Dietz 1996) we showed that the 2CP transcript amount increases in barley leaf blades in response to ozone, if biosynthesis of the buffering low molecular weight antioxidant glutathione was blocked by application of BSO (buthionine sulfoximine). Based on these observations and the response to ascorbate and glutathione, we concluded that 2CPA expression is redox-responsive in the background of strong developmental regulation. When we used more sensitive tools, such as reporter genes whose read-outs accumulate regulation over time, the 2CPA promoter responded also to wounding and to H_2O_2 mediated oxidative stress (Baier et al. 2004).

Array-based comparison (see: compare mode of eFP browser: http://bar.utoronto.ca/efp/cgi-bin/efpWeb.cgi, Winter et al. 2007) and cycle number optimization in semi-quantitative RT-PCR and qPCR analysis demonstrated that the expression level is similar to that of actin-2 (At5g09810) in young Arabidopsis tissues under control conditions. The strong expression intensity results in high protein levels. In barley, 2CP is one of the most abundant chloroplast proteins with 0.6% ± 0.2% of total soluble leaf protein (König et al. 2003). The high background expression masks the regulation in many quantitative analytic techniques focusing on the regulation amplitudes, such as quantitative transcriptomics.

Regulation of 2CP expression was analysed first using inhibitors of signal transduction cascades (Baier and Dietz 1996, 1997; Horling et al. 2001) and later by mapping the *cis*-regulatory elements in the promoter and screening for trans-regulators (Baier et al. 2004). Studies in the liverwort *Riccia fluitans* demonstrated that the ascorbate application strongly decreases 2CP transcript abundance. The response can be blocked by application of the serine/threonine-kinase inhibitor staurosporin, but not by the protein phosphatase inhibitor okadaic acid, the G-protein activator mastoperan and the Ca^{2+}-chelator EGTA (Horling et al. 2001), indicating that the antioxidant suppression is kinase-regulated. Using staurosporin, it was shown later in *Arabidopsis thaliana* that kinase-mediated signaling mechanisms control ascorbate-dependent 2CPA transcription (Baier et al. 2004).

By the use of promoter deletion constructs, the *cis*-regulatory domain of the 2CPA promoter was mapped to a 216 bp domain, designated "redox-box" (Baier et al. 2004). A subsequent yeast-one-hybrid screen resulted in identification of a redox-sensitive transcription factor, Rap2.4a (Shaikhali et al. 2008), which binds in its dimeric form to a CE3-like element within the redox-box. Dimerization takes place at a midpoint potential of −269 mV and, thus, is in the range of physiological redox poise control of the cellular glutathione system (mid-point potential: −230 mV) and the thioredoxin system (mid-point potential: −290 to −300 mV) (Aslund et al. 1997, Collin et al. 2003). Upon severe redox imbalances, Rap2.4a oligomerizes (Shaikhali et al. 2008) and 2CPA promoter activity decreases (Shaikhali et al. 2008,;Shaikhali and Baier 2009).

Low ascorbate levels, such as in the ascorbate-biosynthetic mutant *vtc1* (Conklin et al. 1999), hardly affect 2CPA regulation under control conditions (Wormuth et al. 2006). In contrast, limitation in the thiol availability, such as by application of the thiol-blocking agent diamide (Horling et al. 2003) or application of the glutathione biosynthesis blocker BSO (Baier and Dietz 1996), decreases 2CPA transcript amount. In addition with the recent observation, that the ascorbate response is light-dependent, while the glutathione response is not (Shaikhali and Baier 2009), we concluded that the 2CPA regulation is controlled by the redox state of the cellular thiol pool in a photosynthesis-regulated mechanism. This conclusion is consistent with the previous observation that 2CPA promoter activity depends on the acceptor availability at photosystem I (Baier et al. 2004). Photosynthetic electron transport also controls glutathione reduction (Asada 2000) and, therefore, the redox poise of the cellular thiol pools (Foyer and Halliwell 1977). According to the transcriptome study performed by Piippo et al. (2006), response to the acceptor availability at photosystem I may be the major regulator of genes encoding chloroplast proteins and could trigger the common regulatory mechanism postulated by Biehl et al. (2005) and MacLean et al. (2008) for plastid-nucleus signaling.

Besides redox signals, the 2CPA promoter is responsive to abscisic acid (ABA) (Baier et al. 2004). The sensitive *cis*-element was mapped to the same promoter region as the redox-box. However, while Rap2.4a activation upon slight redox shifts induces the 2CPA promoter (Baier et al. 2004; Shaikhali et al. 2008), ABA suppresses it (Baier et al. 2004). In various studies, including analysis of APx2 regulation (Rossel et al. 2006), ABA and photooxidative stress have been described as positively interacting signals: First, ABA application increases H_2O_2 accumulation (Jiang et al. 2003; Pastori et al. 2003; Hu et al. 2005), and second, H_2O_2 inhibits the phosphatases ABI1 and ABI2, which negatively regulate the ABA-triggered MAPK-signaling cascade (Meinhard and Grill 2001; Meinhard et al. 2002). Since in 2CPA regulation ABA and H_2O_2 trigger inverse responses, independent redox and H_2O_2-signaling have to be assumed (Baier et al. 2004). Binding of Rap2.4a to a *cis*-element, which shows similarities to ABA-regulated CE3-elements, suggested gene regulation through transcription factor competition for common binding sites (Shaikhali et al. 2008). However, the transcription factor Rap2.4b (At1g78080), whose DNA-binding domain is almost identical to that of Rap2.4a (Nakano et al. 2006), is induced by osmotic stress and protects plants from ABA-triggered salt stress (Lin et al. 2008). Consequently, an effect of Rap2.4-transcription factors on either positioning or activation of ABA-regulated transcription factors by the non-conserved C- and N-terminal domains is more likely. Putative mechanisms may be transcription factor interference by protein–protein interactions or protein-dependent modeling of the transcription factor contact surface.

In addition, a mutant screen for 2CPA promoter activators (Heiber et al. 2007) indicates at least six additional regulators. Five of the isolated mutants, designated *rimb* for the redox-imbalance of gene expression regulation, are actually affected in redox regulation of gene expression (Heiber et al. 2007). *Rimb1* and *rimb2* are impaired in central elements of the signaling pathways regulating expression of various chloroplast antioxidant enzymes, including sAPx, tAPx and Csd2 (Heiber et al. 2007), which are also miss-regulated in Rap2.4a knockout lines (Shaikhali et al. 2008). Crude map positions indicate independence from Rap2.4a (H. Hiltscher et al., unpublished data). The affected genes may be *trans*-regulators of Rap2.4a or independent gene activators. In contrast, gene expression analysis and comparison of the antioxidant signature of *rimb5*, *rimb6* and *rimb7* indicates effects on indirect 2CPA expression regulation, such as by increased protection by ascorbate and induction of extra-plastidic antioxidant defense mechanisms (Heiber et al. 2007).

Comparison of gene expression regulation of plastid and extra-plastidic antioxidant defenses in Rap2.4a knockout lines (Shaikhali et al. 2008) and the *rimb*-mutants (Heiber et al. 2007) demonstrated that at least expression of the chloroplast ascorbate peroxidases sAPx and tAPx, Csd2 and 2CPA is controlled independently from ROS-responsive pathways, such as the ZAT10-APx2-pathway. Array-data collected under severe oxidative stress conditions (Davletova et al. 2005; Gadjev et al. 2006) suggest that the induction of genes encoding chloroplast antioxidant enzymes is already inhibited when the cellular redox poise shifts to oxidized levels, which can fully activate the cytosolic antioxidant defense. This is in consistence with the redox-kinetics of Rap2.4a (Shaikhali et al. 2008).

3.3.2 Co-regulation of Chloroplast and Extraplastidic Genes by NDPK2

Gene regulation via the nucleoside diphosphate kinase NDPK2 is an example for a redox-responsive link in regulation of plastid and extraplastidic antioxidant enzymes. Nucleoside diphosphate kinases (NDPK) sense and control the phosphorylation state of nucleosides and desoxynucleosides in cells by catalyzing the reversible transfer of the 5′-terminal phosphate groups from triosphosphate nucleosides and desoxynucleosides to (desoxy-)nucleoside diphosphates and nucleoside disphosphates, except ADP (Park and Agarwal 1973).

NDPK2 was shown to be part of the MPK signaling cascade (Moon et al. 2003). It interacts with the redox-/H_2O_2-sensitive MAPK3 and MAPK6 and enhances the activity of MAPK3 (Moon et al. 2003). Overexpression induces expression of various genes encoding antioxidant and redox-active proteins, such as catalase-3, several cytosolic peroxidases, 2-Cys peroxiredoxin-A and B, thioredoxin, thioredoxin reductase, tAPx and glutathione transferase in *Arabidopsis* (Yang et al. 2003). Besides this, it triggers induction of peroxidase, ascorbate peroxidase and catalase activities in sweetpotato plants (*Ipomoea batatas*) (Kim et al. 2009b). Future work will have to show, which of the putative target genes predicted from array experiments (Yang et al. 2003) is directly activated by NDPK2 or activated via NDPK2-regulated transcription factors.

Fukamatsu et al. (2003) showed that the pea homologue of *Arabidopsis* NDPK1, namely NDK-P1, binds to all three tested catalase isoforms in yeast-two-hybrid approaches. It also co-migrates with catalases on two-dimentional non-denaturing polyacrylamide gels, suggesting that the function of NDPKs in regulation of the cellular antioxidant system includes also post-translational effects.

3.3.3 Glutathione Peroxidase Expression in Arabidopsis

Recently, Soitamo et al. (2008) demonstrated that chloroplast GPx1 and cytosolic GPx6 are induced by the combination of cold and light, which increases cellular ROS formation. The cold induction is light specific, since no significant difference in transcript abundance was observed in cold and dark-treated plants. Moderate high light increased the transcript level of GPx6, but decreased that of GPx1.

In summary, the data demonstrate that cold inverts the light responsiveness of GPx1. For GPx7 the highest regulation amplitudes of all studied genes for chloroplast antioxidant enzymes have been reported in array studies (AtGenExpress Consortium). As mentioned above (Section 3.1), the gene is transiently induced 6.22 and 10.59-fold upon cold and UV-B treatment, respectively. Under control conditions, it is barely active. Low background expression activity is very unusual for a gene encoding a chloroplast antioxidant enzyme. When the gene was first described, in contrast to its paralog GPx1 no ESTs were reported in the databases (Rodriguez Milla et al. 2003). Rodriguez Milla et al. (2003) demonstrated by RT-PCR low, but similar transcript levels in dry seeds, 10-day old seedlings, rosette leaves, stems, cauline leaves and flowers indicating constitutive background expression. In the GPx7-promoter they identified several light regulated I- and T-boxes (Donald and Cashmore 1990; Chan et al. 2001) besides an ABA-sensitive Myc-ATRD22 motif (Abe et al. 1997). None of these elements explains the transient strong induction, since ABA-triggerd osmotic stress results in decreased GPx1 expression (Fig. 2).

The either cytosolic or mitochondrial GPx6 is the strongest stress inducible GPx gene of *Arabidopsis thaliana*. Its transcript level is low in most vegetative tissues and induced by NaCl and heat according to Rodriguez Milla et al. (2003) and oxidative stress, osmotic stress and UV-B according to data provided by the AtGenExpress (Fig. 2). In the GPx6 promoter, stress sensitive motifs such as W-boxes (Yu et al. 2001), similar to those found in the promoter of the TGA-regulatory redox-sensitive protein NPR1 (Despres et al. 2003; Shearer et al. 2009), drought-responsive elements (DRE; Kizis and Pages 2002) and Myb2AT (Urao et al. 1993; Abe et al. 1997) were reported (Rodriguez Milla et al. 2003). In *Lotus japonicus*, the GPx1 and GPx6 promoters have motifs with similarity to redox-regulated antioxidant response elements (ARE) of maize catalase2 (GGTGACCTTGC) (Scandalios 2005; Guan et al. 1996) in their promoters (Ramos et al. 2009). Lotus-GPx1 is homologous to *Arabidopsis* GPx6 and Lotus-GPx6 to chloroplast-targeted *Arabidopsis* GPx1 and GPx7 (Ramos et al. 2009). The homology of these GPx genes suggests similar regulatory elements in *Arabidopsis*.

4 Post-transcriptional Regulation of Superoxide Dismutase Expression by Small RNAs, Chaperones and Heterocomplex Formation

In oxidative defense systems, superoxide dismutases (SOD) constitute the first line of defense against superoxide radicals by rapidly converting superoxide to H_2O_2 (Fridovich 1995). Quiet early, Kliebenstein et al. (1998) systematically compared transcript abundance regulation of superoxide dismutases in Arabidopsis. Based on Northern-Blot hybridization, they demonstrated co-induction of cytosolic Csd1 with chloroplastic Csd2 and inverse regulation of chloroplast Fsd2 compared to presumably extra-plastidic Fsd1 in response to photoinhibitory high light irradiation. Recent work demonstrates that post-transcriptional and also post-translational events, such as control of Cu^{2+}-binding, play crucial roles in regulating the transcript abundances of superoxide dismutase encoding genes.

4.1 The Cu availability Controls Csd1 and Csd2 Transcript Levels of Small RNAs

Gene expression analysis demonstrated that small non-coding 21- to 25-nt long RNAs can strongly impact on transcript stability (Carrington and Ambros 2003; Bartel and Chen 2004; He and Hannon 2004). The small RNAs can be classified into two classes (Bartel and Chen 2004): (i) short endogenous interfering RNAs (siRNAs). The siRNAs are synthesized from long double-stranded RNAs (Borsani et al. 2005). RNAs encoding specific gene functions interact locally with each other, such as in case of SRO5 and P5CDH and, presumably, PrxIIC and GAD2 (Section 3.2.4; Fig. 7). (ii) In contrast microRNAs (miRNAs; μRNAs) are small non-coding RNAs. They are processed from longer hairpin precursors by the Ribonuclease III-like enzyme Dicer (Bernstein et al. 2001; Ketting et al. 2001).

The biochemical machinery involved in siRNA and miRNA formation consists of four Dicer-like (DCL) proteins (Schauer et al. 2002), six predicted RDRs (Mourrain et al. 2000), and ten predicted Argonaute proteins (Morel et al. 2000) in Arabidopsis. siRNA guided cleavage of primary transcripts processes 24-nt trans-acting siRNAs (ta-siRNA) involving RNA-Dependent RNA Polymerase-6 (RDR6) and Suppressor of Gene Silencing-3 (SGS3) (Peragine et al. 2004; Vazquez et al. 2004). The 24-nt ta-siRNA directs then the biogenesis of 21-nt natural antisense transcript-derived siRNAs (nat-siRNAs) by DCL1. These nat-siRNAs guide cleavage of, for example, P5CDH mRNAs (Sunkar et al. 2006). In another case, siRNAs (24 nt) is generated by DCL3, RDR2 and NRPD1A by processing RNAs from transposons, 5S rRNA genes and other repeats (Sunkar et al. 2006). Compared to siRNAs, miRNAs are at first incorporated into an RNA induced silencing complex (RISC) (Hannon 2002; Meister and Tuschl 2004), and induce sequence-specific translational inhibition or transcript cleavage by targeting complementary mRNAs (Bartel and Chen 2004).

Recent studies indicated that miR398 regulates two Cu/Zn superoxide dismutases, cytosolic CSD1 (At1g08830) and chloroplastic CSD2 (At2g28190) (Fig. 8). miR398 was discovered in *Arabidopsis thaliana* and *Oryza sativa* and is conserved in other flowering plants, suggesting that miR398 regulation is conserved over a wide range of plants.

The miR398 family is encoded by three loci in *Arabidopsis thaliana* (Bonnet et al. 2004; Jones-Rhoades and Bartel 2004; Sunkar and Zhu 2004). The two expressed variants bind to a complementary site in the coding region of CSD1 and CSD2 transcripts and cause post-transcriptional cleavage of the mRNAs (Sunkar et al. 2006). In oxidative stress or under the influence of high Cu^{2+} concentrations, miR398 is down-regulated (Sunkar et al. 2006). In response, CSD1 and CSD2 transcripts accumulate, the encoded proteins bind Cu^{2+}-ions and protect plants from oxidative stress. Sugars induce miR398 expression and, therefore, result in decreased CSD transcript levels (Dugas and Bartel 2008). Post-translational regulation by miR398 can explain the co-regulation of Csd1 and Csd2 transcript abundances upon stress as reported by Kliebenstein et al. (1998).

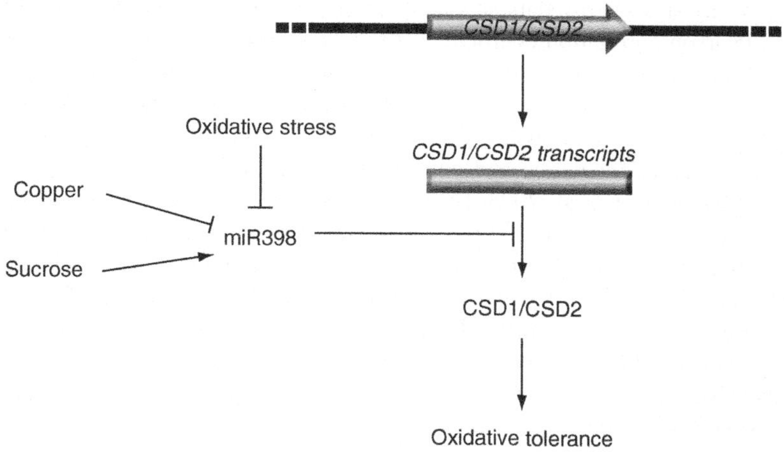

Fig. 8 Transcript abundance of Csd1 and Csd2 is regulated by miR398 in response to oxidative stress, copper availability and sucrose

Recently, Yamasaki et al. (2009) demonstrated that miR398 expression is under control of SQUAMOSA promoter binding protein 7 (SPL7). SPL7 binds to a GTAC motif in the miR398c promoter under low copper conditions. Since the protein also regulates miR397, miR408 and miR857 it is postulated to be a master regulator of miRNA-controlled copper proteins.

Besides the two Csd genes, miR398 regulates Cox5b-expression post-transcriptionally. However, while in Csd-transcripts the complementary sites are highly conserved between Arabidopsis, pea, poplar, rice and *Glycine max*, conservation of Cox5b-binding sites are more variable indicating less conserved post-transcriptional regulation (Dugas and Bartel 2008).

4.2 The Metallo-Chaperone CCS Links Cu Regulation of Csd1 and Csd2 Expression

In *Arabidpopsis thaliana*, the transcript abundance of Csd2 is strictly correlated with the transcript levels of the metallo-chaperone CCS (At1g12520). The small CCS protein delivers Cu to chloroplasts and controls integration of copper into Csd (Abdel-Ghany et al. 2005; Chu et al. 2005). *Arabidopsis thaliana* encodes one CCS gene. The transcript has two in-frame ATG-start sites, which give either a cytosolic CCS or a CCS-preprotein with an organellar targeting signal (Chu et al. 2005). In 10 out of 13 tested 5′-RACE products, the transcript started only one nucleotide upstream of the first ATG suggesting transcript variability. One nucleotide upstream of an ATG may be too short for ribosome binding (Kozak 1991), thus, avoid translation of the chloroplast targeting signal. Without detailed experimental proofing, Chu et al (2005) and Wintz and Vulpe (2002) concordantly suggest alternative transcription initiation.

From 125 stress array data sets of *Arabidopsis thaliana* generated in different labs, a Pearson correlation coefficient of 0.99 was calculated for the relative transcript abundances of Csd1/Csd2 and CCS (M. Baier, unpublished data). This strong co-regulation indicates fine-controlled transcriptional and/or post-transcriptional regulation. However, computational mRNA analysis did not show a miR398 target element in *Arabidopsis*, rice and poplar (Bonnet et al. 2004; Jones-Rhoades and Bartel 2004; Sunkar et al. 2005; Fahlgren et al. 2007). In an experimentally validated approach, Cohu et al (2009) recently identified a 21-nt sequence stretch with high similarity to miR398 in the CCS transcript. The hypothesis on post-transcriptional regulation of CCS by miR398 was confirmed by comparison of CCS transcript levels in constitutive miR398 over-expressor lines and miR398a and miR398c double knockout lines (Cohu et al. 2009). In addition, Cohu et al (2009) showed that the levels of Csd1 and Csd2 are strongly decreased in CCS-knockout lines, which suggests additional post-transcriptional cross-talk.

4.3 The Cu Regulation of Csd1 and Csd2 Expression Is Affected by Plastocyanin Expression

Since plastocyanin is the most prominent Cu-binding protein in plants, the copper homeostasis is altered in plastocyanin-deficient mutants. In response, the mutants accumulate Csd1, Csd2 and CCS (Abdel-Ghany 2009) demonstrating a strong Cu-mediated link between Csd1, Csd2 and plastocyanin expression.

Plastocyanin is an electron carrier protein at the luminal site of thylakoids. It transfers electrons from the cytochrome b_6f complex to photosystem I. According to Schöttler et al. (2004), the concentration of plastocyanin controls the electron transfer efficiency and, in consequence, the assimilation capacity of photosynthesis. If the photosynthetic electron transfer efficiency decreases, more activation energy and electrons can escape from the photosynthetic electron transport chain and activate oxygen (Mehler 1951; Fridovich 1995). Thus, regulation of Csd1 and Csd2 expression by the plastocyanin-controlled Cu availability links protection against superoxide accumulation to photosynthetic electron transport efficiency and photooxidative superoxide formation. In summary, Cu-controlled miRNA regulation provides an ideal system to control Csd availability in chloroplasts and in the cytosol relative to the risks caused by photooxidative ROS generation.

4.4 Regulation of Fsd Expression Is Controlled by Redox Signals, Its Activity by Heterocomplex Formation

Various authors report that stromal FeSOD activity is proportionally up-regulated if Csd1 and Csd2 are suppressed under Cu-limited conditions (Abdel-Ghany et al. 2005; Cohu and Pilon 2007; Yamasaki et al. 2007). Since no miRNA mechanism similar to Csd1 and Csd2 regulation can be assumed for Fsd regulation, one may follow the suggestion by Cohu and Pilon (2007) that Fsd expression is up-regulated to replace Csd activity. In other words, this reciprocal regulation of Csd and Fsd expression indicates that Fsd expression is under control of Csd expression. So far, no molecular mechanisms have described. Since Cohu et al. (2009) recently showed co-suppression of Csd1, Csd2 and Fsd1 in CCS-knockout lines, a direct link between Csd and Fsd1 expression can be excluded. Most likely, Fsd1 responds to oxidative stress signals, such as signals related to ROS accumulation (Apel and Hirt 2004), and, consequently, more sensitive to the redox status and the concentration of ascorbate and glutathione than Csd expression. Comparison of transgenic with increased Zat10 levels (Mittler et al. 2006), suggests regulation of Fsd1 via the MAPK4-Zat-pathway depicted in Fig. 5.

Decreased transcript levels of chloroplast Fsd2 upon photooxidative stress (Kliebenstein et al. 1998) and reciprocal regulation of Fsd1 and Fsd2 suggest additional levels of regulation controlling the SOD activity. Unfortunately, Kliebenstein excluded Fsd3 from their gene-specific expression analysis, since the protein did not cluster with other plant Fsd2 in their phylogenetic analysis. Recent studies demonstrated that Fsd2 and Fsd3 proteins form hetero-complexes in chloroplast nucleosides (Myouga et al. 2008). Transgenic Arabidopsis over-expressing Fsd2 and Fsd3 were more tolerant to oxidative stress than single Fsd-over-expressing lines, although the single proteins retained SOD activity on their own. In Fsd2 over-expressors, the Fsd3 transcripts were decreased, in Fsd3 over-expressors Fsd2 transcripts and in Csd2 over-expressors Fsd2 and Fsd3 indicating that Fsd2 and Fsd3 transcript abundance is negatively correlated with superoxide dismutase activity. This correlation is unidirectional, since over-expression of Fsd2 and Fsd3 resulted in higher Csd2 transcript levels.

The transcript levels of Fsd2 and Fsd3 rapidly accumulated in response to methyl viologen, which promotes superoxide formation at the thylakoid membrane (Fujibe et al. 2004), and other photooxidative stress promoting treatments (Myouga et al. 2008). It is indicated that the two genes are most likely co-regulated by redox singals. Various signal transduction pathways have been postulated for chloroplast-to-nucleus redox signaling (Baier and Dietz 2005; Pfannschmidt et al. 2007; Kleine et al. 2009). However, the precise molecular mechanisms actually regulating Fsd2 and Fsd3 expression still have to be investigated. Hetero-complex formation of Fsd2 and Fsd3 proteins demands for co-regulation and suggests complexities including post-transcriptional and post-translational regulation.

5 Redox-Regulation of Protein Import into Chloroplasts

Compared to extra-organellar antioxidant enzymes, the chloroplast antioxidant enzymes have to pass the two envelope membranes prior to comprising their activity. Compartmentalization and availability of chloroplast antioxidant enzymes can be regulated by post-translational processes (Section 3.2.1), by transcriptional and, like Csd1 and Csd2 (Section 4.1), by post-transcriptional regulation.

So far little attention has been paid on specific regulation of organellar import of the antioxidant enzymes within the group of approximately 3,000 nuclear encoded chloroplast proteins (Leister and Schneider 2003). Various alternative cytosolic factors, such as the differentiation of 14-3-3/Hsp70 (May and Soll 2000) and Hsp90 guidance complexes (Qbadou et al. 2006), signal recognition particles which bind mRNA to translation-import complexes (St Johnston 2005), and vesicle based protein sorting (Nanjo et al. 2006) can interact with pre-proteins in the cytosol, stabilize them and guide them into distinct protein import pathways. The variability implicates that protein import can be differentially regulated and indicates a function of protein import in inter-compartment signaling pathways (Kessler and Schnell 2009). Recently, Kakizaki et al. (2009) demonstrated with protein-import affected Arabidopsis lines that nuclear gene expression is coupled to plastid protein import. Defects in the protein import machinery (caused by silencing of TOC159) resulted in down-regulation of the nuclear expression of photosynthesis-related genes. Based on transcriptome analysis, they suggested that the G-box-binding factor GLK1 may be involved in coordination of gene expression. Consistent with their assumption, over-expression of GLK1 partially restored the gene expression defect in TOC159-silenced lines.

With respect of antioxidant function, redox regulation of protein import may be an important regulatory mechanism. Chloroplast protein import takes place via two translocon machineries present in the outer (OEM) and inner envelope membrane (IEM). They are dedicated TOC (translocon at the outer envelope of chloroplasts) and TIC (translocon at the inner envelope of chloroplasts) (Inaba and Schnell 2008; Jarvis 2008; Stengel et al. 2008; Benz et al. 2009). For most chloroplast proteins, pre-proteins with a N-terminal transit peptide are translated at cytosolic ribosomes. The transit peptides are recognized by receptor proteins on the organellar surface of the outer chloroplast envelope. Via TIC and TOC, the proteins are translocated through both envelope membranes. At the stromal site of the inner chloroplast envelope, a stromal processing peptidase (SPP) cleaves the transit peptide off, thereby generating the mature form of the protein (Bruce 2000).

Translocation across the envelope is a bottleneck in the transport process (Benz et al. 2009). For correct supply of chloroplasts with proteins, protein traffic needs to be tightly regulated. The TOC and TIC translocons impose a regulatory control of incoming pre-proteins. Members of the TIC complex are integrated in the inner chloroplast membrane and are, therefore, closest to the origin of the signal. Via connection to TOC, TIC can transmit the signals to the outer membrane and control channel activity of the TIC–TOC-complex.

In the TIC complex, eight proteins have been implicated with respect to pre-protein import at the IEM of higher plant chloroplasts. For each component a contact with the precursor proteins or a close interaction with one of the established TIC core proteins has been demonstrated (Benz et al. 2009). It is postulated that Tic110 and Tic20 form the translocon channel (Chen et al. 2002; Heins et al. 2002; Teng et al. 2006). The heat shock protein 93 (Hsp93)/caseinolytic protease C

(ClpC) is a putative constituent of the import motor (Chou et al. 2003). Furthermore, the co-chaperone Tic40 is part of the complex (Chou et al. 2003). Tic32, Tic55 and Tic62 are translocon-associated, while Tic22 is an associated soluble protein of inter-membrane space and might participate in TIC–TOC supercomplex formation by interaction with TOC proteins (Balsera et al. 2009a) (Fig. 9). Tic21 is the most recently added component of the TIC (Teng et al. 2006).

TIC proteins sense at least two types of signals, which may be important for linking import of chloroplast antioxidant enzymes to the chloroplast redox state: (a) redox-signals, mediated by the stromal $NADP^+$/NADPH ratio (and possibly also thioredoxins (Bartsch et al. 2008) and glutathione (Oreb et al. 2008), might provide information about the metabolic state of the chloroplast (Chigri et al. 2006; Stengel et al. 2008) and (b) calcium signals, which are controlled by chloroplast calmodulin (CaM) (Chigri et al. 2006) and can be perceived as second messenger in redox regulation (Gomez et al. 2004). Up to now four putative targets have been identified for redox-regulation of protein import: Tic62, Tic32, Tic55 and recently Tic110 (Fig. 9a and b):

1. Tic62 and Tic32 are classified as short-chain dehydrogenases/reductases (SDRs). They are peripherally attached to the stromal side of the inner envelope and share a NADP(H)-binding site. Tic62 can interact with ferredoxin-NADP-oxidoreductase (FNR) (Kuchler et al. 2002; Stengel et al. 2008), which transfers redox power from reduced ferredoxin to $NADP^+$. Consequently, Tic62 represents a good candidate for being the $NADP^+$/NADPH sensor of the TIC–TOC-protein import machinery.

   Tic32 comprises a calmodulin-binding domain (Hormann et al. 2009), indicating sensing of Ca^{2+}-signals. If the $NADP^+$/NADPH ratio shifts to more oxidized values, Tic32 and Tic62 are closely associated with TIC, while reduced conditions lead to translocon disassembly (Chigri et al. 2006; Stengel et al. 2008). Tic62 also shifts between a soluble stromal and a membrane-attached form, depending on the chloroplast redox state (Stengel et al. 2008). Since the soluble form of Tic62 binds NADPH and attaches more strongly to FNR (Balsera et al. 2007), Tic62 regulation is likely to support protein up-take under oxidizing conditions. Under these conditions, better import of antioxidant enzymes may support protection against photooxidative damage.
2. Tic55 is anchored to the inner envelope membrane and exposes a large soluble domain to the stroma. This domain contains a Rieske-type iron–sulfur centre and a mononuclear iron-binding site (Caliebe et al. 1997). The iron–sulfur centre has been suggested to serve as a sensor for photosynthesis-coupled redox-active compounds like glutathione or thioredoxin (Oreb et al. 2008). In fact, it has recently been identified in a screen for thioredoxin targets (Bartsch et al. 2008). Thus, translocation could also be linked to thiol-mediated regulative processes (Balsera et al. 2009b).
3. The channel protein Tic110 has six conserved cysteines that potentially could be redox-controlled by stromal thioredoxins (Balsera et al. 2009a). Besides this, the N-terminal region of Tic110 interacts with the redox-regulated subunit Tic32

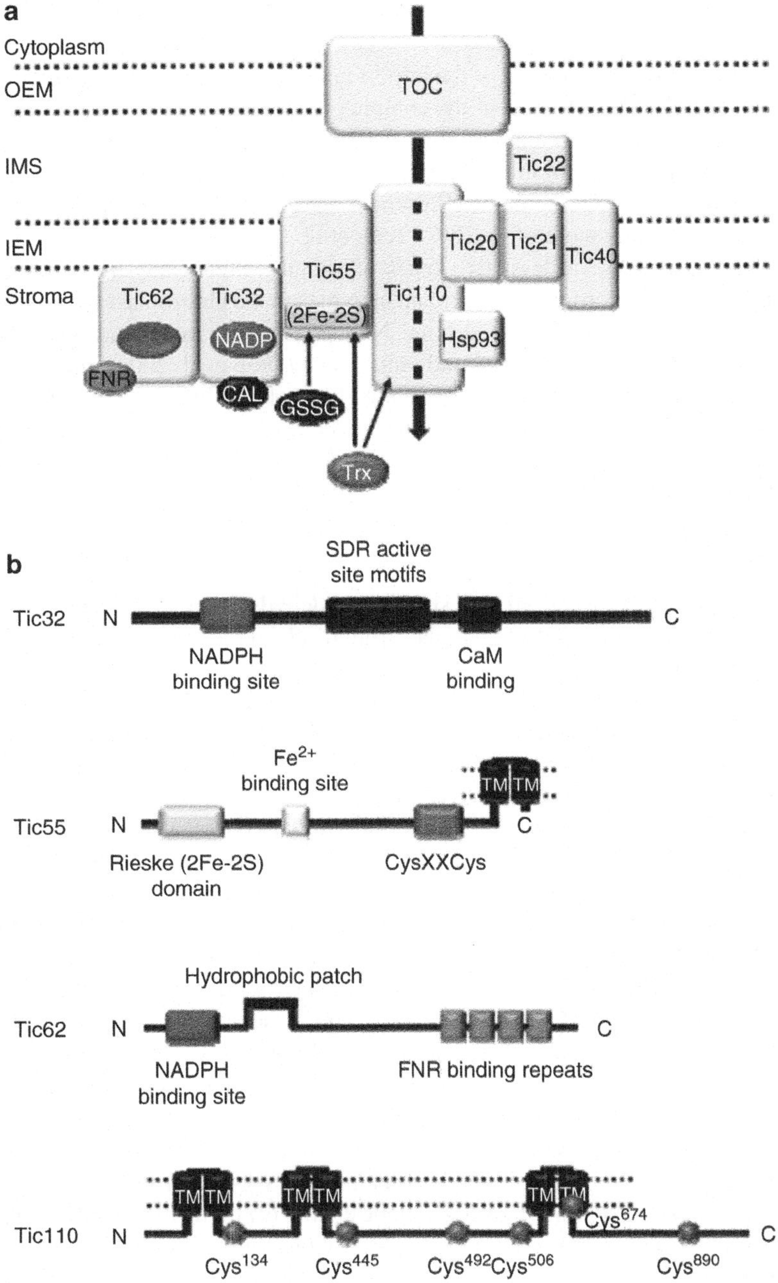

Fig. 9 Redox regulation of protein import. (**a**) The redox-regulated subunits of the Tic translocon. Under oxidizing conditions, FNR assembles with Tic62 activating the translocon. Oxidized glutathione (GSSG) and thioredoxin are postulated to influence the iron–sulfur cluster of Tic55. Under reducing conditions, NADPH induces the release of calmodulin (Cal) from Tic32 and the release of Tic32 and possibly Tic62 from the translocation complex and finally of the translocon. Adapted from Oreb et al. (2008). (**b**) Schematic overview showing predicted functional domains and topology of Tic32, Tic55, Tic62 and Tic110. Transmembrane domains are marked with TM. Regions involved in NADPH binding are indicated in *red*. Cysteine residues possibly involved in regulation by thioredoxins are indicated in *brown* (Adapted from Benz et al. 2009)

(Hormann et al. 2009). For at least two chloroplast proteins, the non-photosynthetic ferredoxin (Fd3) and the FNR isoform 2 (FNR2), differentially regulated import has been shown so far (Hirohashi et al. 2001), which may be depend on different redox states caused by photosynthetic activity (Balsera et al. 2009b). Consequently, in the light the unprocessed precursors of Fd3 and FNR2 were miss-localized to the inter-membrane space for chloroplast envelopes. In the dark, they were efficiently imported into the stroma and processed to its mature form.

As shown by application of different types of disulfide-reducing agents (Stengel et al. 2009), also the formation and reduction of intermolecular disulfide bridges in the TOC receptors and TOC translocation channel strongly influence the import of pre-proteins. Redox regulation of the chloroplast protein import machinery might be a final control step in the expressional control of chloroplast antioxidant and redox-active enzymes. In the green algae *Chlamydomonas reinhardtii*, import control is partially overcome by coupling protein import into chloroplasts to translation (localized translation control; Uniacke and Zerges 2009). With respect to chloroplast antioxidant enzymes, translational import control can be assumed for the 2-Cys peroxiredoxin genes and organellar type-II peroxiredoxin (N.T. Pitsch, B. Witsch, M. Baier (2010) submitted).

With special attention on chloroplast antioxidant enzymes and cellular redox regulation, the future biochemical analysis of the protein import machinery in context of the cellular redox metabolism will show more refined pattern, which adjust chloroplast redox control function to the dangers of photooxidative stress. Additional control mechanisms can be expected from analysis of translational and post-translational regulation. In theory, separation of translation and function between chloroplasts and the cytosol provides a high potential to discover additional regulatory mechanisms, which support coordination of extra-plastidic gene expression to chloroplast function.

6 Conclusions and Perspective

Transcriptomics and specific analyses of gene expression regulation by studying transcript abundance regulation and promoter activities demonstrated that individual regulatory mechanisms have evolved to regulate expression of genes encoding chloroplast and extra-plastid antioxidant enzymes. Exemplarily, the regulatory pathways have been studied in detail for individual genes. The most advanced studies arose from characterization of the ROS-responsive ZAT10/12-pathway regulating expression of cytosolic APx1 and APx2 expression under control of the MAPK-signaling cascade (Kovtun et al. 2000; Mittler et al. 2006) and from RAp2.4a – ABA-antioxidant regulation of chloroplast 2-Cys peroxiredoxin-A by the acceptor availability at photosystem I (Baier et al. 2004). The regulation amplitudes differ by a factor of more than 100 and the preferred directions of regulation are inverted. Links, such as the co-activation by MAPK3/MAPK6 (Yang et al. 2003; Kovtun et al. 2000), demonstrate coordination. In contrast, distinct subsequent signaling pathways, such as ZAT10/ZAT12 and Rap2.4a-dependent regulation allow individual responsiveness, such as the induction of APx2, but suppression of 2CPA in Rap2.4a knockout lines (Shaikhali et al. 2008). Recent analysis of superoxide dismutase expression (Section 4) and NDPK2-regulation (Section 3.3.2) demonstrated that regulation of the chloroplast and extra-plastidic antioxidant defense systems shares transcriptional and post-transcriptional links. The exemplary studies summarized in this review let assume a complex regulatory network, in which master switches are controlled by gene-specific regulatory mechanisms on all levels of gene expression regulation.

Further analysis of signaling cascades and detailed characterization of promoter regulation of additional genes will provide the basis for exploring this signal transduction networks. The examples described here let assume, that we are just at the beginning of understanding the mechanisms which act in concert to protect plants from oxidative stress. There still is a highly complex stet of undiscovered signaling and regulatory pathways controlling the expression of the extra-plastid and plastid antioxidant defense system.

References

Abdel-Ghany SE (2009) Contribution of plastocyanin isoforms to photosynthesis and copper homeostasis in *Arabidopsis thaliana* grown at different copper regimes. Planta 229:767–779

Abdel-Ghany SE, Burkhead JL, Gogolin KA, Andres-Colas N, Bodecker JR, Puig S, Penarrubia L, Pilon M (2005) AtCCS is a functional homolog of the yeast copper chaperone Ccs1/Lys7. FEBS Lett 579:2307–2312

Abe H, Yamaguchi-Shinozaki K, Urao T, Iwasaki T, Hosokawa D, Shinozaki K (1997) Role of *Arabidopsis* MYC and MYB homologs in drought- and abscisic acid-regulated gene expression. Plant Cell 9:1859–1868

Ahlfors R, Lang S, Overmyer K, Jaspers P, Brosche M, Taurianinen A, Kollist H, Tuominen H, Belles-Boix E, Piippo M, Inze M, Palva ET, Kangasjarvi J (2004) *Arabidopsis* radical-induced cell death1 belongs to the WWE protein-protein interaction domain protein family and modulates abscisic acid, ethylene, and methyl jasmonate responses. Plant Cell 16:1925–1937

Alscher RG, Erturk N, Heath LS (2002) Role of superoxide dismutases (SODs) in controlling oxidative stress in plants. J Exp Bot 53:1331–1341

Anthony RG, Henriques R, Helfer A, Meszaros T, Rios G, Testerink C, Munnik T, Deak M, Koncz C, Bogre L (2004) A protein kinase target of a PDK1 signalling pathway is involved in root hair growth in Arabidopsis. EMBO J 23:572–581

Aono M, Saji H, Kondo N, Tanaka K (1997) Paraquat tolerance of transgenic tobacco plants with altered activity of glutathione reductase. Phyton-Annal Rei Botanicae 37:13–18

Apel K, Hirt H (2004) Reactive oxygen species: metabolism, oxidative stress, and signal transduction. Annu Rev Plant Biol 55:373–379

Aravind L, Koonin EV (1999) Fold prediction and evolutionary analysis of the POZ domain: structural and evolutionary relationship with the potassium channel tetramerization domain. J Mol Biol 285:1353–1361

Arisi AC, Noctor G, Foyer CH, Jouanin L (1997) Modification of thiol contents in poplars (*Populus tremula x P. alba*) overexpressing enzymes involved in glutathione synthesis. Planta 203:362–372

Asada K (2000) The water-water cycle as alternative photon and electron sinks. Philos Trans R Soc Lond B Biol Sci 355:1419–1431

Asai T, Tena G, Plotnikova J, Willmann MR, Chiu WL, Gomez-Gomez L, Boller T, Ausubel FM, Sheen J (2002) MAP kinase signalling cascade in Arabidopsis innate immunity. Nature 415:977–983

Aslund F, Berndt KD, Holmgren A (1997) Redox potentials of glutaredoxins and other thiol-disulfide oxidoreductases of the thioredoxin superfamily determined by direct protein-protein redox equilibria. J Biol Chem 272:30780–30786

Babajani G, Effendy J, Plant AL (2009) Sl-SRO(l)under-bar1 increases salt tolerance and is a member of the radical-induced cell death 1-similar to RCD1 gene family of tomato. Plant Sci 176:214–222

Baier M (1993) Zelltyp- und organspezifische Genexpression in Gerste. Universität Würzburg Julius-von-Sachs-Institut, Institut für Botanik, Germany

Baier M (1997) Das 2-Cys-Peroxiredoxin BAS1: Klonierung, Charakterisierung und Funktionsanalyse einer neuartigen, präferentiell in der Zellteilungs- und Zellstreckungszone des Gerstenblatts exprimierten pflanzlichen Peroxidase. Dissertation. Universität Würzburg, Würzburg

Baier M, Dietz KJ (2005) Chloroplasts as source and target of cellular redox regulation: a discussion on chloroplast redox signals in the context of plant physiology. J Exp Bot 56:1449–1462

Baier M, Dietz KJ (1996) Primary structure and expression of plant homologues of animal and fungal thioredoxin-dependent peroxide reductases and bacterial alkyl hydroperoxide reductases. Plant Mol Biol 31:553–564

Baier M, Dietz KJ (1997) The plant 2-Cys peroxiredoxin BAS1 is a nuclear encoded chloroplast protein: its expressional regulation, phylogenetic origin, and implications for its specific physiological function in plants. Plant J 12:179–190

Baier M, Dietz KJ (1999) The costs and benefits of oxygen for photosynthesizing plant cells. Prog Bot 60:282–314

Baier M, Noctor G, Foyer CH, Dietz KJ (2000) Antisense suppression of 2-cysteine peroxiredoxin in Arabidopsis specifically enhances the activities and expression of enzymes associated with ascorbate metabolism but not glutathione metabolism. Plant Physiol 124:823–832

Baier M, Ströher E, Dietz KJ (2004) The acceptor availability at photosystem I and ABA control nuclear expression of 2-Cys peroxiredoxin-A in *Arabidopsis thaliana*. Plant Cell Physiol 45:997–1006

Baker CJ, Orlandi EW, Deahl KL (2000) Oxygen metabolism in plant/bacteria interactions: characterization of the oxygen uptake response of plant suspension cells. Physiol Mol Plant Pathol 57:159–167

Ball L, Accotto GP, Bechtold U, Creissen G, Funck D, Jimenez A, Kular B, Leyland N, Mejia-Carranza J, Reynolds H, Karpinski S, Mullineaux PM (2004) Evidence for a direct link between glutathione biosynthesis and stress defense gene expression in Arabidopsis. Plant Cell 16:2448–2462

Balsera M, Goetze TA, Kovacs-Bogdan E, Schurmann P, Wagner R, Buchanan BB, Soll J, Bolter B (2009a) Characterization of Tic110, a channel-forming protein at the inner envelope membrane of chloroplasts, unveils a response to Ca^{2+} and a stromal regulatory disulfide Bridge. J Biol Chem 284:2603–2616

Balsera M, Soll J, Bolter B (2009b) Protein import machineries in endosymbiotic organelles. Cell Mol Life Sci 66:1903–1923

Balsera M, Stengel A, Soll J, Bolter B (2007) Tic62: a protein family from metabolism to protein translocation. BMC Evolutionary Biology 7. doi:10.1186/1471-2148-7-43

Baniwal SK, Chan KY, Scharf KD, Nover L (2007) Role of heat stress transcription factor HsfA5 as specific repressor of HsfA4. J Biol Chem 282:3605–3613

Bartel DP, Chen CZ (2004) Micromanagers of gene expression: the potentially widespread influence of metazoan microRNAs. Nat Rev Genet 5:396–400

Bartsch S, Monnet J, Selbach K, Quigley F, Gray J, von Wettstein D, Reinbothe S, Reinbothe C (2008) Three thioredoxin targets in the inner envelope membrane of chloroplasts function in protein import and chlorophyll metabolism. Proc Nat Acad Sci USA 105:4933–4938

Benz JP, Soll J, Bolter B (2009) Protein transport in organelles: the composition, function and regulation of the Tic complex in chloroplast protein import. FEBS J 276:1166–1176

Bernstein E, Caudy AA, Hammond SM, Hannon GJ (2001) Role for a bidentate ribonuclease in the initiation step of RNA interference. Nature 409:363–366

Bhattacharjee S (2005) Reactive oxygen species and oxidative burst: roles in stress, senescence and signal transduction in plants. Curr Sci 89:1113–1121

Biehl A, Richly E, Noutsos C, Salamini F, Leister D (2005) Analysis of 101 nuclear transcriptomes reveals 23 distinct regulons and their relationship to metabolism, chromosomal gene distribution and co-ordination of nuclear and plastid gene expression. Gene 344:33–41

Bonnet E, Wuyts J, Rouze P, Van de Peer Y (2004) Detection of 91 potential in plant conserved plant microRNAs in *Arabidopsis thaliana* and *Oryza sativa* identifies important target genes. Proc Natl Acad Sci USA 101:11511–11516

Borsani O, Zhu JH, Verslues PE, Sunkar R, Zhu JK (2005) Endogenous siRNAs derived from a pair of natural *cis*-antisense transcripts regulate salt tolerance in Arabidopsis. Cell 123:1279–91

Bowler C, Alliotte T, De Loose M, Van Montagu M, Inze D (1989) The induction of manganese superoxide dismutase in response to stress in *Nicotiana plumbaginifolia*. EMBO J 8:31–38

Bowler C, Slooten L, Vandenbranden S, De Rycke R, Botterman J, Sybesma C, Van Montagu M, Inze D (1991) Manganese superoxide dismutase can reduce cellular damage mediated by oxygen radicals in transgenic plants. EMBO J 10:1723–1732

Bruce BD (2000) Chloroplast transit peptides: structure, function and evolution. Trends Cell Biol 10:440–447

Caliebe A, Grimm R, Kaiser G, Lubeck J, Soll J, Heins L (1997) The chloroplastic protein import machinery contains a Rieske-type iron-sulfur cluster and a mononuclear iron-binding protein. EMBO J 16:7342–7350

Cao H, Bowling SA, Gordon AS, Dong XN (1994) Characterization of an Arabidopsis mutant that is nonresponsive to inducers of systemic acquired-resistance. Plant Cell 6:1583–1592

Cao H, Glazebrook J, Clarke JD, Volko S, Dong XN (1997) The Arabidopsis NPR1 gene that controls systemic acquired resistance encodes a novel protein containing ankyrin repeats. Cell 88:57–63

Carrington JC, Ambros V (2003) Role of microRNAs in plant and animal development. Science 301:336–338

Chan CS, Guo L, Shih MC (2001) Promoter analysis of the nuclear gene encoding the chloroplast glyceraldehyde-3-phosphate dehydrogenase B subunit of *Arabidopsis thaliana*. Plant Mol Biol 46:131–1341

Chang CC, Slesak I, Jorda L, Sotnikov A, Melzer M, Miszalski Z, Mullineaux PM, Parker JE, Karpinska B, Karpinski S (2009) Arabidopsis chloroplastic glutathione peroxidases play a role in cross talk between photooxidative stress and immune responses. Plant Physiol 150:670–683

Chen XJ, Smith MD, Fitzpatrick L, Schnell DJ (2002) *In vivo* analysis of the role of atTic20 in protein import into chloroplasts. Plant Cell 14:641–654

Chen Z, Young TE, Ling J, Chang SC, Gallie DR (2003) Increasing vitamin C content of plants through enhanced ascorbate recycling. Proc Natl Acad Sci USA 100:3525–3530

Chew O, Whelan J, Millar AH (2003) Molecular definition of the ascorbate-glutathione cycle in Arabidopsis mitochondria reveals dual targeting of antioxidant defenses in plants. J Biol Chem 278:46869–46877

Chigri F, Hormann F, Stamp A, Stammers DK, Bolter B, Soll J, Vothknecht UC (2006) Calcium regulation of chloroplast protein translocation is mediated by calmodulin binding to Tic32. Proc Natl Acad Sci USA 103:16051–16056

Choi HW, Kim YJ, Lee SC, Hong JK, Hwang BK (2007) Hydrogen peroxide generation by the pepper extracellular peroxidase $CaPO_2$ activates local and systemic cell death and defense response to bacterial pathogens. Plant Physiol 145:890–904

Chou ML, Fitzpatrick LM, Tu SL, Budziszewski G, Potter-Lewis S, Akita M, Levin JZ, Keegstra K, Li HM (2003) Tic40, a membrane-anchored co-chaperone homolog in the chloroplast protein translocon. EMBO J 22:2970–29780

Chu CC, Lee WC, Guo WY, Pan SM, Chen LJ, Li HM, Jinn TL (2005) A copper chaperone for superoxide dismutase that confers three types of copper/zinc superoxide dismutase activity in Arabidopsis. Plant Physiol 139:425–436

Ciftci-Yilmaz S, Mittler R (2008) The zinc finger network of plants. Cell Mol Life Sci 65:1150–1160

Ciftci-Yilmaz S, Morsy MR, Song LH, Coutu A, Krizek BA, Lewis MW, Warren D, Cushman J, Connolly EL, Mittler R (2007) The EAR-motif of the Cys2/His2-type zinc finger protein Zat7 plays a key role in the defense response of Arabidopsis to salinity stress. J Biol Chem 282:9260–9268

Cohu CM, Pilon M (2007) Regulation of superoxide dismutase expression by copper availability. Physiol Plant 129:747–755

Cohu CM, Abdel-Ghany SE, Reynolds KAG, Onofrio AM, Bodecker JR, Kimbrel JA, Niyogi KK, Pilon M (2009) Copper delivery by the copper chaperone for chloroplast and cytosolic copper/zinc-cuperoxide dismutases: regulation and unexpected phenotypes in an Arabidopsis mutants. Mol Plant 2:1336–1350

Collin V, Issakidis-Bourguet E, Marchand C, Hirasawa M, Lancelin JM, Knaff DB, Miginiac-Maslow M (2003) The Arabidopsis plastidial thioredoxins: new functions and new insights into specificity. J Biol Chem 278:23747–23752

Cona A, Cenci F, Cervelli M, Federico R, Mariottini P, Moreno S, Angelini R (2003) Polyamine oxidase, a hydrogen peroxide-producing enzyme, is up-regulated by light and down-regulated by auxin in the outer tissues of the maize mesocotyl. Plant Physiol 131:803–813

Conklin PL, Norris SR, Wheeler GL, Williams EH, Smirnoff N, Last RL (1999) Genetic evidence for the role of GDP-mannose in plant ascorbic acid (vitamin C) biosynthesis. Proc Natl Acad Sci USA 96:4198–4203

Czarnecka-Verner E, Pan SQ, Salem T, Gurley WB (2004) Plant class BHSFs inhibit transcription and exhibit affinity for TFIIB and TBP. Plant Mol Biol 56:57–75

Czarnecka-Verner E, Yuan CX, Scharf KD, Englich G, Gurley WB (2000) Plants contain a novel multi-member class of heat shock factors without transcriptional activator potential. Plant Mol Biol 43:459–471

Davletova S, Schlauch K, Coutu J, Mittler R (2005) The zinc-finger protein Zat12 plays a central role in reactive oxygen and abiotic stress signaling in Arabidopsis. Plant Physiol 139:847–856

Dayer R, Fischer BB, Eggen RI, Lemaire SD (2008) The peroxiredoxin and glutathione peroxidase families in *Chlamydomonas reinhardtii*. Genetics 179:41–57

Despres C, Chubak C, Rochon A, Clark R, Bethune T, Desveaux D, Fobert PR (2003) The Arabidopsis NPR1 disease resistance protein is a novel cofactor that confers redox regulation of DNA binding activity to the basic domain/leucine zipper transcription factor TGA1. Plant Cell 15:2181–2191

Despres C, DeLong C, Glaze S, Liu E, Fobert PR (2000) The Arabidopsis NPR1/NIM1 protein enhances the DNA binding activity of a subgroup of the TGA family of bZIP transcription factors. Plant Cell 12:279–290

Dietz KJ, Horling F, Konig J, Baier M (2002) The function of the chloroplast 2-cysteine peroxiredoxin in peroxide detoxification and its regulation. J Exp Bot 53:1321–1329

Donald RG, Cashmore AR (1990) Mutation of either G box or I box sequences profoundly affects expression from the Arabidopsis rbcS-1A promoter. EMBO J 9:1717–1726

Dugas DV, Bartel B (2008) Sucrose induction of Arabidopsis miR398 represses two Cu/Zn superoxide dismutases. Plant Mol Biol 67:403–417

Elstner EF (1990) Der Sauerstoff: biochemie, biologie. Medizin. BI-Wiss.-Verl, Mannheim, Wien, Zürich

Englbrecht CC, Schoof H, Bohm S (2004) Conservation, diversification and expansion of C2H2 zinc finger proteins in the *Arabidopsis thaliana* genome. BMC Genomics 5(1):39

Eshdat Y, Holland D, Faltin Z, BenHayyim G (1997) Plant glutathione peroxidases. Physiol Plant 100:234–240

Eulgem T, Rushton PJ, Robatzek S, Somssich IE (2000) The WRKY superfamily of plant transcription factors. Trens Plant Sci 5:199–206

Eulgem T, Somssich IE (2007) Networks of WRKY transcription factors in defense signaling. Curr Opin Plant Biol 10:366–371

Fan WH, Dong XN (2002) *In vivo* interaction between NPR1 and transcription factor TGA2 leads to salicylic acid-mediated gene activation in Arabidopsis. Plant Cell 14:1377–1389

Fahlgren N, Howell MD, Kasschau KD, Chapman EJ, Sullivan CM, Cumbie JS, Givan SA, SA LTF, Grant SR, Dangl JL, Carrington JC (2007) High-throughput sequencing of *Arabidopsis* microRNAs: evidence for frequent birth and death of MIRNA genes. PLoS ONE 2:e219

Finkemeier I, Goodman M, Lamkemeyer P, Kandlbinder A, Sweetlove LJ, Dietz KJ (2005) The mitochondrial type II peroxiredoxin F is essential for redox homeostasis and root growth of *Arabidopsis thaliana* under stress. J Biol Chem 280:12168–12180

Fita I, Rossmann MG (1985) The active-center of catalase. J Mol Biol 185:21–37

Foyer CH, Halliwell B (1977) Presence of glutathione and glutathione reductase in chloroplasts - proposed role in ascorbic-acid metabolism. Planta 133:21–25

Foyer CH, Lelandais M, Kunert KJ (1994) Photooxidative stress in plants. Physiol Plant 92:696–717

Fridovich I (1995) Superoxide radical and superoxide dismutases. Ann Rev Biochem 64:97–112

Fryer MJ, Ball L, Oxborough K, Karpinski S, Mullineaux PM, Baker NR (2003) Control of ascorbate peroxidase 2 expression by hydrogen peroxide and leaf water status during excess light stress reveals a functional organisation of Arabidopsis leaves. Plant J 33:691–705

Fryer MJ, Oxborough K, Mullineaux PM, Baker NR (2002) Imaging of photo-oxidative stress responses in leaves. J Exp Bot 53:1249–1254

Fujibe T, Saji H, Arakawa K, Yabe N, Takeuchi Y, Yamamoto KT (2004) A methyl viologen-resistant mutant of Arabidopsis, which is allelic to ozone-sensitive rcd1, is tolerant to supplemental ultraviolet-B irradiation. Plant Physiol 134:275–285

Fukamatsu Y, Yabe N, Hasunuma K (2003) Arabidopsis NDK1 is a component of ROS signaling by interacting with three catalases. Plant Cell Physiol 44:982–929

Gadjev I, Vanderauwera S, Gechev TS, Laloi C, Minkov IN, Shulaev V, Apel K, Inze D, Mittler R, Van Breusegem F (2006) Transcriptomic footprints disclose specificity of reactive oxygen species signaling in Arabidopsis. Plant Physiol 141:436–445

Gao CJ, Xing D, Li LL, Zhang LR (2008) Implication of reactive oxygen species and mitochondrial dysfunction in the early stages of plant programmed cell death induced by ultraviolet-C overexposure. Planta 227:755–767

Garreton V, Carpinelli J, Jordana X, Holuigue L (2002) The as-1 promoter element is an oxidative stress-responsive element and salicylic acid activates it via oxidative species. Plant Physiol 130:1516–1526

Gomez LD, Noctor G, Knight MR, Foyer CH (2004) Regulation of calcium signalling and gene expression by glutathione. J Exp Bot 55:1851–1859

Gou P, Hanke GT, Kimata-Ariga Y, Standley DM, Kubo A, Taniguchi I, Nakamura H, Hase T (2006) Higher order structure contributes to specific differences in redox potential and electron transfer efficiency of root and leaf ferredoxins. Biochemistry 45:14389–14396

Guan LQ, Polidoros AN, Scandalios JG (1996) Isolation, characterization and expression of the maize Cat2 catalase gene. Plant Mol Biol 30:913–924

Guan LQM, Zhao J, Scandalios JG (2000) *Cis*-elements and *trans*-factors that regulate expression of the maize Cat1 antioxidant gene in response to ABA and osmotic stress: H_2O_2 is the likely intermediary signaling molecule for the response. Plant J 22:87–95

Halliwell B (1994) How to characterize an antioxidant: an update. Biochem Soc Symp 61:73–101

Hannon GJ (2002) RNA interference. Nature 418:244–251

Hansch R, Lang C, Riebeseel E, Lindigkeit R, Gessler A, Rennenberg H, Mendel RR (2006) Plant sulfite oxidase as novel producer of H_2O_2. Combination of enzyme catalysis with a subsequent non-enzymatic reaction step. J Biol Chem 281:6884–6888

He L, Hannon GJ (2004) MicroRNAs: small RNAs with a big role in gene regulation. Nat Rev Genet 5:522–531

Heiber I, Ströher E, Raatz B, Busse I, Kahmann U, Bevan MW, Dietz KJ, Baier M (2007) The redox imbalanced mutants of Arabidopsis differentiate signaling pathways for redox regulation of chloroplast antioxidant enzymes. Plant Physiol 143:1774–1788

Heins L, Mehrle A, Hemmler R, Wagner R, Kuchler M, Hormann F, Sveshnikov D, Soll J (2002) The preprotein conducting channel at the inner envelope membrane of plastids. EMBO J 21:2616–2625

Hiner ANP, Raven EL, Thorneley RNF, Garcia-Canovas F, Rodriguez-Lopez JN (2002) Mechanisms of compound I formation in heme peroxidases. J Inorg Biochem 91:27–34

Hirohashi T, Hase T, Nakai M (2001) Maize non-photosynthetic ferredoxin precursor is missorted to the intermembrane space of chloroplasts in the presence of light. Plant Physiol 125:2154–2163

Horling F, Baier M, Dietz KJ (2001) Redox-regulation of the expression of the peroxide-detoxifying chloroplast 2-Cys peroxiredoxin in the liverwort *Riccia fluitans*. Planta 214:304–313

Horling F, Konig J, Dietz KJ (2002) Type II peroxiredoxin C, a member of the peroxiredoxin family: its expression and activity in comparison with other peroxiredoxins. Plant Physiol Biochem 40:491–499

Horling F, Lamkemeyer P, Konig J, Finkemeier I, Kandlbinder A, Baier M, Dietz KJ (2003) Divergent light-, ascorbate-, and oxidative stress-dependent regulation of expression of the peroxiredoxin gene family in Arabidopsis. Plant Physiol 131:317–325

Hormann F, Kuchler M, Sveshnikov D, Oppermann U, Li Y, Soll J (2009) Tic32, an essential component in chloroplast biogenesis. J Biol Chem 284:29240

Hu XL, Jiang MY, Zhang AY, Lu J (2005) Abscisic acid-induced apoplastic H_2O_2 accumulation up-regulates the activities of chloroplastic and cytosolic antioxidant enzymes in maize leaves. Planta 223:57–68

Ichimura K, Casais C, Peck SC, Shinozaki K, Shirasu K (2006) MEKK1 is required for MPK4 activation and regulates tissue-specific and temperature-dependent cell death in Arabidopsis. J Biol Chem 281:36969–36976

Ichimura K, Mizoguchi T, Irie K, Morris P, Giraudat J, Matsumoto K, Shinozaki K (1998) Isolation of ATMEKK1 (a MAP kinase kinase kinase) - Interacting proteins and analysis of a MAP kinase cascade in Arabidopsis. Biochem Biophys Res Comm 253:532–543

Ichimura K, Shinozaki K, Tena G, Sheen J, Henry Y, Champion A, Kreis M, Zhang SQ, Hirt H, Wilson C, Heberle-Bors E, Ellis BE, Morris PC, Innes RW, Ecker JR, Scheel D, Klessig DF, Machida Y, Mundy J, Ohashi Y, Walker JC, Grp MAPK (2002) Mitogen-activated protein kinase cascades in plants: a new nomenclature. Trends Plan Sci 7:301–308

Iida A, Kazuoka T, Torikai S, Kikuchi H, Oeda K (2000) A zinc finger protein RHL41 mediates the light acclimatization response in Arabidopsis. Plant J 24:191–203

Inaba T, Schnell DJ (2008) Protein trafficking to plastids: one theme, many variations. Biochem J 413:15–28

Iqbal A, Yabuta Y, Takeda T, Nakano Y, Shigeoka S (2006) Hydroperoxide reduction by thioredoxin-specific glutathione peroxidase isoenzymes of *Arabidopsis thaliana*. FEBS J 273:5589–5597

Iwamoto M, Higo H, Higo K (2000) Differential diurnal expression of rice catalase genes: the 5′-flanking region of CatA is not sufficient for circadian control. Plant Sci 151:39–46

Iwamoto M, Higo H, Higo K (2004) Strong expression of the rice catalase gene CatB promoter in protoplasts and roots of both a monocot and dicots. Plant Physiol Biochem 42:241–249

Jarvis P (2008) Targeting of nucleus-encoded proteins to chloroplasts in plants. New Phytol 179:257–285

Jespersen HM, Kjaersgard IVH, Ostergaard L, Welinder KG (1997) From sequence analysis of three novel ascorbate peroxidases from *Arabidopsis thaliana* to structure, function and evolution of seven types of ascorbate peroxidase. Biochem J 326:305–310

Jiang J, An GY, Wang PC, Wang PT, Han JF, Jin YB, Song CP (2003) MAP kinase specifically mediates the ABA-induced H_2O_2 generation in guard cells of *Vicia faba* L. Chin Sci Bull 48:1919–1926

Jimenez A, Hernandez JA, Pastori G, Del Rio LA, Sevilla F (1998) Role of the ascorbate-glutathione cycle of mitochondria and peroxisomes in the senescence of pea leaves. Plant Physiol 118:1327–1335

Johnson C, Boden E, Arias J (2003) Salicylic acid and NPR1 induce the recruitment of *trans*-activating TGA factors to a defense gene promoter in Arabidopsis. Plant Cell 15:1846–1858

Johnson C, Boden E, Desai M, Pascuzzi P, Arias J (2001) *In vivo* target promoter-binding activities of a xenobiotic stress-activated TGA factor. Plant J 28:237–243

Jonak C, Okresz L, Bogre L, Hirt H (2002) Complexity, cross talk and integration of plant MAP kinase signalling. Curr Opin Plant Biol 5:415–424

Jones-Rhoades MW, Bartel DP (2004) Computational identification of plant MicroRNAs and their targets, including a stress-induced miRNA. Mol Cell 14:787–799

Jürgens G (2004) Membrane trafficking in plants. Ann Rev Cell Dev Biol 20:481–504

Kakizaki T, Matsumura H, Nakayama K, Che F-S, Terauchi R, Inaba T (2009) Coordination of plastid protein import and nuclear gene expression by plastid-to-nucleus retrograde signaling. Plant Physiol 151:1339–5133

Kangasjarvi J, Jaspers P, Kollist H (2005) Signalling and cell death in ozone-exposed plants. Plant Cell Envir 28:1021–1036

Kangasjarvi S, Lepisto A, Hannikainen K, Piippo M, Luomala EM, Aro EM, Rintamaki E (2008) Diverse roles for chloroplast stromal and thylakoid-bound ascorbate peroxidases in plant stress responses. Biochem J 412:275–285

Karpinski S, Escobar C, Karpinska B, Creissen G, Mullineaux PM (1997) Photosynthetic electron transport regulates the expression of cytosolic ascorbate peroxidase genes in Arabidopsis excess light stress. Plant Cell 9:627–40

Karpinski S, Reynolds H, Karpinska B, Wingsle G, Creissen G, Mullineaux P (1999) Systemic signaling and acclimation in response to excess excitation energy in Arabidopsis. Science 284:654–657

Kazan K (2006) Negative regulation of defence and stress genes by EAR-motif-containing repressors. Trends Plant Sci 11:109–112

Kessler F, Schnell D (2009) Chloroplast biogenesis: diversity and regulation of the protein import apparatus. Curr Opin Cell Biol 21:494–500

Ketting RF, Fischer SEJ, Bernstein E, Sijen T, Hannon GJ, Plasterk RHA (2001) Dicer functions in RNA interference and in synthesis of small RNA involved in developmental timing in *C. elegans*. Genes Dev 15:2654–2659

Kilian J, Whitehead D, Horak J, Wanke D, Weinl S, Batistic O, D'Angelo C, Bornberg-Bauer E, Kudla J, Harter K (2007) The AtGenExpress global stress expression data set: protocols, evaluation and model data analysis of UV-B light, drought and cold stress responses. Plant J 50:347–363

Kim SY, Jang HH, Lee JR, Sung NR, Bin Lee H, Lee DH, Park DJ, Kang CH, Chung WS, Lim CO, Yun DJ, Kim WY, Lee KO, Lee SY (2009a) Oligomerization and chaperone activity of a plant 2-Cys peroxiredoxin in response to oxidative stress. Plant Sci 177:227–232

Kim YH, Lim S, Yang KS, Kim CY, Kwon SY, Lee HS, Wang X, Zhou ZL, Ma DF, Yun DJ, Kwak SS (2009b) Expression of Arabidopsis NDPK2 increases antioxidant enzyme activities and enhances tolerance to multiple environmental stresses in transgenic sweetpotato plants. Mol Breed 24:233–244

Kirchsteiger K, Pulido P, Gonzalez M, Cejudo FJ (2009) NADPH thioredoxin reductase C controls the redox status of chloroplast 2-Cys peroxiredoxins in *Arabidopsis thaliana*. Mol Plant 2:298–307

Kitajima S (2008) Hydrogen peroxide-mediated inactivation of two chloroplastic peroxidases, ascorbate peroxidase and 2-Cys peroxiredoxin. Photochem Photobiol 84:1404–1409

Kizis D, Pages M (2002) Maize DRE-binding proteins DBF1 and DBF2 are involved in rab17 regulation through the drought-responsive element in an ABA-dependent pathway. Plant J 30:679–689

Kleine T, Voigt C, Leister D (2009) Plastid signalling to the nucleus: messengers still lost in the mists? Trends Genet 25:185–190

Kliebenstein DJ, Monde RA, Last RL (1998) Superoxide dismutase in Arabidopsis: an eclectic enzyme family with disparate regulation and protein localization. Plant Physiol 118:637–650

König J, Baier M, Horling F, Kahmann U, Harris G, Schürmann P, Dietz KJ (2002) The plant-specific function of 2-Cys peroxiredoxin-mediated detoxification of peroxides in the redox-hierarchy of photosynthetic electron flux. Proc Natl Acad Sci USA 99:5738–5743

König J, Lotte K, Plessow R, Brockhinke A, Baier M, Dietz KJ (2003) Reaction mechanism of plant 2-Cys peroxiredoxin. Role of the C terminus and the quaternary structure. J Biol Chem 278:24409–24420

Kovtun Y, Chiu WL, Tena G, Sheen J (2000) Functional analysis of oxidative stress-activated mitogen-activated protein kinase cascade in plants. Proc Natl Acad Sci USA 97:2940–5

Kozak (1991) Structural features in eukaryotic mRNAs that modulate the initiation of translation. J Biol Chem 266:19867–19870

Kubo K, Sakamoto A, Kobayashi A, Rybka Z, Kanno Y, Nakagawa H, Nishino T, Takatsuji H (1998) Cys(2)/His(2) zinc-finger protein family of petunia: evolution and general mechanism of target-sequence recognition. Nucl Acid Res 26:608–615

Kuchler M, Decker S, Hormann F, Soll J, Heins L (2002) Protein import into chloroplasts involves redox-regulated proteins. EMBO J 21:6136–6145

Kumari M, Taylor GJ, Deyholos MK (2008) Transcriptomic responses to aluminum stress in roots of *Arabidopsis thaliana*. Mol Genet Genomics 279:339–357

Laloi C, Mestres-Ortega D, Marco Y, Meyer Y, Reichheld JP (2004) The Arabidopsis cytosolic thioredoxin h5 gene induction by oxidative stress and its W-box-mediated response to pathogen elicitor. Plant Physiol 134:1006–1016

Lam E, Lam YKP (1995) Binding-Site requirements and differential representation of TGA factors in nuclear Asf-1 activity. Nucl Acid Res 23:3778–3785

Lee H, Jo J, Son D (1998) Molecular cloning and characterization of the gene encoding glutathione reductase in *Brassica campestris*. Biochim Biophys Acta-Gene Struct Expr 1395:309–314

Leister D, Schneider A (2003) From genes to photosynthesis in *Arabidopsis thaliana*. Int Rev Cytol 228:31–83

Li J, Brader G, Palva ET (2004) The WRKY70 transcription factor: a node of convergence for jasmonate-mediated and salicylate-mediated signals in plant defense. Plant Cell 16:319–331

Lin RC, Park HJ, Wang HY (2008) Role of Arabidopsis RAP2.4 in regulating light- and ethylene-mediated developmental processes and drought stress tolerance. Mol Plant 1:42–57

Lu H, Han RL, Jiang XN (2009) Heterologous expression and characterization of a proxidomal ascorbate peroxidase from *Populus tomentosa*. Mol Biol Rep 36:21–7

Lu J, Holmgren A (2009) Selenoproteins. J Biol Chem 284:723–727

MacLean D, Jerome CA, Brown APC, Gray JC (2008) Co-regulation of nuclear genes encoding plastid ribosomal proteins by light and plastid signals during seedling development in tobacco and Arabidopsis. Plant Mol Biol 66:475–490

Mano S, Yamaguchi K, Hayashi M, Nishimura M (1997) Stromal and thylakoid-bound ascorbate peroxidases are produced by alternative splicing in pumpkin. FEBS Lett 413:21–26

Martinez-Hernandez A, Lopez-Ochoa L, Arguello-Astorga G, Herrera-Estrella L (2002) Functional properties and regulatory complexity of a minimal RBCS light-responsive unit activated by phytochrome, cryptochrome, and plastid signals. Plant Physiol 128:1223–1233

May T, Soll J (2000) 14-3-3 proteins form a guidance complex with chloroplast precursor proteins in plants. Plant Cell 12:53–63

Mckersie BD, Chen YR, Debeus M, Bowley SR, Bowler C, Inze D, Dhalluin K, Botterman J (1993) Superoxide-dismutase enhances tolerance of freezing stress in transgenic alfalfa (*Medicago sativa* L). Plant Physiol 103:1155–1163

Mehler AH (1951) Studies on reactions of illuminated chloroplasts: I. Mechanism of the reduction of oxygen and other hill reagents. Arch Biochem Biophys 33:65–77

Meinhard M, Grill E (2001) Hydrogen peroxide is a regulator of ABI1, a protein phosphatase 2C from Arabidopsis. FEBS Lett 508:443–446

Meinhard M, Rodriguez PL, Grill E (2002) The sensitivity of ABI2 to hydrogen peroxide links the abscisic acid-response regulator to redox signalling. Planta 214:775–782
Meister A (1994) Glutathione, ascorbate, and cellular protection. Cancer Res 54:S1969–1975
Meister G, Tuschl T (2004) Mechanisms of gene silencing by double-stranded RNA. Nature 431:343–349
Miao Y, Laun TM, Smykowski A, Zentgraf U (2007) Arabidopsis MEKK1 can take a short cut: it can directly interact with senescence-related WRKY53 transcription factor on the protein level and can bind to its promoter. Plant Mol Biol 65:63–76
Miao Y, Smykowski A, Zentgraf U (2008) A novel upstream regulator of WRKY53 transcription during leaf senescence in *Arabidopsis thaliana*. Plant Biol 10:110–120
Miller G, Mittler R (2006) Could heat shock transcription factors function as hydrogen peroxide sensors in plants? Ann Bot 98:279–288
Miller G, Shulaev V, Mittler R (2008) Reactive oxygen signaling and abiotic stress. Physiol Plant 133:481–489
Mittler R (2002) Oxidative stress, antioxidants and stress tolerance. Trends Plant Sci 7:405–410
Mittler R, Kim Y, Song LH, Coutu J, Coutu A, Ciftci-Yilmaz S, Lee H, Stevenson B, Zhu JK (2006) Gain- and loss-of-function mutations in Zat10 enhance the tolerance of plants to abiotic stress. FEBS Lett 580:6537–6542
Mittler R, Vanderauwera S, Gollery M, Van Breusegem F (2004) Reactive oxygen gene network of plants. Trends Plant Sci 9:490–498
Miyake C, Asada K (1996) Inactivation mechanism of ascorbate peroxidase at low concentrations of ascorbate: hydrogen peroxide decomposes compound I of ascorbate peroxidase. Plant Cell Physiol 37:423–430
Moon H, Lee B, Choi G, Shin S, Prasad DT, Lee O, Kwak SS, Kim DH, Nam J, Bahk J, Hong JC, Lee SY, Cho MJ, Lim CO, Yun DJ (2003) NDP kinase 2 interacts with two oxidative stress-activated MAPKs to regulate cellular redox state and enhances multiple stress tolerance in transgenic plants. Proc Natl Acad Sci USA 100:358–363
Morel JB, Mourrain P, Beclin C, Vaucheret H (2000) DNA methylation and chromatin structure affect transcriptional and post-transcriptional transgene silencing in Arabidopsis. Curr Biol 10:1591–1594
Morell S, Follmann H, De Tullio M, Häberlein I (1997) Dehydroascorbate and dehydroascorbate reductase are phantom indicators of oxidative stress in plants. FEBS Lett 414:567–570
Morgante CV, Rodrigues RAO, Marbach PAS, Borgonovi CM, Moura DS, Silva-Filho MC (2009) Conservation of dual-targeted proteins in Arabidopsis and rice points to a similar pattern of gene-family evolution. Mol Gen Genomics 281:525–538
Mou Z, Fan WH, Dong XN (2003) Inducers of plant systemic acquired resistance regulate NPR1 function through redox changes. Cell 113:935–944
Mourrain P, Beclin C, Elmayan T, Feuerbach F, Godon C, Morel JB, Jouette D, Lacombe AM, Nikic S, Picault N, Remoue K, Sanial M, Vo TA, Vaucheret H (2000) Arabidopsis SGS2 and SGS3 genes are required for posttranscriptional gene silencing and natural virus resistance. Cell 101:533–542
Murakami S, Takahara H, Shiraiwa M (2007) Purification and characterization of three neutral extracellular isoperoxidases from rye leaves. Phytochemistry 68:777–784
Myouga F, Hosoda C, Umezawa T, Iizumi H, Kuromori T, Motohashi R, Shono Y, Nagata N, Ikeuchi M, Shinozaki K (2008) A heterocomplex of iron superoxide dismutases defends chloroplast nucleoids against oxidative stress and is essential for chloroplast development in Arabidopsis. Plant Cell 20:3148–3162
Nakano T, Suzuki K, Fujimura T, Shinshi H (2006) Genome-wide analysis of the ERF gene family in Arabidopsis and rice. Plant Physiol 140:411–432
Nanjo Y, Oka H, Ikarashi N, Kaneko K, Kitajima A, Mitsui T, Munoz FJ, Rodriguez-Lopez M, Baroja-fernandez E, Pozueta-romero J (2006) rice plastidial N-gycosylated nucleotide pyrophosphatase/Phosphodiesterase is transported from the ER-Golgi to the chloroplast through the secretory pathway. Plant Cell 18:2582–2592

Navrot N, Collin V, Gualberto J, Gelhaye E, Hirasawa M, Rey P, Knaff DB, Issakidis E, Jacquot JP, Routhier N (2006) Plant glutathione peroxidases are functional peroxiredoxins distributed in several subcellular compartments and regulated during biotic and abiotic stresses. Plant Physiol 142:1364–1379

Neill S, Desikan R, Hancock J (2002) Hydrogen peroxide signalling. Curr Opin Plant Biol 5:388–395

Nishizawa A, Yabuta Y, Yoshida E, Maruta T, Yoshimura K, Shigeoka S (2006) Arabidopsis heat shock transcription factor A2 as a key regulator in response to several types of environmental stress. Plant J 48:535–547

Noctor G, De Paepe R, Foyer CH (2007) Mitochondrial redox biology and homeostasis in plants. Trends Plant Sci 12:125–134

Noctor G, Foyer CH (1998) Ascorbate and glutathione: keeping active oxygen under control. Annu Rev Plant Physiol Plant Mol Biol 49:249–279

Noctor G, Gomez L, Vanacker H, Foyer CH (2002) Interactions between biosynthesis, compartmentation and transport in the control of glutathione homeostasis and signalling. J Exp Bot 53:1283–1304

Noctor G, Strohm M, Jouanin L, Kunert KJ, Foyer CH, Rennenberg H (1996) Synthesis of glutathione in leaves of transgenic poplar overexpressing [gamma]-glutamylcysteine synthetase. Plant Physiol 112:1071–1078

Noctor G, Veljovic-Jovanovic S, Foyer CH (2000) Peroxide processing in photosynthesis: antioxidant coupling and redox signalling. Philos Trans R Soc Lond B Biol Sci 355:1465–1475

Obara K, Sumi K, Fukuda H (2002) The use of multiple transcription starts causes the dual targeting of Arabidopsis putative monodehydroascorbate reductase to both mitochondria and chloroplasts. Plant Cell Physiol 43:697–705

Obinger C, Regelsberger G, Furtmuller PG, Jakopitsch C, Ruker F, Pircher A, Peschek GA (1999) Catalase-peroxidases in cyanobacteria: similarities and differences to ascorbate peroxidases. Free Radic Res 31:S243–249

Ohta M, Matsui K, Hiratsu K, Shinshi H, Ohme-Takagi M (2001) Repression domains of class II ERF transcriptional repressors share an essential motif for active repression. Plant Cell 13:1959–1968

Oreb M, Tews I, Schleiff E (2008) Policing Tic 'n' Toc, the doorway to chloroplasts. Trends Cell Biol 18:19–27

Otterhag L, Gustavsson N, Alsterfjord M, Pical C, Lehrach H, Gobom J, Sommarin M (2006) Arabidopsis PDK1: identification of sites important for activity and downstream phosphorylation of S6 kinase. Biochimie 88:11–21

Panchuk II, Volkov RA, Schoffl F (2002) Heat stress- and heat shock transcription factor-dependent expression and activity of ascorbate peroxidase in Arabidopsis. Plant Physiol 129:838–853

Park RE, Agarwal RP (1973) Nucleoside disphosphokinases. In: Boyer PD (ed) The enzymes. Academic, New York, pp 307–333

Passardi F, Bakalovic N, Teixeira FK, Margis-Pinheiro M, Penel C, Dunand C (2007) Prokaryotic origins of the non-animal peroxidase superfamily and organelle-mediated transmission to eukaryotes. Genomics 89:567–579

Pastori GM, Kiddle G, Antoniw J, Bernard S, Veljovic-Jovanovic S, Verrier PJ, Noctor G, Foyer CH (2003) Leaf vitamin C contents modulate plant defense transcripts and regulate genes that control development through hormone signaling. Plant Cell 15:939–951

Pena-Ahumada A, Kahmann U, Dietz KJ, Baier M (2006) Regulation of peroxiredoxin expression versus expression of Halliwell-Asada-Cycle enzymes during early seedling development of *Arabidopsis thaliana*. Photosynth Res 89:99–112

Peragine A, Yoshikawa M, Wu G, Albrecht HL, Poethig RS (2004) SGS3 and SGS2/SDE1/RDR6 are required for juvenile development and the production of *trans*-acting siRNAs in Arabidopsis. Genes Dev 18:2368–2379

Petersson UA, Kieselbach T, Garcia-Cerdan JG, Schroder WP (2006) The Prx Q protein of *Arabidopsis thaliana* is a member of the luminal chloroplast proteome. FEBS Lett 580:6055–6061

Pfannschmidt T, Brautigam K, Wagner R, Dietzel L, Schroter Y, Steiner S, Nykytenko A (2007) Abiotic stresses: potential regulation of gene expression in photosynthetic cells by redox and energy state. Photosynth Res 91:S252

Piippo M, Allahverdiyeva Y, Paakkarinen V, Suoranta UM, Battchikova N, Aro EM (2006) Chloroplast-mediated regulation of nuclear genes in *Arabidopsis thaliana* in the absence of light stress. Physiol Genomics 25:142–152

Pitcher LH, Brennan E, Hurley A, Dunsmuir P, Tepperman JM, Zilinskas BA (1991) Overproduction of Petunia chloroplastic copper/zinc superoxide dismutase does not confer ozone tolerance in transgenic tobacco. Plant Physiol 97:452–455

Pitcher LH, Zilinskas BA (1996) Overexpression of copper/zinc superoxide dismutase in the cytosol of transgenic tobacco confers partial resistance to ozone-induced foliar necrosis. Plant Physiol 110:583–588

Pitzschke A, Djamei A, Bitton F, Hirt H (2009) A major role of the MEKK1-MKK1/2-MPK4 pathway in ROS signalling. Mol Plant 2:120–137

Pnueli L, Liang H, Rozenberg M, Mittler R (2003) Growth suppression, altered stomatal responses, and augmented induction of heat shock proteins in cytosolic ascorbate peroxidase (Apx1)-deficient Arabidopsis plants. Plant J 34:185–201

Polidoros AN, Scandalios JG (1999) Role of hydrogen peroxide and different classes of antioxidants in the regulation of catalase and glutathione S-transferase gene expression in maize (*Zea mays* L.). Physiol Plant 106:112–120

Popescu SC, Popescu GV, Bachan S, Zhang ZM, Gerstein M, Snyder M, nesh-Kumar SP (2009) MAPK target networks in *Arabidopsis thaliana* revealed using functional protein microarrays. Genes Dev 23:80–92

Price AH, Taylor A, Ripley SJ, Griffiths A, Trewavas AJ, Knight MR (1994) Oxidative signals in tobacco increase cytosolic calcium. Plant Cell 6:1301–1310

Qbadou S, Becker T, Mirus O, Tews I, Soll J, Schleiff E (2006) The molecular chaperone Hsp90 delivers precursor proteins to the chloroplast import receptor Toc64. EMBO J 25: 1836–1847

Qiu JL, Zhou L, Yun BW, Nielsen HB, Fiil BK, Petersen K, MacKinlay J, Loake GJ, Mundy J, Morris PC (2008) Arabidopsis mitogen-activated protein kinase kinases MKK1 and MKK2 have overlapping functions in defense signaling mediated by MEKK1, MPK4, and MKS1. Plant Physiol 148:212–222

Queval G, Issakidis-Bourguet E, Hoeberichts FA, Vandorpe M, Gakiere B, Vanacker H, Miginiac-Maslow M, Van Breusegem F, Noctor G (2007) Conditional oxidative stress responses in the Arabidopsis photorespiratory mutant cat2 demonstrate that redox state is a key modulator of daylength-dependent gene expression, and define photoperiod as a crucial factor in the regulation of H_2O_2-induced cell death. Plant J 52:640–657

Queval G, Thominet D, Vanacker H, Miginiac-Maslow M, Gakiere B, Noctor G (2009) H_2O_2-activated up-regulation of glutathione in Arabidopsis involves induction of genes encoding enzymes involved in cysteine synthesis in the chloroplast. Mol Plant 2:344–356

Ramos J, Matamoros MA, Naya L, James EK, Rouhier N, Sato S, Tabata S, Becana M (2009) The glutathione peroxidase gene family of *Lotus japonicus*: characterization of genomic clones, expression analyses and immunolocalization in legumes. New Phytol 181:103–114

Raven EL, Lad L, Sharp KH, Mewies M, Moody PCE (2004) Defining substrate specificity and catalytic mechanism in ascorbate peroxidase. Free Radic: Enzymol, Signal Dis 71:27–38

Reddy RA, Kumar B, Reddy PS, Mishra RN, Mahanty S, Kaul T, Nair S, Sopory SK, Reddy MK (2009) Molecular cloning and characterization of genes encoding *Pennisetum glaucum* ascorbate peroxidase and heat-shock factor: interlinking oxidative and heat-stress responses. J Plant Physiol 166:1646–1659

Rentel MC, Lecourieux D, Ouaked F, Usher SL, Petersen L, Okamoto H, Knight H, Peck SC, Grierson CS, Hirt H, Knight MR (2004) OXI1 kinase is necessary for oxidative burst-mediated signalling in Arabidopsis. Nature 427:858–861

Rizhsky L, Davletova S, Liang HJ, Mittler R (2004) The zinc finger protein Zat12 is required for cytosolic ascorbate peroxidase 1 expression during oxidative stress in Arabidopsis. J Biol Chem 279:11736–11743

Robert G, Melchiorre M, Racca R, Trippi V, Lascano HR (2009) Apoplastic superoxide level in wheat protoplast under photooxidative stress is regulated by chloroplast redox signals: effects on the antioxidant system. Plant Sci 177:168–174

Rodriguez Milla MA, Maurer A, Rodriguez HA, Gustafson JP (2003) Glutathione peroxidase genes in Arabidopsis are ubiquitous and regulated by abiotic stresses through diverse signaling pathways. Plant J 36:602–615

Rossel JB, Walter PB, Hendrickson L, Chow WS, Poole A, Mullineaux PM, Pogson BJ (2006) A mutation affecting ASCORBATE PEROXIDASE 2 gene expression reveals a link between responses to high light and drought tolerance. Plant Cell Environ 29:269–281

Rossel JB, Wilson PB, Hussain D, Woo NS, Gordon MJ, Mewett OP, Howell KA, Whelan J, Kazan K, Pogson BJ (2007) Systemic and intracellular responses to photooxidative stress in Arabidopsis. Plant Cell 19:4091–4110

Rouhier N, Gelhaye E, Jacquot JP (2002) Glutaredoxin-dependent peroxiredoxin from poplar: protein-protein interaction and catalytic mechanism. J Biol Chem 277:13609–13614

Rushmore TH, Morton MR, Pickett CB (1991) The antioxidant responsive element: activation by oxidative stress and identification of the DNA consensus sequence required for functional-activity. J Biol Chem 266:11632–11639

Rushton PJ, Torres JT, Parniske M, Wernert P, Hahlbrock K, Somssich IE (1996) Interaction of elicitor-induced DNA-binding proteins with elicitor response elements in the promoters of parsley PR1 genes. EMBO J 15:5690–5700

Sakamoto H, Maruyama K, Sakuma Y, Meshi T, Iwabuchi M, Shinozaki K, Yamaguchi-Shinozaki K (2004) Arabidopsis Cys2/His2-type zinc-finger proteins function as transcription repressors under drought, cold, and high-salinity stress conditions. Plant Physiol 136:2734–2746

Scandalios JG (2005) Oxidative stress: molecular perception and transduction of signals triggering antioxidant gene defenses. Braz J Med Biol Res 38:995–1014

Scarpeci TE, Zanor MI, Carrillo N, Mueller-Roeber B, Valle EM (2008) Generation of superoxide anion in chloroplasts of *Arabidopsis thaliana* during active photosynthesis: a focus on rapidly induced genes. Plant Mol Biol 66:361–378

Schauer SE, Jacobsen SE, Meinke DW, Ray A (2002) DICER-LIKE1: blind men and elephants in Arabidopsis development. Trends Plant Sci 7:487–491

Schöttler MA, Kirchhoff H, Weis E (2004) The role of plastocyanin in the adjustment of the photosynthetic electron transport to the carbon metabolism in tobacco. Plant Physiol 136:4265–4274

Schramm F, Ganguli A, Kiehlmann E, Englich G, Walch D, von Koskull-Doring P (2006) The heat stress transcription factor HsfA2 serves as a regulatory amplifier of a subset of genes in the heat stress response in Arabidopsis. Plant Mol Biol 60:759–772

Shaikhali J, Baier M (2009) Ascorbate regulation of 2-Cys peroxiredoxin-A promoter activity is light-dependent. J Plant Physiol 167(6):461–467

Shaikhali J, Heiber I, Seidel T, Stroher E, Hiltscher H, Birkmann S, Dietz KJ, Baier M (2008) The redox-sensitive transcription factor Rap2.4a controls nuclear expression of 2-Cys peroxiredoxin A and other chloroplast antioxidant enzymes. BMC Plant Biology 8. doi:10.1186/1471-2229-8-48

Shearer HL, Wang LP, DeLong C, Despres C, Fobert PR (2009) NPR1 enhances the DNA binding activity of the Arabidopsis bZIP transcription factor TGA7. Botany-Botanique 87:561–70

Soitamo AJ, Piippo M, Allahverdiyeva Y, Battchikova N, Aro EM (2008) Light has a specific role in modulating Arabidopsis gene expression at low temperature. BMC Plant Biol 8. doi:10.1186/1471-2229-8-13

Spoel SH, Koornneef A, Claessens SMC, Korzelius JP, Van Pelt JA, Mueller MJ, Buchala AJ, Metraux JP, Brown R, Kazan K, Van Loon LC, Dong XN, Pieterse CMJ (2003) NPR1 modulates cross-talk between salicylate- and jasmonate-dependent defense pathways through a novel function in the cytosol. Plant Cell 15:760–770

Srivastava V, Srivastava MK, Chibani K, Nilsson R, Rouhier N, Melzer M, Wingsle G (2009) Alternative splicing studies of the reactive oxygen species gene network in Populus reveal two isoforms of high-isoelectric-point superoxide dismutase. Plant Physiol 149:1848–1859

Stacy RAP, Munthe E, Steinum T, Sharma B, Aalen RB (1996) A peroxiredoxin antioxidant is encoded by a dormancy-related gene, *Per1*, expressed during late development in the aleurone and embryo of barley grains. Plant Mol Biol 31:1205–1216

St Johnston D (2005) Moving messages: the intracellular localization of mRNAs. Nat Rev Mol Cell Biol 6:363–375

Stengel A, Benz JP, Buchanan BB, Soll J, Bolter B (2009) Preprotein import into chloroplasts via the Toc and Tic complexes is regulated by redox signals in *Pisum sativum*. Mol Plant 2:1181–1197

Stengel A, Benz P, Balsera M, Soll J, Bolter B (2008) TIC62 redox-regulated translocon composition and dynamics. J Biol Chem 283:6656–6567

Stevens RG, Creissen GP, Mullineaux PM (1997) Cloning and characterisation of a cytosolic glutathione reductase cDNA from pea (*Pisum sativum* L.) and its expression in response to stress. Plant Mol Biol 35:641–654

Storozhenko S, De Pauw P, Van Montagu M, Inze D, Kushnir S (1998) The heat-shock element is a functional component of the Arabidopsis APX1 gene promoter. Plant Physiol 118:1005–1014

Sunkar R, Gierke T, Jain PK, Zhu JK (2005) Cloning and characterization of microRNAs from rice. Plant Cell 17:1397–1411

Sunkar R, Kapoor A, Zhu JK (2006) Posttranscriptional induction of two Cu/Zn superoxide dismutase genes in Arabidopsis is mediated by downregulation of miR398 and important for oxidative stress tolerance. Plant Cell 18:2051–2065

Sunkar R, Zhu JK (2004) Novel and stress-regulated microRNAs and other small RNAs from Arabidopsis. Plant Cell 16:2001–2019

Teixeira FK, Menezes-Benavente L, Galvao VC, Margis R, Margis-Pinheiro M (2006) Rice ascorbate peroxidase gene family encodes functionally diverse isoforms localized in different subcellular compartments. Planta 224:300–314

Teng YS, Su YS, Chen LJ, Lee YJ, Hwang I, Li HM (2006) Tic21 is an essential translocon component for protein translocation across the chloroplast inner envelope membrane. Plant Cell 18:2247–2257

Toppo S, Vanin S, Bosello V, Tosatto SCE (2008) Evolutionary and structural insights into the multifaceted glutathione peroxidase (Gpx) superfamily. Antioxid Redox Signal 10:1501–1513

Torres MA, Jones JDG, Dangl JL (2005) Pathogen-induced, NADPH oxidase-derived reactive oxygen intermediates suppress spread of cell death in *Arabidopsis thaliana*. Nat Genet 37:1130–1134

Uniacke J, Zerges W (2009) Chloroplast protein targeting involves localized translation in Chlamydomonas. Proc Natl Acad Sci USA 106:1439–1444

Urao T, Yamaguchishinozaki K, Urao S, Shinozaki K (1993) An Arabidopsis Myb homolog is induced by dehydration stress and its gene-product binds to the conserved Myb recognition sequence. Plant Cell 5:1529–1539

Vadassery J, Tripathi S, Prasad R, Varma A, Oelmuller R (2009) Monodehydroascorbate reductase 2 and dehydroascorbate reductase 5 are crucial for a mutualistic interaction between *Piriformospora indica* and Arabidopsis. J Plant Physiol 166:1263–1274

Vaida V (2005) Sunlight initiated atmospheric photochemical reactions. Int J Photoenergy 7:61–70

Van Breusegem F, Slooten L, Stassart JM, Moens T, Botterman J, Van Montagu M, Inze D (1999) Overproduction of *Arabidopsis thaliana* FeSOD confers oxidative stress tolerance to transgenic maize. Plant Cell Physiol 40:515–523

Van Verk MC, Gatz C, Linthorst HJM (2009) Transcriptional regulation of plant defense responses. Plant Innate Immun 51:397–438

VanCamp W, Capiau K, VanMontagu M, Inze D, Slooten L (1996) Enhancement of oxidative stress tolerance in transgenic tobacco plants overproducing Fe-superoxide dismutase in chloroplasts. Plant Physiol 112:1703–1714

Vandenbroucke K, Robbens S, Vandepoele K, Inze D, de Peer YV, Van Breusegem F (2008) Hydrogen peroxide-induced gene expression across kingdoms: a comparative analysis. Mol Biol Evol 25:507–516

Vanderauwera S, Zimmermann P, Rombauts S, Vandenabeele S, Langebartels C, Gruissem W, Inze D, Van Breusegem F (2005) Genome-wide analysis of hydrogen peroxide-regulated gene expression in Arabidopsis reveals a high light-induced transcriptional cluster involved in anthocyanin biosynthesis. Plant Physiol 139:806–821

Vazquez F, Vaucheret H, Rajagopalan R, Lepers C, Gasciolli V, Mallory AC, Hilbert JL, Bartel DP, Crete P (2004) Endogenous *trans*-acting siRNAs regulate the accumulation of Arabidopsis mRNAs. Mol Cell 16:69–79

Vogel JT, Zarka DG, Van Buskirk HA, Fowler SG, Thomashow MF (2005) Roles of the CBF2 and ZAT12 transcription factors in configuring the low temperature transcriptome of Arabidopsis. Plant J 41:195–211

Vranova E, Van Breusegem F, Dat J, Belles-Boix E, Inze D (2002) The role of active oxygen species in plant signal transduction. In: Scheel D, Wasternack C (eds) Plant signal transduction. Oxford University Press, Oxford, pp 45–73

Willekens H, Chamnongpol S, Davey M, Schraudner M, Langebartels C, Van Montagu M, Inzé D, Van Camp W (1997) Catalase is a sink for H_2O_2 and is indispensable for stress defence in C_3 plants. EMBO J 16:4806–4816

Winter D, Vinegar B, Nahal H, Ammar R, Wilson GV, Provart NJ (2007) An "Electronic fluorescence pictograph" browser for exploring and analyzing large-scale data sets. PLoS ONE. doi:10.1371/journal.pone.0000718

Wintz H, Vulpe C (2002) Plant copper chaperones. Biochem Soc Trans 30:732–5

Wormuth D, Baier M, Kandlbinder A, Scheibe R, Hartung W, Dietz KJ (2006) Regulation of gene expression by photosynthetic signals triggered through modified CO_2 availability. BMC Plant Biol 6:15. doi:10.1186/1471-2229-6-15

Wu C (1995) Heat shock transcription factors: structure and regulation. Ann Rev Cell Dev Biol 11:441–469

Xiang CB, Miao ZH, Lam E (1997) DNA-binding properties, genomic organization and expression pattern of TGA6, a new member of the TGA family of bZIP transcription factors in *Arabidopsis thaliana*. Plant Mol Biol 34:403–415

Yabuta Y, Maruta T, Yoshimura K, Ishikawa T, Shigeoka S (2004) Two distinct redox signaling pathways for cytosolic APX induction under photooxidative stress. Plant Cell Physiol 45:1586–1594

Yakimova ET, Yordanova ZP, Slavov S, Kapchina-Toteva VM, Woltering EJ (2009) Alternate AT toxin induces programmed cell death in tobacco. J Phytopathol 157:592–601

Yamasaki H, Abdel-Ghany SE, Cohu CM, Kobayashi Y, Shikanai T, Pilon M (2007) Regulation of copper homeostasis by micro-RNA in Arabidopsis. J Biol Chem 282:16369–16378

Yamasaki H, Hayashi M, Fukazawa M, Kobayashi Y, Shikanai T (2009) SQUAMOSA promoter binding protein-Like7 is a central regulator for copper homeostasis in Arabidopsis. Plant Cell 21:347–361

Yang KA, Moon H, Kim G, Lim CJ, Hong JC, Lim CO, Yun DJ (2003) NDP kinase 2 regulates expression of antioxidant genes in Arabidopsis. Proc Jpn Acad B Phys Biol Sci 79:86–91

Yoshimura K, Yabuta Y, Tamoi M, Ishikawa T, Shigeoka S (1999) Alternatively spliced mRNA variants of chloroplast ascorbate peroxidase isoenzymes in spinach leaves. Biochem J 338:41–48

Yu DQ, Chen CH, Chen ZX (2001) Evidence for an important role of WRKY DNA binding proteins in the regulation of NPR1 gene express ion. Plant Cell 13:1527–1539

Zamocky M, Janecek S, Koller F (2000) Common phylogeny of catalase-peroxidases and ascorbate peroxidases. Gene 256:169–182

Zarra I, Sanchez M, Queijeiro E, Pena MJ, Revilla G (1999) The cell wall stiffening mechanism in *Pinus pinaster* Aiton: regulation by apoplastic levels of ascorbate and hydrogen peroxide. J Sci Food Agric 79:416–420

Zheng ZY, Mosher SL, Fan BF, Klessig DF, Chen ZX (2007) Functional analysis of Arabidopsis WRKY25 transcription factor in plant defense against *Pseudomonas syringae*. BMC Plant Biol 7. doi:10.1186/1471-2229-7-2

Zimmermann P, Heinlein C, Orendi G, Zentgraf U (2006) Senescence-specific regulation of catalases in *Arabidopsis thaliana* (L.) Heynh. Plant Cell Environ 29:1049–1060

Chapter 14
The Peroxisomal Ascorbate–Glutathione Pathway: Molecular Identification and Insights into Its Essential Role Under Environmental Stress Conditions

Sigrun Reumann and Francisco J. Corpas

Abstract Peroxisomes are unique organelles with intricate biochemical machinery involved in oxidative stress management and signaling. Peroxisomal metabolism is characterized by its plasticity because the organelle's enzyme composition varies depending on the cell type, stage of development, and environmental conditions. The present chapter reviews recent progress in understanding the metabolism of reactive oxygen species (ROS) in plant peroxisomes. The ascorbate–glutathione (AsA–GSH) cycle is the second line of defense to cope with hydrogen peroxide (H_2O_2) generated by peroxisomal flavin oxidases particularly under stress conditions when catalase becomes inactivated. The cycle has been studied comprehensively at the biochemical and physiological level and its role been elucidated under biotic and abiotic stress conditions. The last two members of the peroxisomal AsA–GSH cycle, glutathione reductase and dehydroascorbate reductase, have recently been identified by proteome analysis of Arabidopsis leaf peroxisomes and been established as peroxisomal proteins by in vivo targeting analysis. The identification of all cycle members now opens new doors with which to study the function of the cycle in vivo, including reverse genetics. The release of ROS signaling molecules, such as H_2O_2, from plant peroxisomes adds new functions in the cross-talk events among organelles and cells under physiological and stress conditions.

Keywords Peroxisome • stress • ascorbate glutathione cycle • reactive oxygen species • subcellular targeting • signaling

---

S. Reumann (✉)
Faculty of Science and Technology, Centre for Organelle Research (CORE), University of Stavanger, Stavanger, Norway
e-mail: sigrun.reumann@uis.no

F.J. Corpas
Departamento de Bioquímica, Biología Celular y Molecular de Plantas, Estación Experimental del Zaidín (EEZ), CSIC, Granada, Spain

N.A. Anjum et al. (eds.), *Ascorbate-Glutathione Pathway and Stress Tolerance in Plants*, DOI 10.1007/978-90-481-9404-9_14, © Springer Science+Business Media B.V. 2010

1 Overview of Peroxisomal Reactive Oxygen Species Metabolism

Peroxisomes are subcellular organelles with a single boundary membrane and a rather simple morphological structure, but with intricate biochemical machinery involved in oxidative stress management and signaling. Peroxisomal metabolism and physiological functions are plastic as the organelle's enzyme composition is modulated depending on the cell type, stage of development, and prevailing environmental conditions (del Río et al. 1992, 2002; Johnson and Olsen 2001; Hayashi and Nishimura 2006). With reference to their major physiological functions, plant peroxisomes have been classified as glyoxysomes in germinating seedlings, leaf peroxisomes in photosynthetic tissue, gerontosomes in senescing tissue, and unspecialized peroxisomes (Johnson and Olsen 2001; Kamada et al. 2003; Hayashi and Nishimura 2006). Since the enzymatic differences among some peroxisome variants and intermediate forms might be subtle, the term "peroxisome" is recommended to commonly refer to all plant peroxisomal variants (Pracharoenwattana and Smith 2008).

Plant peroxisomes perform a plethora of functions, including photorespiration, lipid and nitrogen metabolism, detoxification reactions, synthesis of plant hormones, back-conversion of polyamine, metabolism of reactive oxygen and nitrogen species (ROS and RNS), and urate catabolism (Beevers 1979; Hayashi and Nishimura 2003; del Río et al. 2006; Moschou et al. 2008; Kaur et al. 2009; Palma et al. 2009). Newly uncovered functions of plant peroxisomes include, for instance, their involvement in the biosynthesis of phylloquinone (Babujee et al. 2010). Apart from these diverse metabolic functions, plant peroxisomes play essential roles in photomorphogenesis and plant–pathogen interactions (Hu et al. 2002; Taler et al. 2004; Koh et al. 2005; Lipka et al. 2005; McCartney et al. 2005; Bednarek et al. 2009). Peroxisome functions are also essential for embryogenesis and seed germination, as demonstrated by lethal plant phenotypes of knock-out mutants deficient in peroxisome biogenesis proteins (PEX, Hu et al. 2002; Rylott et al. 2003; Schumann et al. 2003; Sparkes et al. 2003; Fan et al. 2005).

One of the major functions of plant peroxisomes is the compartmentalization of metabolic reactions that produce harmful by-products such as hydrogen peroxide (H_2O_2) and their immediate detoxification at the site of production. These functions are common to all peroxisome variants. Massive amounts of H_2O_2 are produced during photosynthesis in peroxisomes by glycolate oxidase as part of the photorespiratory cycle, and in oilseeds, such as Arabidopsis seeds, during germination by acyl-CoA oxidase as part of fatty acid β-oxidation (Beevers 1979; Tolbert 1980, 1981; Graham and Eastmond 2002). Other peroxisomal H_2O_2-producing enzymes include superoxide dismutase, polyamine oxidase, uricase, and sulfite oxidase (for review see del Río et al. 2002; Kaur et al. 2009).

Peroxisomal H_2O_2 is scavenged primarily by catalase, which is the most abundant peroxisomal protein in the matrix. Despite being the key enzyme in H_2O_2 metabolism, catalase is very inefficient in removing H_2O_2 at low concentrations because of its high Km value for H_2O_2 (~1 M, Huang et al. 1983). Moreover, catalase is inactivated and degraded under high light conditions (Feierabend and Engel 1986; Feierabend et al. 1992; Hertwig et al. 1992).

The ascorbate (AsA)–GSH cycle is one of the major lines of defense in most cell compartments to cope with H_2O_2. The cycle includes the cooperative activities of ascorbate peroxidase (APX), monodehydroascorbate reductase (MDAR), dehydroascorbate reductase (DHAR), and glutathione reductase (GR) (Foyer and Noctor 2003). The cycle is initialized by APX, which catalyzes the reduction of H_2O_2 to water, coupled to the oxidation of ascorbate to monodehydroascorbate (Asada 1992). Monodehydroascorbate is converted back into ascorbate by MDAR in an NADH-dependent manner (Hossain and Asada 1984; Asada and Takahashi 1987). Dehydroascorbate is produced as a side-product either by a non-enzymatic reaction of ascorbate with superoxide or H_2O_2 or by the spontaneous disproportionation of monodehydroascorbate. Dehydroascorbate is reduced to ascorbate by DHAR using reduced glutathione as a co-factor (Hossain and Asada 1984). Oxidized glutathione is reduced by GR with NADPH as the reductant. In plant cells, the AsA–GSH cycle is present in several cell compartments, including the chloroplast stroma, mitochondria, peroxisomes, and the cytosol (Asada 1999).

In peroxisomes, the AsA–GSH cycle is the second line of defense to cope with high concentrations of H_2O_2 generated by peroxisomal flavin oxidases, particularly under stress conditions when catalase becomes inactivated. The peroxisomal cycle includes all four enzymes, which are partly soluble and partly membrane associated (Yamaguchi et al. 1995; Bunkelmann and Trelease 1996; Jiménez et al. 1997; Mullen et al. 1999; Karyotou and Donaldson 2005; Leterrier et al. 2005; Lisenbee et al. 2005). The peroxisomal AsA–GSH cycle has been studied comprehensively at the biochemical and physiological level in pea (del Río et al. 2002) and partly at the molecular level in Arabidopsis, as summarized below (Section 2). In this book chapter we also summarize the literature regarding the regulation of peroxisomal AsA–GSH cycle enzymes and the organelle's redox state under environmental conditions (Sections 3 and 4), as well as the available data on the release of peroxisomal H_2O_2 as a signaling molecule (Section 5).

2 Molecular Identification of the Arabidopsis Enzymes

Despite the peroxisomal isozymes of the AsA–GSH cycle having been identified by biochemical or cell biology techniques in different plant species (Corpas et al. 2001), their identification at the molecular level in a model plant species such as Arabidopsis is important to understand these metabolic pathways and enzymatic functions in sufficient detail. Isoform redundancy in several cell compartments long prevented the cloning of three enzymes of the peroxisomal AsA–GSH cycle. The possibility that biochemical data on peroxisomal enzymes were compromised by the co-purification of plastidic, mitochondrial, peroxisomal, or cytosolic contaminants, could eventually be excluded by cDNA identification and in vivo peroxisome targeting confirmation. Proteome research of Arabidopsis peroxisomes was proven to be a particularly important new technology to identify candidate isoforms, upon

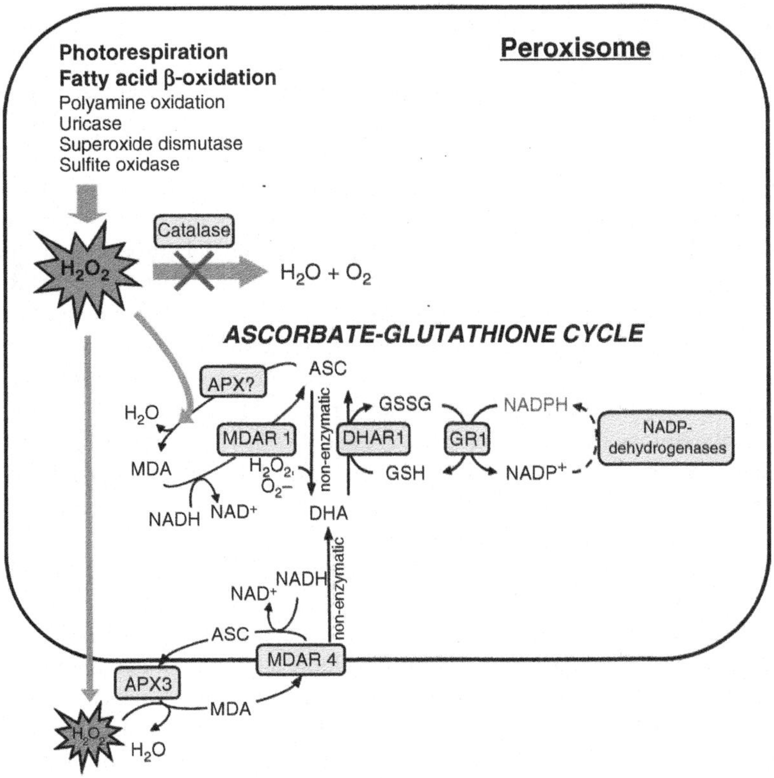

Fig. 1 Model of the sub-compartmental localization of the ascorbate–glutathione cycle in Arabidopsis peroxisomes. Abbreviations: APX, ascorbate peroxidase; ASC, reduced ascorbate; DHA, dehydroascorbate; DHAR, dehydroascorbate reductase; GR, glutathione reductase; GSH, reduced glutathione; GSSG, oxidized glutathione; MDA, monodehydroascorbate; MDAR, monodehydroascorbate reductase

which their peroxisome targeting could then be verified by straight-forward in vivo analyses (Reumann et al. 2007, 2009) (Fig. 1).

As the starting point of the AsA–GSH cycle, APX catalyzes the reduction of H_2O_2 to water coupled to the oxidation of ascorbate to monodehydroascorbate (Asada 1992). In Arabidopsis, the peroxisome-associated isoform is ascorbate peroxidase 3 (APX3, Table 1), which is integral to the peroxisomal membrane with the catalytic domain facing the cytosol (Zhang et al. 1997; del Río et al. 2002). In concert with membrane-associated MDAR4 both enzymes are thought to detoxify H_2O_2 as it leaks out of peroxisomes, thereby preventing oxidative damage to other cell compartments (Yamaguchi et al. 1995; Mullen and Trelease, 1996) (Fig. 2). The high affinity of APX for H_2O_2 (spinach: Km = 74 μM) makes the enzyme suitable to fulfil this scavenging function (Mittler and Zilinskas 1991; Ishikawa et al. 1998). Overexpression of APX3 has been reported to increase plant protection against oxidative stress (Wang et al. 1999). In a reverse genetic study, APX3 loss of function

Table 1 Arabidopsis Asc–glutathione cycle enzymes localized in peroxisomes

Enzyme	Acronym	AGI code	Intraperoxisomal localization	PTS	Reference
Ascorbate peroxidase isoform 3	APX3	At4g35000	Integral membrane protein exposed to the cytosolic side	mPTS	Zhang et al. 1997
Monodehydroascorbate reductase isoform 1	MDAR1	At3g52880	Soluble protein	AKI>	Lisenbee et al. 2005
Monodehydroascorbate reductase isoform 4	MDAR4	At3g27820	Integral membrane protein exposed to the matrix side	mPTS	Lisenbee et al. 2005
Dehydroascorbate reductase isoform 1	DHAR1	At1g19570	Soluble protein	Unknown	Reumann et al. 2009
Glutathione reductase isoform 1	GR1	At3g24170	Soluble protein	TNL>	Kataya and Reumann 2010

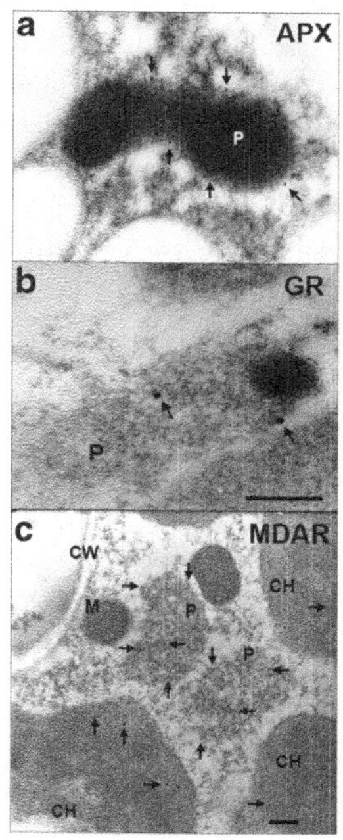

Fig. 2 Peroxisomal immunogold electron microscopy localizations of several enzymes of the ascorbate–glutathione cycle. Panel (**a**): electron micrograph showing the immunocytochemical localization of APX in peroxisomes of cucumber cotyledons (Reproduced, with permission, from Corpas et al. 1994). Panel (**b**): electron micrograph showing the immunocytochemical localization of GR in pea leaves peroxisomes (Reproduced, with permission, from Romero-Puertas et al. 2006). Panel (**c**): electron micrograph showing the immunocytochemical localization of MDAR in pea leaves peroxisomes (Reproduced, with permission, from Leterrier et al. 2005). CW, cell wall; P, peroxisome; M, mitochondrion; CH, chloroplast. Bar = 0.5 μm

did not affect Arabidopsis growth and development suggesting that its peroxidase function could be compensated for by other antioxidant enzymes (Narendra et al. 2006).

It currently remains unknown whether a soluble APX isoform is also part of the peroxisomal AsA–GSH cycle, similar to MDAR (see below). Arabidopsis is predicted to have eight to nine APXs (Jespersen et al. 1997; Panchuk et al. 2002;

Chew et al. 2003; Narendra et al. 2006). Based on similarities in sequence and secondary structure, and the presence of predicted peroxisome targeting signals type 1 or type 2 (PTS1/2), APX5 and APX4 (LKL>) have been proposed to be integral to the peroxisome membrane and localized in the matrix, respectively (Panchuk et al. 2002). However, experimental evidence for their association with peroxisomes is not yet reported.

The expression and activity of catalase and APX (and possibly MDAR, DHAR, and GR) must be coordinated to control intraperoxisomal H_2O_2 levels generated under physiological or adverse environmental stress conditions, as well as the release of H_2O_2 release from peroxisomes. However, data yielding insight into the regulation of the Asc–glutathione cycle at the transcriptional or post-translational level are extremely scarce. The five genes involved in the peroxisomal AsA–GSH cycle are constitutively expressed at significant levels at all developmental stages, indicating a constitutive detoxification function under standard growth conditions (Fig. 3). In addition, several genes are reported to be further induced by abiotic and biotic stress (see Section 3).

In addition to gene induction, the peroxisomal AsA–GSH cycle might be regulated at the post-translational level. An isoform of the 14-3-3 protein family, G-box regulating factor 6 (GRF6), also annotated as GF14λ, has been detected in a yeast two-hybrid screen and characterized to interact with APX3 (Yan et al. 2002). 14-3-3 proteins generally bind to phosphorylated proteins and regulate signal transduction and protein activity (DeLille et al. 2001), suggesting that APX3 activity or protein–protein interactions might be regulated by reversible phosphorylation. The same 14-3-3 protein (GF14λ) has been identified, along with APX3 itself, in the proteome of Arabidopsis leaf peroxisomes. Together with the absence of

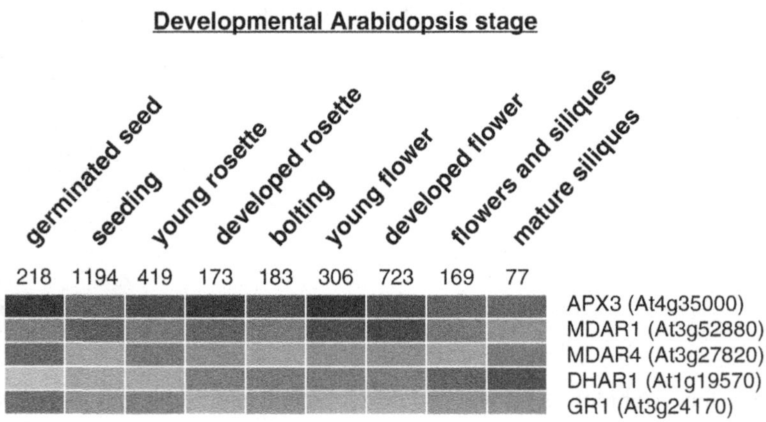

Fig. 3 Gene expression analysis of the five Arabidopsis enzymes of the peroxisomal AsA–GSH cycle at different developmental plant stages. The expression data derive from microarray experiments and were retrieved using Genevestigator (www.genevestigator.com; Zimmermann et al. 2004, 2005). High and low expression levels are reflected semi-quantitatively by *dark* and *light blue* coloring, respectively. The numbers indicate the number of underlying microarrays

predicted matrix targeting signals, the data strongly suggest that GF14λ associates with leaf peroxisomes by its physical interaction with the cytosolic domain of APX3 (Reumann et al. 2009). In vivo validation of such an indirect protein targeting mechanism to peroxisomes might not be trivial and remains to be reported.

Monodehydroascorbate generated by APX is converted back into ascorbate by MDAR using NAD(P)H as the reductant (Hossain and Asada 1984; Asada and Takahashi 1987). Arabidopsis peroxisomes possess two MDARs; MDAR1 is a PTS1-containing matrix enzyme (AKI>), whereas MDAR4 is an integral membrane protein with its catalytic domain facing the matrix side (Lisenbee et al. 2005) (Fig. 2). The differential sub-compartmental localization of the two enzymes indicates slightly divergent roles in AsA recycling, i.e., detoxification of H_2O_2 escaping from peroxisomes (MDAR4, see above) and clearance of intra-peroxisomal H_2O_2 (MDAR1). By application of a forward genetic screen, Eastmond (2007) demonstrated that MDAR4 is essential to prevent oil bodies from incurring oxidative damage, when in close proximity to peroxisomes, and triacylglycerol lipase from inactivation. The loss-of-function mutant is conditionally seedling-lethal because its seeds are unable to break down storage oil to provide carbon skeletons and energy for early seedling growth.

Dehydroascorbate, produced nonenzymatically from AsA or monodehydroascorbate, is reduced to ascorbate by DHAR using reduced glutathione as the reducing co-factor (Hossain and Asada 1984). Arabidopsis encodes five DHAR isoforms that are either shown to be or predicted to be cytosolic or mitochondrial/plastidic (Chew et al. 2003). DHAR1 (At1g19570) was detected in Arabidopsis leaf peroxisomes by proteomics and confirmed to be peroxisome-localized by in vivo subcellular targeting analysis (Reumann et al. 2009). Similar to a number of newly identified matrix proteins, DHAR1 does not carry a predicted PTS neither type 1 nor type 2, and is thus predicted to be imported by a novel import pathway.

Glutathione reductase (GR) is a major enzyme of the antioxidative defense system and plays an important physiological role in maintaining and regenerating reduced glutathione in response to biotic and abiotic stresses in plants (Romero-Puertas et al. 2006). GR mediates the reduction of GSSG to GSH by using NADPH as an electron donor. In contrast to the large gene families of APX, MDAR, and DHAR in Arabidopsis, only two GR isoforms are encoded in the Arabidopsis genome. Proteomic experiments identified GR1 (At3g24170) in Arabidopsis peroxisomes from both mature leaves and suspension-cultured cells (Eubel et al. 2008; Reumann et al. 2007, 2009). GR1 orthologs were previously reported to be cytosolic. However, expression of GR1 as an N-terminal fusion protein with a fluorescent reporter protein, GR1 has validated its peroxisomal localization (Kataya and Reumann 2010). The efficiency of peroxisome targeting, however, was weak upon expression from a strong promoter, consistent with the idea that the enzyme is dually targeted to peroxisomes and the cytosol in vivo. The C-terminal tripeptide of GR1, TNL>, was shown to represent a novel, albeit weak, PTS1 (Kataya and Reumann 2010).

Arabidopsis GR1 is the first plant protein that is dually targeted to peroxisomes and the cytosol. Strikingly, Arabidopsis GR2 is likewise dually targeted, namely to chloroplasts and mitochondria (Creissen et al. 1995). Nonetheless, the underlying regulatory mechanism of dual targeting to chloroplast and mitochondria remains unknown. Dual targeting of GR1 orthologs to both peroxisomes and the cytosol appears to be conserved across Viridiplantae and might partly be determined by the weak peroxisome targeting efficiency of TNL>. Additionally, in light of the importance of the peroxisomal AsA–GSH cycle under sudden conditions of catalase inactivation (Feierabend and Engel 1986), the ratio of GR1 distribution between peroxisomes and the cytosol may be dynamic and adjustable depending on H_2O_2 overproduction in the matrix by post-transcriptional and/or post-translational mechanisms (Kataya and Reumann 2010). Since GR1 provides reduced glutathione not only for DHAR but also for several glutathione-dependent enzymes that have recently been discovered in plant peroxisomes (see below), the physiological function of peroxisomal GR can hardly be overestimated.

3 Peroxisomal AsA–Glutathione Enzymes Under Environmental Stress Conditions

Plants are exposed to continuous environmental changes. In many cases, these changes can induce oxidative stress that provokes cellular damage. Therefore, plants have developed a battery of antioxidative enzymatic and non-enzymatic mechanisms to attenuate or overcome the negative effects of ROS. In this context, the AsA–GSH cycle is, together with catalase, superoxide dismutase (SOD) and other peroxidases, a first line of defense against the deleterious effect of ROS. Usually, the antioxidant systems are located close to the place of ROS generation, including cytosol, chloroplasts, mitochondria and peroxisomes. Under environmental stress conditions, H_2O_2 may be overproduced in the peroxisomal matrix, causing oxidation of proteins, lipids, metabolites, and co-factors in the matrix, the peroxisome membrane and likely, upon diffusion across the membrane, other cell compartments. Thus, the presence of all components (enzymatic and non-enzymatic) of the AsA–GSH cycle in plant peroxisomes is an auxiliary strategy for plant cells to maintain the H_2O_2 level in this cell compartment, which could have a relevant significance under adverse stress conditions.

3.1 Abiotic Stress Conditions

The term abiotic stress includes many environmental factors that can cause an increased formation of ROS, including high light, extreme temperature, heavy metals, salt, and drought. Knowledge of the specific contribution of peroxisomes in the mechanism of response against these stresses is very limited. Even less is known about the AsA–GSH cycle, considering that the description of this cycle in peroxisomes is relatively new. Table 2 summarizes the available data of some of the components of the AsA–GSH cycle from isolated peroxisomes from different

Table 2 Peroxisomal Asc–glutathione enzyme activities and gene expression under environmental stress conditions

Plant species	Abiotic or biotic stress condition	Peroxisomal enzyme	Reference
Pea (*Pisum sativum*)	Dark induced senescent leaves	Decrease in APX and MDAR activities	Jiménez et al. 1998
	Cadmium (50 μM)	Increase in APX and GR activities	Romero-Puertas et al. 1999, 2006
	Wounding, low temperature (8°C), 2,4-dichlorophenoxyacetic acid (2,4-D)	Increase in *MDAR* expression	Leterrier et al. 2005
	Nitrate-fed and nodulated plants at two different growth stages (9 and 12 weeks)	Decrease in APX and MDAR activities, increase in DHAR and GR activities	Palma et al. 2006
Barley (*Hordeum vulgare*)	Heat, salt, and abscisic acid treatments	Increase in the transcript level of *APX*	Shi et al. 2001
Grey mangrove (*Avicennia marina*)	Light stress (500 μE $m^{-2} s^{-1}$), NaCl (500 mM), H_2O_2 (90 mM), Fe(III) citrate (1.0 mM)	Increase in the transcript level of *APX*	Kavitha et al. 2008
Tobacco (*Nicotiana tobacum*)	Oxidative stress damage caused by aminotriazole	Overexpression of Arabidopsis APX3	Wang et al. 1999
Pepper (*Capsicum annuum*)	Fruit ripening during the transition from green to red fruit	Decrease in GR activity	Mateos et al. 2003
Tomato (*Lycopersicon esculentum*)	NaCl (100 mM)	Increased APX and MDAR activities and reduced GR activity	Mittova et al. 2003, 2004
	Botrytis	Decrease in APX, GR, MDAR, and DHAR activities	Kuzniak and Skłodowska 2005a
Wheat (*Triticum aestivum*)	Powdery mildew	Enhanced *APX* expression	Chen et al. 2006

APX, ascorbate peroxidase; MDAR, monodehydroascorbate reductase; DHAR, dehydroascorbate reductase; GR, glutathione reductase

organs and plant species. Peroxisomal isoenzymes of the AsA–GSH cycle must be investigated in isolated peroxisomes to avoid an overlap with the non-peroxisomal isoenzymes, which points to some of the difficulties in analyzing peroxisomal enzymes and their activities.

3.2 Biotic Stress Conditions

During pathogen infection, it has been proposed that peroxisomes are involved mechanistically in systemic defense responses (Kimura et al. 2001). Peroxisomes are involved in the production of lipid-based signaling molecules, such as jasmonic acid, as well as in the generation of ROS and the modulation of antioxidative enzymes (Koo et al. 2006; del Río et al. 2006). For example, it has been reported that pathogens can induce rapid changes in H_2O_2 leading to a variety of physiological responses in plants (Bolwell 1999; Shetty et al. 2007). It remains to be demonstrated to what extent peroxisomes contribute to the overall oxidative burst. Catalase has been reported to be activated by calcium/calmodulin (Yang and Poovaiah 2002). This regulatory mechanism might play a role in pathogen-induced ROS production. However, to our knowledge, little information is available on how the AsA–GSH pathway is affected by pathogen attack. For example, tomato peroxisomes exhibit a decrease in the activities of APX, GR, MDAR, and DHAR when infected by the fungus *Botrytis cinerea*, which was also accompanied by a significant decrease in AsA and GSH concentrations (Kuzniak and Skłodowska 2005a).

4 Peroxisomal Redox Status of Ascorbate, Glutathione and NADPH Under Environmental Stress Conditions

For the function of the AsA–GSH cycle, there is a close interplay among AsA, GSH, and NADPH. These multifunctional redox metabolites play essential roles in both the AsA–GSH cycle and in mediating signaling processes. Moreover, the concentrations of these metabolites can be greatly modified in response to a variety of environmental factors, mainly those that cause increased oxidative stress. Several of these molecules have been clearly detected in plant peroxisomes by different approaches, but the available information remains incomplete (Donaldson 1982; Jiménez et al. 1997).

Glutathione (γ-Glu-Cys-Gly) is one of the major, soluble, low molecular weight antioxidants, as well as the major non-protein thiol in plant cells (0.2–10 mM, Foyer and Noctor 2003). Glutathione not only fulfils multiple metabolic functions, such as detoxification of xenobiotics, ROS, and heavy metals (Tausz et al. 2004), but also acts as an independent redox-signaling molecule (Foyer and Noctor 2005). In this sense, GSH can also react with nitric oxide (NO) to form S-nitrosoglutathione (GSNO), which may function as an intracellular NO reservoir. GSNO may also act as a vehicle for NO throughout the cell, as a connection between ROS and RNS (Corpas et al. 2008). Glutathione has a wide distribution in the different subcellular compartments including mitochondria, nuclei, chloroplasts, cytosol and also in peroxisomes (Müller et al. 2004; Fernández-García et al. 2009). The peroxisomal localization of GSH has been demonstrated by either HPLC in isolated peroxisomes or by electron-microscopic immunogold cytochemistry in different plant species under normal and adverse conditions (Jiménez et al. 1997; Mateos et al. 2003; Müller et al. 2004; Zechmann et al. 2008; Fernández-García et al. 2009; Kolb et al. 2009). In pumpkin plants infected by zucchini yellow mosaic virus (ZYMV), a general rise in GSH has been described in older leaves (as evaluated by electron microscopic immunogold cytochemistry) with GSH levels 1.7-fold higher in peroxisomes, 1.6-fold in the cytosol, 1.4-fold in mitochondria, and 1.2-fold in nuclei when compared to control cells (Zechmann et al. 2007). Even beyond its function in the peroxisomal AsA–GSH cycle, glutathione is now recognized as a major co-factor of metabolic reactions in plant peroxisomes. For instance, several glutathione-S transferases have recently been discovered in peroxisomes (Reumann et al. 2007, 2009; Dixon et al. 2009).

Ascorbate is the most abundant non-thiol (10–100 mM) small molecule antioxidant in plants (Foyer and Noctor 2003). Ascorbate has also been detected in plant peroxisomes by HPLC, and its concentrations were shown to be dynamic and affected by different growth conditions. For example, during natural and induced leaf senescence, the peroxisomal Asc content increased (Jiménez et al. 1998; Palma et al. 2006). Similar antioxidant changes have been described in peroxisomes isolated from pepper fruits where the concentration of peroxisomal AsA increased during fruit ripening (conversion of green to red peppers), while the concentration of peroxisomal GSH remained constant (Mateos et al. 2003). In contrast, the infection of tomato plants by *Botrytis cinerea* caused a significant reduction in the concentration of both AsA and GSH in peroxisomes (Kuzniak and Skłodowska 2005b). Likewise, a reduction in the concentration of Asc in root tomato peroxisomes has been reported under salinity stress (100 mM NaCl, Mittova et al. 2004).

NADPH is the specific donor of reducing equivalents to GR. Therefore, the production of this pyridine nucleotide inside of peroxisomes is required for a functional AsA–GSH cycle. The presence of NADP(H) in peroxisomes was reported by Donaldson (1982), where isolated glyoxysomes from castor beans contained 16% of the total NADP(H); although, to our knowledge, there is not any additional data. It has been reported that the recycling of NADPH from NADP can be carried out in plant peroxisomes by at least three dehydrogenases: glucose-6-phosphate dehydrogenase (G6PDH), 6-phosphogluconate dehydrogenase (6PGDH), and isocitrate dehydrogenase (ICDH) (Corpas et al. 1998, 1999, 2001). Specific Arabidopsis isoforms of both 6PGDH and ICDH have been localized to peroxisomes in vivo (Reumann et al. 2007; Eubel et al. 2008). Consequently, any factors that could affect the activity of these peroxisomal dehydrogenases may also affect the function of the AsA–GSH cycle.

5 Release of Peroxisomal H_2O_2 and ROS Signaling Molecules

Hydrogen peroxide is relatively stable and has been shown to act as a signaling molecule in different physiological and phytopathological processes (Alvarez et al. 1998; Bolwell 1999; López-Huertas et al. 2000; Orozco-Cárdenas et al. 2001; Shetty et al. 2007). A theoretical model, based on the latency of peroxisomal catalase from rat liver, concludes that the peroxisomal membrane represents a barrier to the free diffusion of H_2O_2. However, it could not be excluded that around 2% of the H_2O_2 produced in peroxisomes can diffuse out of the organelle

(Poole 1975). In support of this prediction, Mueller et al. (2002) have demonstrated the release of H_2O_2 using a chemiluminescence method combined with in vitro analysis of isolated rat liver peroxisomes. Interestingly, the diffusion of H_2O_2 out of peroxisomes appears to be controlled not only by the compartmentalization function of the peroxisomal membrane, but also at second level where the highly packed peroxisomal matrix itself seems to act as an additional diffusion barrier (Fritz et al. 2007). While catalase generally removes matrix H_2O_2 efficiently, its spatial separation from urate oxidase prevents the removal of H_2O_2 derived from urate oxidation. It has been demonstrated that urate oxidase can directly release H_2O_2 into the cytoplasm via 5-nm primary tubules in crystalline cores (Fritz et al. 2007).

Plant peroxisomes can be considered more efficient in keeping H_2O_2 under control as compared to mammalian and fungi peroxisomes because they have catalase in the matrix and APX in the membrane (Fig. 2). However, under stress conditions, the generation of ROS can overcome the capacity of the peroxisomal antioxidative system. Consequently, it has been hypothesized that some ROS, mainly H_2O_2, can be released to the cytosol where they may trigger some signal transduction pathways (Corpas et al. 2001). In plants, there is some evidence that support this hypothesis. For example, in catalase-deficient Arabidopsis plants exposed to high light intensity (1,600–1,800 µmol m^{-2} s^{-1} for 3–8 h), there is a correlated increase of cellular H_2O_2 levels with the up-regulation of 349 transcripts and the down-regulation of 88 genes. Under these high light conditions, H_2O_2 seems to play a key role in the transcriptional up-regulation of small heat shock proteins (Vandenabeele et al. 2004; Vanderauwera et al. 2005). On the other hand, electron microscopic analysis of leaves from cadmium-treated pea plants showed a significant H_2O_2 accumulation in the membrane of peroxisomes when in close contact with chloroplasts and the tonoplast (Romero-Puertas et al. 2004).

6 Conclusions

Considering that metabolism of peroxisomes is very dynamic because it is adapted to many physiological and adverse conditions, the Asc–glutathione pathway must be studied in great detail at the molecular level and in relation with other peroxisomal components, such as the NADPH-recycling system and RNS. At the enzymatic level, reverse genetics must be applied to all genes to investigate the exact physiological function of all enzymes, including pathway regulation at the transcriptional, post-transcriptional and post-translational level and the regulation of dual targeting. Our understanding of sub-compartmentalization of the peroxisomal AsA–GSH cycle into membrane-bound and matrix-located pathways remains fragmentary and might require the identification of additional isoforms (e.g., soluble APX). A significant gap concerns the transfer of pathway intermediates across the peroxisome membrane. Future research must also focus on interactions among the different molecules present in these organelles, such as H_2O_2, AsA, GSH, nitric oxide (NO) and S-nitrosoglutathione (GSNO), and their metabolic enzymes. All of these molecules are also involved in signaling processes and consequently in the cross-talk among the different cell compartments.

Acknowledgements SR and FJC were supported by UiS funding and grants from the Ministry of Education and Science (BIO2009-12003-C02-01), respectively.

References

Alvarez ME, Pennell RI, Meijer PJ, Ishikawa A, Dixon RA, Lamb C (1998) Reactive oxygen intermediates mediate a systemic signal network in the establishment of plant immunity. Cell 92:773–784

Asada K (1992) Ascorbate peroxidase: a hydrogen peroxide-scavenging enzyme in plants. Physiol Plant 85:235–241

Asada K (1999) The water-water cycle in chloroplasts: scavenging of active oxygens and dissipation of excess photons. Annu Rev Plant Physiol Plant Mol Biol 50:601–639

Asada K, Takahashi M (1987) Production and scavenging of active oxygen species in photosynthesis. In: Kyle D, Osmond C, Arntzen C (eds) Photoinhibition. Elsevier Science, New York, pp 227–287

Babujee L, Wurtz V, Ma C, Lueder F, Soni P, van Dorsselaer A, Reumann S (2010) The proteome map of spinach leaf peroxisomes indicates partial compartmentalization of phylloquinone (vitamin K1) biosynthesis in plant peroxisomes. J Exp Bot 61(5):1441–1453

Bednarek P, Pislewska-Bednarek M, Svatos A, Schneider B, Doubsky J, Mansurova M, Humphry M, Consonni C, Panstruga R, Sanchez-Vallet A, Molina A, Schulze-Lefert P (2009) A glucosinolate metabolism pathway in living plant cells mediates broad-spectrum antifungal defense. Science 323:101–106

Beevers H (1979) Microbodies in higher-plants. Annu Rev Plant Physiol Plant Mol Biol 30:159–193

Bolwell GP (1999) Role of active oxygen species and NO in plant defence responses. Curr Opin Plant Biol 2:287–294

Bunkelmann JR, Trelease RN (1996) Ascorbate peroxidase. A prominent membrane protein in oilseed glyoxysomes. Plant Physiol 110:589–598

Chen Y, Wang H, Wang X, Cao A, Chen P (2006) Cloning and expression of peroxisomal ascorbate peroxidase gene from wheat. Mol Biol Rep 33:207–213

Chew O, Whelan J, Millar AH (2003) Molecular definition of the ascorbate-glutathione cycle in Arabidopsis mitochondria reveals dual targeting of antioxidant defenses in plants. J Biol Chem 278:46869–46877

Corpas FJ, Barroso JB, del Río LA (2001) Peroxisomes as a source of reactive oxygen species and nitric oxide signal molecules in plant cells. Trends Plant Sci 6:145–150

Corpas FJ, Barroso JB, Sandalio LM, Distefano S, Palma JM, Lupiáñez JA, del Río LA (1998) A dehydrogenase-mediated recycling system of NADPH in plant peroxisomes. Biochem J 330:777–84

Corpas FJ, Barroso JB, Sandalio LM, Palma JM, Lupiáñez JA, del Río LA (1999) Peroxisomal NADP-dependent isocitrate dehydrogenase. Characterization and activity regulation during natural senescence. Plant Physiol 121:921–928

Corpas FJ, Bunkelmann J, Trelease RN (1994) Identification and immunochemical characterization of a family of peroxisome membrane proteins (PMPs) in oilseed glyoxysomes. Eur J Cell Biol 65:280–290

Corpas FJ, del Río LA, Barroso JB (2008) Post-translational modifications mediated by reactive nitrogen species: nitrosative stress responses or components of signal transduction pathways? Plant Signal Behav 3:301–303

Creissen G, Reynolds H, Xue Y, Mullineaux P (1995) Simultaneous targeting of pea glutathione reductase and of a bacterial fusion protein to chloroplasts and mitochondria in transgenic tobacco. Plant J 8:167–175

del Río LA, Corpas FJ, Sandalio LM, Palma JM, Gómez M, Barroso JB (2002) Reactive oxygen species, antioxidant systems and nitric oxide in peroxisomes. J Exp Bot 53:1255–1272

del Río LA, Sandalio LM, Corpas FJ, Palma JM, Barroso JB (2006) Reactive oxygen species and reactive nitrogen species in peroxisomes. Production, scavenging, and role in cell signaling. Plant Physiol 141:330–335

del Río LA, Sandalio LM, Palma JM, Bueno P, Corpas FJ (1992) Metabolism of oxygen radicals in peroxisomes and cellular implications. Free Radic Biol Med 13:57–80

DeLille JM, Sehnke PC, Ferl RJ (2001) The Arabidopsis 14-3-3 family of signaling regulators. Plant Physiol 126:35–38

Dixon DP, Hawkins T, Hussey PJ, Edwards R (2009) Enzyme activities and subcellular localization of members of the Arabidopsis glutathione transferase superfamily. J Exp Bot 60:1207–1218

Donaldson RP (1982) Nicotinamide cofactors (NAD and NADP) in glyoxysomes, mitochondria, and plastids isolated from castor bean endosperm. Arch Biochem Biophys 215:274–279

Eastmond PJ (2007) MONODEHYROASCORBATE REDUCTASE4 is required for seed storage oil hydrolysis and postgerminative growth in Arabidopsis. Plant Cell 19:1376–1387

Eubel H, Meyer EH, Taylor NL, Bussell JD, O'Toole N, Heazlewood JL, Castleden I, Small ID, Smith SM, Millar AH (2008) Novel proteins, putative membrane transporters, and an integrated metabolic network are revealed by quantitative proteomic analysis of Arabidopsis cell culture peroxisomes. Plant Physiol 148:1809–1829

Fan J, Quan S, Orth T, Awai C, Chory J, Hu J (2005) The Arabidopsis PEX12 gene is required for peroxisome biogenesis and is essential for development. Plant Physiol 139:231–239

Feierabend J, Engel S (1986) Photoinactivation of catalase in vitro and in leaves. Arch Biochem Biophys 251:567–576

Feierabend J, Schaan C, Hertwig B (1992) Photoinactivation of catalase occurs under both high- and low-temperature stress conditions and accompanies photoinhibition of Photosystem II. Plant Physiol 100:1554–1561

Fernández-García N, Martí MC, Jiménez A, Sevilla F, Olmos E (2009) Sub-cellular distribution of glutathione in an Arabidopsis mutant (vtc1) deficient in ascorbate. J Plant Physiol 166:2004–2012

Foyer CH, Noctor G (2005) Redox homeostasis and antioxidant signaling: a metabolic interface between stress perception and physiological responses. Plant Cell 17:1866–1875

Foyer CH, Noctor G (2003) Redox sensing and signaling associated with reactive oxygen in chloroplasts, peroxisomes and mitochondria. Physiol Plant 119:355–364

Fritz R, Bol J, Hebling U, Angermüller S, Völkl A, Fahimi HD, Mueller S (2007) Compartment-dependent management of H_2O_2 by peroxisomes. Free Radic Biol Med 42:119–129

Graham IA, Eastmond PJ (2002) Pathways of straight and branched chain fatty acid catabolism in higher plants. Prog Lipid Res 41:156–181

Hayashi M, Nishimura M (2003) Entering a new era of research on plant peroxisomes. Curr Opin Plant Biol 6:577–582

Hayashi M, Nishimura M (2006) *Arabidopsis thaliana* – a model organism to study plant peroxisomes. Biochim Biophys Acta 1763:1382–1391

Hertwig B, Streb P, Feierabend J (1992) Light dependence of catalase synthesis and degradation in leaves and the influence of interfering stress conditions. Plant Physiol 100:1547–1553

Hossain MA, Asada K (1984) Purification of dehydroascorbate reductase from spinach and its characterization as a thiol enzyme. Plant Cell Physiol 25:85–92

Hu J, Aguirre M, Peto C, Alonso J, Ecker J, Chory J (2002) A role for peroxisomes in photomorphogenesis and development of Arabidopsis. Science 297:405–409

Huang AHC, Moore TS, Trelease RN (1983) Plant peroxisomes. Academic Press, New York

Ishikawa T, Yoshimura K, Sakai K, Tamoi M, Takeda T, Shigeoka S (1998) Molecular characterization and physiological role of a glyoxysome-bound ascorbate peroxidase from spinach. Plant Cell Physiol 39:23–34

Jespersen HM, Kjaersgard IV, Ostergaard L, Welinder KG (1997) From sequence analysis of three novel ascorbate peroxidises from *Arabidopsis thaliana* to structure, function and evolution of seven types of ascorbate peroxidase. Biochem J 326:305–310

Jiménez A, Hernández JA, del Río LA, Sevilla F (1997) Evidence for the presence of the ascorbate-glutathione cycle in mitochondria and peroxisomes of pea leaves. Plant Physiol 114:275–284

Jiménez A, Hernández JA, Pastori G, del Rio LA, Sevilla F (1998) Role of the ascorbate-glutathione cycle of mitochondria and peroxisomes in the senescence of pea leaves. Plant Physiol 118:1327–1335

Johnson TL, Olsen LJ (2001) Building new models for peroxisome biogenesis. Plant Physiol 127:731–739

Kamada T, Nito K, Hayashi H, Mano S, Hayashi M, Nishimura M (2003) Functional differentiation of peroxisomes revealed by expression profiles of peroxisomal genes in *Arabidopsis thaliana*. Plant Cell Physiol 44:1275–1289

Karyotou K, Donaldson RP (2005) Ascorbate peroxidase, a scavenger of hydrogen peroxide in glyoxysomal membranes. Arch Biochem Biophys 434:248–257

Kataya AR, Reumann S (2010) Arabidopsis glutathione reductase 1 is dually targeted to peroxisomes and the cytosol. Plant Signal Behav 5(2):171–175

Kaur N, Reumann S, Hu J (2009) Peroxisome biogenesis and function. In: Somerville CR, Meyerowitz EM (eds) The Arabidopsis book. The American Society of Plant Biologists, Rockville, MD, pp 1–41

Kavitha K, Venkataraman G, Parida A (2008) An oxidative and salinity stress induced peroxisomal ascorbate peroxidase from Avicennia marina: molecular and functional characterization. Plant Physiol Biochem 46:794–804

Kimura A, Takano Y, Furusawa I, Okuno T (2001) Peroxisomal metabolic function is required for appressorium-mediated plant infection by *Colletotrichum lagenarium*. Plant Cell 13:1945–1957

Koh S, Andre A, Edwards H, Ehrhardt D, Somerville S (2005) Arabidopsis thaliana subcellular responses to compatible *Erysiphe cichoracearum* infections. Plant J 44:516–529

Kolb D, Müller M, Zellnig G, Zechmann B (2009) Cadmium induced changes in subcellular glutathione contents within glandular trichomes of *Cucurbita pepo* L. Protoplasma. doi:10.1007/s00709-009-0043-x

Koo AJ, Chung HS, Kobayashi Y, Howe GA (2006) Identification of a peroxisomal acyl-activating enzyme involved in the biosynthesis of jasmonic acid in Arabidopsis. J Biol Chem 281:33511–3320

Kuzniak E, Skłodowska M (2005a) Compartment-specific role of the ascorbate-glutathione cycle in the response of tomato leaf cells to Botrytis cinerea infection. J Exp Bot 56:921–933

Kuzniak E, Skłodowska M (2005b) Fungal pathogen-induced changes in the antioxidant systems of leaf peroxisomes from infected tomato plants. Planta 222:192–200

Leterrier M, Corpas FJ, Barroso JB, Sandalio LM, del Río LA (2005) Peroxisomal monodehydroascorbate reductase. Genomic clone characterization and functional analysis under environmental stress conditions. Plant Physiol 138:2111–2123

Lipka V, Dittgen J, Bednarek P, Bhat R, Wiermer M, Stein M, Landtag J, Brandt W, Rosahl S, Scheel D, Llorente F, Molina A, Parker J, Somerville S, Schulze-Lefert P (2005) Pre- and postinvasion defenses both contribute to non-host resistance in Arabidopsis. Science 310:1180–1183

Lisenbee CS, Lingard MJ, Trelease RN (2005) Arabidopsis peroxisomes possess functionally redundant membrane and matrix isoforms of monodehydroascorbate reductase. Plant J 43:900–914

López-Huertas E, Charlton WL, Johnson B, Graham IA, Baker A (2000) Stress induces peroxisome biogenesis genes. EMBO J 19:6770–6777

Mateos RM, León AM, Sandalio LM, Gómez M, del Río LA, Palma JM (2003) Peroxisomes from pepper fruits (*Capsicum annuum* L.): purification, characterisation and antioxidant activity. J Plant Physiol 60:1507–1516

McCartney AW, Greenwood JS, Fabian MR, White KA, Mullen RT (2005) Localization of the tomato bushy stunt virus replication protein p33 reveals a peroxisome-to-endoplasmic reticulum sorting pathway. Plant Cell 17:3513–3531

Mittler R, Zilinskas BA (1991) Purification and characterization of pea cytosolic ascorbate peroxidase. Plant Physiol 97:962–968

Mittova V, Guy M, Tal M, Volokita M (2004) Salinity up-regulates the antioxidative system in root mitochondria and peroxisomes of the wild salt-tolerant tomato species *Lycopersicon pennellii*. J Exp Bot 55:1105–1113

Mittova V, Tal M, Volokita M, Guy M (2003) Up-regulation of the leaf mitochondrial and peroxisomal antioxidative systems in response to salt-induced oxidative stress in the wild salt-tolerant tomato species *Lycopersicon pennellii*. Plant Cell Environ 26:845–856

Moschou PN, Sanmartin M, Andriopoulou AH, Rojo E, Sanchez-Serrano JJ, Roubelakis-Angelakis KA (2008) Bridging the gap between plant and mammalian polyamine catabolism: a novel peroxisomal polyamine oxidase responsible for a full back-conversion pathway in Arabidopsis. Plant Physiol 147:1845–1857

Mueller S, Weber A, Fritz R, Mütze S, Rost D, Walczak H, Völkl A, Stremmel W (2002) Sensitive and real-time determination of H_2O_2 release from intact peroxisomes. Biochem J 363:483–491

Mullen RT, Trelease RN (1996) Biogenesis and membrane properties of peroxisomes: does the boundary membrane serve and protect? Trends Plant Sci 1:389–394

Mullen RT, Lisenbee CS, Miernyk JA, Trelease RN (1999) Peroxisomal membrane ascorbate peroxidase is sorted to a membranous network that resembles a subdomain of the endoplasmic reticulum. Plant Cell 11:2167–2185

Müller M, Zechmann B, Zellnig G (2004) Ultrastructural localization of glutathione in Cucurbita pepo plants. Protoplasma 223:213–219

Narendra S, Venkataramani S, Shen G, Wang J, Pasapula V, Lin Y, Kornyeyev D, Holaday AS, Zhang H (2006) The Arabidopsis ascorbate peroxidase 3 is a peroxisomal membrane-bound antioxidant enzyme and is dispensable for Arabidopsis growth and development. J Exp Bot 57:3033–3042

Orozco-Cárdenas ML, Narváez-Vásquez J, Ryan CA (2001) Hydrogen peroxide acts as a second messenger for the induction of defense genes in tomato plants in response to wounding, systemin, and methyl jasmonate. Plant Cell 13:179–191

Palma JM, Corpas FJ, del Río LA (2009) Proteome of plant peroxisomes: new perspectives on the role of these organelles in cell biology. Proteomics 9:2301–2312

Palma JM, Jiménez A, Sandalio LM, Corpas FJ, Lundqvist M, Gómez M, Sevilla F, del Río LA (2006) Antioxidative enzymes from chloroplasts, mitochondria, and peroxisomes during leaf senescence of nodulated pea plants. J Exp Bot 57:1747–1758

Panchuk II, Volkov RA, Schoffl F (2002) Heat stress- and heat shock transcription factor-dependent expression and activity of ascorbate peroxidase in Arabidopsis. Plant Physiol 129:838–853

Poole B (1975) Diffusion effects in the metabolism of hydrogen peroxide by rat liver peroxisomes. J Theor Biol 51:149–167

Pracharoenwattana I, Smith SM (2008) When is a peroxisome not a peroxisome? Trends Plant Sci 13:522–525

Reumann S, Babujee L, Ma C, Wienkoop S, Siemsen T, Antonicelli GE, Rasche N, Luder F, Weckwerth W, Jahn O (2007) Proteome analysis of Arabidopsis leaf peroxisomes reveals novel targeting peptides, metabolic pathways, and defense mechanisms. Plant Cell 19:3170–3193

Reumann S, Quan S, Aung K, Yang P, Manandhar-Shrestha K, Holbrook D, Linka N, Switzenberg R, Wilkerson CG, Weber AP, Olsen LJ, Hu J (2009) In-depth proteome analysis of Arabidopsis leaf peroxisomes combined with in vivo subcellular targeting verification indicates novel metabolic and regulatory functions of peroxisomes. Plant Physiol 150:125–143

Romero-Puertas MC, Corpas FJ, Sandalio LM, Leterrier M, Rodríguez-Serrano M, del Río LA, Palma JM (2006) Glutathione reductase from pea leaves: response to abiotic stress and characterization of the peroxisomal isozyme. New Phytol 170:43–52

Romero-Puertas MC, McCarthy I, Sandalio LM, Palma JM, Corpas FJ, Gómez M, del Río LA (1999) Cadmium toxicity and oxidative metabolism of pea leaf peroxisomes. Free Radic Res 31:S25–S31

Romero-Puertas MC, Rodríguez-Serrano M, Corpas FJ, Gómez M, del Río LA, Sandalio LM (2004) Cadmium-induced subcellular accumulation of $O_2^{\cdot-}$ and H_2O_2 in pea leaves. Plant Cell Environ 27:1122–1134

Rylott EL, Rogers CA, Gilday AD, Edgell T, Larson TR, Graham IA (2003) Arabidopsis mutants in short- and medium-chain acyl-CoA oxidase activities accumulate acyl-CoAs and reveal that fatty acid beta-oxidation is essential for embryo development. J Biol Chem 278:21370–21377

Schumann U, Wanner G, Veenhuis M, Schmid M, Gietl C (2003) AthPEX10, a nuclear gene essential for peroxisome and storage organelle formation during Arabidopsis embryogenesis. Proc Natl Acad Sci USA 100:9626–9631

Shetty NP, Mehrabi R, Lütken H, Haldrup A, Kema GH, Collinge DB, Jørgensen HJ (2007) Role of hydrogen peroxide during the interaction between the emibiotrophic fungal pathogen Septoria tritici and wheat. New Phytol 174:637–647

Shi WM, Muramoto Y, Ueda A, Takabe T (2001) Cloning of peroxisomal ascorbate peroxidase gene from barley and enhanced thermotolerance by overexpressing in Arabidopsis thaliana. Gene 273:23–27

Sparkes IA, Brandizzi F, Slocombe SP, El-Shami M, Hawes C, Baker A (2003) An Arabidopsis pex10 null mutant is embryo lethal, implicating peroxisomes in an essential role during plant embryogenesis. Plant Physiol 133:1809–1819

Taler D, Galperin M, Benjamin I, Cohen Y, Kenigsbuch D (2004) Plant eR genes that encode photorespiratory enzymes confer resistance against disease. Plant Cell 16:172–184

Tausz M, Sircelj H, Grill D (2004) The glutathione system as a stress marker in plant ecophysiology: is a stress-response concept valid? J Exp Bot 55:1955–1962

Tolbert NE (1980) Photorespiration. In: Davies DD (ed) The biochemistry of plants. Academic, New York, pp 487–523

Tolbert NE (1981) Metabolic pathways in peroxisomes and glyoxysomes. Annu Rev Biochem 50:133–157

Vandenabeele S, Vanderauwera S, Vuylsteke M, Rombauts S, Langebartels C, Seidlitz HK, Zabeau M, Van Montagu M, Inzé D, Van Breusegem F (2004) Catalase deficiency drastically affects gene expression induced by high light in Arabidopsis thaliana. Plant J 39:45–58

Vanderauwera S, Zimmermann P, Rombauts S, Vandenabeele S, Langebartels C, Gruissem W, Inzé D, Van Breusegem F (2005) Genome-wide analysis of hydrogen peroxide-regulated gene expression in Arabidopsis reveals a high light-induced transcriptional cluster involved in anthocyanin biosynthesis. Plant Physiol 139:806–821

Wang J, Zhang H, Allen RD (1999) Overexpression of an Arabidopsis peroxisomal ascorbate peroxidase gene in tobacco increases protection against oxidative stress. Plant Cell Physiol 40:725–732

Yamaguchi K, Mori H, Nishimura M (1995) A novel isoenzyme of ascorbate peroxidase localized on glyoxysomal and leaf peroxisomal membranes in pumpkin. Plant Cell Physiol 36:1157–1162

Yan J, Wang J, Zhang H (2002) An ankyrin repeat-containing protein plays a role in both disease resistance and antioxidation metabolism. Plant J 29:193–202

Yang T, Poovaiah BW (2002) Hydrogen peroxide homeostasis: activation of plant catalase by calcium/calmodulin. Proc Natl Acad Sci U S A 99:4097–4102

Zechmann B, Mauch F, Sticher L, Müller M (2008) Subcellular immunocytochemical analysis detects the highest concentrations of glutathione in mitochondria and not in plastids. J Exp Bot 59:4017–4027

Zechmann B, Zellnig G, Urbanek-Krajnc A, Müller M (2007) Artificial elevation of glutathione affects symptom development in ZYMV-infected *Cucurbita pepo* L. plants. Arch Virol 152:747–762

Zhang H, Wang J, Nickel U, Allen RD, Goodman HM (1997) Cloning and expression of an Arabidopsis gene encoding a putative peroxisomal ascorbate peroxidase. Plant Mol Biol 34:967–971

Zimmermann P, Hennig L, Gruissem W (2005) Gene-expression analysis and network discovery using Genevestigator. Trends Plant Sci 10:407–409

Zimmermann P, Hirsch-Hoffmann M, Hennig L, Gruissem W (2004) GENEVESTIGATOR. Arabidopsis microarray database and analysis toolbox. Plant Physiol 136:2621–2632

Chapter 15
Identification of Potential Gene Targets for the Improvement of Ascorbate Contents of Genetically Modified Plants

Adebanjo A. Badejo and Muneharu Esaka

Abstract Up to half of the ~6.8 billion people living on planet earth live on less than $3 a day and also suffer from at least one micronutrient deficiency especially in Africa and Southeast Asia. Human lacks the capacity to synthesize vitamin C (ascorbic acid) and its deficiency causes scurvy. As excess vitamin C cannot be stored in the human's body, there is the need to regularly consume fruits and vegetables to supply this essential compound. In plants, it is multifunctional and indispensable. Overexpression of heterogonous genes to introduce novel traits into plants offers an effective way to increase the vitamin content of crops. Although many alternative biosynthesis routes for vitamin C have been proposed, the Smirnoff–Wheeler (L-galactose) pathway has been proven to be the functional pathway in Arabidopsis and many other fruit-bearing plants. Identifying limiting genes in the biosynthesis pathways and overexpression of such genes severally and collectively as well as in combination with genes from ascorbate recycling pathway may lead to the generation of transgenic plants with 'substantial' amount of vitamin C for both human nutrition leading to reduced 'hidden hunger' and agronomic purposes. Proper dissemination of scientifically proven safety information about such transgenic plants will also increase public confidence in selecting and consuming such nutritionally enhanced genetically modified food crops.

Keywords Biosynthesis pathway • Gene expression • Genetic engineering • Genetically modified foods

---

A.A. Badejo
Faculty of Life and Environmental Science, Shimane University, Matsue, Shimane 690-8504, Japan
e-mail: badejoaa1@yahoo.com

M. Esaka (✉)
Graduate School of Biosphere Sciences, Hiroshima University, Kagamiyama, Higashi-Hiroshima, Hiroshima 739-8528, Japan
e-mail: mesaka@hiroshima-u.ac.jp

N.A. Anjum et al. (eds.), *Ascorbate-Glutathione Pathway and Stress Tolerance in Plants*,
DOI 10.1007/978-90-481-9404-9_15, © Springer Science+Business Media B.V. 2010

1 Ascorbic Acid in Living Organisms

According to the Food and Agriculture Organization of the United Nations, close to 50% of the world's population are suffering from diseases caused by lack of essential minerals and vitamins (FAO 2006). The developed societies have taken measures to reduce such diseases by ensuring that fruits and vegetables are included in their diet. These fruits and vegetables are rich sources of antioxidants that have also been found to reduce the risk of certain cancers and cardiovascular diseases (Hancock and Viola 2005a). Unfortunately, malnutrition is widespread in the developing world where the populace depends upon monotonous cereal-rich diets that lacks the essential vitamins and minerals (Timmer 2003).

Many vertebrates synthesize ascorbic acid. Fish, amphibians and reptiles synthesize ascorbic acid in the kidney; mammals on the other hand produce it in the liver (Chatterjee 1973; Moreau and Dabrowski 1998). African trypanosome, *Trypanosoma brucei*, and the American trypanosome, *Trypanosoma cruzi* have been shown to synthesis ascorbic acid in a single-membrane organelle, the glycosome (Wilkinson et al. 2005). Although the biosynthesis capacity of most eukaryotic organisms appeared early in the evolutionary history especially the fish (Moreau and Dabrowski 1998); many including human, have lost the capacity to synthesize the compound. The mutation and non-functional L-gulono-1,4-lactone oxidase, the enzyme catalyzing the last step in the biosynthesis of ascorbic acid, incapacitated human's ability to synthesize the compound (Chatterjee 1973). Most plant species studied so far have been found to synthesize ascorbic acid (Davey et al. 2000; Smirnoff et al. 2001) including the photosynthesizing algae *Euglena gracilis* (Ishikawa et al. 2008). Yeasts are unusual in that they predominantly synthesize erythroascorbate, a five-carbon analogue of ascorbic acid that possesses all of its biochemical properties (Huh et al. 1998).

It took more than a century from the time the Scottish doctor James Lind found that administration of fresh citrus fruits could cure scurvy to the time it was first isolated by Svent Gyorgy (Davey et al. 2000). The studies on the biosynthesis of ascorbic acid in plants have been very interesting as it generates many direct and indirect connecting routes to the final product and has been reviewed by many authors (Smirnoff and Wheeler 2000; Valpuesta and Botella 2004; Hancock and Viola 2005b). It was towards the close of the last century that Wheeler et al. (1998) proposed a functional biosynthesis pathway for ascorbic acid (Fig. 1). The pathway has since been supported by genetic, molecular and biochemical proves (Conklin et al. 2006; Dowdle et al. 2007; Linster et al. 2007).

Ascorbic acid is present in all subcellular compartments such as mitochondria, apoplast (cell wall), cytosol, vacuoles, chloroplasts (Rautenkranz et al. 1994; Smirnoff and Wheeler 2000), with concentration ranging from 20 to 300 mM depending on the location within the plant. Except the last enzyme in the biosynthetic pathway that is localized in the inner mitochondria, the whole pathway is cytosolic (Smirnoff 2000). Under normal condition, the ascorbic acid pool in plant is maintained at close to 90% in the reduced state (Foyer 1993). Ascorbate oxidase and Ascorbate peroxidase are the two enzymes that catalyze the oxidation of ascorbic

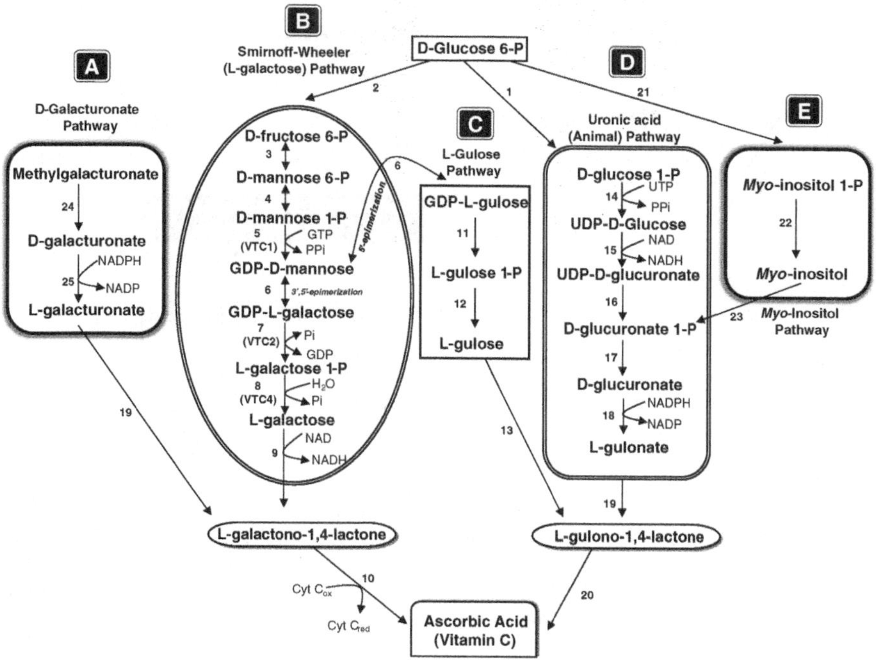

Fig. 1 Biosynthetic pathway of ascorbic acid. The pathway D represent the animal pathway. The flow in B represents the Smirnoff-Wheeler (L-galactose) pathway and the network of other alternative pathways are shown in A, C and E. The reactions are catalyzed by: 1. phosphoglucose mutase; 2. phosphoglucose isomerase (EC 5.3.9.1); 3. phosphomannose isomerase (EC 5.3.1.8); 4. phosphomannose mutase (EC 5.4.2.8); 5. GDP-mannose pyrophosphorylase (VTC1) (EC 2.7.7.22); 6. GDP-mannose-3′,5′-epimerase (EC5.1.3.18); 7. GDP-L-galactose phosphorylase; 8. L-galactose 1-phosphate phosphatase (VTC4); 9. L-galactose dehydrogenase; 10. L-galactono-1,4-lactone dehydrogenase (EC 1.3.2.3); 11. nucleotide pyrophosphatase or sugar-1-P guanylyl-transferase; 12. sugar phosphatase; 13. sugar dehydrogenase; 14. UDP-glucose pyrophosphorylase; 15. UDP-glucose dehydrogenase; 16. UDP-glucuronate pyrophosphorylase; 17. Glucuronate 1-kinase; 18. D-glucuronate reductase; 19. Aldonolactonase (EC 3.1.1.18); 20. L-gulonolactone oxidase or dehydrogenase; 21. *myo*-inositol 1-P synthase; 22. *myo*-inositol 1-P phosphatase; 23. *myo*-inositol oxygenase; 24. (pectin) methylesterase (EC 3.1.1.11); 25. D-galacturonate reductase (EC 1.1.1.203)

acid and formation of monodehydroascorbate (MDHA) radical (Fig. 2). MDHA reductase uses NAD(P)H as a reductant to reduce MDHA back to ascorbic acid in the cytosol or chloroplast. In the chloroplast, thylakoid-associated ferredoxin has been found to reduce MDHA more effectively than MDHAR (Asada 1999). However MDHA emanating from the thylakoid lumen as a result of the action of violaxanthin deepoxidase is not always available as a substrate for both reductants; thus a drop in pH of the lumen causes MDHA to just disproportionate into dehydroascorbate (DHA) and ascorbic acid (Asada 1999). DHA reductase uses glutathione as a reductant to reduce DHA back to ascorbate leading to the ascorbate–glutathione cycle (Fig. 2). In cases where there are not enough DHA reductase

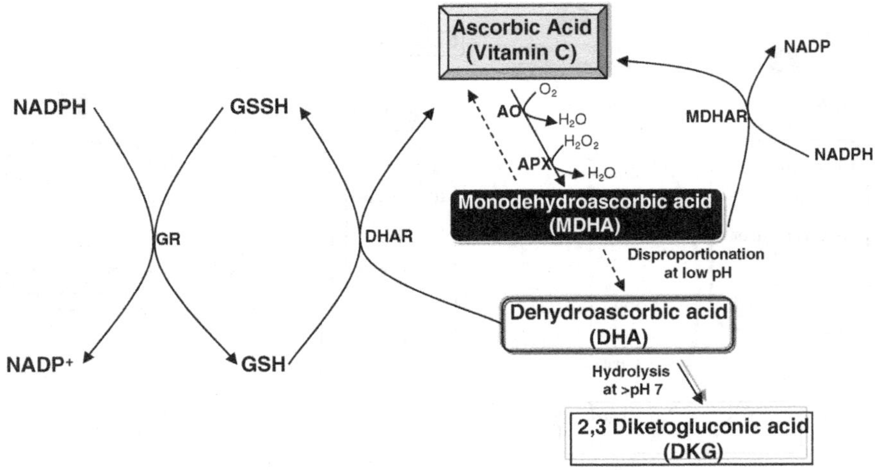

Fig. 2 The ascorbate–glutathione (Halliwell–Foyer–Asada) cycle. The enzymes catalyzing the reactions are as follows; AO: Ascorbate oxidase (EC 1.10.3.3); GR: Glutathione reductase (EC 1.8.1.7); DHAR: Dehydroascorbate reductase (EC 1.8.5.1); MDHAR: Monodehydroascorbate reductase (EC 1.6.5.4); APX: Ascorbate peroxidase (EC 1.11.1.11). GSH: Glutathione; GSSH: Oxidized Glutathione. The reaction mechanism is discussed in the text. The disproportionation of MDHA at low pH is shown with *broken arrows*

(as the apoplast does not have much DHA reductase and glutathione) to reduce DHA to ascorbic acid, the excess DHA which is very unstable above pH7 undergoes irreversible hydrolysis to 2,3-diketogulonic acid. The degradation of ascorbic acid has been extensively discussed (Hancock and Viola 2005b).

Since micronutrients are highly invaluable in human diets and more so billions of people in the developing countries suffer from at least one micronutrient deficiency, till date many multinational companies like Roche, BASF, and Cognis Health and Nutrition are spending millions of euros to increase their vitamin production facilities (Herbers 2003).

2 Functions of Ascorbic Acid

2.1 Antioxidant

Adverse environmental conditions and normal aerobic metabolic processes such as photosynthesis or respiration leads to the generation of reactive oxygen species (ROS) in the form of superoxide, singlet oxygen, ozone, and hydrogen peroxide. ROS are very toxic to the plants and if the amount generated are not controlled it can lead to oxidation of protein resulting in adverse effect on plants (Smirnoff 1996). Ascorbic acid acts as an antioxidant wherein it protects plant by detoxifying the ROS through decomposition (Conklin 2001).

2.2 *Enzyme Co-factor*

Ascorbic acid is highly involved in the regeneration of α-tocopherol commonly called vitamin E from the tocopheroxyl radical thus providing membrane protection (Thomas et al. 1992). About 10–20% of the ascorbate is found in the thylakoid lumen of plant where it functions greatly as a co-factor for the enzymes violaxanthin de-epoxidase (the enzyme catalyzing the formation of the pigment zeaxanthin from violaxanthin and antheroxanthin in the xanthophyll cycle where it is necessary for dissipation of excess excitation energy), and ethylene forming enzyme (necessary for the ripening of various fruits), 2-oxoacid-dependent dioxygenases (important in the synthesis of abscisic acid and gibberellic acid) (Eskling et al. 1997; Davey et al. 2000; Smirnoff 2000; Arrigoni and De Tullio 2002).

2.3 *Cell Division and Growth*

Large concentration of ascorbate is found in the meristematic tissues that are the major sites for cell division (Conklin 2001) and it has been shown to be involved in cell division and expansion (Kato and Esaka 1999, 2000, see review by Davey et al. 2000). Tobacco Bright Yellow-2 cells with antisense L-galactono-1,4-lactone dehydrogenase grew with abnormal shapes as well as a reduction in cell growth because the ascorbate content was only 70% of the wild type (Tabata et al. 2001). On the other hand when the same gene was overexpressed in tobacco BY2 cells the mitotic index was higher in the transgenic cells than the wild type (Tokunaga et al. 2005). Keller et al. (1999) also reported that transgenic potatoes with antisense GDP-mannose pyrophosphorylase are smaller in size, had lower ascorbate content and the leaves undergo senescence faster compared to the wild type. All these strongly support the involvement of ascorbic acid in growth and cell division.

3 One Product (Ascorbate) with Multiple Pathways

3.1 *Inversion and Non-inversion Routes*

The synthesis of ascorbate in plant was thought to be similar to that of animal where the compound was synthesized from UDP-D-Glucose through the “inversion” pathway in which the C1 of glucose becomes C6 of ascorbate and C6 becomes C1 (Smirnoff and Wheeler 2000). Although plants were capable of converting D-galacturonic acid derivatives (Mapson and Isherwood 1956) and L-galactono-γ-lactone to ascorbate (Mapson et al. 1954) just as the animals could convert D-glucuronate into L-gulono-1,4-lactone and finally to ascorbate; when labeled hexoses were fed to plants it disproves the synthesis of ascorbate through "inversion" of carbon (Loewus et al. 1958; Loewus 1963). An alternative pathway involving "non-inversion" of carbon was later proposed in which ascorbate was formed directly from glucose via D-glucosone (D-arabino-hexos-2-ulose) and L-sorbosone (L-xylo-hexo-2-ulose) (Loewus et al. 1990). Although the reaction takes place at a very slow rate, the enzyme catalyzing the last step of the reaction involving NADP-dependent oxidation of L-sorbosone to ascorbic acid was partially purified (Loewus et al. 1990) to justify the existence of the pathway. Unfortunately, there were no strong evidences to support the first two reactions which involve the conversion of D-glucose to D-sorbosone and epimerization of D-glucosone to L-sorbosone, and so the pathway was not regarded as a major route to ascorbate biosynthesis.

Available evidences support L-glactono-1,4-lactone as an effective precursor of ascorbate (Ostergaard et al. 1997; Davey et al. 1999) but there were uncertainties as to whether it is present in plant and also the possibility to synthesizing it without "inversion" of the hexose carbon. Over a decade ago Wheeler et al. (1998) were able to demonstrate the presence of L-galactose dehydrogenase in pea and also the NAD^+ dependent oxidation of L-galactose at C1 to L-galactono-1,4-lactone thus proposing the Smirnoff–Wheeler (L-galactose) pathway. In this newly proposed biosynthetic pathway, GDP-galactose formed from GDP-mannose undergoes two sequential oxidation reaction leading to L-galactono-1,4-lactone and ascorbic acid without the inversion of carbon chain. Recent works on fruit bearing plants such as blackcurrant, peach, kiwifruit, and acerola show correlation between ascorbic acid content of these fruits and gene expression of the enzymes involved in the Smirnoff–Wheeler pathway (Hancock et al. 2007; Imai et al. 2009; Bulley et al. 2009; Badejo et al. 2009a).

In some of the early radiolabel experiments with detached strawberry, it was shown that ascorbic acid could be synthesized in vivo from D-glucuronic acid, D-glucuronolactone, and D-galacturonic acid methyl ester with little redistribution of radiolabel and also without the inversion of carbon chain (Loewus et al. 1958). In linking this to the proposed pathway by the Wheeler group, the D-galacturonic acid methyl ester in plant could possibly be reduced by a non-specific aldo-keto reductase activity, to form L-galactono-1,4-lactone which is the substrate for L-galactono-1,4-lactone dehydrogenase (Davey et al. 2000). Unfortunately, the conversion of D-glucuronolactone, D-glucuronic acid methyl ester and L-gulonolactone to ascorbic acid did not surface in the Smirnoff–Wheeler pathway (Wheeler et al. 1998) thus the suggestion for the possibility of a separate uronic acid pathway (Smirnoff et al. 2001). It was not long before the D-galacturonic acid reductase was cloned and characterized from strawberry fruits thus providing molecular evidence for the uronic acid pathway in which pectin-derived D-galacturonic acid is reduced to L-galactonic acid and later to the terminal substrate in the synthesis of ascorbic acid (Agius et al. 2003). Another key enzyme in this uronic acid pathway is

Aldonolactonase which is highly vital to the synthesis of ascorbate in photosynthesizing algae *Euglena gracilis* (Ishikawa et al. 2008). Till date Aldonolactonase and/or its orthologs have not been cloned from plants neither has it been found in among higher plants data base (Ishikawa et al. 2008).

3.2 Other Alternative Routes

The isolation of Arabidopsis vitamin C (*vtc1*) mutant that encodes GDP-mannose pyrophosphorylase (GMP) lends credence to the Smirnoff–Wheeler pathway (Conklin et al. 1999; Conklin 2001) at the same time opened up yet another possible link to the synthesis of ascorbic acid in higher plant. GMP catalyzes the conversion of D-mannose-1-P to GDP-D-mannose. GDP-mannose-3′,5′ epimerase has been known to catalyze the conversion of GDP-D-mannose to GDP-L-galactose (Wolucka et al. 2001). The same group also found that another product, GDP-L-gulose, is formed alongside GDP-L-galactose in the epimerization reaction (Wolucka and Van Montagu 2003). While GDP-L-galactose is formed through the 3′,5′-epimerization of GDP-D-mannose, GDP-L-gulose is formed through the 5′-epimerization. This reaction product established a new branch of ascorbic acid biosynthesis pathway in plants with a link to the biosynthesis pathway in animals (Wolucka and Van Montagu 2003; Valpuesta and Botella 2004).

Lorence et al. (2004) showed molecular and biochemical evidence for a possible biosynthetic route using *myo*-inositol (MI) as the initial substrate. In the MI pathway, the conversion of D-glucose-6-phosphate to *myo*-inositol 1-phosphate by *myo*-inositol 1-phosphate synthase (IPS) precedes the dephosphorylation of *myo*-inositol-1-phosphate by the enzyme *myo*-inositol-1-phosphate phosphatase to form MI. *Myo*-inositol oxygenase (MIOX) then catalyzes the oxidation of MI to D-glucuronate (Lorence et al. 2004) before undergoing additional steps in which L-gulonate and L-gulono lactone are formed prior to ascorbic acid synthesis. It is worth noting that *myo*-inositol was not considered as a major precursor to ascorbic acid. Furthermore no correlation could be found between the gene expression of MI pathway genes (IPS and MIOX) and high ascorbate accumulation in kiwifruit (Bulley et al. 2009). All these alternative pathways reveal a complex picture of ascorbic acid biosynthesis in *planta* than had been expected (Valpuesta and Botella 2004).

4 Key Ascorbate Biosynthesis Enzymes of Smirnoff–Wheeler Pathway

4.1 Phosphomannose Isomerase (PMI)

On the basis of sequence similarity, three classes of PMI (EC 5.3.1.8) have been defined. Bacteria and human PMIs belong to the type1 (Coulin et al. 1993). PMI activity has been confirmed from ManA (Rv3255c), an enzyme that has been shown to be essential for mycobacterial growth in vitro (Patterson et al. 2003). Unfortunately, PMI is often not expressed in plants (Gao et al. 2005; Zhu et al. 2005; Badejo et al. 2009b) which makes the origin of the D-mannose 6-phosphate unclear. Furthermore, PMI has been used as an antibiotic-free positive selection marker in plant biotechnology (Wenck and Hansen 2005). Recently, two putative PMI genes were identified in *Arabidopsis thaliana* (PMI1, At3g02570; and PMI2, At1g67070), and have been used to show that under normal growth conditions, PMI1 but not PMI2 is essential for the biosynthesis of AsA in *A. thaliana* leaves (Maruta et al. 2008).

4.2 Phosphomannomutase

Phosphomannosemutases (PMM; EC 5.4.2.8) are phosphotransferases belonging to the superfamily of haloalkanoic acid dehalogenases. In the functional AsA biosynthetic pathway, PMM catalyzes the interconversion of mannose 6-phosphate and mannose 1-phosphate, and is required for the synthesis of GDP-D-mannose (Wheeler et al. 1998; Dowdle et al. 2007). It has also been found that PMM is constitutively expressed in both the vegetative and reproductive organs of plants (Qian et al. 2007; Badejo et al. 2009b). PMM is required in the synthesis of D-mannose which is vital not only for AsA biosynthesis but also in post-translational modifications, such as protein glycosylation and glycosylphosphatidylinositol (Hoeberichts et al. 2008). In human a mutation in the PMM gene was reported to result in a fatal clinical disorder that provokes impaired neurological development and increased childhood mortality (Van Schaftingen and Jaeken 1995). Most plants contain a single PMM gene (Qian et al. 2007; Hoeberichts et al. 2008; Badejo et al. 2009b) and till date no PMM knockout mutants have been reported. Efforts to generate knockdown mutants were unsuccessful (Hoeberichts et al. 2008). Two point mutations were located on the first exon of the PMM gene outside the conserved DXDX(T/V) motif characteristic of the phosphotransferases. Sequence comparison with the wild type shows that the first mutation resulted in a GGA to AGA codon change (codon 7), leading to a substitution of a neutral glycine residue with a basic arginine; while in the second a CGA to CAA codon change (codon 37) occurred thereby replacing an arginine with a neutral glutamine (Hoeberichts et al. 2008).

4.3 GDP-D-Mannose Pyrophosphorylase

GDP-D-mannose pyrophosphorylase (GMP; EC 2.7.7.22), localized to the cytosol, is the enzyme that catalyzes the synthesis of GDP-mannose (the activated form of mannose), an important step in the formation of all guanosine-containing sugar nucleotides in plants. Prokaryotes and eukaryotes use this GDP-mannose in the synthesis of complex structural proteins. It is highly essential for the synthesis of three different structural carbohydrates in the plant cell wall – mannans, L-fucose and L-galactose (Bewley et al. 1997; Bonin et al. 1997; Wheeler et al. 1998). The vitamin C deficient Arabidopsis mutant VTC1 locus encodes GMP (Conklin et al. 2000). The *vtc1-1* mutant which is ozone-sensitive was isolated from ethylmethane-sulfonate-mutagenized pools and is incapable of properly converting glucose and mannose to ascorbate thus possessing only about a quarter of the wild-type ascorbate content (Conklin et al. 1999; Conklin 2001).

4.4 GDP-D-Mannose-3′,5′-Epimerase

GDP-D-mannose-3′,5′-epimerase (GME; EC 5.1.3.18) represent the intersection between the pathway that leads to AsA synthesis and that of the cell wall polysaccharides and glycoprotein in plants. Upon the formation of GDP-D-Mannose by GMP, GME catalyzes the formation of GDP-L-galactose from GDP-D-mannose. It is one of the most highly conserved proteins involved in the biosynthesis of AsA (Wolucka and Van Montagu 2007). In reporting an alternative pathway for the biosynthesis of AsA, Wolucka and Van Montagu (2003) identified another epimerization product, GDP-L-gulose, released by the catalytic action of GME in addition to the already established GDP-L-galactose. GME is unique as it performs three chemical reactions-oxidation, epimerization and reduction- at the same active site (Major et al. 2005). The conserved NAD binding domain in rice GME compared to that of Arabidopsis reflected on the reaction it catalyzed; as GME was inhibited by GDP and strongly assisted by NAD^+ (Watanabe et al. 2006).

4.5 GDP-L-Galactose Phosphorylase

L-galactose-1-P has been described as the first metabolite in the Smirnoff–Wheeler pathway that is dedicated wholly to ascorbate biosynthesis. The enzyme GDP-L-galactose phosphorylase (GGP) catalyzes the conversion of GDP-L-galactose to L-galactose-1-phosphate (Laing et al. 2007). Vitamin C deficient Arabidopsis mutants were first identified as ozone sensitive, with characteristically low vitamin C contents and later mapped to a locus on chromosome 4 that was designated VTC2 (vitamin C 2) (Conklin et al. 2000; Conklin 2001). The mutated gene identified as At4g26850 was found to encode a novel protein. At4g26850 shows high similarity to another *Arabidopsis* gene At5g55120 both of which were uncharacterized (Laing et al. 2007). In attempt to characterize the protein sequence predicted for At4g26850, it was found by BLAST (Basic Logic Alignment Search Tool) that it shared similarities with HIT (histidine triad) protein family with the characteristic HXHXH (where X is a hydrophobic amino acid) motif (Laing et al. 2007). The HIT enzymes share a lot in common with the VTC gene. All the known HIT enzymes use the second His residue (His238 in *Arabidopsis* VTC2) of the HIT motif in a covalent nucleotidylylation reaction where the intermediate enzyme can then be broken down by simple hydrolysis, phosphrolysis or a specific phosphorylated compound (Linster and Clarke 2008).

VTC2 and kiwifruit homolog were found to specifically convert GDP-L-galactose to L-galactose-1-P and to require the presence of a guanylyl acceptor other than water to catalyze the reaction (Linster et al. 2007; Dowdle et al. 2007; Laing et al. 2007). Linster et al. (2008) went on to provide evidence for covalent guanylylation of the second His residue of the HIT motif of Arabidopsis VTC2 thus proving that VTC2 is a member of the phosphorylasetransferase (but not hydrolase) branch of the HIT protein superfamily.

Laing et al. (2007) named the Arabidopsis At4g26850 gene as GDP-L-galactose-D-mannose-1-phosphate guanyltransferase proving with an indirect enzyme coupled assay that the gene is able to use a range of hexose-1-P sugars. The inability of the enzyme to effectively transfer guanosine monophosphate from GDP-L-galactose-1-phosphate to either pyrophosphate or phosphate led to the conclusion that it is neither a pyrophosphorylase nor a phosphorylase (Laing et al. 2007). One the other hand two other independent groups (Dowdle et al. 2007, Linster et al. 2007) referred to the same gene as GDP-L-galactose pyrophosphorylase because they were able to find high VTC2 activities in the presence of inorganic phosphate. This little but highly controversial enzyme activity led to the proposed VTC2 cycle (Wolucka and Van Montagu 2007). Further works by the Linster group was able to clarify that VTC2 acts as a phosphorylase forming GDP, and not a transferase in which GDP-hexose is formed at the end of the reaction (Fig. 3) because the VTC2 has phosphorylase activity that was 100-fold higher than the transferase activity (Linster et al. 2008; Linster and Clarke 2008).

At5g55120 was also found to be GGP and was designated VTC5 (Dowdle et al. 2007). The kinetic data and sequence identity of VTC5 is highly similar to those of VTC2 but the gene expression of VTC2 in reproductive and vegetative tissues of Arabidopsis was far greater than those of VTC5 (Dowdle et al. 2007).

4.6 L-Galactose-1-Phosphate Phosphatase

L-Galactose-1-phosphate phosphatase (GPP) was previously classified as *myo*-inositol-1-phosphate phosphatase but phylogenetic analyses as well biochemical characterizations have been used to clarify the differences (Laing et al. 2004; Conklin et al. 2006). GGP purified from kiwifruit is highly specific in its hydrolysis of L-galactose-1-phosphate in a reaction that is highly dependent on Mg^{2+} with concentration less than 2 mM to form L-galactose (Laing et al. 2004). Ascorbate deficient Arabidopsis mutant VTC4 encodes the gene GPP. Alignment shows high identity between Arabidopsis and kiwifruit GPP and both are highly active against L-galactose-1-phosphate compared to *myo*-inositol-1-phosphate (Laing et al. 2004).

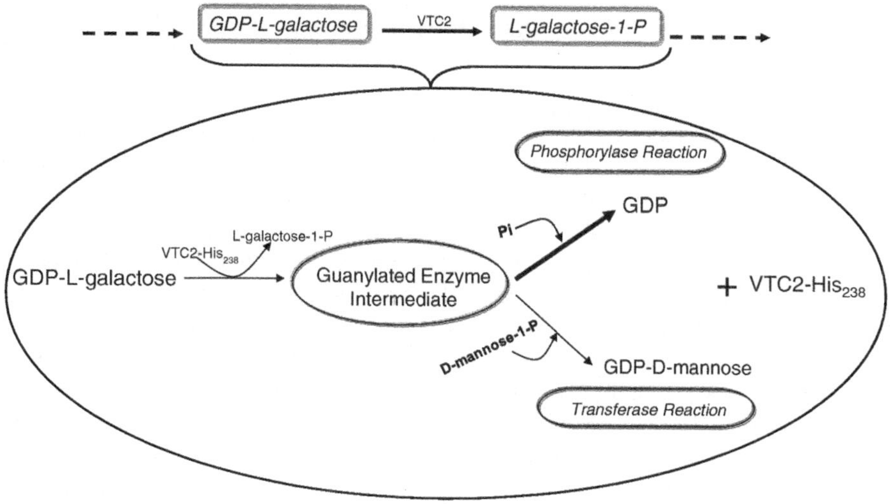

Fig. 3 The proposed reactions catalyzed by VTC2 adapted from Linster and Clarke (2008). The reaction undergoes two sequential steps in which L-galactose-1-P is formed when guanylated enzyme (guanylylated active site His) intermediate emerges upon the use of a-phosphate of GDP-L-galactose. In the latter reaction the dominant phosphorylase reaction (depicted with *thick arrow*) generates GDP while the less dominant transferase reaction (depicted with the thin) forms GDP-D-mannose. The *broken arrows* on the left and right represent reactions from D-glucose to GDP-L-galactose and from L-galactose 1P to ascorbic acid respectively

4.7 L-Galactose Dehydrogenase (GDH)

L-Galactose dehydrogenase (GDH; EC 1.1.1.117) catalyzes the penultimate step involving the conversion of L-galactose to L-galactono-1,4-lactone in an irreversible reaction by oxidizing the C1 of L-galactose in the Smirnoff–Wheeler pathway (Wheeler et al. 1998). GDH purified from spinach has high substrate specificity toward L-galactose but none for L-fructose, L-mannose, L-xylose, L-arabinose, D-galactose, D-glucose, L-fucose, D-mannose and D-arabinose (Mieda et al. 2004). Among the characteristics of this cytosol localized gene is that it has a *K*m of 0.43 and 3.7 mM for L-galactose and L-gulose respectively in pea and can be inhibited up to 65% and 90% by 5 mM oxidized glutathione and 1 mM *N*-ethylmaleimide respectively (Gatzek et al. 2002). The spinach GDH activity was enhanced by more than 15% in 1 mM DTT and 1 mM glutathione (Mieda et al. 2004). Antisense suppression of GDH activity in *A. thaliana* reduces foliar ascorbate concentration but was not thought to have pleiotropic effect on other enzymes required for regenerating ascorbate from its oxidized state; however overexpression of the gene in plant was only able to increase GDH activity but not the ascorbate content (Gatzek et al. 2002).

4.8 L-Galactono 1,4 Lactone Dehydrogenase (GalLDH)

L-Galactono 1,4 lactone dehydrogenase (GalLDH; EC 1.3.2.3) catalyzes the last step in the biosynthesis pathway for ascorbate in plants. It is localized to the inner membrane of the mitochondria (Siendone et al. 1999) and uses cytochrome *c* as an electron acceptor (Ostergaard et al. 1997; Imai et al. 1998). GalLDH is incapable of catalyzing the synthesis of ascorbate in the presence of potassium cyanide, an inhibitor of cytochrome *c* oxidase (Bartoli et al. 2000). It is one of the enzymes in the Smirnoff–Wheeler pathway (Wheeler et al. 1998) whose activity in plant has been known for a long time (Mapson et al. 1954). The enzyme has been purified from a number of plant species including potato and cauliflower (Oba et al. 1995; Ostergaard et al. 1997). Unlike the animal L-gulono-1,4-lactone oxidase that can utilize L-gulono-1,4-lactone and L-galactono-1,4-lactone, GalLDH uses L-galactono-1,4-lactone as a specific substrate (Oba et al. 1995; Ostergaard et al. 1997).

5 Factors Affecting Ascorbate Biosynthesis and Accumulation

Researches have shown that many factors control the ascorbic acid contents in plants. It has been shown earlier that ascorbate content of plant changes with the light intensity available to the plant (Tabata et al. 2002). The AsA content is diurnally regulated in a number of plant species where the peak of accumulation is recorded when daylight intensity is maximal and this correlates with the photosynthetic activity (Tamaoki et al. 2003; Chen and Gallie 2004). Total foliar ascorbate of potato leaves was found to be dependent on the age of the leaves in the plant (Bartoli et al. 2000). The ascorbate content of the young leaves was ten times more than those of leaves undergoing senescence (Bartoli et al. 2000). Ascorbate is also found in different concentrations in the vegetative and reproductive organs of plants. The highest concentrations was recorded in the fruits of acerola (Badejo et al. 2007) while the flowers and roots of Arabidopsis has the highest and lowest concentrations respectively (Lorence et al. 2004). Various cell compartments in plants also accumulate ascorbate to different degrees. The chloroplast and cytoplasm have relatively higher concentration compared to the vacuole of plant cell (Foyer et al. 1983; Davey et al. 2000). Two major substrates in the biosynthesis pathway, namely GDP-D-mannose and L-galactose, are of great importance in the control of ascorbate concentration (Wheeler et al. 1998). GDP-D-mannose and the major epimerization product, GDP-L-galactose, are channeled not only to ascorbate biosynthesis but also to cell-wall polysaccharide (xyloglucan and rhamnogalacturonans) and/or glycoprotein biosynthesis (Roberts 1971; Rayon et al. 1999). The diversion to cell-wall polysaccharide and protein glycosylation will put pressure on the de novo GDP-D-mannose available for ascorbate biosynthesis. The steps from L-galactose-1-P are fully committed to ascorbate biosynthesis as L-galactose is only a minor component of cell wall polysaccharide (Roberts 1971).

6 Targeted Genes for Ascorbate Improvement

6.1 Genes from Smirnoff–Wheeler Pathway

The overexpression and virus-induced gene silencing knockdown experiments performed on Arabidopsis and tobacco (*Nicotiana benthamiana*), respectively, indicated that PMM is playing an active role in ascorbic acid biosynthesis (Hoeberichts et al. 2008). With viral-vector mediated ectopic expression in *N. benthamiana*, PMM expression increased thereby resulting in an increase in the ascorbate content of the transgenic plant by 20~50% compared to the wild type (Qian et al. 2007). In the same research the Qian team also performed the transgenic expression of *Arabidopsis thaliana* PMM–GFP fusion protein in Arabidopsis which also increased ascorbate content by 25–33%. Transgenic tobacco overexpressing acerola (*Malpighia glabra*) PMM was found to have ascorbate level that was threefold that of the wild type plant (Badejo et al. 2009b).

It has been shown that biosynthesis in situ rather than translocation contributed to the accumulation of ascorbate in blackcurrant fruits with metabolic and genetic proofs that ascorbate is synthesized through the Smirnoff–Wheeler pathway (Hancock et al. 2007). The gene expression studies on kiwifruit (*Actinidia* spp.) showed that there is a correlation between the rise in ascorbate concentration and the expression of the Smirnoff–Wheeler pathway biosynthetic genes especially GDP-L-galactose guanyltransferase (GDP-L-galactose phosphorylase (GGP)) (Bulley et al. 2009). In the fruits of *Actinidia eriantha*, not only GGP was found to correlate with the increase in ascorbate concentration, the expression of GDP-mannose-3,5-epimerase (GME) was also found to be vital in regulating the ascorbate concentration (Bulley et al. 2009). When GGP gene was over expressed in Arabidopsis and transiently in tobacco it was found to elevate ascorbate level by more than threefold (Laing et al. 2007; Bulley et al. 2009). Transgenic tobacco transformed with GME and GGP, showed an increase in the ascorbate content that was twice the amonut observed with either genes over expressed alone (Bulley et al. 2009). The ascorbate content of tomatoes in which GME gene was silenced by RNAi were found to be reduced by 20–40% in the fruits 20 days post anthesis, and by 40–60% in the young fully expanded leaves (Gilbert et al. 2009). Analysis of the fruit pericarp of the RNAi transgenic tomatoes showed a significant reduction in both the cell number and cell size (Gilbert et al. 2009).

With acerola fruits having a high ascorbate content, it was found that the ascorbate concentration of the fruit is positively correlated with the genes in the Smirnoff–Wheeler pathway especially GDP-D-mannose pyrophosphorylase (GMP), GME and GDP-L-galactose phosphorylase (GGP) (Badejo et al. 2008, 2009a). In attempt to understand why acerola has such high ascorbate content, the transcript expressions of the biosynthesis genes were compared to that of the model plant *A. thaliana* and it found that transcript expression of acerola was 5–700-fold higher (Badejo et al. 2009a). Antisense inhibition of GMP activity in potato resulted in a decrease in the ascorbate content of the transgenic potato as well as the mannose content of the cell wall polysaccharide (Keller et al. 1999), but when acerola GMP gene was overexpressed in tobacco it resulted in two- to threefold increase in the ascorbate content of the transgenic tobacco (Badejo et al. 2008).

The Bulley team found it difficult to clone the last gene in the ascorbate biosynthesis pathway, L-galactono-1,4-lactone dehydrogenase (GalLDH) with degenerate primers. The expression of GalLDH could not be measured because there were no kiwifruit orthologues in the HortResearch Actinidia EST database. In acerola however GalLDH was cloned and its transcript level in the leaves was found to be the lowest among the five genes involved in the Smirnoff–Wheeler pathway that were analyzed (Badejo et al. 2009a). Although GalLDH expression have been found to correlate with ascorbate content (Badejo et al. 2009a; Pateraki et al. 2004; Tamaoki et al. 2003), it does not seem to exert strong control on ascorbate flux neither did the penultimate gene GDH (Gatzek et al. 2002). When GalLDH was knocked out in tomato, no change was recorded in the total ascorbate content of the young leaves and fruits (Alhagdow et al. 2007).

6.2 Genes from D-Galacturonate Pathway

Agius et al. (2003) have been able to show that ascorbate could be synthesized through D-galacturonic acid derived from the pectin of ripening strawberry fruits. The D-galacturonic acid which is also an abundant component in the cell wall of all plants is reduced to L-galactonic acid by the enzyme D-galacturonate reductase (Smirnoff 2003). This gene has expression correlating with ascorbate content in ripening strawberry fruits (Agius et al. 2003). Overexpression of the gene in *Arabidopsis thaliana* resulted in two- to threefold increase in the ascorbate content of the transgenic plants (Table 1). Although Aldonolactonase has not been reported in higher plants, RNAi silencing of Aladonolactonase in photosynthetic algae *Euglena gracilis* halted the cell growth, only to be restored on supplementation with one of L-galactono-1,4-lactone, D-galacturonate or D-glucuronate (Ishikawa et al. 2008).

6.3 Genes from Animal Pathway and myo-Inositol Route

At the beginning of the millennium, Jain and Nessler (2000) showed that ascorbate content of plants could be increased with overexpression of the rat gene L-gulono-γ-lactone oxidase (GLO) (Table 1). Interestingly the same rat gene was also able to rescue the ascorbate content of the Arabidopsis vitamin C mutants including *vtc1* when overexpressed in them (Radzio et al. 2003). However, plants overexpressing rat GLO still uses and prefers L-galactono-γ-lactone of the Smirnoff–Wheeler pathway as a precursor over L-gulono-γ-lactone (from the animal pathway) when fed artificially to the tobacco leaves (Jain and Nessler 2000). In the animal pathway

Table 1 Changes in the ascorbate content of some transgenic plants

Engineered gene	Gene source	Plant engineered	Effect on AsA pool	Reference
GalLDH	Tobacco	Tobacco BY2 cells	1.5–2.0-Fold increase	Tokunaga et al.(2005)
GLO	Rat	Lettuce	4.0–7.0-Fold increase	Jain and Nessler (2000)
MIOX	Arabidopsis	Arabidopsis	2.0–3.0-Fold increase	Lorence et al. (2004)
PMM	Arabidopsis	Arabidopsis	0.25–0.33-Fold increase	Qian et al. (2007)
PMM	Acerola	Tobacco	~2.0-Fold increase	Badejo et al. (2009b)
GMP	Acerola	Tobacco	2.0–3.0-Fold increase	Badejo et al. (2008)
GGP	Acerola	Tobacco	2.0–3.0-Fold increase	Badejo et al. (2009c)
GGP	Kiwifruit	Tobacco	3.0-Fold increase	Laing et al. (2007)
GGP	Kiwifruit	Arabidopsis	4.0-Fold increase	Bulley et al. (2009)
GalUAR	Strawberry	Arabidopsis	2.0–3.0-Fold increase	Agius et al. (2003)
DHAR	Wheat	Maize	~1.9-Fold increase	Chen et al. (2003)
GME/GGP	Kiwifruit	Tobacco	7.0-Fold increase	Bulley et al. (2009)

Genetically modified plants with increased ascorbate content. The genes engineered are: GalLDH, L-galactono-1,4-lactone dehydrogenase; GLO, L-gulono-γ-lactone oxidase; MIOX, *myo*-inositol oxygenase; PMM, phosphomannomutase; GMP, GDP-D-mannose pyrophosphorylase; GGP, GDP-L-galactose phosphorylase; GalUAR, D-galacturonic acid reductase; DHAR, Dehydroascorbate reductase; GME, GDP-D-mannose-3′,5′-epimerase glucuronate reductase catalyzes the formation of L-gulonic acid from L-glucuronic acid and the latter is acted upon by aldono lactonase to form the penultimate substrate to ascorbate synthesis (Fig. 1). The D-glucuronic acid can be formed through the action of *myo*-inositol oxygenase (Fig. 1). Lorence et al. (2004) showed that ectopic expression of *myo*-inositol oxygenase doubled the ascorbate content of transgenic *Arabidopsis thaliana* (Table 1). Also overexpression of purple acid phosphatase gene AtPAP15 which is involved in *myo*-inositol metabolism was also found to double the ascorbate content of transgenic *A. thaliana* (Zhang et al. 2008).

6.4 Genes from Ascorbate Recycling Pathway

Poplar tree hybrids expressing ascorbate recycling gene glutathione reductase showed slight increase in ascorbate concentration and were able to withstand oxidative stress (Foyer et al. 1995). When *A. thaliana* MDHA reductase that is essential for maintaining reduced ascorbate in plant was overexpressed in tobacco, it resulted in about twofold higher ascorbate contents in the transgenic plant compared to the wild type (Eltayeb et al. 2007). DHAR, which regenerates the reduced form of ascorbate, elevated the ascorbic acid content of transgenic tobacco in which it was overexpressed (Chen et al. 2003).

7 Factors for and Against Genetic Engineering

Since most staple foods are low in ascorbate content, consumption of the recommended dietary allowance is being achieved by adding chemically synthesized vitamin C to foods (fortification), or by taking vitamin C pills (supplementation). Supplementation and fortification have been plagued with many demerits such as the increasing resistance among consumers to the use of additives in foods and a tightening of the legislation on foods and drinks especially in the developed world (Hancock and Viola 2005a). The recurrent cost, inadequate distribution system and lack of industrial food system makes implementation very difficult in developing countries (Bouis 2002). All these pose a great challenge to supplementation and fortification of foods. Biofortification, which is the process of enriching the nutrient content of crops as they grow by employing genetic engineering of genes in the plants, is a promising and viable option to nutrient enhancement in plants (Naqvi et al. 2008).

Genetic modifications in the past decade have been used to elucidate the biosynthetic pathways for important health components of crops. Research progress made in the area of genomics of food science and nutrition has improved the understanding of how common dietary chemicals affect human and animal health by altering the expression and/or structure of the genome (van Ommen 2004). Genetic engineering certainly increases the nutritional contents of plants. One important factor that has deterred the progress of genetic engineering technology when considered in the past decade is insufficient support from the upstream and downstream sectors and the importance of genetically engineered crops/foods have been watered down in many literatures lacking scientific facts (Vain 2006). Bioavailability of the nutrient in the transgenic plants is very vital to its acceptability, as increased levels of nutrients are not necessarily correlated with enhanced bioavailability (Jeong and Guerinot 2008). Further scientific research into the bioavailability of nutritionally enhanced genetically modified crops/foods will either debunk or strengthen the non-science based speculations and fears about the acceptance of transgenic crops. Also, there has been the need to cope with meager funding, the hostile market conditions combined with the tight regulatory frame work surrounding transgenic crops.

8 Genetic Engineering-Solution to Food Security and Malnutrition

'Hidden hunger', described as the micronutrient malnutrition inherent in human diets that are adequate in calories but lack vitamins and/or mineral elements (White and Broadley 2009), is ravaging a significant portion of the world population. Genetic engineering is one area of science that offers the opportunity to exponentially increase the rate of food production by avoiding losses due to diseases and pests and by increasing tolerance under adverse conditions (Bhalla 2006) as in the case of ascorbic acid that play roles in resistance to environmental stresses such as ozone, high temperature and UV radiation (Conklin et al. 1996; Sanmartin et al. 2003; Chen and Gallie 2005). Genetic engineering is also laden with potentials to breed crops with novel desirable characteristics, such as reduced allergenicity (Herman et al. 2003), improved nutritional qualities (Paine et al. 2005; Diaz de la Garza et al. 2007; Naqvi et al. 2008; Badejo et al. 2008, 2009b, c), hybrid seed production (Bhalla and Smith 1998) and improved plant productivity (Sakamoto et al. 2005). Many of the crops to which genetic engineering is being applied are the same as those being targeted for many years by plant breeding (Dale 1993). Genetically engineered foods have the potential to reach relatively remote rural populations that conventional interventions are not now reaching and with increased agricultural productivity (Bouis 2002).

Most diets rich in fruits and vegetables are associated with decreased risk of certain cancers and cardiovascular diseases, and this has been attributed to the high concentrations of antioxidants such as ascorbic acid in such foods. Production of these fruits and vegetables could also be enhanced through the application of mineral fertilizers. There is an increasing use of nitrogen fertilizers on agricultural soils worldwide resulting in the accumulation of ammonium in those soils (Britto et al. 2001). It was reported recently that mutation in GMP, one of the vital genes in the biosynthesis of vitamin C and essential for the synthesizing of D-mannose, confer hypersensitivity to ammonium (Qin et al. 2008). Defective protein glycosylation in the root rather than decreased vitamin C content in the mutant was reported to account for the hypersensitivity to ammonium (Qin et al. 2008). Genetic manipulation of GMP through over expression in plant has resulted in an increase in vitamin C content (Badejo et al. 2008). D-mannose formed through the activity of GMP is not channeled to vitamin C biosynthesis only but also to protein glycosylation and many other cellular processes. GDP-L-fucose is a constituent of glycoprotein and it is also synthesized from GDP-mannose (Bonin et al. 1997). The deficiency and/or a null mutant of GMP gene may be lethal (Conklin et al. 1999). Any genetic manipulation that increases the de novo synthesis of D-mannose or its activated form, GDP-mannose, may possibly increase ascorbate content as well as improve on protein glycosylation. Theoretically the same manipulation should doubtless reduce the sensitivity to ammonium in the natural and industrial fertilizers thereby enhancing the growth of plants. Also biofortification of crops through the application of mineral fertilizers, combined with breeding varieties with an increased ability to acquire mineral elements, has been advocated for incorporating essential mineral elements into the staple foods commonly consumed worldwide (White and Broadley 2009).

9 Concluding Remarks

The first decade of this millennium has witnessed an unprecedented progress in the study relating to improving ascorbate concentration in plants. Most of the overexpression of ascorbate biosynthesis genes have been performed on the leafy 'vegetables' and 'unfortunately' it is only the rat gene L-gulono-γ-lactone oxidase overexpressed in lettuce that yielded sevenfold increase (Jain and Nessler 2000). Albeit attractive, incorporation of a rat gene in readily consumable vegetable may not be a great choice for a willing consumer. Overexpression of genes from a high ascorbic acid containing plant, acerola, did not increase ascorbic acid content of plants beyond threefold (Badejo et al. 2008, 2009b, c) neither has the genes from Arabidopsis raised ascorbate content exponentially (Ishikawa et al. 2006). Thus, integrating increased biosynthesis with increased recycling capacity by overexpressing not only single but a combination of plant genes from 'all' ascorbate biosynthesis pathways and the ascorbate–glutathione recycling pathway may aid in tailoring plants with improved ascorbate contents at will and this is going to be of great benefit to many people worldwide. It will also ensure the generation of plants that is highly resistance to pathogens and environmental stress. While it is worth noting that genetic engineering only is not a panacea for all nutritional deficiencies, it can be a promising and cost-effective additional tool, complementing existing micronutrient interventions. It is still very difficult to discuss genetically modified foods without dividing people into different schools of thought as the expected benefits are given less credence than the feared risks, however a ban on genetically modified crops, as Conner et al (2003) described, could limit the options of farmers and be imprudent rather than precautionary especially to the developing world where the technology will have a lot to offer.

References

Agius F, Gonzalez-Lamothe R, Caballero JL, Munoz-Blanco J, Botella MA, Valpuesta V (2003) Engineering increased vitamin C levels in plants by over-expression of a D-galacturonic acid reductase. Nat Biotechnol 21:177–181

Alhagdow M, Mounet F, Gilbert L, Nunes-Nesi A, Garcia V, Just D, Petit J, Beauvoit B, Fernie AR, Rothan C, Baldet P (2007) Silencing of the mitochondrial ascorbate synthesizing enzyme L-galactono-1, 4-lactone dehydrogenase affects plant and fruit development in tomato. Plant Physiol 145:1408–1422

Arrigoni O, De Tullio MC (2002) Ascorbic acid: much more than just an antioxidant. Biochim Biophys Acta 1569:1–9

Asada K (1999) The water-water cycle in chloroplasts: scavenging of active oxygens and dissipation of excess photons. Annu Rev Plant Physiol Plant Mol Biol 50:601–639

Badejo AA, Jeong ST, Goto-Yamamoto N, Esaka M (2007) Cloning and expression of GDP-D-mannose pyrophosphorylase gene and ascorbic acid content of acerola (*Malpighia glabra* L.) fruit at ripening stages. Plant Physiol Biochem 45:665–672

Badejo AA, Tanaka N, Esaka M (2008) Analysis of GDP-D-mannose pyrophosphorylase gene promoter from acerola (*Malpighia glabra*) and increase in ascorbate content of transgenic tobacco expressing the acerola gene. Plant Cell Physiol 49:126–132

Badejo AA, Fujikawa Y, Esaka M (2009a) Gene expression of ascorbic acid biosynthesis related enzymes of the Smirnoff-Wheeler pathway in acerola (*Malpighia glabra*). J Plant Physiol 166:652–660

Badejo AA, Eltelib HA, Fukunaga K, Fujikawa Y, Esaka M (2009b) Increase in ascorbate content of transgenic tobacco plants overexpressing the acerola (*Malpighia glabra*) phosphomannomutase gene. Plant Cell Physiol 50:423–428

Badejo AA, Eltelib HA, Fujikawa Y, Esaka M (2009c) Genetic manipulation for enhancing vitamin C content in tobacco expressing acerola (*Malpighia glabra*) GDP-L-galactose phosphorylase gene. Hort Environ Biotech 50:329–333

Bartoli CG, Pastori GM, Foyer GM (2000) Ascorbate biosynthesis in mitochondria is linked to the electron transport chain between complexes III and IV. Plant Physiol 123:335–343

Bewley JD, Burton RA, Morohashi Y, Fincher GB (1997) Molecular cloning of a cDNA encoding a (1→4)-β-mannan endohydrolase from the seeds of germinated tomato (*Lycopersicon esculentum*). Planta 203:454–459

Bhalla PL, Smith N (1998) *Agrobacterium tumefaciens*-mediated transformation of Australian cultivars of cauliflowers, *Brassica oleracea* var. botrytis. Mol Breed 4:531–541

Bhalla PL (2006) Genetic engineering of wheat – current challenges and opportunities. Trends Biotech 24:305–311

Britto DT, Siddiqi MY, Glass AD, Kronzucker HJ (2001) Futile transmembrane NH_4^+ cycling: a cellular hypothesis to explain ammonium toxicity in plants. Proc Natl Acad Sci USA 98:4255–4258

Bonin CP, Potter I, Vanzin GF, Reiter WD (1997) The MUR1 gene of Arabidopsis thaliana encodes an isoform of GDP-D-mannose-4, 6-dehydratase, catalyzing the first step in the de novo synthesis of GDP-L-fucose. Proc Natl Acad Sci USA 94:2085–2090

Bouis HE (2002) Plant breeding: a new tool for fighting micronutrient malnutrition. J Nutr 132:491S–494S

Bulley SM, Rassam M, Hoser D, Otto W, Schunemann N, Wright M, MacRae E, Gleave A, Laing W (2009) Gene expression studies in kiwifruit and gene over-expression in Arabidopsis indicates that GDP-L-galactose guanyltransferase is a major control point of vitamin C biosynthesis. J Exp Bot 60:765–778

Chatterjee IB (1973) Evolution and the biosynthesis of ascorbic acid. Science 182:1271–1272

Chen Z, Young TE, Ling J, Chang S, Gallie DR (2003) Increased vitamin C content of plants through enhanced ascorbate recycling. Proc Natl Acad Sci USA 100:3525–3530

Chen Z, Gallie DR (2004) The ascorbic acid redox state controls guard cell signalling and stomatal movement. Plant Cell 16:1143–1162

Chen Z, Gallie DR (2005) Increasing tolerance to ozone by elevating foliar ascorbic acid confers greater protection against ozone than increasing avoidance. Plant Physiol 138:1673–1689

Conklin PL (2001) Recent advances in the role and biosynthesis of ascorbic acid in plants. Plant Cell Environ 24:383–394

Conklin PL, Norris SR, Wheeler GL, Williams EH, Smirnoff N, Last RL (1999) Genetic evidence for the role of GDP-mannose in plant ascorbic acid (vitamin C) biosynthesis. Proc Natl Acad Sci USA 96:4198–4203

Conklin PL, Williams EH, Last RL (1996) Environmental stress sensitivity of an ascorbic acid-deficient Arabidopsis mutant. Proc Natl Acad Sci USA 93:9970–9974

Conklin PL, Saracco SA, Norris SR, Last RL (2000) Identification of ascorbic acid-deficient *Arabidopsis thaliana* mutants. Genetics 154:847–856

Conklin PL, Gatzek S, Wheeler GL, Dowdle J, Raymond MJ, Rolinski S, Isupov M, Littlechild JA, Smirnoff N (2006) *Arabidopsis thaliana* VTC4 encodes L-Galactose-1-P phosphatase, a plant ascorbic acid biosynthetic enzyme. J Biol Chem 281:15662–15670

Conner A, Glare TR, Nap JP (2003) The release of genetically modified crops into the environment. Plant J 33:19–46

Coulin F, Magnenat E, Proudfoot AEI, Payton MA, Scully P, Wells TNC (1993) Identification of Cys-150 in the active site of phosphomannose isomerase from *Candida albicans*. Biochemistry 32:14139–14144

Dale PJ (1993) The release of transgenic crops into agriculture. J Agric Sci Camb 120:1–5

Davey MW, Gilot C, Persiau G, Ostergaard J, Han Y, Bauw GC, Van Montagu MC (1999) Ascorbate biosynthesis in Arabidopsis cell suspension culture. Plant Physiol 121:535–543

Davey MW, Van Montagu M, Inze D, Sanmartin M, Kanellis A, Smirnoff N, Benzie IJJ, Strain JJ, Favell D, Fletcher J (2000) Plant L-ascorbic acid: chemistry, function, metabolism, bioavailability and effects of processing. J Sci Food Agric 80:825–860

Diaz de la Garza RI, Gregory JF III, Hanson AD (2007) Folate biofortification of tomato fruit. Proc Natl Acad Sci USA 104:4218–4222

Dowdle J, Ishikawa T, Gatzek S, Rolinski S, Smirnoff N (2007) Two genes in *Arabidopsis thaliana* encoding GDP-L-galactose phosphorylase are required for ascorbate biosynthesis and seedling viability. Plant J 52:673–689

Eltayeb AE, Kawano N, Badawi GH, Kaminaka H, Sanekata T, Shibahara T, Inanaga S, Tanaka K (2007) Overexpression of monodehydroascorbate reductase in transgenic tobacco confers enhanced tolerance to ozone, salt and polyethylene glycol stresses. Planta 225:1255–1264

Eskling M, Arvidsson P, Akerlund HE (1997) The xanthophyll cycle, its regulation and components. Physiol Plant 100:806–816

FAO (2006) State of food insecurity in the world. Food and Agriculture Organization of the United Nations, Rome

Foyer CH (1993) Ascorbic acid. In: Alscher RG, Hess JL (eds) Antioxidants in higher plants. CRC Press, Boca Raton, FL, pp 31–58

Foyer CH, Rowell J, Walker D (1983) Measurements of the ascorbate concentration of spinach leaf protoplasts and chloroplasts during illumination. Planta 157:239–244

Foyer CH, Souriau N, Perret S, Lelandais M, Kunert K, Pruvost C, Jouanin L (1995) Overexpression of glutathione reductase but not glutathione synthetase leads to increases in antioxidant capacity and resistance to photoinhibition in poplar trees. Plant Physiol 109:1047–1057

Gao Z, Xie X, Ling Y, Muthukrishnan S, Liang GH (2005) *Agrobacterium tumefaciens* -mediated sorghum transformation using a mannose selection system. Plant Biotech J 3:591–599

Gatzek S, Wheeler GL, Smirnoff N (2002) Antisense suppression of L-galactose dehydrogenase in *Arabidopsis thaliana* provides evidence for its role in ascorbate synthesis and reveals light modulated L-galactose synthesis. Plant J 30:541–553

Gilbert L, Alhagdow M, Nunes-Nesi A, Quemener B, Guillon F, Bouchet B, Faurobert M, Gouble B, Page D, Garcia V, Petit J, Stevens R, Causse M, Fernie AR, Lahaye M, Rothan C, Baldet P

(2009) GDP-D-mannose 3, 5-epimerase (GME) plays a key role at the intersection of ascorbate and non-cellulosic cell-wall biosynthesis in tomato. Plant J 60:499–508

Hancock RD, Viola R (2005a) Improving the nutritional value of crops through enhancement of L-ascorbic acid (vitamin C) content: rationale and biotechnological opportunities. J Agric Food Chem 53:5248–5257

Hancock RD, Viola R (2005b) Biosynthesis and catabolism of L-ascorbic acid in plants. Crit Rev Plant Sci 24:167–188

Hancock RD, Walker PG, Pont SDA, Marquis N, Vivera S, Gordon SL, Brennan RM, Viola R (2007) L-ascorbic acid accumulation in fruit of *Ribes nigrum* occurs by in situ biosynthesis via the L-galactose pathway. Funct Plant Biol 34:1080–1091

Herbers K (2003) Vitamin production in transgenic plants. J Plant Physiol 160:821–829

Herman EM, Helm RM, Jung J, Kinney AJ (2003) Genetic modification removes an immunodominant allergen from soybean. Plant Physiol 132:36–43

Hoeberichts FA, Vaeck E, Kiddle G, Coppens E, van de Cotte B, Adamantidis A, Ormenese S, Foyer CH, Zabeau M, Inze D, Perilleux C, Van Breusegem F, Vuylsteke M (2008) A temperature-sensitive mutation in the *Arabidopsis thaliana* phosphomannomutase gene disrupts protein glycosylation and triggers cell death. J Biol Chem 283:5708–57018

Huh WK, Lee BH, Kim ST, Kim YR, Rhie GE, Baek YW, Hwang CS, Lee JS, Kang SO (1998) D-Erythroascorbic acid is an important antioxidant molecule in *Saccharomyces cerevisiae*. Mol Microbiol 30:895–903

Imai T, Ban Y, Terakami S, Yamamoto T, Moriguchia T (2009) L-ascorbate biosynthesis in peach: cloning of six L-galactose pathway-related genes and their expression during peach fruit development. Physiol Plant 136:139–149

Imai T, Karita S, Shiratori G, Hattori M, Nunome T, Oba K, Harai M (1998) L-galactono-γ-lactone dehydrogenase from sweet potato: purification and cDNA sequence analysis. Plant Cell Physiol 39:1350–1358

Ishikawa T, Dowdle J, Smirnoff N (2006) Progress in manipulating ascorbic acid biosynthesis and accumulation in plants. Physiol Plant 126:343–355

Ishikawa T, Nishikawa H, Gao Y, Sawa Y, Shibata H, Yabuta Y, Maruta T, Shigeoka S (2008) The pathway via D-galacturonate/L-galactonate is significant for ascorbate biosynthesis in *Euglena gracilis*. J Biol Chem 283:31133–1141

Jain AK, Nessler CL (2000) Metabolic engineering of an alternative pathway for ascorbic acid biosynthesis in plants. Mol Breed 6:73–78

Jeong J, Guerinot ML (2008) Biofortified and bioavailable: the gold standard for plant-based diets. Proc Natl Acad Sci USA 105:1777–1778

Kato N, Esaka M (1999) Changes in ascorbate oxidase gene expression and ascorbate levels in cell division and cell elongation in tobacco cells. Physiol Plant 105:321–329

Kato N, Esaka M (2000) Expansion of transgenic tobacco protoplasts expressing pumpkin ascorbate oxidase is more rapid than that of wild-type protoplasts. Planta 210:1018–1022

Keller R, Springer F, Renz A, Kossmann J (1999) Antisense inhibition of the GDP-mannose pyrophosphorylase reduces the ascorbate content in transgenic plants leading to developmental changes during senescence. Plant J 19:131–141

Laing WA, Bulley S, Wright M, Cooney J, Jensen D, Barraclough D, MacRae E (2004) A highly specific L-galactose-1-phosphate phosphatase on the path to ascorbate biosynthesis. Proc Natl Acad Sci USA 101:16976–16981

Laing WA, Wright MA, Cooney J, Bulley SM (2007) The missing step of the L-galactose pathway of ascorbate biosynthesis in plants, an L-galactose guanyltransferase, increases leaf ascorbate content. Proc Natl Acad Sci USA 104:9534–9539

Linster CL, Adler LN, Webb K, Christensen KC, Brenner C, Clarke SG (2008) A Second GDP-L-galactose Phosphorylase in Arabidopsis en Route to Vitamin C. J Biol Chem 283:18483–18492

Linster CL, Clarke SG (2008) L-ascorbate biosynthesis in higher plants: the role of VTC2. Trend Plant Sci 13:567–573

Linster CL, Gomez TA, Christensen KC, Adler LN, Young BD, Brenner C, Clarke SG (2007) Arabidopsis VTC2 encodes a GDP-L-galactose phosphorylase, the last unknown enzyme in the Smirnoff-Wheeler pathway to ascorbic acid in plants. J Biol Chem 282:18879–18885

Loewus FA (1963) Tracer studies of ascorbic acid formation in plants. Phytochemistry 2:109–128

Loewus MW, Bedgar DL, Saito K, Loewus FA (1990) Conversion of L-sorbosone to L-ascorbic acid by a NADP-dependent dehydrogenase in bean and spinach leaf. Plant Physiol 94:1492–1495

Loewus FA, Jang R, Seegmiller CG (1958) The conversion of C^{14} -labelled sugars to L-ascorbic acid in ripening strawberries. J Biol Chem 232:533–541

Lorence A, Chevone BI, Mendes P, Nessler CL (2004) Myo-inositol oxygenase offers a possible entry point into plant ascorbate biosynthesis. Plant Physiol 134:1200–1205

Major LL, Wolucka BA, Naismith JH (2005) Structure and function of GDP-mannose-3', 5'-epimerase: an enzyme which performs three chemical reactions at the same active site. J Am Chem Soc 127:18309–18320

Mapson LW, Isherwood FA (1956) Biological synthesis of L-ascorbic acid: the conversation of derivatives of D-galacturonic acid to L-ascorbate in plant extracts. Biochem J 64:13–22

Mapson LW, Isherwood FA, Chen YT (1954) Biological synthesis of L-ascorbic acid: the conversion of L-galactono-γ-lactone into L-ascorbic acid by plant mitochondria. Biochem J 56:21–28

Maruta T, Yonemitsu M, Yabuta Y, Tamoi M, Ishikawa T, Shigeoka S (2008) Arabidopsis phosphomannose isomerase 1, but not phosphomannose isomerase 2, is essential for ascorbic acid biosynthesis. J Biol Chem 283:28842–28851

Mieda T, Yabuta Y, Rapolu M, Motoki T, Takeda T, Yoshimura K, Ishikawa T, Shigeoka S (2004) Feedback inhibition of spinach L-galactose dehydrogenase by L-ascorbate. Plant Cell Physiol 45:1271–1279

Moreau R, Dabrowski K (1998) Body pool and synthesis of ascorbic acid in adult sea lamprey (*Petromyzon marinus*): an agnathan fish with gulonolactone oxidase activity. Proc Natl Acad Sci USA 95:10279–10282

Naqvi S, Zhu C, Farre G, Ramessar K, Bassie L, Breitenbach J, Conesa DP, Ros G, Sandmann G, Capell T, Christou P (2008) Transgenic multivitamin corn through biofortification of endosperm with three vitamins representing three distinct metabolic pathways. Proc Natl Acad Sci USA 106:7762–7767

Oba K, Ishikawa S, Nishikawa M, Mizuno H, Yamamoto T (1995) Purification and properties of L-galactono-γ-lactone dehydrogenase, a key enzyme for ascorbic acid biosynthesis, from sweet potato roots. J Biochem 117:120–124

Ostergaard J, Persiau G, Davey MW, Bauw G, Van Montagu M (1997) Isolation of a cDNA coding for L-galactono-γ-lactone dehydrogenase, an enzyme involved in the biosynthesis of ascorbic acid in plants. J Biol Chem 272:30009–30016

Paine JA, Shipton CA, Chaggar S, Howells RM, Kennedy MJ, Vernon G, Wright SY, Hinchliffe E, Adams JL, Silverstone AL, Drake R (2005) Improving the nutritional value of Golden Rice through increased pro-vitamin A content. Nat Biotech 23:482–487

Pateraki I, Sanmartin M, Kalamaki MS, Gerasopoulos D, Kenallis AK (2004) Molecular characterization and expression studies during melon fruit development and ripening of L-galactono-1, 4-lactone dehydrogenase. J Exp Bot 55:1623–1633

Patterson JH, Waller RF, Jeevarajah D, Billman-Jacobe H, McConville MJ (2003) Mannose metabolism is required for mycobacterial growth. Biochem J 372:77–86

Qian W, Yu C, Qin H, Liu X, Zhang A, Johansen IE, Wang D (2007) Molecular and functional analysis of phosphomannomutase (PMM) from higher plants and genetic evidence for the involvement of PMM in ascorbic acid biosynthesis in Arabidopsis and *Nicotiana benthamiana*. Plant J 49:399–413

Qin C, Qian W, Wang W, Wu Y, Yu C, Jiang X, Wang D (2008) GDP-mannose pyrophosphorylase is a genetic determinant of ammonium sensitivity in *Arabidopsis thaliana*. Proc Natl Acad Sci USA 105:18308–18313

Radzio JA, Lorence A, Chevone BI, Nessler CL (2003) L-Gulono-1, 4-lactone oxidase expression rescues vitamin C-deficient Arabidopsis (vtc) mutants. Plant Mol Biol 53:837–844

Rautenkranz AAF, Li L, Machler F, Martinoia E, Oertli JJ (1994) Transport of ascorbic and dehydroascorbic acids across protoplast and vacuole membranes isolated from barley (*Hordeum vulgare* L. cv Gerbel) leaves. Plant Physiol 106:187–193

Rayon C, Cabanes-Macheteau M, Loutelier-Bourhis C, Salliot-Maire I, Lemoine J, Reiter WD, Lerouge P, Faye L (1999) Characterization of N-glycans from Arabidopsis. Application to a fucose-deficient mutant. Plant Physiol 119:725–733

Roberts RM (1971) The metabolism of D-mannose-^{14}C to polysaccharide in corn roots. Specific labelling of L-galactose, D-mannose, and L-fucose. Arch Biochem Biophys 145:685–692

Sakamoto T, Morinaka Y, Ohnishi T, Sunohara H, Fujioka S, Ueguchi-Tanaka M, Mizutani M, Sakata K, Takatsuto S, Yoshida S, Tanaka H, Kitano H, Matsuoka M (2005) Erect leaves caused by brassinosteroid deficiency increase biomass production and grain yield in rice. Nat Biotech 24:105–109

Sanmartin M, Drogoudi PD, Lyons T, Pateraki I, Barnes J, Kanellis AK (2003) Over-expression of ascorbate oxidase in the apoplast of transgenic tobacco results in altered ascorbate and glutathione redox states and increased sensitivity to ozone. Planta 216:918–928

Siendone E, Gonzalez-Reyes JA, Santos-Ocana C, Navas P, Cordoba F (1999) Biosynthesis of ascorbic acid in kidney bean. L-galactono-γ-lactone dehydrogenase is an intrinsic protein located at the mitochondrial inner membrane. Plant Physiol 120:907–912

Smirnoff N (1996) The function and metabolism of ascorbic acid in plants. Ann Bot 78:661–669

Smirnoff N (2000) Ascorbic acid: metabolism and functions of a multi-facetted molecule. Curr Opin Plant Biol 3:229–235

Smirnoff N (2003) Vitamin C booster. Nat Biotech 21:134–6

Smirnoff N, Conklin PL, Loewus FA (2001) Biosynthesis of ascorbic acid in plants: a renaissance. Annu Rev Plant Physiol Plant Mol Biol 52:437–467

Smirnoff N, Wheeler GL (2000) Ascorbic acid in plants: biosynthesis and function. Crit Rev Biochem Mol Biol 35:291–314

Tabata K, Oba K, Suzuki K, Esaka M (2001) Generation and properties of ascorbic acid-deficient transgenic tobacco cells expressing antisense RNA for L-galactono-1, 4-lactone dehydrogenase. Plant J 27:139–148

Tabata K, Takaoka T, Esaka M (2002) Gene expression of ascorbic acid-related enzymes in tobacco. Phytochemistry 61:631–635

Tamaoki M, Mukai F, Asai N, Nakajima N, Kubo A, Aono M, Saji H (2003) Light-controlled expression of a gene encoding L-galactono-γ-lactone dehydrogenase which affects ascorbate pool size in *Arabidopsis thaliana*. Plant Sci 164:1111–1117

Thomas CE, McLean LR, Parker RA, Ohlweiler DF (1992) Ascorbate and phenolic antioxidant interactions in prevention of liposomal oxidation. Lipids 27:543–550

Timmer CP (2003) Biotechnology and food systems in developing countries. J Nutr 133:3319–3322

Tokunaga T, Miyahara K, Tabata K, Esaka M (2005) Generation and properties of ascorbic acid-overproducing transgenic tobacco cells expressing sense RNA for L-galactono-1, 4-lactone dehydrogenase. Planta 220:854–863

Vain P (2006) Global trends in plant transgenic science and technology (1973–2003). Trend Biotech 24:206–211

Valpuesta V, Botella MA (2004) Biosynthesis of L-ascorbic acid in plants: new pathways for an old antioxidant. Trend Plant Sci 9:573–577

Van Ommen B (2004) Nutrigenomics: exploiting systems biology in the nutrition and health arenas. Nutrition 20:4–8

Van Schaftingen E, Jaeken J (1995) Phosphomannomutase deficiency is a cause of carbohydrate-deficient glycoprotein syndrome type I. FEBS Lett 377:318–320

Watanabe K, Suzuki K, Kitamura S (2006) Characterization of a GDP-D-mannose 3", 5"-epimerase from rice. Phytochemistry 67:338–346

Wenck A, Hansen G (2005) Positive selection. Meth Mol Biol 286:227–236

Wheeler GL, Jones MA, Smirnoff N (1998) The biosynthetic pathway of vitamin C in higher plants. Nature 393:365–369

White PJ, Broadley MR (2009) Biofortification of crops with seven mineral elements often lacking in human diets – iron, zinc, copper, calcium, magnesium, selenium and iodine. New Phytol 182:49–84

Wilkinson SR, Prathalingam SR, Taylor MC, Horn D, Kelly JM (2005) Vitamin C biosynthesis in trypanosomes: a role for the glycosome. Proc Natl Acad Sci USA 102:11645–11650

Wolucka BA, Persiau G, van Doorsselaere J, Davey MW, Demol H, Vandekerckhove J, van Montagu M, Zabeau M, Boerjan W (2001) Partial purification and identification of GDP-mannose 3", 5"-epimerase of Arabidopsis thaliana, a key enzyme of the plant vitamin C pathway. Proc Natl Acad Sci USA 98:14843–14848

Wolucka BA, Van Montagu M (2003) GDP-mannose 3', 5'-epimerase forms GDP-L-gulose, a putative intermediate for the de novo biosynthesis of vitamin C in plants. J Biol Chem 278:47483–47490

Wolucka BA, Van Montagu M (2007) The VTC2 cycle and the *de novo* biosynthesis pathways for vitamin C in plants: an opinion. Phytochemistry 68:2602–2613

Zhang W, Gruszewski HA, Chevone BI, Nessler CL (2008) An arabidopsis purple acid phosphatase with phytase activity increases foliar ascorbate. Plant Physiol 146:431–440

Zhu YJ, Agbayani R, McCafferty H, Albert HH, Moore PH (2005) Effective selection of transgenic papaya plants with the PMI/man selection system. Plant Cell Rep 24:426–432

Index

A

ABA and ABA/giberelin acid (GA) signaling pathways, 31
ABA metabolism, 241
Abiotic and biotic stresses, 2, 3, 8, 9, 15, 28–31, 103, 142, 165, 178, 182, 196, 199, 232, 265–290, 304–306, 315–317, 323–330, 354, 392–396
Abiotic stress conditions, 265–290, 394–395
Abiotic stress-driven oxidative stress, 268
Abiotic stresses, 25, 28–31, 103, 130, 155, 160, 165, 178, 182, 196, 199, 232, 265–290, 304–306, 311, 315, 316, 323–330, 354, 393, 394
Abnormal meristems, 239, 245
Abscisic acid synthesis pathway, 19
Abutilon theophrasti, 201
ACC. *See* 1-Aminocyclopropane-1-carboxylase
Acclimation, 5, 24–26, 102, 138, 176, 177, 179, 273, 276–278, 283, 311, 314
ACCO. *See* 1-Aminocyclopropane-1-carboxylate oxidase
Acer saccharinum, 143, 171
Action of ascorbate and glutathione, 13–16
Activation of defence mechanisms, 265–290
Adaptation, 15, 28, 67, 142, 160, 169, 210, 215, 221–223, 304, 329, 341
ADP-ribose polymers, 67
Adventitious root formation, 59, 77–80
Adventitious rooting, 77–79
Adverse abiotic stress conditions, 265
Agricultural productivity, 272, 290, 421
Agrobacterium rhizogenes, 57
Air pollutants/strong light, 2, 9, 17, 284, 287
Air pollution, 15
Aldolase, 27, 75
Allium, 166
Allium schoenoprasum, 143, 166
Allocasuarina luehmannii, 169
Alternative splicing, 341, 344–345
Al-tolerant, 281
Aluminium (Al), 281–282, 358, 359
AM. *See* Arbuscular mycorrhizal
1-Aminocyclopropane-1-carboxylase (ACC), 72, 242, 255
1-Aminocyclopropane-1-carboxylate oxidase (ACCO), 255
Aminotriazole, 16, 84, 103, 395
Anaerobic environment, 194
Androgenesis, 233, 241
Angiosperm, 233, 241, 246, 345
Animal pathway, 407, 418–420
Anoda, 166
Antheraxanthin (Ax), 17
Antioxidant compounds, pools of, 23
Antioxidant defense system, 2, 170, 210–211, 222, 269, 275, 279, 337–372
Antioxidant enzymes, 11, 30, 62, 81, 99, 101, 129, 140, 142, 156, 158–160, 162, 164–168, 171, 173–177, 179–183, 220, 223, 269, 273, 275–277, 282, 284, 289, 290, 304, 309, 313, 337–372, 391
Antioxidant enzymes in higher plants, cellular localization, 340
Antioxidants, 2, 59, 99, 115, 138, 197, 210, 240, 251, 266, 304, 324, 339, 391, 406
Antioxidative enzymes, 9, 28, 30, 123, 127, 182, 210, 222, 306, 307, 310, 325, 394, 396
Antioxidative protection, 13, 156, 222, 267, 290, 307, 342, 359, 361, 369
Antioxidative system, components, 2, 13–16, 23
Antiperoxidative element (ARE), 95
Antisense constructs, 354, 358
AO. *See* Ascorbate oxidase
Apoplasm, 340

Apoplast, 4, 8–10, 24, 26, 32, 60, 68, 69, 116–122, 124, 125, 127, 128, 138, 142, 196, 218, 219, 267, 271, 287–289, 306, 308, 309, 311, 312, 338, 339, 406, 408
  oxidative metabolism, 117–122
Apoplastic AA pool, 10
Apoplastic ascorbate/phenolic/peroxidase system, 119
Apoplastic ascorbate pool, 117, 118, 120, 219
Apoptosis, 64, 215, 338
  signaling, 215
APX. *See* Ascorbate peroxidase
APX1-deficient mutant, 314
Arabidopsis, 6, 8, 12, 13, 16, 28, 29, 58, 65–67, 76, 94, 98, 104, 124, 127, 168, 175, 179, 194, 196, 211–213, 215–217, 235, 239, 240, 255–259, 273, 274, 278, 281–284, 286, 288, 289, 306, 311, 313, 315, 330, 342, 344, 345, 347, 352–360, 362–368, 391–393, 395, 397, 398, 411, 413, 414, 416–419, 422
Arabidopsis mutant, 65, 194, 216, 255, 257, 278, 288, 315, 413, 414
Arabidopsis thaliana (A. thaliana), 4–8, 11, 14, 16, 29, 65, 95, 97–99, 102–104, 118, 123–125, 129, 143, 166, 173, 196, 198, 211, 216, 217, 223, 271–273, 279, 281, 282, 288, 289, 313, 314, 330, 341–343, 350, 354, 356, 358, 360, 363–366, 412, 415, 417, 418, 420
Arbuscular mycorrhizal (AM), 159–161
Arbutus unedo, 143, 170
ARE. *See* Antiperoxidative element
ARGONAUTE (AGO), 75, 239
ARGONAUTE 1 (AGO 1), 76, 239
AsA-and GSH-mediated functions, 130
AsA/DHA and GSH/GSSG ratios, 15, 26, 28, 126, 127, 164, 172, 198–199, 235–239, 245, 265–290, 309, 310
AsA–GSH cycle-related mechanism, 122
AsA-requiring enzymes, 255
Asc and GSH and Asc/DHA or GSH/GSSG redox couples, 28, 31
Asc/DHA
  and GSH/GSSG redox states, 62
  ratio, 26, 70, 82
  redox state, 70
Asc–GSH cycle, 8, 12, 13, 73, 82
Ascorbate (AsA/Asc)
  biosynthesis, 66, 71, 254
  biosynthetic pathway, 6
  consumption, 22, 23
  content, 66, 70, 71, 79, 80, 82, 405–422
  deficient mutants, 4, 11, 118, 284, 414
  developmental stages, 6
  forms, 10, 20
  metabolism, 70, 84
  oxidised form of, 20
  pool, 4, 10, 14, 23, 31, 61, 64, 70, 117, 118, 120, 125, 172, 219, 233, 235, 244, 269, 273
  recycling, 268, 271, 288, 393
  recycling pathway, 270, 288, 420
  redox state, 9, 119, 120, 157, 231–246, 259, 268, 271, 275, 277, 284, 288
  signalling module, 253–256
  synthesis, 245, 254, 255, 273, 280, 284, 325, 413, 420
  transporter, 22
Ascorbate-and glutathione-dependent biochemical systems, 84
Ascorbate-deficient *(vtc),* 4, 11, 118, 284, 414
Ascorbate–glutathione (AsA–GSH)
  coordinated role of, 218–223, 323–330
  cycle, 4–14, 17–28, 31–33, 81, 91–106, 116, 118, 120–126, 130, 137–184, 210, 220, 252–254, 258–260, 267, 268, 270, 273, 275, 276, 279, 282, 286, 303–318, 324, 326–327, 389–392, 394–398
  interaction of, 59–62
  pathway components, 1–33, 100, 130, 171, 222, 223, 254, 258, 275, 387–399
  regulation, 246, 290
Ascorbate-metabolising enzymes, 70
Ascorbate oxidase (AO), 9, 10, 32, 118–120, 219, 242, 268, 271, 274, 289, 406, 408
Ascorbate peroxidase (APX)
  characteristics of, 92–96
  encoding genes, 95–96
  family, 94
  functions, 91–92
  isoforms, 93–95, 155, 267, 391
  protein content, 101
Ascorbic acid, 21, 22, 183, 210, 218, 219, 221, 222, 243, 251–253, 255, 256, 305, 308, 324, 325, 328, 406–411, 415–417, 420–422
  in living organisms, 406–408
  functions of, 324, 408–409
  synthesis, 325, 406, 410, 411
Ascorbic acid free radical (AFR), 243–245
a-tochopherol, 267
ATP-dependent reactions, 8
ATP-sulfurylase activity, 222
Atrazine, 192, 198, 200–202
Auxin, 10, 57–63, 69, 77–80, 192, 217, 234, 236, 311
Ax. *See* Antheraxanthin

B

Benzylaminopurine (BAP), 57, 62, 83
Betula pendula, 288
Biochemical dysfunctions, 10
Biochemical machinery, 364, 388
Bioinformatic analysis, 344
Biomolecular regulation, 303–318
Bioreactors, 81
Biosynthesis, 4, 6, 11, 12, 14, 30, 65, 66, 71, 72, 118, 169, 179, 180, 192, 195–196, 202, 213, 216, 217, 219, 221, 223, 235, 236, 241–243, 253–255, 258, 279–282, 308, 324, 325, 340, 357, 359, 360, 388, 406, 410–417, 421, 422
Biosynthetic pathway of ascorbic acid, 243, 253, 284, 407
Biotic and abiotic stress, 2, 3, 8, 9, 15, 31, 103, 142, 165, 178, 182, 196, 199, 232, 265–290, 304–306, 315–317, 323–330, 354, 393–396
Biotic and abiotic stress response, 28–31
Biotic stress conditions, 125, 395–396
Botrytis cinerea (B. cinerea), 118, 119, 121–124, 127, 396, 397
Brassica
  B.campestris, 98, 222, 281
  B.juncea, 16, 202, 222, 279
  B.napus, 76, 77, 233–234, 237, 239–242, 246
BSO-treated embryos, 72, 73, 77
Buffering, 9, 24, 68, 117, 118, 120, 272, 359
Bupleurum, 166
Bupleurum chinense, 143, 166

C

Cadmium (Cd), 12, 16, 32, 101, 102, 104, 221, 222, 278–281, 395, 398
  stress, 16, 102, 103, 221, 279–281
  tolerance, 221, 279
  tolerance, tomato, 16
Calendula officinalis, 273
Callus, 56–58, 66, 78, 82–84
Calmodulin, 311, 369, 370, 396
Calvin cycle enzymes, 27, 129, 258
C_3 and C_4 plant, 142, 163, 172
Capparis ovata (C. ovata), 143, 161
Capsicum annuum, 219, 395
cAPXs. *See* Cytosolic ascorbate peroxidases
Ca^{2+} signals, 104, 105, 351, 369
Casuarina, 99
Catharanthus, 166
Catharanthus roseus, 144, 167, 181
CCP. *See* Cytochrome *c* peroxidase
Cd-hyperaccumulating ecotype, 221
CDK. *See* Cyclin-dependent kinases
CDNB. *See* 1-Chloro-2,4-dinitrobenzene
Cell compartments, 2, 24, 155, 211, 258, 259, 308, 389, 390, 394, 399, 416
Cell cycle, 58, 62, 64, 66, 67, 236, 238, 244, 257, 325, 350
Cell death, 2, 11, 32, 33, 102, 120, 123, 124, 142, 222, 237, 238, 266, 273, 287, 357
Cell division, 7, 32, 62–73, 77, 84, 217, 235, 236, 240, 241, 243, 244, 409
Cell division and growth, 409
Cell growth, 68–69, 71, 325, 350, 409, 418
Cell proliferation, 58, 66–68, 71–74, 76, 235, 237, 244, 255
Cell redox buffers, 25
Cell suspension cultures, 57, 66
Cellular antioxidant, 32, 222, 346, 362
Cellular antioxidant defense system, component of, 222
Cellular homeostasis, 240, 266, 267, 280
Cellular information-rich redox buffers, 24
Cellular proliferation, 235
Cellular receptors, 312
Cellular redox balance, 28, 212, 215, 267, 271, 309
Cellular redox environment, 126–128, 232, 245, 310
Cellular redox homeostasis, 23, 24, 126, 215, 280, 325–326
Cellular redox level, 25
Cellular redox regulation, 290, 325, 371
Cellular redox state, 23–28, 116, 126, 127, 215, 220, 232, 233, 240, 245, 259, 268, 274, 281, 282
Cellular redox systems, 77
Cereals, 162–165, 199, 286, 406
Chaperones, 342, 363
Chelation of heavy metals, 278
Chilling, 2, 15, 16, 19, 26, 99, 100, 102, 146, 161, 164, 165, 216, 266, 275, 276, 306
Chilling-low temperature stress, 275–277
Chilling tolerance, maize, 16, 165
Chitinase, 30
Chlamydomonas reinhardtii, 96, 97, 341, 342, 371
Chloracetanilide herbicides, 16
1-Chloro-2,4-dinitrobenzene (CDNB), 202
Chlorophyll fluorescence, 161
Chlorophyta, 17
Chloroplastic DHAR (chlDHAR), 270, 315
Chloroplastic/mitochondrial GR genes, 98, 213
Chloroplasts
  antioxidants, 337–372
  redox signals, 339, 369

Chrysopogon zizanioides, 201
Cicer arietinum, 329
9-*cis*-epoxycarotenoid dioxygenase gene (*SgNCED1*), 180
Cistus clusii, 144, 170
CLAVATA (CLV), 58
CLAVATA 1 (CLV 1), 75
CLAVATA 3, 77, 239
Cleavage polyembryony, 238
C, N and S, 11
Coffea canephora, 144, 157
Cold treatments, 16, 32, 346
Compartmentalization, 125, 195, 201, 287, 367, 388, 398
Compartmentalization of metabolic reactions, 388
Compartments, 2, 4, 8, 10, 11, 24, 26, 31, 93, 103, 105, 116–118, 120, 122–126, 130, 155, 183, 199, 210, 211, 213, 235, 256–259, 267, 269, 308, 309, 339, 340, 343, 368, 389, 390, 394, 396, 399, 406, 416
Component of plant antioxidant networks, 215
Conifer, 233, 246
Controlling cellular redox state, 23–28
Copper (Cu), 15, 16, 32, 139, 277, 281, 283, 284, 309
  availability, 364–366
Co-regulation, 124, 362, 364, 366, 367
Crop productivity, 269, 329
Cross-talk, 8, 23, 26, 27, 123, 129, 355, 366, 399
Cucumis sativus, 145, 180
Cucurbita pepo, 281–282
CUPSHAPED COTYLEDONS, 239
Cuscuta reflexa, 18
CuZn-superoxide dismutase, 30, 139
Cyclin-dependent kinases (CDK), 58
Cysteines, 8, 27, 28, 65, 97, 104, 128, 129, 199, 210, 214–216, 235, 256–258, 275–277, 324, 339, 353–355, 359, 369, 370
Cytochrome *c* peroxidase (CCP), 93, 94, 183
Cytokinin, 57, 58, 61–62, 69, 181, 234
  levels, 61
Cytosol, 4, 8, 10, 11, 24, 91, 92, 97, 98, 116, 118–120, 122, 124, 127–129, 139, 140, 155, 168, 179, 184, 195, 196, 211–213, 258, 259, 267, 279, 287, 288, 308, 309, 314, 324, 326, 340, 342, 343, 355, 368, 371, 389, 390, 393, 394, 396–398, 406, 407, 412, 415
Cytosolic antioxidant enzymes, 346, 348–359, 362
Cytosolic ascorbate peroxidases (cAPXs), 93–95, 101–103, 154, 278, 283, 285, 341, 349–350

D
DDE. *See* Diadinoxanthin de-epoxidases
Ddx. *See* Diadinoxanthin
Dedifferentiation, 55–84, 218, 236, 238–240, 243, 308, 368
De-epoxidases, 17, 19–22, 139, 219, 267, 308, 409
De-epoxidation, 19–23, 157, 164
  of epoxy xanthophylls, 17, 21
Defense mechanisms, 15, 28, 31, 175, 182, 198, 232, 306–309, 361
Defense pathways, 2
Defense-related genes, 28, 30, 256, 354
Degradation of glutathione, 195–196
Dehydrascorbate reductase, 310
Dehydroascorbate (DHA)
  AsA redox potential, 259
  ascorbate redox couples, 13
Dehydroascorbate reductase (DHAR), 3, 4, 11, 14, 30, 60, 63, 76, 79, 81, 82, 92, 116, 119, 122–127, 129, 138–142, 144–146, 149, 151–154, 157, 163, 164, 168, 171, 172, 175, 176, 180, 220, 222, 268, 270, 271, 273, 274, 277, 278, 280, 282–286, 288, 289, 307, 310, 315, 326, 340, 343, 358, 389–396, 408, 419, 420
Depletion of GSH pool, 280
Desertification, 305
Desert plants, 100, 176
Detoxification
  heavy metals, 32, 215, 219
  H_2O_2, 15, 70, 92, 138, 267, 280, 286, 339, 388, 393
  xenobiotics, 65, 126, 139, 197, 198, 200–201, 396
Developmental pathways, 56, 58
DHA. *See* Dehydroascorbate
DHAR. *See* Dehydroascorbate reductase
Diadinoxanthin (Ddx), 17–19, 21, 23
Diadinoxanthin de-epoxidases (DDE), 18–23
Diatoms, 17, 22, 23
Diatoxanthin (Dtx), 17, 19, 21, 23
Differentiation, 55–84, 218, 236, 238–240, 243, 308, 368
2,4-Diphenoxyacetic acid (2,4-D), 57, 59, 192, 193, 200, 236, 395
Diphenyloiodonium (DPI), 78, 81
Dismutation, 4, 210
Distal meristem (DM), 63, 153
Dithiol disulfide exchange, 210
D-mannose/L-galactose pathway, 308
D-mannose-1-phosphate, 4, 414
DNA-coded molecule, 195
DNA segment (T-DNA), 57

DPI. *See* Diphenyloiodonium
Drought
  stress, 2, 15, 99–101, 137–184, 219, 269–271, 314, 328–329, 348
  stress and desiccation, 2
  stressed, 159, 160, 167, 168, 175
  stressed plants, 142–170, 173–177, 179–183
  tolerance, 100, 142, 155, 157–162, 165, 173, 177–183, 269–271, 315
Drought-mediated oxidative stress, 141, 155, 156, 183
Dry mass, 170, 222, 223
Dtx. *See* Diatoxanthin
Dual-targeting, 97, 123, 313, 344–345, 394, 398

E
EBR. *See* 2,4-Epibrassinosteroid
Electrolyte leakage, 156, 157, 163, 165, 171
Electron donor, 11, 91–93, 119, 140, 141, 218, 308, 324, 326, 327, 342, 393
Elicitation, 81–82
Embryo
  conversion, 75, 243–245
  development, 57, 72, 73, 75, 77, 231–246
  formation, 56, 57, 71, 75–77, 83, 232, 233
Embryogenesis, 56–59, 69–76, 82–84, 213, 233, 234, 236–238, 240–242, 244–246, 257, 327, 388
Endogenous electron, 20
Endogenous metabolites, 16
Enhanced affinity, 22
Environment
  conditions, 11, 15, 26, 30, 121, 123, 290, 305, 338, 389, 408
  factors, 9, 67, 138, 232, 394, 396
  fluctuations, 116
  irradiance, 158, 172
  stimuli, 24, 338
Environment stress
  conditions, 105, 176, 341, 387–399
  factors, 312, 314
Enzymatic and non-enzymatic antioxidant pathways, 266
Enzymatic and non-enzymatic integrated mechanisms, 303–318
Enzymatic and non-enzymatic molecules, 266
Enzymatic antioxidants, 10, 121, 123, 138, 140, 156, 158, 160, 163, 168, 172, 179, 304, 310
Enzymatic conjugation, 193
Enzymatic mechanisms, 117, 259
Enzymatic reactions, 222
Enzyme
  ascorbate–glutathione cycle, 81, 139, 140, 143, 155, 164, 171–173, 183, 184, 210, 270, 273, 282, 286, 303–318, 391
  catalysed reactions, 15
  co-factor, 409
  Smirnoff–Wheeler pathway, 410–416
2,4-Epibrassinosteroid (EBR), 220
Escherichia coli, 98, 99, 177, 194, 213, 215, 342
Ethylene, 31, 64, 70, 72, 118, 241, 242, 254, 255, 353, 409
Eucalyptus globulus, 145, 156
Eukaryotes, 98, 211, 212, 253, 257, 406, 412
Eupatorium adenophorum, 100, 171
Eupatorium odoratum, 100
Euphorbia, 165
Euphorbia esula, 16, 166, 270
Eustoma grandiflorum, 65
Evolution
  metabolic network, 259–260
  metabolism, 195
Expressional regulation of antioxidant enyzmes, 346–363
External applications, 180–183
External environment, 8, 126
Extra-plastidic antioxidant defenses, 337–372
Extra-plastidic antioxidant systems, 339–345
Extraprotoplastic matrix, 8
Extreme abiotic stress conditions, 266
Extreme temperatures, 15, 176, 394

F
Fabaceae, 194
Factors affecting ascorbate biosynthesis and accumulation, 416
Factors for and against genetic engineering, 420–421
Fagus sylvatica, 145, 170
FAO. *See* Food and Agriculture Organization
Fence model, 254
Fenchlorazole, 16
Ferredoxin-thioredoxin system, 311
Ferulic acid peroxidase (FPOX), 71
First line of defence, 4, 287, 363, 394
Fluxes of H_2O_2 production, 4
Food and Agriculture Organization (FAO), 406
Food security and malnutrition, 421–422
Foyer–Halliwell-cycle, 343
FPOX. *See* Ferulic acid peroxidase
Functional proteins, 174

G
GA. *See* Gibberellic acid
Galactono-γ-lactone (GalL), 66, 71, 79, 80, 280

Gamma-glutamylcysteine synthetase (γ-ECS), 7, 8, 31, 65, 195, 196, 217, 219, 236
GDP-D-mannose, 415
GDP-D-mannose-3′,5′-epimerase, 413
GDP-D-mannose pyrophosphorylase, 412–413, 417
GDP-L-galactose, 415
GDP-L-galactose phosphorylase, 407
Gene expression
analysis, 59, 361, 364, 367, 392
coordination of, 368
influence, 28–31, 355
patterns of, 30, 59, 75–77, 175
and protein expression, 266
General stress response, 327–328
Genes
from D-galacturonate pathway, 418
from Smirnoff–Wheeler pathway, 417–418
Genes encoding, enzymes of arcobate glutathione cycle, 313
Genetically modified foods, 421, 422
Genetically modified plants, 405–422
Genetic engineering, 193, 218, 284, 288, 315, 317, 420–422
Genetic manipulation, 270, 275, 276, 316, 422
Genetic transformation, 56, 276
Genotype and phenotype, 7
Germination, 70, 71, 75, 147, 168, 171, 213, 217, 231–246, 257, 275, 388
GHS/GSSG redox status, 72
Gibberellic acid (GA), 10, 181, 409
Glomus versiforme, 159
Glucose-6-phosphate dehydrogenase, 124, 180, 397
Glutaredoxins (GRXs), 28, 128–130, 210, 211, 214–217, 259, 342
Glutathione (GSH)
administration, 75
and ascorbate redox state, 231–246
biosynthesis, 65, 66, 213, 216, 217, 221, 258
deficiency, 16, 257
deficient mutant, 6, 7
dependent changes, 73–75
functions of, 197
homeostasis, 120
metabolism, 12, 30, 31, 59, 76, 81, 195, 212, 220, 235, 236, 243, 257, 267, 285–287
oxidation and transport, 199
oxidation ratio, 169
peroxidase expression, 362–363
pool, 12, 15, 24, 26, 27, 30, 32, 72, 73, 76, 78, 79, 92, 125–127, 141, 142, 172, 173, 194, 223, 235, 240–242, 245, 269, 270, 273, 285–287, 355
pool size of, 15
redox state, 72, 126, 235–236, 241, 269, 275, 278, 285
redox status, 13, 15
reductase-encoding genes, 98–99
reductase family, 98
signalling module, 257–258
synthesis, 8, 12, 15, 16, 30–32, 196, 199, 201, 211–213, 217, 221, 235, 237, 257, 272, 275, 277, 279, 287, 324
synthesizing capacity, 277
Glutathione (γ-L-glutamyl-L-cysteinyl glycine; GSH), 4
Glutathione-and ascorbate-mediated redox regulation, 126, 128
Glutathione-and ascorbate-related redox regulations, 126–130
Glutathione disulfide (GSSG), 3, 4, 9, 12, 13, 15, 21, 26, 27, 30, 32, 63–67, 71–74, 76, 78, 79, 92, 96, 97, 119, 120, 123–125, 128, 139, 141, 143, 145, 146, 149, 151, 152, 155, 163, 164, 169, 172, 182, 197, 199, 200, 210, 211, 213, 221, 232, 240, 242, 245, 309, 314, 324, 327, 370, 390, 393
Glutathione disulfide-glutathione (GSH/GSSG) couple, 26, 27, 32, 128, 199, 232, 245, 267, 269
Glutathione disulfide/glutathione (GSSG/2GSH) ratio, 13
Glutathione-induced changes, 75–77
Glutathione peroxidase (GPX), 3, 11, 12, 16, 81, 93, 130, 139, 163, 179, 197, 210, 221, 222, 267, 277, 290, 307, 329, 330, 339, 340, 342
Glutathione reductase (GR), 3, 11, 61, 65, 91–106, 116, 119, 139, 140, 195, 197, 210, 211, 220–222, 235, 268, 306, 310, 324, 326, 328, 329, 343, 344, 359, 389–391, 393, 408, 420
characteristics of, 96–99
Glutathione–short historical survey and evolution, 194–195
Glutathione-S-transferase (GST), 163, 307, 358
significance of, 201–202
Glutathione synthetase (GSHS), 8, 30, 31, 195, 196, 202, 324, 330
Glutathionylation, 27, 28, 68, 129, 214, 253, 258
Glutharedoxins, 232, 342
Glycine, 8, 12, 196, 257, 412
Glycine max, 146, 161, 194, 365
Gossypium hirsutum, 146, 166

G1 phase, 63, 66, 67, 236
GPOX. *See* Guaiacol peroxidase
GPX. *See* Glutathione peroxidase
GR. *See* Glutathione reductase
GR isoforms, 96, 155, 179, 393
GRXs. *See* Glutaredoxins
GSH. *See* Glutathione
GSH/GSH + GSSG ratio, 72
GSHS. *See* Glutathione synthetase
GSH1 transcripts, 16
G1-S phase, 244
GSSG. *See* Glutathione disulfide
GSSG-treated embryos, 72, 76, 242
GST. *See* Glutathione-S-transferase
GST-mediated herbicide resistance, 201
Guaiacol peroxidase (GPOX), 71, 81, 93, 197, 210, 221, 222

H

Haem-containing enzyme, 93
Hairy root culture, 57, 82
Half-life, 93, 192
Halliwell–Asada cycle, 4, 79, 84, 198, 232, 243
Halliwell–Foyer–Asada cycle, 408
H^+-ATPase, 68
Heat shock, 2, 15, 95, 233, 266, 277, 314, 352
Heat shock factors, (HSFs) 105, 350, 353
Heatshock proteins (HSPs), 278, 312, 327, 368, 398
Heat shock transcription factors (HSFs), 30, 283
Heat stress, 16, 102, 277–278, 313, 314, 348, 352, 356
Heavy metals, 2, 7, 16, 32, 101, 139, 192, 211, 215, 219, 223, 235, 258, 266, 278, 279, 305, 308, 326, 394, 396
Heavy metal stress, 15, 223, 278–283
Helianthus annuus, 100, 146, 172, 181, 272, 329
Hemiparasitic plants, 18
Herbaceous species, 165–168
Herbicide Resistance Action Committee, 191
Herbicides, 16, 192–194, 197–202, 258, 304
  classifications, 192–193
  conjugation, 193, 197, 199–200, 258
  defense system, 197–198
  resistance, 191–202
  stress, 192, 199–200, 202
Heterocomplex formation, 363–367
Hexuronic acid, 252
High light (HL), 5, 6, 11, 14, 19, 22, 23, 99, 102, 103, 156, 170, 173, 179, 266, 278, 283–285, 304, 311, 313, 316, 338, 353, 354, 362, 363, 389, 394, 398
High light stress, 2, 173, 284, 285, 350
Histodifferentiation, 77, 241, 245
Holoparasitic, 18
Homo-glutathiones, 194, 195, 202, 211, 324
Hordeum vulgare, 101, 395
Horizontal connection, 253, 254, 258–259
HRGPs. *See* Hydroxyproline rich glycoproteins
HSFs. *See* Heat shock factors; Heat shock transcription factors
HSPs. *See* Heatshock proteins
Hydrilla verticillata, 221
Hydrohymethylglutathione, 194
Hydrophilic redox buffer, 14, 116
Hydroxyproline, 64, 67, 120, 244, 255
Hydroxyproline-reach proteins, 244
Hydroxyproline rich glycoproteins (HRGPs), 67, 120, 255, 256
Hydroxyproline synthesis, 64
Hypoxia, 286

I

IBA. *See* Indole butyric acid
Imbalance between ROS and antioxidant defense, 115, 222, 223, 304
Immunogold electron microscopy, 391
Indole alkaloid content, 167
Indole butyric acid (IBA), 57, 78, 83
Influence, 3, 10–12, 14, 19, 25, 26, 28–32, 58, 67, 81, 119, 166, 167, 169, 180, 181, 192, 196, 218, 223, 284, 308, 329, 340, 355, 364, 370, 371
Inga sapindoides, 18
Inversion and non-inversion routes, 409–411
In vitro cultures, 55–84, 194
  cells, 56–59, 61, 66, 84
  organs, 56–59
  tissues, 56–60
In vitro embryogenesis, 233, 236–238
Ionic toxicity, 272, 329
Ipomoea batatas, 362
Irradiance, 19, 158–160, 166, 172, 176, 177, 304
Isoenzyme patterns, 289, 341–344
Isoforms and subcelluar localization, 93–94
  of glutathione reductase, 97

J

Jasmonic acid (JA), 6, 16, 128, 223, 396

K

Ketoconazole (KCZ), 181
Knockout, 8, 30, 95, 212, 341, 352–354, 359, 361, 366, 367, 372

Knockout mutant, 7, 196, 412
KNOTTED-like homeobox (KNOX), 76

L
Late embryogenesis abundant (LEA) proteins, 75, 242, 327
Laurus azorica, 146, 156, 157
Lavandula spica, 161
L-buthione sulfoximine, 65–67, 72, 73, 76, 77, 79, 235, 237, 239–242, 359, 360
Leaf water potential, 157, 222
LEAFY COTYLEDON 1 (LEC1), 59
L-galactono 1,4 lactone dehydrogenase, 169, 253, 254, 284, 407, 409, 410, 416, 418, 419
L-galactose dehydrogenase, 284, 407, 410, 415, 418
L-galactose-1-phosphate, 4, 413, 414
L-galactose-1-phosphate phosphatase, 407, 414
Light stress, 33, 95, 102, 283–285, 352, 395
Lignin peroxidases (LiPs), 94
Ligustrum vulgare, 146, 160, 161
Lipid peroxidation, 5, 15, 61, 82, 123, 139, 157, 158, 161, 162, 164, 168, 182, 183, 197, 223, 266, 275, 281, 282, 289, 329
Lipid peroxides, 16, 223, 305, 342
Lower affinity, 22
Low light (LL), 14, 19, 22, 169, 179, 283, 284
Low oxygen stress, 285–287
Lumen, 11, 17, 19, 21–23, 138, 139, 259, 342, 407, 409
Lutein epoxide (Lx), 17, 18
Lycium barbarum, 83
Lycopersicon esculentum, 146, 219, 395
Lycopersicon peruvianum, 100
Lycorine, 71, 243, 245, 254

M
Maize, 15, 16, 26, 30, 96, 100, 163–165, 171, 192, 197, 199, 202, 217, 219, 220, 256, 270, 275–277, 283, 306, 314, 328, 343, 356, 363, 419
Malondialdehyde (MDA), 81, 156, 160, 162, 163, 165, 172, 174, 175, 220, 222, 223, 390
Malpighia glabra, 417
Malus domestica, 146, 169
Manganesedependent peroxidases (MnPs), 94
Mantoniella squamata, 17
MAPK. *See* Mitogen-activated protein kinase
Markers of plant stress, 13, 28
Mass spectrometry, 176
MDA. *See* Malondialdehyde
MDEs. *See* Microspore-derived embryos
MDHA. *See* Monodehydroascorbate
MDHAR. *See* Monodehydroascorbate reductase
Mechanical stress, 2, 103
Mehler reaction, 23, 31, 283
Membrane-bound antioxidant compounds, 139
Membrane lipid peroxidation, 266
Membrane stability, 158, 162, 182
Menadione, 16
Meristems, 7, 56–58, 62–66, 70–77, 83, 84, 213, 216–218, 239–242, 244, 245, 256, 257
  reactivation, 71, 72, 231–246
Mesophyll cells, 16, 30, 171, 180, 236, 288, 338
Metabolic activity, 10, 138, 259
Metabolic electron consumption, 338
Metabolic homogeneity, 66
Metabolic processes, 14, 24, 25, 27, 28, 31–33, 117, 218, 232, 235, 245, 304, 308, 310, 408
Metabolic reactions, 324, 338, 388, 397
Metabolic regulation, 286, 310–315
Metabolism, 2, 6, 12, 13, 16, 23–25, 29–31, 59, 61, 64, 65, 67, 70, 73–76, 79, 84, 115–122, 125, 138, 141, 142, 159, 166, 168, 177, 180, 192, 193, 195–197, 200–202, 210, 212, 214, 215, 218, 220, 221, 223, 233, 235–236, 238, 240, 241, 243–245, 252, 253, 257, 266, 268–271, 275, 278, 283, 306, 308, 311, 315, 326, 330, 371, 388–389, 398, 420
Metabolism and modulation of growth and development, 235–236
Metabolite synthesis, 81–82
Metallo-Chaperone, 365–366
Metal pollution, 278
Metal stress, 220, 222, 223
Metaphase, 244
Methionine sulfoxide reductases, 210
Methylglyoxal (MG), 16, 32, 275
Methyl viologen, 102, 103, 215, 314, 346, 367
Microarray experiments, 240, 256, 392
Microarray studies, 75
Microbodies, 92–94, 101, 102, 138, 340, 341
MicroRNAs, 364
Microspore-derived and somatic embryos, 234
Microspore-derived embryogenesis system, 234, 237
Microspore-derived embryos (MDEs), 233, 234, 237, 239–242
Mineral ion toxicity, 304
Mineral nutrients deficiency, 304
miRNA mechanism, 366

Mitochondria, 2, 4, 10, 12–13, 24, 32, 64, 91–95, 97, 98, 116, 122–125, 138, 164, 173, 211, 213, 214, 218, 219, 253, 257, 259, 266, 267, 282, 289, 306–310, 313, 314, 324–326, 340–345, 363, 389, 393, 394, 396, 397, 406, 416
Mitochondrial and peroxisomal ascorbate–glutathione cycle, 12
Mitogen-activated protein kinase (MAPK), 10, 350
  cascade, 104, 105, 311, 350–356, 361, 371
  pathway, 60, 311
Mitosis, 142, 325
Mitotic activity, 62, 63, 67, 237, 243, 244
Mitotic index, 66, 67, 409
Modulation
  of ascorbate–glutathione cycle, 311–313
  of redox state, 10, 219, 232, 235–236, 243–245
  of ROS-antioxidant interaction, 310
Molecular identification, 387–399
Molecular oxygen, 4, 283, 330
Monodehydroascorbate (MDHA), 3, 4, 8–11, 24, 68, 92, 119, 139–141, 145, 154, 172, 218, 219, 241, 243, 256, 268, 271, 273, 308, 309, 326, 340, 389, 390, 393, 407, 408, 420
Monodehydroascorbate reductase (MDHAR), 4, 8, 9, 11, 13, 24, 29, 60, 63, 73, 76, 79, 81, 92, 116, 118, 119, 122–124, 138–142, 145, 147, 149, 151, 153, 154, 157, 164, 168, 171, 172, 175, 180, 218, 270, 273, 280, 282–286, 288, 289, 307–310, 313, 314, 326, 340, 344, 345, 357–359, 389–391, 395, 407, 408
Monoepoxide, 17
Morphogenesis, 56, 58, 233, 236, 241, 245
Morphogenetic pathways, 56, 57, 59
mRNA level, 16, 59, 76, 95, 99, 101–103, 140, 178, 255, 256, 270, 279, 284, 286, 288, 289, 314, 315, 357, 364, 366, 368
Multifunctional compounds, 116
Mutants, 4–8, 11, 13, 26, 29–31, 58, 65, 66, 118, 123, 125, 173, 194, 196, 212, 213, 216, 217, 236, 240, 255–257, 278, 284, 288, 314, 315, 346, 360, 361, 366, 388, 393, 411–414, 418, 421, 422
Myo-inositol route, 418–420
Myrica faya, 156

N

β-Naphtoxyacetic (NAA), 57, 60, 78
NAD(P), 24
NADH oxidation, 24, 64, 67, 68, 389
NAD/NADH redox pair, 67–68
NADPH-dependent reaction, 11
NAD pool, 38
NAM. *See* NO APICAL MERISTEM
Natural and anthropogenic phenomena, 192
Natural metabolism, 192
Near-isogenic lines (NILs), 277
Necrotrophic pathogens, 120, 123, 129
Nernst equation, 126, 199
Net CO_2 assimilation rate, 169, 183, 285
Net photosynthesis rates, 164, 314
Nickel (Ni), 282–283
Nicotiana tabacum, 10, 98, 147, 179, 271, 289, 314
NILs. *See* Near-isogenic lines
Nitric oxide, 16, 26, 31, 77, 80, 81, 180, 182, 272–274, 306, 396, 399
N, N′-diethyldithiocarbonate, 84
NO APICAL MERISTEM (NAM), 75
Non-enzymatic and enzymatic integration, 303–318
Non-enzymatic antioxidant pools, 142–170
Non-enzymatic antioxidants, 2, 10, 123, 128, 138, 142–170, 172, 180, 183, 210, 214, 222, 223, 266, 305, 307, 309, 310
Non-enzymatic antioxidant system, 138, 222, 223, 304, 339
Non-enzymatic compounds, 307–309, 330
Nonexpressor of pathogen related genes (NPR1), 30, 104, 105, 127–129, 355, 363
Non-heme peroxidases, 341
Non-photochemical quenching, 11, 157, 283
Non-protein thiol glutathione, 222
Non-protein thiols, 211, 308, 396
NPR1. *See* Nonexpressor of pathogen related genes
Nucleotide metabolism, 67, 73–75, 244
Nucleus, 68, 104, 124, 127, 128, 340, 355, 361, 367

O

O-acetylserine, 12
2-ODDs. *See* 2-Oxoacid-dependent dioxygenases
Olea europaea, 158
OPP. *See* Oxidative pentose phosphate pathway
Organic sulphur-hydrogenating substance, 252
Organogenesis, 55–84, 244
Organ primordia, 58
Oryza sativa, 98, 99, 165, 314, 364
Osmolarity, 166
Osmotic balance, 272

Oxalic acid (OxA), 31
Oxalic and L-tartaric acids, 1
Oxidation, 4, 9, 10, 12, 13, 15, 24–28, 30, 32, 60, 61, 63, 64, 67–69, 73, 78, 79, 82, 93, 104, 118–121, 123, 124, 126, 169, 170, 199, 200, 214, 215, 220, 223, 243, 256, 266–270, 272, 273, 275, 278, 280, 282, 285, 286, 289, 305, 307, 311, 315, 324, 326, 338, 339, 388–390, 394, 398, 406, 409–411, 413
Oxidation and reduction of protein thiols, 214
Oxidation of glutathione pool, 73, 79
Oxidative burst, 9, 117, 120, 124, 125, 306, 311, 396
Oxidative damages, 12, 14, 16, 81, 82, 116, 138, 159–161, 163, 164, 168, 171, 179–183, 198, 199, 210, 214–216, 270, 272, 274, 278, 286, 304, 305, 308, 315, 324, 329, 390, 393
Oxidative pentose phosphate pathway (OPP), 124
Oxidative stress, 11, 64, 101, 115, 138, 196, 211, 258, 266, 304, 326, 338, 388, 420
Oxidative stress-promoting chemical agents, 271
Oxidising agents, 55
OxA. *See* Oxalic acid
2-Oxoacid-dependent dioxygenases (2-ODDs), 255, 409
Oxygen scavengers and chelators, 305
Ozone, 12, 15, 16, 25, 26, 99, 101, 102, 167, 170, 235, 255, 266, 287–290, 304, 313–315, 338, 359, 408, 413, 421
Ozone-sensitive mutants, 255
Ozone stress, 266, 287–290

P

Paclobutrazol (PBZ), 181
PAL. *See* Phenylalanine ammonium lyase
Panax ginseng, 81, 82
Panax quinqefolium, 82
PAPS. *See* Phosphoadenylyl sulfate
PAR. *See* Photosynthetically active radiation
Paraquat, 95, 99, 102, 103, 180, 197, 198, 284
PARP-coding genes, 67
Partial drying treatment (PDT), 70, 75
Pathogen attack, 2, 15, 117, 124, 125, 192, 304, 306, 311, 338, 355, 357, 396
Pathogenesis-related (PR)
  genes, 28–30, 118, 349, 355, 359
  proteins, 28–30
Pathogens, 9, 25, 30, 31, 117–129, 142, 157, 213, 216, 338, 396, 422
Pathogen-triggered cell death, 123
PATOGENESIS RELATED PROTEIN 1 (PR1), 61
PBZ. *See* Paclobutrazol
PC. *See* Phytochelatins
PCD. *See* Programmed cell death
PDT. *See* Partial drying treatment
Peptide ligands, 278
Peroxiredoxin regulation, 356–357
Peroxiredoxins, 29, 130, 210, 339–342, 356, 357, 359
Peroxisomal ascorbate–glutathione pathway, 12, 387–399
Peroxisomal reactive oxygen species metabolism, 388–389
Peroxisomal redox status of ascorbate, glutathione and NADPH, 396–397
Peroxisomes, 2, 4, 12–13, 91, 94, 97, 116, 122, 124, 125, 266, 306, 307, 309, 326, 338, 388–398
Persea indica, 156, 157
Pesticide, 180, 192
Phaeodactylum tricornutum, 17
Phaeophyta, 17
Phaseolus vulgaris, 123, 178, 279, 287
Phenological stage, 166, 275
Phenotype of ascorbate (vtc) comaparison, 5
Phenylalanine ammonium lyase (PAL), 30, 120, 285
Phloem glutathione, 170
Phosphoadenylyl sulfate (PAPS), 211
Phospholipid hydroperoxide glutathione peroxidase, 139, 307
Phosphomannomutase (PMM), 412, 417
Phosphomannose isomerase (PMI), 411–412
Phosphorylation, 74, 311, 350, 392
Photochemical, 10, 275
Photoinhibition, 141, 157, 284, 285
Photo-oxidative stress, 101, 276, 283, 284, 330, 334, 361, 367, 371
Photoprotection mechanism, 17
Photoprotective pigments, 169
Photoprotective processes, 23
Photorespiration, 10, 12, 25, 266, 306, 307, 330, 388
Photorespiration pathway, 12
Photosynthesis, 10, 12, 13, 25, 29, 31, 61, 65, 139, 158, 161, 164, 167, 172, 177, 179, 182, 192, 214, 216, 223, 266, 271, 287, 288, 313–315, 325, 338, 360, 366, 368, 369, 388, 408
Photosynthetic activity, 275, 288, 371
Photosynthetically active radiation (PAR), 156, 158
Photosynthetic apparatus, 141

Photosynthetic cells, 171
Photosynthetic components, 275
Photosynthetic electron transport, 10, 23, 24, 283, 285, 338, 339, 360, 366
Photosynthetic electron transport chain, 23, 24, 285, 338, 339, 366
Photosynthetic operation, 13, 326
Photosynthetic organisms, 308
Photosynthetic performance, 164
Photosynthetic pigments, 166, 182, 183
Photosynthetic potential, 222, 223
Photosystem II (PSII), 157, 177, 192, 193, 216, 283, 284, 314, 330
Physcomitrella patens, 341
Physiological functions, 25, 195, 330, 388, 394, 398
Physiological responses, 2, 396
Physiological status, 12
Phytochelatins (PCs), 16, 32, 139, 196, 213, 217, 220–223, 257, 278, 279, 281, 282
  biosynthesis, 217, 279, 281
Phytohormones, 58, 65, 241
Phytophthora brassicae, 30
Phytoremediation, 82, 202, 217
Picea
  P. abies, 237
  P. asperata, 156
  P. glauca, 69, 233–234, 237, 246
Pinus canariensis, 169
Piriformospora indica, 175, 315, 358
Pisum sativum, 98, 102, 173, 182
Plant defense response, metabolic processes, 31–33
Plant–pathogen interactions, 115–130, 388
Plants
  defense mechanisms, 306–309
  development, 62, 64–66, 179, 212–216, 235, 236, 257, 325
  peroxisomes, 388, 394, 397, 398
  regeneration, 58, 84, 236–238
  resistance to herbicide stress, 199–200
  stress tolerance, 1–33, 105
Plasma membrane, 2, 8, 21, 63, 68, 119, 120, 266, 267, 306, 312, 338, 339
Plastid and extra-plastid antioxidants, 346–349
Plastids, 8, 196, 212, 213, 259, 285, 305, 307, 316, 324, 326, 330, 346–349, 361, 362, 368, 371, 372
Plastid transcription kinase (PTK), 285
Plastochinon pool, 26
Plastocyanin, 366
Plastocyanin expression, 366
PM. *See* Proximal meristem
PMI. *See* Phosphomannose isomerase (PMI)
PMM. *See* Phosphomannomutase (PMM)
Polyacrylamide gel electrophoresis, 176
Poly (ADP-ribose) polymerase (PARP) activity, 67, 68
Poncirus trifoliata, 159
Populus accutifolius, 270
Populus euramericana, 100
Populus kangdingensis, 155
Populus trichocarpa, 98, 314
Post-transcriptional regulation, 357, 363–368
Potential gene targets, 405–422
Precursors, 6, 31, 32, 65, 66, 69, 73, 74, 79, 139, 194–196, 213, 218, 223, 238, 243, 245, 257, 279, 280, 308, 324, 325, 350, 359, 364, 368, 371, 410, 411, 418
Priformospora indica, 306
Primary symptoms of oxidative burst, 9
Production of reactive oxygen species, 115, 304–305
Programmed cell death (PCD), 11, 33, 102, 118, 121, 123, 124, 237, 238, 306
Properties, 10, 19, 84, 91–93, 99, 101, 122, 165, 181, 210, 275, 304, 305, 316, 344, 406
Protection, 2, 10, 13–15, 17, 19, 31, 120, 125, 138, 156–158, 199, 210, 219, 222, 223, 258, 267, 269, 278, 289, 290, 305, 307, 308, 316, 341–344, 354, 359, 361, 366, 369, 390, 409
Protectors, 209–223
Protein
  folding, 26, 199, 256
  glutathonylation, 258
  import, 313, 367–371
  induced by drought stress, 174
  labeling studies, 344
  oxidation, 214, 266
  protein interaction, 77, 239, 355, 361, 392
  synthesis, 75, 99, 102, 105, 197, 314
Proteomic analysis, 176, 281
Proteomic approaches, 344
Proteomic studies, 176–177
Proton donor, 20
Proton-electrochemical gradient, 218, 308
Proximal meristem (PM), 63
Pryrimidine precursors, 245
PSII. *See* Photosystem II
Pseudomonas aeruginosa, 98, 99
Pseudomonas syringae, 10, 125
Pteris vittata, 216
PTK. *See* Plastid transcription kinase
Pyrimidine
  metabolism, 73
  nucleotides, 73, 74, 238, 244

Q
Quercus species, 18
Quiescent center (QC), 62–64, 216, 217, 235, 244

R
RAC. *See* ROOTING AUXIN CASCADE (RAC)
Ratio GSHred/GSHox, 10
Reactions catalyzed by VTC2, 415
Reactive nitrogen species (RNS), 16, 214, 388, 396, 398
Reactive oxygen and nitrogen species, 388
Reactive oxygen species (ROS)
  detoxification, 15, 116, 286, 339, 357
  generation, 24, 117, 339, 366, 394
  production, 2, 24, 27, 32, 60, 82, 92, 104, 125, 274, 304, 306, 312, 352, 396
  scavenging, 2, 9, 11, 92, 103–105, 116, 124, 130, 155, 177, 182, 215, 222, 223, 258, 267, 273, 279, 280, 310, 312, 314, 315, 328
  signaling molecules, 397–398
  signal transduction cascade, 104–105
  signal transduction pathway, 105
Real time-PCR, 173
Reaumuria soongorica, 100
Redox
  active components, 29
  active elements, ferredoxin, NADPH and GSH, 24
  buffers, 24, 25, 116, 130, 139, 219
  flux, 116, 210, 269
  homeostasis, 4, 23, 24, 117, 125, 126, 215, 325–326
  imbalances, 23, 339, 351, 360, 361
  metabolites, 23, 24, 396
  metabolites, NAD(P), GSH and Asc, 24
  pairs, 23, 123, 251–260, 326
  potentials, 10, 19, 23, 99, 116, 123, 126, 199, 210, 215, 258, 259, 311, 338
  reactions, 26, 307
  sensitive proteins, 268, 269
  signaling, 2, 26, 28, 117, 120, 126, 127, 130, 198, 212, 214, 266, 286, 289, 309, 342, 367, 396
  signals, 13, 23, 115–130, 312, 326, 339, 350, 361, 366–367, 369
  state, 2, 9, 10, 12, 15, 23–28, 30–33, 60, 62, 70–73, 116, 118–124, 126–128, 157, 169, 170, 173, 197, 199, 213, 217, 219, 220, 231–246, 257–259, 268, 269, 271, 272, 274, 275, 277, 278, 281, 282, 284, 285, 287–289, 315, 339, 343, 355, 360, 369, 371, 389
  state of glutathione, 15, 27, 33, 123, 126, 213, 285, 343
  systems, 24, 25, 27, 68, 77, 119
Redox active compounds, 23, 117, 126, 369
  pools of, 23
Redox regulation
  central role of, 304, 310
  2-Cys Peroxiredoxin-A, 359–362
  protein import, 367–371
Redox couples: Asc/DHA, GSH/GSSG, and NADPH/NADP+, 13
Redox-regulated pathways, 26
Redox-sensitive cellular processes, 216
Redox-sensitive receptors, 25, 104, 105
Reduced ascorbate environment, 245
Reduced ascorbate redox state, 245
Reduced glutathione environment, 237, 238
Reduced glutathione pool, 15, 223, 285
Reduced photosystem I, 338
Reductase, 3, 60, 91, 116, 139, 197, 210, 235, 253, 268, 306, 324, 340, 389
Reduction, 11, 14, 24, 28, 32, 57, 60, 62, 70, 72–74, 83, 99, 118, 119, 126–129, 139, 141, 169, 179, 181, 182, 198–200, 210–217, 219, 238, 240, 241, 243, 244, 258, 266–271, 273–275, 281, 283, 285–287, 304, 307, 309, 310, 315, 317, 318, 324, 326, 327, 340, 343, 355, 361, 371, 389, 390, 393, 397, 409, 413, 417
Regenerating, 11, 56–60, 63, 76, 79, 124, 169, 218, 219, 280, 285, 328, 393, 415
Regeneration, 13, 24, 28, 55–84, 118, 120, 124, 127, 129, 141, 215, 236–239, 242, 267, 273, 283, 308, 314, 339, 340, 342, 343, 409
Regulation
  of catalase genes, 356
  of cytosolic MDHAR, 357–359
  of defense genes, 355
  of expression genes, 6
  of extra-plastidic antioxidant defense system, 337–372
  of Fsd expression, 366–367
  of genes encoding chloroplast antioxidant enzymes, 337–372
  of nucleotide synthesis, 55
  of transcript abundance, 348, 358
  of transcript abundance of genes, 348
  via HSFs, 352–353
Regulators, 28, 57, 59, 62–65, 76, 77, 116, 127, 139, 173, 181, 197, 233, 241, 254, 258, 266, 271, 352, 354, 355, 360, 361, 365

Regulatory amplitudes of genes, comparison of 348
Regulatory factor, 65–66
Regulatory proteins, 27, 267, 356
Relative leaf water content, 170, 179
Relative water content (RWC), 156–158, 160–162, 166, 169, 170, 172, 182
Repression/activation, 2
Retama raetam, 176, 177
Rhodophyta, 17
Ribonucleotide reductase (RNR), 64, 210, 211, 244
Ricinus communis, 314, 315
RNA blot analyses, 271
RNAi construct, 314
RNR. *See* Ribonucleotide reductase
RNS. *See* Reactive nitrogen species
Roles of Asc and GSH, 82
Root
  apical meristem, 62–64, 66, 73, 75, 256
  cultures, 57, 81–82, 286
  glutathione, 170
  primordium, 77
  regeneration, 13, 24, 28, 55–84, 118, 120, 124, 127, 129, 141, 215, 236–238, 242, 267, 273, 283, 308, 314, 339, 340, 342, 343, 409
Root-colonizing endophytic fungus, 175
Root-inducing (Ri), 57
ROOTING AUXIN CASCADE (RAC), 59
ROOT MERISTEMLESS 1 (RML1) gene, 65
ROS. *See* Reactive oxygen species
Rosa hybrida, 174
ROS-scavenging mechanisms, 2
RUBISCO (P*ssu-ipt*), 61, 177, 345
RWC. *See* Relative water content (RWC)

S
S-adenosylmethionine, 242
Salicylic acid (SA), 10, 32, 127–129, 180, 182, 216, 274, 306, 355
Salicylic acid (SA) biosynthesis, 11
Salinity, 9, 15, 31, 32, 99, 101, 103, 192, 222, 266, 272–275, 304, 305, 307, 315–318, 327, 329
Salinity stress, 99, 103, 192, 222, 223, 272–275, 305–307, 315, 316, 329, 354, 397
Salinization, 305, 317, 318
Salt dependent-oxidative cell death, 273
Salt-sensitive genotype, 100
Salt stress, 2, 14, 99–101, 162, 178, 180, 272–275, 305, 306, 314, 329–330, 346, 348, 361
Salt-tolerant maize, 100
SAM. *See* Shoot apical meristem
SCARECROW (SCR) gene, 75
Scavenging, 2, 4, 22, 67, 79, 84, 92, 96, 116, 124, 138–140, 155, 177, 180–182, 218, 240, 258, 271, 274, 280, 308, 390
Scheme of ascorbate–glutathione (AsA-GSH) cycle, 3, 92
Sclerotinia sclerotiorum, 121
Secondary messengers, 2, 24, 142, 266, 369
Secondary metabolite production, 82
Secondary redox-reactions, 338
Sedum alfredii, 221, 222
Selaginella moellendorffii, 341
Senescence, 12, 61, 62, 221, 281, 287, 338, 356, 358, 397, 409, 416
Sensors of reactive oxygen species-perception, 104
Sequestration of toxic metals, 31
SERK. *See* Somatic embryogenesis receptor-like kinase
Setaria italica, 99, 272, 329
Shoot apical meristem (SAM), 58, 71–73, 75–77, 240, 244, 245
  architecture, 239, 242
SHOOT MERISTEMELESS (STM), 58, 76, 77, 239, 240
Shoot organogenesis, 56–58, 82–84, 244
Shoot regeneration, 84
Shoot–root translocation of glutathione, 170
Shrub species, 160–161
Signal
  molecules of oxidative stress, 103–104
  regulation, 312
  transduction, 9, 11, 23, 26, 28, 32, 101, 104–106, 142, 174, 184, 219, 251–260, 267, 311, 342, 349, 350, 352, 356, 360, 372, 392
  transduction pathways, 2, 23–28, 104–106, 120, 311, 346, 367, 398
Signaling/Signalling
  cascades, 23, 33, 327, 361, 362, 371, 372
  molecules, 82, 116
  pathways, 10, 26, 31, 32, 59–61, 77, 80, 103, 117, 118, 121, 127, 128, 130, 212, 276, 290, 309, 311, 313, 325, 326, 350, 359, 361, 368, 371–372
Signalling-related implications, 253
Signal-transducing processes, 8
siRNAs, 357, 358, 364
Small RNAs, 363–367
Smirnoff–Wheeler pathway, 253, 410–418
Smirnoff–Wheeler (L-galactose) pathway, 407, 410, 415

S-nitrosoglutathione, 16, 26, 396, 399
SOD. *See* Superoxide dismutase
Sodium nitroprusside (SNP), 31, 81, 180, 182, 183, 272
Solanum lycopersicum, 272
Somatic embryo formation, 56, 57, 69, 75–77, 83
Somatic embryogenesis, 56–59, 69–76, 82–84, 233, 236, 240, 242, 244
Somatic embryogenesis receptor-like kinase (SERK), 59
Somatic embryos, 56, 57, 69–73, 76, 233, 234, 237, 238, 241, 242, 245
Sorghum bicolor, 151, 172
STM. *See SHOOT MERISTEMELESS*
Stomatal conductance, 160, 167, 169, 271, 275
Strategies to improve the antioxidant levels in crops, 315–318
Stress
  conditions, 4–13, 15, 24, 27, 28, 31, 64, 67, 81, 95, 96, 99–101, 103, 105, 106, 125, 164, 169, 176–179, 192, 213, 235, 257, 265–290, 304, 316, 318, 326, 341, 344, 362, 387–399
  factors, 9, 12, 19, 24, 31, 103, 117, 183, 192–194, 197, 266, 312, 314
  tolerance, 1–33, 91–106, 162, 163, 168, 174, 182, 219, 275, 278, 284, 287, 306, 314, 316, 329, 330, 357
Stress-response signal transduction pathway, 2
Stromal proteins, 342
Structural, physiological and molecular changes, 238–243
Suaeda salsa, 95, 100, 101
Subcellular localization, 94, 97, 340–344
  of antioxidant system, 339–340
Subcellular targeting, 393
Sub-compartmental localization, 390, 393
Superoxide dismutase (SOD), 3, 9, 30, 61, 81–84, 140, 156, 157, 161, 163–168, 171, 172, 174, 175, 180, 181, 183, 210, 220–222, 270, 279, 282, 283, 286, 306, 307, 313, 328, 340, 343–345, 356, 363–367, 372, 388, 394
Superoxide radical, 2, 220, 221, 266, 304, 306, 339, 363
Symplast, 8–10, 100, 275, 289
Synergistic antioxidants, 305
Systemic defence, 121, 130

T

Tagetes erecta, 80
Targeted genes for ascorbate improvement, 417–420
TDM. *See* Triadimefon
T-DNA insertions, 8, 175, 358
TE. *See* Tracheary elements
T. goesingense, 282
Thiol
  buffer, 10, 325
  concentrations, 194
Thiol/disulfide-containing proteins, 232
Thiol-disulfide interconversions, 257
Thiol-disulphide, 27, 68, 120, 129, 257
Thiol-group, 26, 104, 129, 199, 211, 267, 308, 311, 324, 326, 357
Thiol (–SH) group, 324, 326
Thiol-mediated signalling, 129
Thioredoxins (TRX), 24, 30, 123, 128–130, 210, 211, 214–217, 232, 256, 311, 339, 342, 355, 360, 362, 369, 370
  pathway, 311
Thylakoid-associated ferredoxin, 407
Thylakoid lumen, 17, 21–23, 342, 407, 409
Thylakoid membrane, 10, 11, 17, 19, 21, 22, 93, 138, 139, 283, 338, 344, 367
Thymidine, 245
Tobacco, 10, 32, 59–61, 66, 83, 97, 102, 119, 124, 177–180, 202, 212, 243, 259, 268, 271, 273–275, 279, 282–284, 288, 289, 309, 329, 341, 343, 344, 352, 395, 409, 417–420
  callus culture, 82
Tolerance, 1–33, 72, 91–106, 142, 155–163, 165, 168, 173, 174, 178–183, 192, 199–202, 217–223, 241, 269, 271–279, 281–284, 287, 288, 304–306, 314–317, 327, 329, 330, 354, 357, 421
Tomato, 15, 16, 18, 95, 96, 99, 100, 104, 118, 122, 124–126, 182, 219, 272, 275, 276, 289, 305–307, 316, 317, 329, 395–397, 418
Tomato seedling, 59, 60, 78–80, 277
Tortula ruralis, 100
Toxic compounds, 192, 193, 197
Toxic metabolic products, 1
Trace metals, 278
Tracheary elements (TE), 65, 236
Transcription
  chloroplast isoforms, 345
  control, 327, 344, 353
  factors, 30, 59, 104, 105, 127–130, 178, 210, 215, 256, 258, 267, 269, 311, 312, 350, 351, 354–357, 360–362
  regulation, 173–175, 242, 315, 346, 359
Transcripts
  abundance, 30, 102, 286, 346, 356–360, 362–367

abundance regulation of genes, 346–348, 356, 357, 359, 363, 371
factors, 104, 105
levels, 13, 72, 75, 95, 101–104, 118, 173–175, 178, 270, 285, 289, 346, 347, 349, 352–354, 356–359, 362–367, 395, 418
Transgenic plants, 4, 61, 62, 121, 125, 177–180, 183, 202, 215, 271–275, 279, 282, 284, 288–290, 307, 314, 417–421
Transgenic plants lines, 4
Transgenic poplars, 16
Transgenic tobacco, 10, 61, 124, 178–180, 202, 268, 271, 273–275, 279, 282–284, 288, 289, 329, 417, 418, 420
Transmembrane domain, 94, 96, 283, 370
Tree species, 142, 155–160
Triadimefon (TDM), 181
Trifolium repens, 153, 167
Triose phosphate isomerase, 28
Triticum aestivum, 152, 153, 162–164, 169, 182, 194, 286, 288, 315, 395
TRX. *See* Thioredoxins*Trypanosoma cruzi,* 93, 406
Two-dimensional polyacrylamide gel electrophoresis, 176

U
Ubiquitous thiol, 32, 235
Ultraviolet radiation, 2
UMP. *See* Uridine monophosphate
United Nations, 406
Up-regulation of sulphur assimilation, 12, 287
Uridine, 73, 74, 245
Uridine monophosphate (UMP), 73, 74, 238, 245
UV-B radiation, 155, 180, 183, 285

V
Vaccinium, 165
Vaccinium myrtillus, 153, 166
Vernalization-induced, 65
Vicinal cysteine residues, 256
Vigna radiata, 281
Violaxanthin (Vx), 11, 17, 18, 138, 409
Vitis vinifera, 177, 315

W
Water deficit, 100, 142, 156, 157, 164, 166, 170, 172, 173, 177, 179, 181, 350
Water-deficit stress, 167, 174, 175, 177, 182, 270
Water-soluble antioxidants, 11, 219
Water–water cycle, 11, 14, 218, 258, 283, 308
Whole ascorbate–glutathione cycle, 171–173
Winning two pair, 251–260
Wolffia arrhiza, 223
WUSHEL (WUS), 58, 59

X
Xanthophyll cycle, photorespiration, 10
Xanthophyll cycles activity, regulation, 17–23
Xenobiotics, 12, 15, 16, 65, 95, 139, 192, 196, 200, 201, 258
Xenobitic detoxification, 65, 126, 139, 197, 198, 200–201, 396

Z
ZAT10-APx2-pathway, 362
ZAT factors, 353–355
Zea mays, 154, 163–165, 182, 220, 270, 283, 315
Zeaxanthin, 5, 6, 11, 138, 157, 163, 170, 218, 267, 308, 409
Zinc hyperaccumulation, 221
Zinnia, 65
Zinnia elegans, 236
ZWILLE, 77, 240
Zx epoxidase (ZE), 17

The manufacturer's authorised representative in the EU is Springer Nature Customer Service Centre GmbH, Europaplatz 3, 69115 Heidelberg, Germany. If you have any concerns regarding our products, please contact ProductSafety@springernature.com

Printed and bound by CPI Group (UK) Ltd, Croydon, CR0 4YY
22/09/2026
02220985-0009